REVERSE OSMOSIS MEMBRANE RESEARCH

REVERSE OSMOSIS MEMBRANE RESEARCH

Based on the symposium on "Polymers for Desalination"
held at the 162nd National Meeting of the
American Chemical Society in Washington, D.C., September 1971

Edited by

H. K. Lonsdale
Alza Research
Palo Alto, California

and

H. E. Podall
Office of Saline Water
U.S. Department of the Interior
Washington, D. C.

PLENUM PRESS • NEW YORK-LONDON • 1972

Library of Congress Catalog Card Number 72-87518
ISBN 0-306-30710-3

A Division of Plenum Publishing Corporation
227 West 17th Street, New York, N. Y. 10011

United Kingdom edition published by Plenum Press, London
A Division of Plenum Publishing Company, Ltd.
Davis House (4th Floor), 8 Scrubs Lane, Harlesden, London,
NW10 6SE, England

Printed in the United States of America

To the pioneers of reverse osmosis

Charles E. Reid
Sidney Loeb
Srinivasa Sourirajan

PREFACE

This book had its origin in the symposium on "Polymers for Desalination" sponsored by the Division of Polymer Chemistry of the American Chemical Society and held in September, 1971 in Washington D. C. at the 162nd national meeting of the Society. However, the book is not simply the proceedings of that symposium. A number of additional papers were contributed by other workers in the field, and the original papers presented at the symposium have, for the most part, been expanded. The book thus represents a broad cross section of membrane research and development activities in the United States and abroad within the field of reverse osmosis. The purposes of the book are to bring attention to important new developments in this field, to suggest what the next generation of reverse osmosis equipment may look like, and to indicate where further research and development are needed. The vast majority of the papers collected here represent work supported by the Office of Saline Water of the United States Department of the Interior, and the emphasis here is clearly on the application of the reverse osmosis process to water purification. However, many of the concepts, methods, and conclusions are expected to be useful in other areas of membrane science and technology.

We would like to thank Dr. W. S. Gillam of the Office of Saline Water for his encouragement in the initial planning of the symposium and in the publication of this monograph. We also wish to thank Mrs. Helen Ryan and Mrs. Pat Sebilian for their excellent job in typing the manuscript.

June, 1971

H. K. Lonsdale, Palo Alto, California
H. E. Podall, Washington, D. C.

CONTENTS

INTRODUCTION

In this introductory section, we will attempt to provide a general description of the reverse osmosis process and briefly summarize the present state-of-the-art, including problem areas to which much of the research and development work presented here is addressed. We will also present a generalized outline of the book along with a brief introduction to each of the papers.

STATE-OF-THE-ART

Reverse osmosis is a membrane permeation process for separating relatively pure water (or other solvent) from a less pure solution. The solution is passed over the surface of an appropriate semipermeable membrane at a pressure in excess of the effective osmotic pressure of the feed solution. The permeating liquid is collected as the product and the concentrated feed solution is generally discarded. It is immediately obvious that the membrane must be highly permeable to water, highly impermeable to solutes, and capable of withstanding the applied pressure without failure.

There are other equally important constraints on the membrane but these do not become so obvious until the details of the process and, most importantly, the costs of the process are introduced. Consider the example of water desalination. All of the salt solution must be pressurized but only a fraction can be recovered as product because the osmotic pressure of the concentrated feed solution exceeds practical operating pressures at some point. However, the pressure-volume work put into the feed solution represents the principal operating cost, and the cost of disposal of the concentrated brine can also be significant under some circumstances; both factors favor high water recovery. In order to achieve high recovery and maintain acceptable quality of the product water, it is necessary to use membranes of high permselectivity. Completely solute-impermeable membranes would, in fact, be optimum for most applications. Because the membrane represents a significant fraction of the capital cost of the process, it is essential that water flux be maximized. This means that the membrane material must be highly water permeable and that the fabricated membrane, or more precisely speaking the solute-impermeable barrier, must be as thin as possible, consistent with strength requirements. Further, because membrane replacement is an important operating cost, membrane life must be maximized. This means that the membrane must be resistant to chemical and microbiological attack, and that its transport and mechanical characteristics should be invariant after long exposure to high pressure. Finally, the fact that the cost of the pressure vessel used to confine the membrane is a significant fraction of the capital cost adds another constraint: the membrane should be capable of being fabricated into shapes that offer high packing densities, i.e., high surface-to-volume ratios.

We can summarize these criteria as follows:

1. The membrane material must be highly permeable to water and highly impermeable to solutes.

2. The membrane barrier must be as thin as possible consistent with strength requirements.

3. The membrane should be chemically inert, mechanically strong, and creep resistant.

4. The membrane should be capable of being fabricated into configurations of high surface-to-volume ratio.

The first researches into the application of the reverse osmosis process to water desalination were made by Prof. C. E. Reid and his students at the University of Florida in the mid-1950's [1,2]. They discovered that cellulose acetate satisfied the first and third of the above criteria. Loeb and Sourirajan subsequently discovered [3] how to cast asymmetric cellulose acetate membranes, which are effectively extremely thin, thus satisfying the second criterion. In the mid-1960's several system designs evolved with the objective of producing modularized membrane packages with minimum cost and maximum performance.

The first commercial reverse osmosis units were made available to the public in the mid-1960's and the interval from demonstration of feasibility to reduction to commercial practice thus took about a decade. The growth of the reverse osmosis industry since then has been explosive: it has been estimated that as many as 1000 reverse osmosis units of capacity 1000 gallons per day or greater have been built as of 1972 and the capacity of the largest installed plant has also risen dramatically such that plants approaching one million gallons-per-day capacity are now being assembled*. Because

*Most of the U.S. suppliers of reverse osmosis processing equipment are listed in the annual "Pollution Control Directory" of Environmental Science and Technology, published by the American Chemical Society.

of limitations in membrane performance, virtually all of these units are being operated on feed waters not more concentrated than 2000-3000 parts per million of total dissolved solids. There are no commercially available single stage seawater desalination units as yet, although recent improvements in membrane performance (which are discussed in this book) indicate that such units will be available in the near future.

Only two types of membranes are presently being offered in commercial reverse osmosis equipment. The first of these is the asymmetric or "skinned" cellulose acetate membrane made in flat sheet or tubular form, generally according to the recipe originally proposed by Loeb and Sourirajan or improvements in their formulation[4,5]. The properties of these membranes can be tailord over a significant range by annealing them in water at temperatures up to about 90°C. The highly annealed membrane has a dense surface layer or skin approximately 0.2 μ thick. The remainder of the membrane, which is typically 100 μ thick, is a highly porous structure which serves essentially as a finely porous support for the thin skin[6,7]. The pores are interconnected and the major permeation barrier resides in the thin skin. Transport within this barrier apparently proceeds by a simple solution-diffusion process [8], and the basis for the high performance of these membranes is their extreme effective thinness and the fact that the permeability of dense cellulose acetate films to water is greater than their permeability to most salts by several orders of magnitude[9,10]. Incompletely annealed asymmetric membranes are not as permselective and the mechanism of water and salt transport is not as well established. While tailoring is commonly practiced, it has its limitations. Thus, increased annealing temperature improves permselectivity or "salt rejection" (which is defined as unity minus the ratio of product concentration to feed concentration, usually expressed as a percentage) but at the expense of reduced water flux at a fixed pressure. Further, increasingly rigorous annealing reaches the point of diminishing returns: as the annealing temperature is increased beyond about 90°C, flux continues to decline without further improvement in salt rejection (beyond about 98% for sodium chloride under typical conditions). The important point is that these membranes are not suitable for single-pass seawater desalination. In this application, reduction of the dissolved solids to potable limits (500 ppm in the U.S.) requires a salt rejection generally in excess of 99%. Two-pass seawater desalination has been demonstrated with these membranes but it does not appear to be competitive with the distillation process at this time.

Typical performance characteristics for these membranes under field conditions with brackish water are as follows: At 40-50 atm operating pressure, water flux is $0.5\text{-}1 \times 10^{-3}$ cm^3/cm^2-sec (10-20 gallons/ft^2-day). Rejection of sodium chloride is 95-96%. Rejection of divalent ions is virtually quantitative but the rejection of certain ions (nitrate, for example) is lower, and the rejection of several low molecular weight non-ionic species (boric acid and phenol, for example) is quite low. Salt rejections are a strong function of membrane fabrication conditions and, for trace ionic species, rejection can be a strong function of solution composition[11].

One unfortunate property of these membranes is their tendency to exhibit declining water flux with time. This flux decline is the result of two phenomena, one is an essentially irreversible reduction in membrane permeability commonly referred to as "membrane compaction", and the other is membrane fouling. The mechanism of membrane compaction is not completely clear at this time; the changes in performance are consistent with a thickening of the dense skin, apparently the result of creep. It is known, however, that flux decline due to compaction increases with increasing operating pressure and it is more severe with less highly annealed membranes. Reverse osmosis systems are thus generally designed and operated under conditions where compaction is minimized: operating pressures are generally limited to 50 atm and the selection of membrane annealing conditions is made consistent with the designed operating pressure. Compaction rates are typically reported as the slope of a plot of log (water flux) against log (time)[12]. Fouling is a phenomenon common to all membrane separation processes. A number of suspended and colloidal materials, both inorganic and organic, have been identified as contributing to membrane fouling in reverse osmosis. Several methods for minimizing fouling and for removing fouled layers have been reported[13,14]; water pretreatment including filtration is widely practiced and appears to be reasonably successful. Fouling is not discussed further in this book, but it is important to note that reverse osmosis is essentially limited to the processing of waters which, after pretreatment, are very low in suspended matter.

The other membrane material which is now used in commercial reverse osmosis equipment is an aromatic polyamide or polyamide-hydrazide. In commercial installations, the membrane is used only in the form of hollow fine fibers. The composition of these membranes and their performance in flat sheet form has been reported in a patent[15]. Beyond that, little had been published concerning these membranes prior to the two chapters appearing in this book. The hollow fibers are reported to have the asymmetric structure typical of the Loeb-Sourirajan cellulose acetate membrane. According to company literature, typical performance characteristics for these hollow fiber systems are as follows: At 27 atm operating pressure, water flux is $0.05\text{-}0.1\times10^{-3}$ cm^3/cm^2-sec (1-2 gal/ft^2-day) and the rejection of sodium chloride is approximately 95%. Flux decline characteristics under field conditions are not reported.

This review of the state-of-the-art can be brought to a close with a summary of the limitations of present reverse osmosis membranes. These limitations form the basis for the continued research and development described in the remainder of this book.

1. Membrane selectivity is inadequate for single pass seawater desalination.

2. For brackish water applications, reduction in cost could be attained if the water flux could be substantially increased without loss in salt rejection or other properties.

3. Flux decline is serious with high flux membranes.

4. Certain species are inadequately rejected.

5. For certain applications, existing membranes are not sufficiently resistant to chemical or microbiological attack, or their mechanical or thermal stability is inadequate.

6. Practical in situ membrane regeneration methods do not exist.

ORGANIZATION OF THIS BOOK

The several papers have been arranged in six sections, as follows:

I. Fundamentals

II. Cellulosic Membranes

III. Non-Cellulosic Membranes

IV. Hollow Fibers and Small Tubes

V. Composite Membranes

VI. Summary and Future Trends

Section I is devoted to a review of the fundamentals of the transport processes in reverse osmosis (or hyperfiltration), both within the membrane and in the high pressure feed solution immediately adjacent to the membrane. In the opening paper, O. Kedem of the Weizmann Institute of Science reviews the important theoretical developments in the general field of pressure-induced membrane transport. There have been three models used to describe transport in uncharged membranes which appear to have withstood the test of time (at least for a decade or so). There is the model evolving from the thermodynamics of irreversible processes (sometimes referred to as the Kedem-Katchalsky model), the solution-diffusion model, and the frictional model. Each has its own set of strengths and limitations and each provides its own specific set of insights into membrane function and the interactions between solute, solvent, and membrane. Unfortunately, the models are sufficiently different conceptually that a complete discription of membrane properties according to one model does not assure an adequate (or even useful) description according to either of the other models. For this reason most of the workers in the field continue to describe membrane properties according to their "traditional" model, even though it may be either unnecessarily complex or too simple. This problem does not seem to be capable of early resolution but, fortunately, the important performance parameters of water flux and salt rejection are universally understood. While commercial reverse osmosis membranes are non-ionic, some highly favorable results have been obtained in recent years with membranes which contain ionogenic groups and thus carry fixed charges over much of the pH range. (Some of these membranes are described in Sections III and V.) Dr. Kedem also treats the question of co-ion exclusion and limiting salt rejection in such membranes.

The subject of concentration polarization is treated in the next paper by W. Pusch of the Max-Planck-Institut für Biophysik. As water is transported through the membrane, salt tends to accumulate at the water-membrane interface. This high local salt concentration has several consequences, all of which have a negative effect on process efficiency. First, the local osmotic pressure is increased reducing the effective driving force for water flow. Second, the concentration gradient of salt across the membrane is increased, thereby increasing salt flow. Third, the high interfacial concentration promotes both inorganic and organic fouling due to precipitation of sparingly soluble salts and gel formation, respectively. While these boundary layer phenomena are independent of membrane transport properties per se, their effects can be quite significant, particularly in the case of high water flux, and overlooking these effects can lead to incorrect conclusions about membrane properties. Dr. Pusch presents the theoretical treatment of concentration polarization in both the turbulent and laminar flow situations and compares the results with experiments.

Section II deals in the main with cellulosic membranes. The first two papers are concerned with the mechanism of formation of asymmetric membranes, of which the Loeb-Sourirajan membrane is probably the foremost example. First, Elias Klein and James K. Smith of Gulf South Research Institute describe the use of solubility parameters in selecting solvent systems for asymmetric membrane formation. The method is useful both in providing a rationale for the several solvent/non-solvent systems which can be used in casting the asymmetric cellulose acetate membrane and in predicting useful solvent systems for other polymers, as the authors illustrate in their work with ethyl cellulose.

A different approach is described next by Moshe A. Frommer and Doron Lancet of Hydronautics, Inc. With the aid of optical and scanning electron microscopy, the authors have defined and evaluated the important kinetic parameters in asymmetric membrane formation. The importance of solvent evaporation and solvent/non-solvent interdiffusion during coagulation to membrane structure and asymmetry and to void formation are described. The research presented in these first two papers in this section and in other recent papers by these authors provides not only an understanding of the mechanism of formation of present asymmetric membranes but also a basis for rationally extrapolating to other polymer systems.

Cellulose acetate membranes are subjected to a variety of mechanical stresses during their manufacture, when they are incorporated into modules, and in use at high pressure. The effects of stress on molecular orientation and on the transport properties of dense cellulose acetate membranes is described by Eberhard A. Meinecke and Devendra V. Mehta of The University of Akron. The effect of strain rate, total strain, and temperature on these properties was examined.

The transport properties of cellulose acetate membranes are known to depend on the degree of acetylation, with increasing permselectivity accompanying increased acetylation. W. M. King, D. L. Hoernschemeyer, and C. W. Saltonstall, Jr. of Envirogenics Co. describe the beneficial effect of blending cellulose diacetate with cellulose triacetate to produce a membrane with an overall degree of acetylation of 2.6-2.7. With the aid of special solvent systems, a broad spectrum of membranes has been achieved, ranging from highly efficient "brackish water" membranes to membranes providing adequate salt retention for single-pass seawater desalination. The blend membranes also exhibit improved flux stability over conventional membranes.

Next, D. L. Hoernschemeyer, R. W. Lawrence, C. W. Saltonstall, Jr., and O. S. Schaeffler of Envirogenics Co. describe the preparation and reverse osmosis properties of a crosslinkable polymer system, cellulose acetate methacrylate. Redox systems were developed for crosslinking these membranes in an aqueous environment and the lightly crosslinked membranes exhibit improved mechanical properties and flux stability, even in the case of high flux membranes.

Another approach to improving the mechanical and flux stability properties of cellulose acetate membranes is described by F.

Kimura-Yeh, H. B. Hopfenberg, and V. T. Stannett of North Carolina State University. These authors graft-copolymerized styrene to dense and asymmetric cellulose acetate membranes and also prepared membranes from blends of homopolymer styrene and cellulose acetate. Significant improvement in tensile creep behavior resulted. The effects of grafting conditions and the length of the grafted chain on water sorptivity and diffusivity, water flux, salt rejection, and compaction rate are reported.

In the final paper in this section, John L. Richardson, Gilbert Segovia, Wilfred H. Bachle, and H. Andre Parker-Jones of Philco-Ford Corp. describe the development of techniques for the precision casting of tubular cellulose acetate reverse osmosis membranes in lengths up to twenty feet. Based on considerations of the effects of casting variables on membrane performance, a device was designed and developed for the semi-continuous fabrication of tubular membranes with properties comparable to those of batch-cast membranes. This technology has led to a new engineering design in reverse osmosis processing equipment.

Section III deals with non-cellulosic membranes. Since the early work of Reid and coworkers and of Loeb and Sourirajan an intensive search has been carried out in a number of laboratories for polymers with better equilibrium and transport properties than cellulose acetate and related cellulosics. Several systems have been proposed and some of the most important of these are described in this section. In the first paper, Lynn E. Applegate and C. Richard Antonson of duPont Co. describe a new aromatic polyamide-hydrazide membrane which exhibits excellent permeation characteristics. The synthesis of the polymer, casting of asymmetric membranes, and performance of flat sheet membranes in reverse osmosis tests are described. These membranes are now being incorporated into reverse osmosis equipment in hollow fiber form.

Next, R. McKinney, Jr. of Chemstrand Research Center describes the preparation and properties of both aromatic polyamide and polyamide-hydrazide asymmetric membranes. When tested in flat sheet form, the properties of one of these membrane systems were consistent with those reported in the previous paper, i.e., water fluxes of 10 gal/ft^2-day or greater and rejections of seawater well in excess of 99%. The membranes also exhibit good mechanical and chemical stability.

A new polymer system with particularly favorable high temperature properties is described in the following paper by Anthony B. LaConti, Paul J. Chludzinski, and Arnold P. Fickett of General Electric Co. The polymer is sulfonated poly (2,6-dimethyl phenylene oxide) or PPO. This material is an ion exchanger, and the properties of PPO membranes are strongly dependent on the degree of sulfonation or ion exchange capacity. However, even the least permeable of these materials has a water permeability several times that of cellulose 2.5-acetate such that dense membranes 5 to 10 μ thick exhibit fluxes on the order of $0.2\text{-}0.3 \times 10^{-3}$ cm^3/cm^2-sec (4-6 gal/ft^2-day) at moderate pressures. At ambient temperatures, membrane performance is comparable to that of asymmetric cellulose acetate membranes in certain applications (when corrected for thickness differences), but the properties are undiminished even at 75°C while conventional cellulose 2.5-acetate membranes fail in operation well below this temperature. The PPO membranes appear to be exceptionally stable mechanically and chemically.

Another new, potentially useful membrane material, polybenzimidazole (PBI), is described by Frank S. Model and Leonard A. Lee of Celanese Research Co. Asymmetric flat sheet membranes prepared from this material appear to have flux and rejection properties at ambient temperature comparable to those of cellulose acetate. However, the favorable properties are maintained even at temperatures as high as 90°C, consistent with the well known high temperature stability of PBI. Some preliminary results are reported for these membranes in hollow fiber form.

Next, H. Yasuda and A. Shindler of Research Triangle Institute present a model of water flow and salt rejection in reverse osmosis membranes, in which the key parameter in describing homogeneous membrane properties is the equilibrium volume fraction of water. The diffusive and viscous flow permeabilities of both non-ionic and ionic membranes are represented as functions of the water content parameter. Reverse osmosis results in support of the model are presented for a series of ionic polymers. The authors conclude that ionic membranes offer the inherent advantage over non-ionic membranes of higher water flux at comparable salt rejection.

Support for this contention is presented in the concluding paper in this section. Here, G. Lopatin and H. A. Newey of Shell Development Co. present a description of the preparation and properties of another family of ionic membranes, these based on block copolymers of polystyrene and polyisoprene, polybutadiene, or ethylene-propylene. In each case, the ionic groups (sulfonate or trimethylammonium groups) were added to the non-styrene block, producing polymers with capacities in the range of 1-2 meq/gram. Dense sheet membranes of these materials exhibited exceptional transport properties: water permeability is substantially greater than that of cellulose 2.5-acetate while rejections of 0.1M NaCl were in excess of 90%.

<u>Section IV</u> is concerned with hollow fibers and small tubes. Near the outset of this introductory section, it was noted that the ability to fabricate membranes into shapes amenable to high packing

density is an important criterion in membrane optimization. Thus, water production per unit volume of pressure vessel can be an important figure of merit. Depending on relative membrane costs and pressure vessel costs, it can be even more important than water production per unit area of membrane surface. Because, for a given geometry, the surface-to-volume ratio increases with the reciprocal of the minor dimension, very small membrane units (such as microspheres or fine fibers) can offer significant economic advantages over units of more conventional size. Another advantage of such systems is that they can withstand rather large external pressures without additional mechanical support. In early attempts to capitalize on this concept, dense hollow fibers were used. Wall thicknesses were small (20-50 μ) but still substantially greater than the effective thickness of the asymmetric cellulose acetate membrane (about 0.2 μ), so that with fiber materials of equal water permeability, water fluxes were exceedingly low (generally less than $10^{-5}cm^3/cm^2$-sec). Specific water fluxes (gallons/ft^3-day) tended to be less than was already attainable with the more conventional modular designs. Recently, workers in the field have succeeded in preparing asymmetric or "skinned" hollow fine fibers and specific water fluxes have risen sharply. In the first paper in this section, Mark E. Cohen, Michael A. Grable, and Billy M. Riggleman of Hercules, Inc. describe the preparation and performance of skinned hollow fibers of cellulose acetate. These fibers can be tailored by varying the annealing conditions. Water fluxes up to $0.3x10^{-3}cm^3/cm^2$-sec (6 gal/ft^2-day) and NaCl rejections as high as 96% at 40 atm are reported.

A method for preparing larger fibers or "tubules" is described in the next paper by A. Gollan and M. P. Tulin of Hydronautics, Inc. Small tubular, skinned cellulose acetate membranes on the order of 1 mm O.D. are cast on the inside of supporting tubes by a hydrocasting process. The method offers the additional advantage that in situ membrane regeneration is possible. Unsupported tubes have been tested at pressures of 10 atm or more and performance characteristics appear to be comparable to those of flat sheet Loeb-Sourirajan membranes.

Section V deals with composite membranes. As noted above, current commercial reverse osmosis membranes as well as those described here up to this point consist of an asymmetric structure where the uppermost layer or "skin" and underlying porous substructure are composed of the same polymer. The mechanical as well as the transport requirements for the skin and substructure are, however, vastly different. Thus, one desires the skin to be moderately hydrophilic but non-porous to minimize coupled flow of salt and water, while the substructure should preferably be hydrophobic, for greater compressive strength, and porous to enable water passage from the barrier. It follows that a preferred structure from the standpoint of transport and mechanical properties is that of a composite, comprised of an ultrathin desalination "barrier" or coating on a microporous support. The barrier and substrate support materials may then each be selected for their requisite properties, rather than to seek a compromise with a single polymer in the case of a homogeneous membrane. The composite approach also permits the selection of rational methods for the preparation of each of the components, i.e., the barrier and the porous support. Finally, the composite membrane system offers a more versatile approach to the design of membranes for different applications. In the opening paper in this section, James S. Johnson, Jr. of Oak Ridge National Laboratory describes a dynamic method for preparing one kind of composite membrane. A hydrous metal oxide sol is passed under pressure over the surface of a finely porous supporting membrane and a thin, charged membrane is thus deposited in the pores. This membrane is in itself capable of desalinating, but performance with natural saline waters is improved if an outer layer of polyacrylic acid is laid down over the metal oxide membrane by the same procedure. The properties of these promising membranes are described in detail by Dr. Johnson, including the effects of salt concentration, pressure, and molecular weight of the polyacrylic acid used on membrane performance. Operating data on several natural waters and waste streams are also presented.

Composite polyacrylic acid (PAA) membranes have been prepared by a different means by C. E. Milstead and M. Tagami of Gulf General Atomic Co., as reported in the following paper. Here, an aqueous solution of high molecular weight polyacrylic acid was cast directly on the surface of a finely porous support and allowed to dry. Properties can be altered somewhat by annealing the membrane in air. When the PAA is cast on a finely porous polysulfone substrate, the desalination properties of the resulting membranes are comparable to those of anisotropic cellulose acetate membranes.

Composite cellulose acetate membranes are described in the next two papers. First, L. T. Rozelle, J. E. Cadotte, A. J. Senechal, W. L. King, and B. R. Nelson of North Star Research and Development Institute report on their development of a method for casting ultrathin films on the inside surface of microporous polysulfone tubular substrates in situ. The thin films of cellulose 2.5-acetate are only a few hundred angstroms thick, and the pores in the substrate are sufficiently small to support these films at pressures in excess of 50 atm without apparent rupture. Water fluxes greater than $1\times10^{-3} cm^3/cm^2$-sec (20 gal/ft^2-day) and rejections of 1% NaCl of 97-98% are reported.

Next, R. L. Riley, G. Hightower, and C. R. Lyons of Gulf General Atomic Co. describe the preparation and properties of cellulose triacetate composite membranes. These membranes are prepared by dip-coating an ultrathin film onto the surface a microporous substrate. The substrate is first coated with a layer of polyacrylic acid to prevent intrusion of the casting solution into the pores. These composite membranes exhibit water fluxes of $0.5 \times 10^{-3} cm^3/cm^2$-sec (10 gal/$ft^2$-day) and greater than 99.5% rejection of 3.5% sodium chloride at 100 atm. They appear to be promising candidates for single pass seawater desalination.

An X-ray diffraction study of the structure and ordering within ultrathin cellulose acetate films is described in the next paper by S. Krishnamurthy and D. McIntyre of The University of Akron. The effect of annealing and pressure on the crystallinity is described. The presence of a small fraction of an ordered polymer phase in ultrathin films, which is not present in thick films, is suggested by the results.

A novel approach to preparing finely porous tubular substrates to be used in preparing composite membranes is described in the concluding paper in this section by Bernard Baum, R. A. White, and W. H. Holley, Jr. of DeBell and Richardson, Inc. "Tubulets", 700-1500 μ O.D., have been prepared from a number of thermoplastic polymers including polyvinyl chloride and certain copolymers and polymer blends of vinyl chloride. These tubulets are subsequently rendered microporous by aqueous extraction of water-soluble materials. Tensile, porosity, and water permeability properties of these substrates are described.

Section VI is devoted to a summing up of these recent research and development activities, and a look into the future. This concluding paper was prepared by W. Sherman Gillam, Assistant Director, Research of the Office of Saline Water. During Dr. Gillam's nearly two decades at the OSW, the reverse osmosis process has evolved from a laboratory curiosity to the point where it forms the basis for a thriving new industry. It is widely considered today as the favored process for many desalting applications and new uses for the process are still evolving. The future of the process may be even more dramatic than its past, however, and Dr. Gillam points out the important areas where continued research is needed in order to fully exploit the opportunities inherent in the process.

CONCLUDING REMARKS

As extensive as is the coverage of the field of reverse osmosis membrane research presented in this book, it is by no means complete. A good deal of additional work is in progress. An excellent source for current Office of Saline Water programs in this area is the series of OSW annual Saline Water Conversion Reports, available from the Superintendent of Documents, U.S. Government Printing Office. There is also a substantial amount of research in progress elsewhere of which that at the University of California at Los Angeles under the direction of Prof. Joseph McCutchan[16] and at the National Research Council of Canada under the direction of Dr. S. Sourirajan[17] can be cited as two examples. There is, in addition, much activity in this area overseas and much of this appears in the journal Desalination, published by Elsevier.

A closing note of apology is in order for the inconsistency in the use of symbols in this book. The only solace we can offer the reader is that the authors have been reasonably careful to define their symbols, and in those chapters in which the greatest number of symbols are used, symbol lists are provided.

REFERENCES

1. E. J. Breton, Jr. "Water and Ion Flow Through Imperfect Osmotic Membranes", Office of Saline Water Research and Development Progress Report No. 16 (1957).

2. C. E. Reid and E. J. Breton, J. Appl. Poly. Sci. 1, 133 (1959).

3. S. Loeb and S. Sourirajan, Advan. Chem. Ser. 38, 117 (1962).

4. S. Manjikian, Ind. Eng. Chem. Prod. Res. Devel. 6, 23 (1967).

5. S. Sourirajan and T. S. Govindan, "Membrane Separation of Some Inorganic Salts in Aqueous Solutions", Proceedings of the First International Symposium on Water Desalination, U.S. Government Printing Office, Washington D.C. (1967).

6. R. L. Riley, J. O. Gardner, and U. Merten, Science 143, 801 (1964).

7. R. L. Riley, U. Merten, and J. O. Gardner, Desalination 1, 30 (1966).

8. H. K. Lonsdale, U. Merten, and R. L. Riley, J. Appl. Poly. Sci. 9, 1341 (1965).

9. R. Blunk "A Study of Criteria for the Semipermeability of Cellulose Acetate Membranes to Aqueous Solutions", UCLA Dept. Eng. Rept. No. 64-28 (1964).

10. H. K. Lonsdale, B. P. Cross, F. M. Graber, and C. E. Milstead, J. Macromol. Sci.-Phys. B(5)1, 167 (1971).

11. T. D. Hodgson, Desalination 8, 99 (1970).

12. U. Merten, H. K. Lonsdale, R. L. Riley, and K. D. Vos, Office of Saline Water Research and Development Progress Report No. 208 (1966).

13. R. G. Sudak, "Test of 10,000 GPD Reverse Osmosis (Spiral Wound) Pilot Plant (San Diego)" Office of Saline Water Research and Development Progress Report No. 453 (1968).

14. C. G. DeHaven, M. A. Jarvis, and C. R. Wunderlich, "Operation of Reverse Osmosis Pilot Plants", Office of Saline Water Research and Development Progress Report No. 356 (1968).

15. J. W. Richter and H. H. Hoehn, U. S. Patent 3,567,632 (1971).

16. See, for example, the series of reports issued by the UCLA Department of Engineering, available through the UCLA Water Resources Center.

17. S. Sourirajan, Reverse Osmosis, Academic Press, New York (1970).

I. FUNDAMENTALS

WATER AND SALT TRANSPORT IN HYPERFILTRATION

O. Kedem
The Weizmann Institute of Science
Rehovot, Israel

INTRODUCTION

The impressive practical achievements of reverse osmosis have stimulated the study of thermodynamics, membrane structure, and transport mechanisms underlying this separation process. In many cases the theoretical analysis was developed for the interpretation of a specific membrane system. It is attempted in this outline to review work done in this laboratory and to stress and correlate common elements in the somewhat different points of view presented by a number of authors.

We shall deal consecutively with the following aspects of reverse osmosis:

1. Definition of membrane performance under hyperfiltration conditions. This will not include polarization, which is, of course, an important factor in plant performance, but not relevant to the particular polymer membrane used.

2. Transport coefficients characterizing a given membrane, and their relation to hyperfiltration performance.

3. The relationships between transport coefficients and membrane structure, and mechanisms of transport.

SALT REJECTION IN HYPERFILTRATION*

A basic feature of reverse osmosis as compared to other membrane methods is the fact that only one solution, the feed, is of given composition (c'_s) while the concentration of the solution on the other side of the membranes is completely determined by the process itself: the concentration of product, c''_s, is equal to the ratio of salt flow, J_s, to total volume flow, J_v:

$$c''_s = J_s/J_v \tag{1}$$

The effectiveness of the hyperfiltration is described by the salt (or solute) rejection, R, or the desalination ratio, S

$$S \equiv c'_s/c''_s = c'_s J_v/J_s$$

$$R \equiv 1 - c''_s/c'_s = 1 - J_s/J_v c'_s = 1 - 1/S \tag{2}$$

Another useful form of equation (2) for the solute rejection is:

$$R = [1 + \frac{J_s/(c'_s - c''_s)}{J_v}]^{-1} \tag{3}$$

It has been consistently observed that R increases with increasing flow (or pressure head) and it is now generally accepted that this need not reflect a change of membrane properties under the influence of pressure. At high flow rates a limiting value, R_∞, is approached. Together with the flow rate at given pressure, R_∞ is the most important criterion for performance and hence a useful transport theory must be able to predict R_∞ from the coefficients characterizing the membrane material. The general features of the rejection curve for a single solute in any membrane can be given, with very few specific assumptions.

If surface processes are fast, and polarization is negligible, transport within the membrane is rate-determining. We may then assume that equilibrium with feed and product solutions at the surface is not disturbed by the process, local equilibrium is maintained everywhere, and the thermodynamic potentials of water and solute change continuously from feed, across the membrane, to product. A homogeneous membrane may thus be represented by

*The formal description of salt rejection given here is based on the treatment by Spiegler and Kedem[1].

a series of infinitesimally thin layers, separated by thin solution compartments of uniform concentration (Fig. 1). The small changes of the thermodynamic potentials, μ_w and μ_s, across each membrane layer may be expressed as changes of pressure, p, and solute concentration, c_s, or osmotic pressure, π, in the hypothetical solution compartments. These are at equilibrium with the adjacent membrane elements, and do not contribute to the flow resistances.

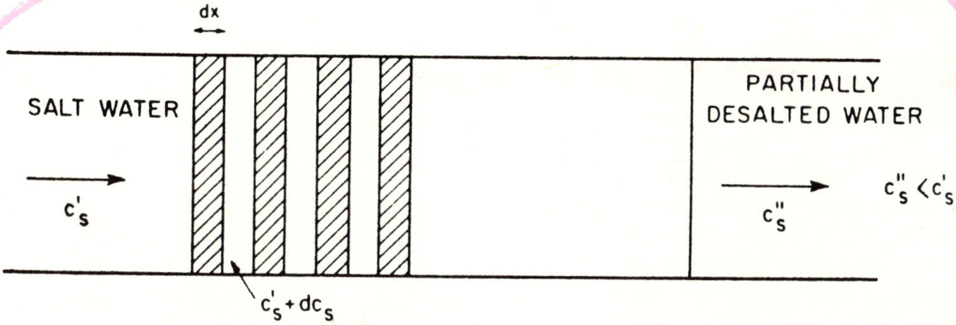

Fig. 1: Schematic Representation of a Hyperfiltration Membrane. The membrane is broken down into differential elements, separated by uniform solution segments which are in equilibrium with the two contiguous membrane faces.

All fluxes going from left to right are counted positive. The gradient of a scalar y, dy/dx, is taken as the value of the scalar on the right (double prime) minus the value on the left (single prime), divided by the distance. On the other hand, the operator Δ is defined with the opposite sign, in order to bring our notation in line with that of previous publications[14,17]: $\Delta c \equiv c' - c''$. Thus $\Delta y \equiv - \int_{'}^{''} dy$. (Reprinted from Reference [1])

From the thermodynamics of solutions:

$$\frac{d\mu_w}{dx} = \bar{V}_w \left(\frac{dp}{dx} - \frac{d\pi}{dx}\right) \tag{4}$$

$\bar{V}_w$ is the molar volume of water and $d\pi/dx$ is proportional to the concentration gradient, dc_s/dx. For the solute:

$$\frac{d\mu_s}{dx} = \frac{1}{c_s}\frac{d\pi}{dx} + \bar{V}_s \frac{dp}{dx} \tag{5}$$

In many cases the pressure dependent part of $d\mu_s/dx$ may be neglected.

Across each differential membrane, solute flow depends on the concentration and pressure gradients. Alternatively, one may express J_s as a function of the concentration gradient and volume flow: it is *a priori* clear that the second choice will lead to a simpler expression for the product concentration, since this is itself directly related to J_v.

For each thin layer we write:

$$J_s = -\bar{P}\frac{dc_s}{dx} + (1-\sigma)c_s J_v \tag{6}$$

where $\bar{P}$ is the local permeability coefficient measured at vanishing volume flow, σ is the reflection coefficient which is a factor smaller than, or equal to, unity and usually positive. Equation (6) (or a closely analogous equation) was derived from the formalism of the thermodynamics of irreversible processes[1, 2,3] and the reflection coefficient was defined previously by Staverman[4] in the context of this theory. However, irreversible thermodynamics is not a necessary basis for this equation, which may also be regarded as an empirical relation, describing a wide variety of membrane systems.* The differences in membrane structure and transport mechanism are reflected in the values of $\bar{P}$ and σ. With σ=0, Eq. (6) is the customary local equation for solute flow in capillary systems, and its integrated form, the so-called Hertzian[6], gives the concentration profile in a capillary connecting two reservoirs of given, different concentrations. With the boundary conditions of hyperfiltration, σ=0 simply means that

*This does not include, e.g., some of the models of Banks and Sharples[5], which are based on diffusion in the solution parallel to the membrane as the rate-determining step, and other unusual features.

there is no salt rejection: no concentration gradient can be created by forced volume flow. At the other end, $\sigma=1$ means that solute flow is totally independent of volume flow. It will be shown below that, strictly speaking, $1-\sigma$ vanishes only for ideally semipermeable membranes, in which also $\overline{P}=0$. However, the influence of J_v on J_s may become practically negligible in not-quite-ideal membranes, if the water does not drag along the solute and if solute permeability is small compared to the hydraulic permeability. This has been shown to hold for the very important case of salt flow through dense cellulose acetate. Equation (6), together with Eq. (1) is then simplified to

$$\frac{J_s}{J_v} = c''_s = -\frac{\overline{P}}{J_v}\frac{dc_s}{dx} \tag{7}$$

If the permeability coefficient is not concentration dependent, the rejection is

$$R = \frac{1}{1 + P/J_v} \tag{8}$$

where P is the overall salt permeability, $P=\overline{P}/\Delta x$, and Δx is the membrane thickness. Equation (8) follows immediately from Eq. (3), if J_s is independent of J_v.

According to Eq. (8) the curve of R vs. J_v is in this case determined by a single coefficient, P. Two membranes of different $\overline{P}$, Δx, and hydraulic permeability but equal $\overline{P}/\Delta x$ should give superimposable curves.

The salt rejection approaches unity at high J_v, and the initial slope is

$$(dR/dJ_v)_{J_v \to 0} = 1/P \tag{9}$$

In the general case, when σ is neither zero nor unity, $1-\sigma$ defines the influence of total flow rate on solute flow. If the coefficients are constant, integration of Eq. (6) with the boundary conditions of hyperfiltration gives:

$$R = \sigma\frac{\exp(BJ_v) - 1}{\exp(BJ_v) - \sigma} \tag{10}$$

where

$$B = \frac{1-\sigma}{P} \tag{11}$$

At very large volume flow, $BJ_v >> 1$, and R tends to its limiting value, σ. The initial slope at very small flow rates is now determined by the ratio between σ and the solute permeability:

$$(dR/dJ_v)_{J_v=0} = \sigma/P \tag{12}$$

If the limiting rejection at very large volume flow is defined as R_∞, then for the extreme cases discussed above, $R_\infty = 1$ for $\sigma=1$, and $dR/dJ_v=0$ for $\sigma=0$, as expected.

After integration of Eq. (6) the concentration profile along the series of differential membrane and solution layers can be evaluated. It is particularly instructive to examine the concentration gradient near the feed interface, at high flow rates. Under these circumstances, $R \rightarrow \sigma$, and the ratio between solute flow and volume flow is:

$$J_s/J_v = c''_s = c'_s(1-R) = c'_s(1-\sigma) \tag{13}$$

Substituting this into the local flow equation, Eq. (6), and remembering that at $x=0$, $c_s=c'_s$, it is obvious that the concentration gradient within the membrane near the feed interface must vanish at high flow rates. While constant coefficients are used here, it has been shown in at least one case that this flattening of the profile will ensue also with concentration dependent coefficients[7] and may be assumed to exist as long as there is no drastic variation of the coefficients with concentration.* The derivation of R_∞ for a given model is thus greatly simplified: one only has to calculate J_s/J_v, in the absence of a concentration gradient, in a membrane element which is in equilibrium with a solution of known composition: the feed.

Reverse osmosis membranes are either prepared as an anisotropic structure, the Loeb-Sourirajan membrane[8] or its more recent analogs, or put together to form a composite structure. The effect of a porous substructure or support on membrane performance must therefore be considered[9]. For every point x, in the support as well as the denser selective part, an equation of the form of Eq. (6) can be written; the coefficients are different in both parts. Assuming that anisotropic as well as

*It should be emphasized that solutions of a single solute are considered here, and the approximation of zero gradient at x=0 will not, in general, be valid for multicomponent systems.

composite membranes may be regarded as two-layered structures, with the coefficients constant in each part separately, the local flow equation can then readily be integrated. The flows are constant throughout, and the concentration c_s is continuous at the plane of contact between the layers. We shall not discuss here in full the resulting equation, given and tested in reference 9. For the important simple case when the porous part is completely unselective, it shows that the salt rejection is equal to that of the dense layer facing the feed solution at the same volume flow. In many membranes the contribution of the support to the hydraulic resistance is not negligible (i.e., the hydraulic permeability of the support is not infinite compared to that of the dense part). The rejection at the same pressure is then lower than that of the dense layer itself. The limiting rejection of the two-layered structure is equal to that of the "up-stream" layer, even if the second layer is selective.

VOLUME FLOW IN HYPERFILTRATION

The major driving force for volume flow through a thin membrane element is the pressure gradient, partly opposed by the gradient of osmotic pressure created in hyperfiltration.

$$J_v = - \bar{A} \frac{dp}{dx} + C \frac{d\pi}{dx} \tag{14}$$

$\bar{A}$ is the local hydraulic permeability. C is not another independent coefficient; as a consequence of Onsager's reciprocal relations C must be equal to $\sigma\bar{A}$, where σ is the reflection coefficient appearing in Eq. (6) (cf. Appendix). Therefore

$$J_v = - \bar{A}\left(\frac{dp}{dx} - \sigma \frac{d\pi}{dx}\right) \tag{15}$$

If $\bar{A}$ and σ are constant, differences may be substituted for the gradients and

$$J_v = A(\Delta p - \sigma\Delta\pi) \tag{16}$$

where

$$A = \bar{A}/\Delta x \tag{17}$$

In a reverse osmosis experiment, the change of J_v with Δp contains the dependence of $\Delta\pi$ on Δp and the slope of J_v vs. Δp-$\Delta\pi$ decreases as long as R increases.

At pressure gradients higher than those required for the maximal salt rejection, the volume flow of course continues to increase with pressure, and the slope of J_v vs. the pressure head, Δp, in this region is the hydraulic permeability of the membrane.

TRANSPORT COEFFICIENTS

The transport coefficients which define the reverse osmosis performance of a membrane should be measurable by different standard methods, without applying high pressure. Measurement by these alternative methods is necessary mainly for two purposes: a The direct determination of the highest salt rejection a material can give requires high volume flow and thus very thin membranes - exploratory testing of materials prior to thin membrane development may be valuable. b The comparison between the "statically" measured coefficients and the reverse osmosis results measures the change caused by the pressure - the short term compaction effects.

According to Eq. (6), $\bar{P}$ is the local solute permeability, measured at zero volume flow. This can readily be done in a simple chamber, with one compartment closed or flow kept at zero by other means. If measurements are carried out with small concentration differences, concentration dependence can be estimated by variation of the average concentration.

Considering the second coefficient, σ, as it appears in Eq. (6), one should measure solute flow caused solely by a small volume flow between identical solutions. This is a very difficult task, though it has been done in a few cases. It is much easier to use Eq. (16) for the volume flow, and to measure the pressure at which volume flow is zero for given osmotic pressure difference across the membrane.

$$\sigma = (\Delta p/\Delta\pi) \quad \text{at} \quad J_v = 0 \tag{18}$$

In practice one often substitutes an osmotic pressure gradient caused by an impermeable solute for the pressure gradient.

The application of Onsager's relation to reverse osmosis is thus: If membrane properties are not changed by the application of pressure, the limiting salt rejection for a given membrane material is equal to the effective osmotic pressure ratio, $(\Delta p/\Delta\pi)_{J_v=0}$, measured in a static experiment.

An important qualification has to be added here: it was emphasized above that in a two-layer membrane, with $\sigma_b=0$, R_∞ is equal to σ of the dense part, σ_a. The reflection coefficient for the two-layer membrane, determined in an osmotic measurement, is equal to σ_a only if the salt permeability of the loose part is very large as compared to that of layer a; otherwise a correction has to be applied[9].

For a comparison of the hydraulic permeability at high pressure with that measured at low pressure in a "static" cell, this measurement must be carried out at $\Delta\pi=0$. Since volume flow through a selective membrane must create an osmotic gradient, effective stirring on both sides of the membrane is essential. In other words, no support can be used. Volume measuring apparatus for this purpose, sensitive enough to apply quite low pressures, was recently developed[10]. With ideal, or practically ideal membranes it is of course possible to obtain A from osmotic flow experiments. There is surprisingly little information in the literature on an explicit comparison of A from pressure driven flow measurements and osmotic flow measurements with impermeable solutes.

In most cases the volume flow is very nearly equal to the water flow and, moreover, the hydraulic permeability, $\bar{A}$, is very close, though not equal, to the water permeability, $\bar{A}_w$, measured in the absence of permeable solutes.

$$\bar{V}_w J_w = -\bar{A}_w \frac{dp}{dx} = A_w \Delta p \qquad A_w = \bar{A}_w/\Delta x \tag{19}$$

The water permeability depends on the water content and the hydrodynamic resistance. The driving force acting on the water at any point in the membrane is $d\mu_w/dx$, creating a velocity, v_w.

$$v_w = -\frac{1}{f_{wm}} \frac{d\mu_w}{dx} = -m_{wm} \frac{d\mu_w}{dx} \tag{20}$$

where f_{wm} is the hydrodynamic resistance, or friction coefficient, of water in the membrane per mol of water, and m_{wm}, the reciprocal of f_{wm}, is the water mobility. To obtain the flow from the velocity, v_w has to be multiplied by its concentration in the membrane:

$$\bar{V}_w J_w = -\bar{V}_w \bar{c}_w m_{wm} \frac{d\mu_w}{dx} \tag{21}$$

Under all practically interesting circumstances $\bar{c}_w$ changes only slightly, and thus, for constant mobility, we may replace the gradient by differences. Introducing Eqs. (4) and (19),

$$\bar{A}_w = \bar{V}_w \phi_w m_{wm} \qquad \text{and} \qquad A = \frac{\bar{V}_w \phi_w m_{wm}}{\Delta x} \tag{22}$$

ϕ_w is the volume fraction of water. In Eq. (22) A is separated into two factors: the water content, which may be determined in an equilibrium measurement, and a kinetic parameter, m_{wm}. The physical interpretation of each of these depends on the specific system. For example, in a highly porous membrane, ϕ_w is the pore fraction and m_{wm} is related to the pore radius by Poiseuille's equation.

THE SOLUTION-DIFFUSION MODEL

After demonstration of the outstanding performance of the cellulose acetate membrane[8], various mechanisms were suggested to explain its high selectivity, combined with quite high water permeability. Closer study increasingly indicated that water does not pass by viscous flow through pores, but is dissolved in the dense cellulose acetate film and crosses it in very fine, near-molecular, dispersion.

The plausibility of selective permeation by dissolution, long accepted by physiologists, was demonstrated for cellulose acetate in liquid model systems[11]. Also, the extremely small salt permeability and the independence of salt and water flows, observed in various experiments, would be very hard to explain by a pore model[12]. As a further test for porosity the hydraulic permeability of cellulose acetate was compared with the diffusion of tagged water through the same membrane[13]. The "water permeability" derived from flow under pressure or osmotic gradient is usually larger than that derived from isotope exchange. This does not mean that there are two mechanisms for water flow, but simply that one is observing two different processes: in one case flow in one direction relative to the membrane and in the other flow of water molecules past the membrane and other water molecules. Only if the mutual drag between water molecules is negligible are the two equal. This was observed only in some special systems, like the black lipid membranes, which are intensively studied as models for biological membranes, and in some very slightly permeable polymers. For cellulose acetate the ratio is larger than one, but much lower than that obtained for pore flow.

In a solvent-membrane water flow is dependent only on its own thermodynamic potential, since there is no interaction between flows.

Water flow in cm^3 is:

$$\bar{V}_w J_w = A_w \frac{\Delta\mu_w}{\bar{V}_w} = A_w(\Delta p - \Delta\pi) \tag{23}$$

and the water permeability is given by Eq. (22) with the physical interpretation appropriate to the nature of the system: ϕ_w is the solubility of water in the membrane phase and m_{wm} its mobility. Both the thermodynamic factor, ϕ_w, and the kinetic factor, m_{wm}, are determined by membrane-water interactions on a molecular level. The relation between m_{wm} and basic parameters, like amount of crystallinity, mobility of polymer segments and kinetics of absorption-desorption has not yet been analysed. Operationally, m_{wm} is defined by Eqs. (22) and (23) in which everything except the mobility is directly measurable.

Following the general formality of diffusion in solution, $\underline{R}Tm_{wm}$ is often called the diffusion coefficient of water in the membrane. This product has indeed the dimension of a diffusion coefficient and it is illuminating to compare its order of magnitude to that of diffusion coefficients in liquids, but it must not be concluded that water flow during reverse osmosis in cellulose acetate may be correctly described by Fick's law, with a diffusion coefficient given by $\underline{R}Tm_{wm}$.

In ordinary liquid diffusion the only driving force acting on both solute and solvent is the activity gradient. The mutual diffusion coefficient, usually called the diffusion coefficient of the solute, is given by:

$$J_s = -\frac{c_s}{f_s}\frac{d\mu_s}{dx} = -\frac{1}{f_s} c_s \underline{R}T \frac{d\ln a_s}{dx}$$

$$J_s = -\frac{1}{f_s}\underline{R}T\left(1 + \frac{d\ln\gamma}{d\ln c_s}\right)\frac{dc_s}{dx}$$

Hence the diffusion coefficient is:

$$D = \frac{\underline{R}T}{f_s}\left(1 + \frac{d\ln\gamma}{d\ln c_s}\right) = \underline{R}T\, m_s\left(1 + \frac{d\ln\gamma}{d\ln c_s}\right)$$

where f_s is the hydrodynamic resistance of the solute (per mol), and m_s is the mobility and γ the activity coefficient of the solute. Only for ideal solutions is D equal to $\underline{R}Tm_s$.

In a polymer film in which an external pressure head causes water to move relative to the matrix, the relation between $d\mu_w/dx$ and a water concentration gradient is not at all clear. It may not even be meaningful to discuss the presence or absence of a hydrostatic pressure gradient in the polymer phase, and there is no reason to expect an ideal solution of water in, e.g., cellulose acetate in contact with water or a dilute aqueous solution[3].

For the very slightly solute salt, however, Fick's law may be a good approximation. The salt flow is driven by its thermodynamic potential gradient, independent of water flow. In a differential layer:

$$J_s = - m_{sm}\bar{c}_s \frac{d\mu_s}{dx}$$

where $\bar{c}_s$ is the concentration in the membrane at point x in $mols/cm^3$ and m_{sm} is the mobility of the salt. If the influence of the pressure gradient is negligible (cf. Eq. (5))

$$J_s = - m_{sm}\bar{c}_s \frac{d\mu_s}{dx} = - m_{sm}K_s c_s d\mu_s/dx = - m_{sm}K_s d\pi/dx \tag{24}$$

K_s is the distribution coefficient and c_s the concentration in a solution at equilibrium with the membrane at point x. For constant distribution and mobility:

$$J_s = \frac{m_{sm}K_s}{\Delta x} \Delta\pi \tag{25}$$

From Eqs. (3) and (25), the salt rejection of a uni-univalent salt is obtained as a function of J_v:

$$R = [1 + \frac{2\underline{R}Tm_{sm}K_s}{\Delta x\ J_v}]^{-1} \qquad (\Delta\pi \simeq 2\underline{R}T\Delta c_s) \tag{26}$$

R versus the pressure gradient is [3], from Eqs. (22), (23) and (26):

$$R = [1 + \frac{2\underline{R}Tm_{sm}K}{\bar{V}_w\ m_{wm}(\Delta p-\Delta\pi)}]^{-1} \tag{27}$$

$K = K_s/\phi_w$ is the solute distribution coefficient per cm^3 of water in the membrane.

At high enough pressure gradient R tends towards unity, with the pressure driven water flow overtaking the slow salt diffusion.

One may ask whether the possibility of ideal separation by hyperfiltration is guaranteed by the fact that water and solute pass the membrane independently: is $\sigma<1$ only if the solute is dragged along by the water? The following consideration shows that for complete rejection, $\sigma=1$, it is required in addition that solute permeability be low, relative to water permeability.

When dealing with a highly permeable solute which does not drag water, the contribution of J_s to the volume flow cannot be neglected in the measurement of the effective osmotic pressure. In the stationary state volume flow is zero, but water flow is not:

$$J_w\bar{V}_w = -J_s\bar{V}_s \tag{28}$$

Introducing Eqs. (23) and (25) in Eq. (28):

$$(-\bar{A}_w + \bar{V}_s K_s m_{sm})\Delta\pi + \bar{A}_w \Delta p = 0$$

and therefore

$$\sigma = (\Delta p/\Delta\pi)_{J_v=0} = 1 - \frac{\bar{V}_s K_s m_{sm}}{\bar{A}_w} \tag{29}*$$

Equation (29) was given previously[14] and was recently tested in a biological system[15]. For a series of lipid-soluble solutes, reflection coefficients between ∿0.1 and ∿1 were predicted from permeability measurements and observed directly in osmotic experiments.

POROUS MEMBRANES: THE FRICTION MODEL

In the large number of synthetic membrane systems investigated for desalination as well as other applications, independence of water and solute flows is an exception rather than the rule. In most membranes, the velocity of each species is influenced not

*If the contribution of pressure to μ_s is not neglected, one obtains:

$$\sigma = 1 - \frac{\bar{V}_s K_s m_{sm}(1 + c_s\bar{V}_s)}{\bar{A}_w}$$

only by its own driving force but also by the driving forces acting on all other species. The mutual drag may be expressed as friction between the ions and molecules moving at different velocities. At mechanical equilibrium the thermodynamic driving force acting on each particle is exactly balanced by the frictional forces between these particles and all others in the surrounding medium. For one mol of ion i the driving force is the gradient of the electrochemical potential, $-d\tilde{\mu}_i/dx$, and mechanical equilibrium is described by the force-balance equation

$$-\frac{d\tilde{\mu}_i}{dx} = \sum_{ij} f_{ij}(v_i - v_j)$$

where j denotes all other species including the membrane, and f_{ij} is the friction coefficient per mol. It is convenient to choose the membrane as the frame of reference so that $v_m=0$.

From these equations, permeability, transport number, membrane potentials, etc. can be derived in terms of distribution and friction coefficients[6,16-18]. In this and in the following sections we shall analyze exclusively the rejection of a single electrolyte at high flow rates.

The calculation of R_∞ is very simple, if it is remembered that at high volume flow the concentration gradient near the feed boundary vanishes.

The force balance equations for the two ions, 1 and 2, are:

$$\begin{aligned} -d\tilde{\mu}_1/dx &= f_{1w}(v_1 - v_w) + f_{1m}v_1 \\ -d\tilde{\mu}_2/dx &= f_{2w}(v_2 - v_w) + f_{2m}v_2 \end{aligned} \tag{30}$$

where the v's are velocities relative to the membrane and f_{ij} is the friction coefficient of one mol of i with water or membrane. We assume here that the friction between the ions is negligible.

The flow of each ion is the product of its velocity and concentration

$$J_i = \bar{c}_i v_i$$

For simplicity we shall consider only uni-univalent salts. For these, electroneutrality (in the absence of an electric current) requires that

$$J_1 = J_2 \equiv J_s \tag{31}$$

and the thermodynamic potential of the salt is given by the sum of $\tilde{\mu}_1$ and $\tilde{\mu}_2$. Thus in the absence of a concentration gradient

$$\frac{d\mu_s}{dx} = \frac{d\tilde{\mu}_1}{dx} + \frac{d\tilde{\mu}_2}{dx} = 0 \tag{32}$$

Volume flow is practically equal to water flow, so that

$$\frac{J_w}{\bar{\bar{c}}_w} = \frac{J_v}{\phi_w} \tag{33}$$

With the restrictions and approximations expressed in Eqs. (31), (32) and (33), Eqs. (30) give:

$$J_s\left(\frac{f_{1m}+f_{1w}}{\bar{c}_1{}'} + \frac{f_{2m}+f_{2w}}{\bar{c}_2{}'}\right) = \frac{f_{sw}}{\phi_w} J_v \qquad (f_{sw} = f_{1w} + f_{2w}) \tag{34}$$

The concentrations $\bar{c}_1{}'$ and $\bar{c}_2{}'$ are ion concentrations in the membrane, close to the surface in contact with the feed solution.

If the polymer matrix does not carry fixed charges, $\bar{c}_1 = \bar{c}_2$, and the ratio between salt and volume flow is given by:

$$J_s/J_v = \frac{f_{sw}}{f_{sw} + f_{sm}} \frac{K_s c_s}{\phi_w} \tag{35}$$

$$f_{sm} = f_{1m} + f_{2m}$$

Introducing the salt distribution per cm^3 of water in the membrane, the salt rejection is given by:

$$R_\infty = \sigma = 1 - \frac{f_{sw}}{f_{sw} + f_{sm}} K = 1 - \beta K \tag{36}$$

$1-\sigma$ is thus separated into an equilibrium factor, K, and a kinetic factor, β[1,2].

According to Eq. (36) salt rejection may be achieved in an uncharged membrane if

a) salt is at least partly excluded from the water contained in the membrane, and/or

b) the "frictional" interaction between salt and membrane is not negligible compared to the friction salt-water. While ultrafiltration of macromolecules by porous membranes has been traditionally based on these principles, it is very difficult to achieve either salt exclusion or considerable frictional retardation of sodium chloride or similar electrolytes by uncharged porous membranes.

CHARGED MEMBRANES

The Ideal Fixed Charge Model

When the polymer carries charged groups, salt can be excluded to a considerable degree even from porous membranes, by a Donnan equilibrium which is established between membrane phase and aqueous solutions. Salt rejection by the Donnan mechanism is unfortunately, and unavoidably, decreased by increasing salt concentration.

Following Teorell[19] and others, who suggested and analyzed the fixed charge model for ion transport through membranes[6], we shall assume ideality and homogeneous distribution of ions not only in the external solutions, but also in the water-filled pores of the membrane[7]. Electroneutrality requires that the counter-ion concentration, $\bar{c}_1$, balance both the fixed charges and the co-ions:

$$\bar{c}_1 = \bar{c}_2 + X \tag{37}$$

(X = fixed charge concentration in mols per cm^3). The co-ion concentration is then the salt concentration in the membrane and the salt distribution coefficient K is defined by

$$K = \bar{c}_2/\phi_w c_s$$

At equilibrium between membrane and solution

$$\bar{c}_1 \bar{c}_2/\phi_w^2 = c_s^2 \tag{38}$$

The internal counter-ion concentration is larger than the external concentration:

$$\bar{c}_1/\phi_w = c_s/K$$

From (37) and (38) the salt distribution coefficient is readily obtained as a function of the ratio between external salt concentration and membrane charge density.

$$K = \frac{1}{2} \frac{X}{c_s \phi_w} \left(\sqrt{1 + \frac{4 c_s \phi_w^2}{X^2}} - 1\right) \tag{39}$$

In Eq. (34), relating salt and volume flow, we now have to introduce the two different concentrations $\overline{c}_1'$ and $\overline{c}_2'$. This gives:

$$J_s/J_v = K' c_s' \frac{f_{sw}}{(f_{1w}+f_{1m})K'^2 + f_{2w}+f_{2m}} \tag{40}$$

K' is the salt distribution coefficient at the external concentration c_s'. The limiting salt rejection is:

$$R_\infty = \sigma = 1 - \frac{1}{t_1+t_2K'^2} \frac{f_{sw}}{f_{sw}+f_{sm}} K' \tag{41}$$

where

$$t_1 = \frac{f_{2w}+f_{2m}}{f_{sw}+f_{sm}} \quad \text{and} \quad t_2 = 1 - t_1 \tag{42}$$

The kinetic factor is now

$$\beta = \frac{1}{t_1+t_2K'^2} \frac{f_{sw}}{f_{sw}+f_{sm}} \tag{43}$$

Comparison of (36) and (42) shows that in charged membranes β is multiplied by an additional factor, larger than one.

If the friction between ions and membrane matrix is negligible, R_∞ reduces to

$$R_\infty = 1 - \frac{1}{t_1+t_2(K')^2} K' \tag{44}$$

Assuming that the mobilities of the ions in the pores are the same as in equeous solution (except for a tortuosity effect, equal for both), t_1 and t_2 are the transport numbers of counter-ion and co-ion in free solution.*

Equations (36) and (44) show that by a consideration of the salt exclusion alone, as if the membrane were uncharged, one would overestimate the salt rejection. This is not surprising, since K really measures the exclusion of the co-ions, while the concentration of counter-ions is quite large. These tend to be swept along with the water and create an electric field which accelerates the co-ions and retards the counter-ions, maintaining electroneutrality. The rejection-decreasing effect of β is more pronounced when the mobility of the co-ion is relatively large - in extreme cases this may even lead to a negative value of R[7,20,21].

Ion Association in Charged Membranes

The starting point of the ideal fixed charge model is the ideal Donnan equilibrium - clearly an unrealistic assumption. Experience with ion-exchangers and with polyelectrolyte solutions shows that the salt exclusion is much smaller than predicted by Eq. (39). More specifically, the activity coefficient of the counter-ion is much smaller than one, while for the co-ion, activity and concentration are approximately equal.

Detailed calculations of the electrostatic interaction between polyion, or charged matrix, and counter-ion were given[22]. It turns out, however, that a good description of experimental data can be derived from a simple association model[23]. If one assumes that only a fraction, ϕ_p, of the fixed charges are dissociated, and the rest are neutralized by "bound" counter-ions, one arrives at correct predictions of osmotic pressure and salt distribution over a wide range of polyelectrolyte and salt concentrations.

These concepts are readily applied to ion transport in charged membranes[24]. An effective charge density, $\phi_p X$, is substituted for the stoichiometric charge density, and in the concentration of counter-ions, $\bar{c}_1$, only the free counter-ions are included. These changes do not affect the relationship between salt exclusion and salt rejection. The association model can thus be tested in two steps:

*In reference 7 R was calculated by integration of the flow equations, with the appropriate boundary conditions. For R_∞ the relation given in Eq. (44) was obtained.

1. The relation between K and R_∞ should be given by Eq. (44), with the effective concentrations.

2. The ratio between the effective charge density derived from the measured distribution coefficient and the stoichiometric charge density should agree with the values found in corresponding polyelectrolyte solutions.

Results on one type of positive polyelectrolyte membranes gave fair agreement on both points, but the model has yet to be tested for different membranes.

Inhomogeneous Ion Distribution

It was assumed so far that ion concentrations are uniform across the water filled pores. This assumption is justified only if the pore radius is smaller than the thickness of the electric double layer, which may not always be the case in porous membranes. For very large pores, where the thickness of the double layer is a small fraction of the pore radius, transport phenomena have been described by the classical Helmholtz theory. Recently, detailed solutions were presented for stationary and non-stationary flows in highly porous membranes[25]. These solutions, however, are not directly relevant to hyperfiltration: no appreciable salt rejection can be obtained in these systems.

Salt exclusion in charged porous bodies with intermediary capillary dimensions was treated by Dresner[26]. In these systems, like, e.g., a mass of clay particles, the fixed charges are assumed to be located on the surfaces of cylindrical pores or on an assembly of rods, and the mobile ions attain an equilibrium distribution in the pores which is determined by the external concentration, the charge density and the dimensions of the pores. From the Poisson-Boltzmann equation, quantitative results were obtained for the limiting cases of good and bad ion-exclusion and an approximation was given for the whole region. The relation between salt rejection and salt exclusion was not treated explicitly, but the author states that for certain limiting conditions, the concentration of the filtrate is expected to be 3/2 the average concentration of co-ions in the pores. The ratio between the concentration of the product at high flow rates and that of the pore-solution, at equilibrium with the feed, is equal to the kinetic factor, β, defined previously:

$$\frac{c''}{(\bar{c}'/\phi_w)} = \frac{c''}{c'} \frac{c'}{(\bar{c}'/\phi_w)} = \frac{1-R_\infty}{K} = \beta$$

The value given by Dresner for β is lower (i.e., R_∞ is higher) than that expected from Eq. (44) for NaCl in a finely porous membrane at high salt exclusion.

A slightly different model was chosen by R. Simons[27]. The fixed charges are located in ion-exchange walls, separated by water-filled pores or channels, and the ions equilibrate between ion-exchanger and pore-solution. Numerical integration of the Poisson-Boltzmann equation gave the equilibrium ion distribution, and the Navier-Stokes equation the velocity profile in the rectangular pores during hyperfiltration. Limiting salt rejection for given salt exclusion was found to be, rather surprisingly, very close to that given in Eq. (44), even if the ion concentration in the middle of the pore was considerably different from that near the wall. The calculated ion exclusion was found to be, as expected, poorer than for homogeneous distribution of the fixed charges.

The inhomogeneous distribution of co-ions in the pores is represented by a step function in the model of Yasuda and Schindler[28]: close to the charged walls the co-ions do not invade at all, and near the center of the pore their concentration is equal to that of the feed. The ratio between the two transport volumes is thus the average distribution coefficient of the co-ions, per cm^3 of water in the membrane. It is not easy to envisage, however, that the water layer close to the polymer matrix would move as fast as the water in the middle of the pore, as assumed by these authors. For viscous flow in a capillary, the velocity near the wall must be smaller, and if some of the water is bound to the functional groups on the matrix[28], the difference in the velocities should be even more pronounced.

For a more detailed picture of flow in charged porous membranes, allowing for hydration and specific ion-ion interactions, the state of water and ions in the membrane will have to be investigated by a wider variety of physico-chemical methods.

ACKNOWLEDGMENT

The author would like to express thanks to the Office of Saline Water for the support of work on hyperfiltration membranes at the Weizmann Institute.

APPENDIX

The dissipation function per unit volume at point x is:

$$\phi = - J_w \frac{d\mu_w}{dx} - J_s \frac{d\mu_s}{dx} \qquad \text{(A1)}$$

If J_v and J_s are chosen as the flows, their conjugate forces X_v and X_s must be such that the dissipation function remains unchanged.

$$(J_w \bar{V}_w + J_s \bar{V}_s) X_v + J_s X_s = J_w \left(- \frac{d\mu_w}{dx}\right) + J_s \left(- \frac{d\mu_s}{dx}\right) \qquad \text{(A2)}$$

Identification of coefficients for J_w and J_s gives, with Eqs. (4) and (5)

$$X_v = - \frac{1}{\bar{V}_w} \frac{d\mu_w}{dx} = - \left(\frac{dp}{dx} - \frac{d\pi}{dx}\right) \qquad \text{(A3)}$$

$$X_s = - \frac{d\mu_s}{dx} + \frac{\bar{V}_s}{\bar{V}_w} \frac{d\mu_w}{dx} = - \frac{1}{c_s} \frac{d\pi}{dx} \qquad \text{(A4)}$$

$$\phi = - J_v \left(\frac{dp}{dx} - \frac{d\pi}{dx}\right) - J_s \frac{1}{c_s} \frac{d\pi}{dx} \qquad \text{(A5)}$$

Both J_v and J_s are linear functions of X_v and X_s. The Onsager relation requires:

$$(J_s/J_v)_{d\pi/dx=0} = - \left[\left(\frac{dp}{dx} - \frac{d\pi}{dx}\right) / \left(\frac{d\pi}{dx} \frac{1}{c_s}\right)\right]_{J_v=0}$$

$$= c_s \left[1 - \frac{dp}{dx} / \frac{d\pi}{dx}\right]_{J_v=0}$$

From Eq. (6)

$$(J_s/J_v)_{d\pi/dx=0} = (1-\sigma) c_s$$

Hence

$$\left(\frac{dp}{dx}/\frac{d\pi}{dx}\right)_{J_v=0} = \sigma \tag{A6}$$

Pusch and Woermann[29] base their discussion of the relation between salt rejection and the reflection coefficient on another set of fluxes and forces: the volume flow and the chemical flux density χ, defined by

$$\chi' = -\frac{V'}{ac_{av}}\frac{dc'}{dt} \quad \text{or} \quad \chi'' = \frac{V''}{ac_{av}}\frac{dc''}{dt}, \tag{A7}$$

where a is the membrane area, V' and V'' are the volumes of the feed and product compartment, respectively, and

$$c_{av} = \frac{c_s' + c_s''}{2} .$$

The authors assume that $\chi' = \chi''$. A single chemical flux density, χ, is a useful approximation when forces and fluxes are small, but this is in general not the case in practical hyperfiltration.

From the mass balance the concentration changes in both compartments are:

$$\frac{dc'}{dt} = \frac{-aJ_s}{V'} + \frac{ac'J_v}{V'}$$

$$\text{and} \quad \frac{dc''}{dt} = \frac{aJ_s}{V''} - \frac{ac''J_v}{V''} . \tag{A8}$$

Hence, the difference between χ'' and χ' is given by

$$\chi'' - \chi' = \frac{c'J_v - c''J_v}{c_{av}} = \frac{c'RJ_v}{c_{av}} . \tag{A9}$$

At stationary hyperfiltration $V''dc''/dt=0$, but $V'dc'/dt$ does not tend to zero with increasing V', and is related to the flows by equations (A8) and (A9).

From the assumption $\chi' = \chi'' = 0$, the authors obtain that $(\Delta\pi/\Delta p)$ in hyperfiltration is a function of transport coefficients only, which would not lead to the generally observed tendency towards a limiting salt rejection.

LIST OF SYMBOLS

Subscripts 1 and 2 stand for counter-ion and co-ion, respectively. Subscripts w and s stand for water and salt, respectively, subscript m for membrane. Superscripts ' and '' stand for feed and product, respectively.

$\overline{A}$	local hydraulic permeability, $cm^3/watt\text{-}sec^2$
A	hydraulic permeability, $A = \overline{A}/\Delta X$
$\overline{A}_w$	local water permeability
A_w	water permeability
a	membrane area, cm^2
B	cf. Equation (11)
$c_{s,w}$	concentration in solution, mol/cm^3
$\overline{c}_{s,w}$	concentration in membrane, mol/cm^3
D	diffusion coefficient, cm^2/sec
f_{ij}	friction coefficient between species i and j, $watt\text{-}sec^2/cm^2\text{-}mol$
$J_{s,w}$	solute and water flow, $mol/cm^2\text{-}sec$
J_v	volume flow, cm/sec
K_s	solute distribution coefficient, $\overline{c}_s/c_s$
K	solute distribution coefficient, $K=K_s/\phi_w$
m_{sm}, m_{wm}	mobility of solute and water in membrane
$\overline{P}$	local solute permeability, cm^2/sec
P	solute permeability, $P=\overline{P}/\Delta X$
p	pressure, $watt\text{-}sec/cm^3$
$\underline{R}$	gas constant
R	salt rejection, Equation (2)
S	desalination ratio, $S=c'/c''$

$t_{1,2}$	cf. Equation (42)
T	temperature
v_i	linear velocity, cm/sec
$\overline{V}_i$	molar volume, cm^3/mol
x	length coordinate
Δx	membrane thickness
χ	charge density, mol/cm^3
β	kinetic factor, Equation (36)
μ_i	thermodynamic potential, watt-sec/mol
ϕ_p	osmotic factor in salt-free polyelectrolyte solution
ϕ_w	volume fraction of water in membrane, $\phi_w = \overline{c}_w \overline{V}_w$
π	osmotic pressure, watt-sec/cm^3
σ	reflection coefficient

REFERENCES

1. K. S. Spiegler and O. Kedem, Desalination, 1, 311 (1966).

2. J. S. Johnson, Jr., L. Dresner and K. A. Kraus, Chapter 8 in Principles of Desalination, K. S. Spiegler (Ed.), Academic Press, New York (1966).

3. U. Merten, (Ed.), Desalination by Reverse Osmosis, M.I.T. Press, Cambridge, Mass. (1966).

4. A. J. Staverman, Rec. Trav. Chim., 70, 344 (1951).

5. W. Banks and A. Sharples, J. Appl. Chem. 16, 153 (1966).

6. R. Schlögl, Stofftransport durch Membranen, Steinkopf Verlag, Darmstadt, 1964. pp. 66, 73.

7. E. Hoffer and O. Kedem, Desalination, 2, 25 (1967)

8. S. Loeb and S. Sourirajan, Advan. Chem. Ser. 38, 117 (1962).

9. J. Jagur-Grodzinski and O. Kedem, Desalination, 1, 327 (1966).

10. C. Stymans and H. Fischer, Desalination, 6, 203 (1969).

11. K. A. Kraus, R. J. Raridon and W. H. Baldwin, J. Am. Chem. Soc. 86, 2571 (1964).

12. H. K. Lonsdale, U. Merten and R. L. Riley, J. Appl. Polymer Sci. 9, 1341 (1965).

13. G. Thau, R. Bloch and O. Kedem, Desalination, 1, 129 (1966).

14. O. Kedem and A. Katchalsky, Biochim. Biophys. Acta, 27, 229 (1958).

15. P. Smulders and E. M. Wright, J. Membr. Biol., 5, 297 (1971).

16. K. S. Spiegler, Trans. Faraday Soc., 54, 1409 (1958).

17. O. Kedem and A. Katchalsky, J. Gen. Physiol., 45, 143 (1961).

18. P. Meares, D. G. Dawson, A. H. Sutton and J. F. Thain, Ber. Bunsenges., 71, 765 (1967), and references there.

19. T. Teorell, Progr. Biophysics, 3, 305 (1953).

20. E. Hoffer and O. Kedem, Desalination, 5, 167 (1968).

21. W. Pusch and D. Woermann, Naturwissenschaften, 55, 228 (1969).

22. S. Lifson and A. Katchalsky, J. Polymer Sci., 13, 43 (1953).

23. A. Katchalsky, Z. Alexandrowicz and O. Kedem, in Chemical Physics of Ionic Solutions, B. E. Conway and R. G. Barradas (Eds.), Wiley, New York (1966).

24. E. Hoffer and O. Kedem, unpublished results.

25. Y. Kobatake and H. Fujita, J. Chem. Phys. 40, 2212,2219 (1964); P. Meares and K. R. Page, Phil. Trans. Roy. Soc., in press.

26. L. Dresner and K. A. Kraus, J. Phys. Chem., 67, 990 (1963); L. Dresner, J. Phys. Chem., 69, 2230 (1965)

27. R. Simons and O. Kedem, to be published.

28. See the paper by H. Yasuda and A. Schindler in this volume, and references there.

29. W. Pusch and D. Woermann, Ber. Bunsenges., 74, 445 (1970).

CONCENTRATION POLARIZATION IN HYPERFILTRATION SYSTEMS

W. Pusch
Max-Planck-Institut für Biophysik
6 Frankfurt am Main, Germany

INTRODUCTION

When using hyperfiltration membranes for brackish or seawater desalination, an enrichment of the solute (concentration polarization) always occurs at the membrane surface juxtaposed with the brine solution. This is due to the formation of a boundary layer separating the membrane surface from the bulk flow. The thickness of this boundary layer is dependent upon the stirring or streaming velocity. Because the salt transport within the boundary layer is not affected by the bulk flow, the rejected salt at the membrane surface must return to the bulk flow by means of diffusion only. This diffusion of salt into the well stirred or streaming solution is counterbalanced by the convective salt flux (associated with the permeation of water through the membrane), thereby forming a concentration gradient within the boundary layer. The enhanced salt concentration at the membrane surface results in a lower water flux and salt rejection. Thus, the "intrinsic water flux" and the "intrinsic salt rejection", which would be observed without concentration polarization, are always greater than the observed water flux and salt rejection. The effect of concentration polarization upon volume flow, q, and salt rejection, r_{obs}, of a membrane was first investigated by Merten[1]. He showed that the volume flow is limited by concentration polarization because of the effect on local osmotic pressure. Furthermore, Merten et al.[2] determined the effect of streaming velocity on volume flow and salt rejection for NaCl and $CaSO_4$ solutions. As a consequence of the poor solubility of $CaSO_4$ concentration polarization can lead to precipitation of salt near the membrane surface. Banks and Sharples[3] assumed that the asymmetric behavior of Loeb-

Sourirajan type membranes was attributable to concentration polarization. More detailed theoretical and experimental investigations were performed by Sherwood et al.[4] and Raridon et al.[5]. These authors calculated the concentration profile within the unstirred boundary layer for laminar and turbulent flow. The theoretical findings of these authors will be treated in greater detail within the following section.

THEORETICAL TREATMENT OF TURBULENT FLOW

In Figure 1, the flow condition near the membrane surface is illustrated. The treatment follows the ideas presented by Brian[6]. The simplest model of a concentration boundary layer in turbulent flow is the Nernst film model. The boundary layer in this model is considered to be a thin liquid film which separates the membrane surface from the bulk flow. Eddy diffusion is assumed negligible, thus limiting mass transport within the film to diffusion and convection mechanisms. The bulk flow is assumed to be turbulent so as no concentration gradients exist outside of the boundary layer. Furthermore, mass flow, parallel to the membrane surface, is assumed nonexistent. The ultimate result is a one dimensional flow problem. With a steady state established, the salt flux is determined by the following differential equation:

$$\phi_s = - D_s \frac{dc_s}{dy} - c_s q \;, \tag{1}$$

where ϕ_s = salt flux (mol/cm^2-sec),

D_s = diffusion coefficient of salt (cm^2/sec),

q = volume flux (cm/sec),

y = distance coordinate perpendicular to the membrane surface (cm) (See Fig. 1),

c_s = salt concentration (mol/cm^3).

With use of the simplifications from the aforementioned film model, the salt flux ϕ_s becomes constant throughout the boundary layer film. Using the salt rejection $r = (c_s^m - c_s'')/c_s^m$, one obtains the following relation:

$$\phi_s = - (1 - r)\, c_s^m\, q \;, \tag{2}$$

where c_s' = brine concentration of salt,

c_s'' = product concentration of salt,

c_s^m = concentration at the membrane surface.

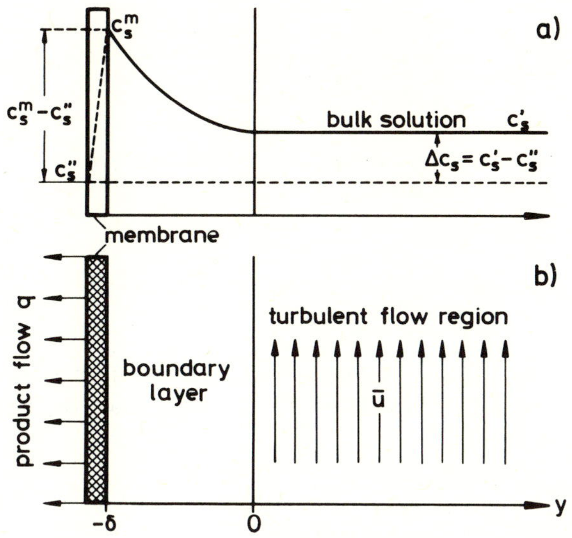

Fig. 1: a) Concentration profile near the membrane surface.
b) Flow conditions near the membrane surface for the film model.

With this relation for ϕ_s, it is assumed that the concentration of water throughout the entire solution is practically constant and nearly 1 gram/cc. Furthermore, the permeation velocity q is assumed constant within the film. Eqs. (1) and (2) together yield:

$$D_s \frac{dc_s}{dy} + c_s q = (1 - r) c_s^m q \ . \tag{3}$$

Using the following boundary conditions:

$$c_s = c_s^m \quad \text{at } y = -\delta,$$

$$c_s = c_s' \quad \text{at } y = 0 \ ,$$

where δ = boundary layer thickness, one can integrate across the film to yield:

$$\frac{c_s^m}{c_s'} = \frac{\exp(q\delta/D_s)}{r + (1 - r)\exp(q\delta/D_s)} \ . \tag{4}$$

Because the actual boundary layer thickness is unknown, it is useful to introduce the mass transfer coefficient which would exist for the same flow system and the same Reynolds number but in the absence of permeation velocity. The mass transfer coefficient, k, is defined as $k \equiv D_s/\delta$. The introduction of the mass transfer coefficient into Eq. (4) yields:

$$\frac{c_s^m}{c_s'} = \frac{\exp(q/k)}{r + (1 - r)\exp(q/k)} . \tag{5}$$

For the special case of complete desalination (r = 1), Eq. (5) simplifies to the following:

$$\frac{c_s^m}{c_s'} = \exp(q/k) .$$

If the Chilton-Colburn factor, defined as $j \equiv (k/\bar{u})N_{Sc}^{2/3}$, is introduced into Eq. (5), the following equation results:

$$\frac{c_s^m}{c_s'} = \frac{\exp[(q/j\bar{u})N_{Sc}^{2/3}]}{r + (1 - r)\exp[(q/j\bar{u})N_{Sc}^{2/3}]} , \tag{6}$$

where $\bar{u}$ is the average velocity of bulk solution past the membrane, N_{Sc} the Schmidt number, defined as $N_{Sc} \equiv \nu/D_s$, and ν is the kinematic viscosity. Introducing the observed salt rejection, $r_{obs} = (c_s' - c_s'')/c_s'$, and the intrinsic salt rejection, $r = (c_s^m - c_s'')/c_s^m$, Eq. (6) can be rearranged to yield:

$$\ln[(1 - r_{obs})/r_{obs}] = (q/j\bar{u})N_{Sc}^{2/3} + \ln[(1 - r)/r] . \tag{7}$$

As Shor et al.[7] have shown, there exists a similar relation between r_{obs}, r, q, and $\bar{u}$ for tubes with cylindrical symmetry:

$$\ln[(1 - r_{obs})/r_{obs}] = k'[(d/\nu)^{1/4} N_{Sc}^{2/3} q/\bar{u}^{3/4}] + \ln[(1 - r)/r] . \tag{8}$$

k' refers to a semiempirical constant in the Chilton-Colburn analogy (see Sherwood[8]) and d is the tube diameter.

With this simple treatment, neither eddy diffusion nor other kinds of convective flow possessing density instabilities as a consequence of concentration gradients (cf. the Benard Problem[9]) are considered. If eddy diffusion is taken into account, the same basic differential equation is obtained but the mass transfer coefficient k will become a function of y.

Therefore, no complete integration of the differential equation is possible. This problem is dealt with in greater detail by Brian[6].

EXPERIMENTAL INVESTIGATIONS OF CONCENTRATION POLARIZATION WITH TURBULENT FLOW

The dependence of r_{obs} on q and $\overline{u}$, established by Eq. (8), was examined experimentally by Shor, Kraus, Johnson, and Smith [7]. Dynamic membranes of Zr(IV) oxide were formed on the outer surface of a carbon tube. The carbon tubes measured 11.0 mm o.d., had a median pore radius in the range 0.18 to 0.34 micron, and possessed a void fraction of 0.22 to 0.28. Using 0.015 to 0.26 molar $MgCl_2$ solutions and 0.0068 to 0.1 molar NaCl solutions, r_{obs} was determined as a function of q and $\overline{u}$. With these fixed-charge membranes it is necessary to take into account the dependence of the intrinsic rejection r on the concentration c_s^m. With the aid of Eq. (8), the authors were able to determine experimentally the value of r of a membrane by plotting $\ln[(1 - r_{obs})/r_{obs}]$ against $q/\overline{u}^{3/4}$ and extrapolating the produced straight line to $\overline{u} \to \infty$ or $1/\overline{u} = 0$. In the measurements cited, the variation of q was from 0.20 cm/min to 0.33 cm/min, whereas $\overline{u}$ was chosen between 110 cm/sec and 670 cm/sec (the higher value corresponding to a Reynolds number of 40,000 at room temperature).

In a recent publication, Pusch et al.[10] discussed the effect of concentration polarization on r_{obs} and the permeation velocity q with turbulent flow conditions. Using 0.05, 0.10, and 0.50 molar NaCl solutions, the dependence of q and r_{obs} on pressure was measured with pressures up to 100 atm. The thickness of the unstirred layer was raised by covering the active surface of the asymmetric cellulose acetate membranes with membrane filters of suitable pore size. In Figure 2, the functions $q(\Delta P)$ and $r_{obs}(\Delta P)$ are diagrammatically shown. Using Eq. (4), the concentration profile within the boundary layer was calculated and the concentration c_s^m estimated. For a boundary layer thickness of 300μ and a permeation velocity $q = 3\times10^{-4}$ cm/sec, the calculated concentration profile is reproduced in Figure 3. In addition, the authors were able to estimate the effective osmotic pressures by two different methods: (1) by use of the relation $q = L_p(\Delta P - \Delta\Pi_{eff})$, and (2) by use of the measured values of r_{obs} and r. In the first case, it was assumed that the diminution of q with increasing boundary layer thickness was caused only by the increased osmotic pressure due to concentration polarization. The agreement between the two determined values for $\Delta\Pi_{eff}$ was good. Furthermore, it was shown that the enrichment of salt leads to values of $c_s^m/c_s^{\prime}$ up to 10 with boundary layer thicknesses between 100μ and 300μ. Simultaneously, it was

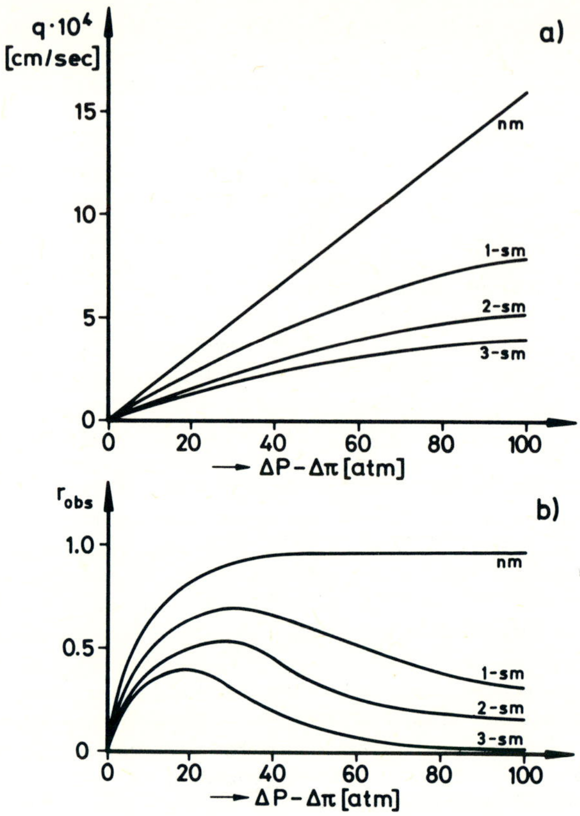

Fig. 2: a) Volume flux, q, as a function of effective pressure (ΔP-$\Delta\Pi$) using 0.1 molar NaCl brine solution; nm: air-dried surface juxtaposed with brine; 1-sm, 2-sm, 3-sm = air-dried surface juxtaposed with brine and covered with one, two, or three membrane filters, respectively. $\Delta\Pi$ = osmotic difference between brine and product. From Pusch et al.[10].

b) Salt rejection, r_{obs}, as a function of effective pressure for an asymmetric cellulose acetate membrane annealed at 82.5°C. From Pusch et al.[10].

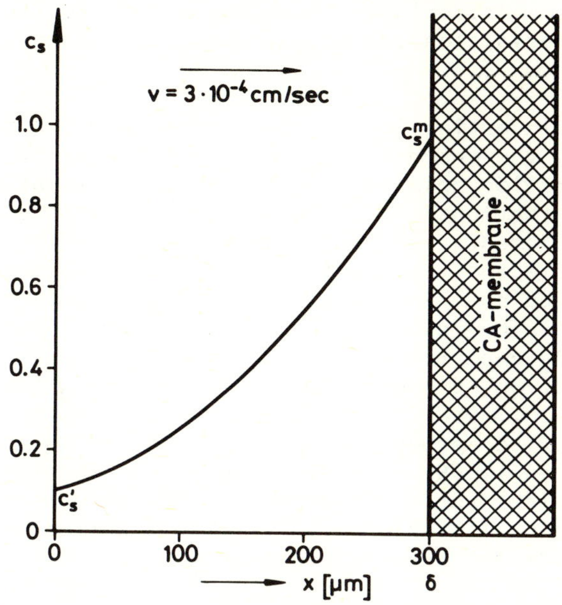

Fig. 3: Calculated concentration profile within the overlying membrane ($c_s' = 0.1$ molar NaCl solution, $D_s = 8 \times 10^{-6} cm^2/sec$). From Pusch et al. [10].

proved experimentally that the asymmetric behavior of Loeb-Sourirajan type membranes is a consequence only of concentration polarization within the porous substructure of the cellulose acetate membranes.

Similar investigations were carried out by Sheppard and Thomas[11] with the sedimentation of fouling products and colloids on the membrane surface in hyperfiltration systems. Sheppard and Thomas covered the surface of the membranes with artificial films in order to examine the effect of deposits on r_{obs} and q. They observed a strong dependence of r_{obs} on the thickness of the covering layer. With a q of about 2.5×10^{-4} cm/sec, r_{obs} reached a maximum with films having thicknesses of 100μ and 200μ and thereafter decreased with increasing pressure. r_{obs} decreased from approximately 93% for an uncovered membrane, to approximately 20% for a membrane covered with a film of 200μ thickness and having a q of about 1.18×10^{-3} cm/sec. In this same connection Thomas et al.[12-16] demonstrated that it is possible to fabricate detached promoters which neither block the membrane surface nor provide stagnant regions where solids can accumulate. At streaming velocities of about 23 cm/sec in their system, these turbulence promoters prevented the sedimentation of fouling materials on the membrane surface, thereby enhancing q and r_{obs}.

With detailed investigations, Strathmann[17] attempted to test experimentally the theoretical predictions of Brian[6] concerning concentration polarization. Using NaCl solutions of different concentrations and cellulose acetate membranes for one series of measurements, and tetrabutylammonium chloride solutions and polyelectrolyte resin membranes for another series, the behavior of turbulent flow hyperfiltration systems at about 7 atm was analysed. The relation of c_s^m/c_s^l to $\overline{u}$ for 100 cm/sec $< \overline{u} <$ 600 cm/sec was experimentally determined by analysing the data for salt rejection, product flow, salt flux, and feed and product concentration. Whereas the experimentally determined values for c_s^m/c_s^l varied between 1.1 and 1.5, the calculated values varied between 1.005 and 1.01. The reason for these differences is not obvious but they may be attributable to the indirect experimental determination of c_s^m/c_s^l. Similar experiments by Bixler and Cross[18] confirm the results of Strathmann.

THEORETICAL TREATMENT OF CONCENTRATION POLARIZATION FOR LAMINAR FLOW

For the theoretical treatment of concentration polarization in turbulent flow, it was unnecessary to solve the differential equations for the flow profile. However, with laminar flow one must first solve the Navier-Stokes equations. The basic solution was given by erman[19]. He solved the Navier-Stokes equations for laminar flow within a tube of rectangular cross section having porous walls. One side of the cross section, representing the distance between the porous walls, is taken to be much smaller than the other. The following conditions are imposed: (1) a steady state prevails, (2) the fluid is imcompressible, (3) no external forces act upon the fluid, (4) the flow is laminar, and (5) the velocity of the fluid leaving the walls of the channel is independent of position. A coordinate system is chosen with origin at the center of the channel. The y axis is perpendicular to the channel walls and the x axis is in a plane parallel to the channel walls. (See Figure 4.) The distance between the walls is taken to be 2h and the channel length is L. Under the assumed conditions and choice of axes, the Navier-Stokes equations are:

$$u \frac{\partial u}{\partial x} + \frac{v}{h} \frac{\partial u}{\partial \lambda} = - \frac{1}{\rho} \frac{\partial p}{\partial x} + \nu \left[\frac{\partial^2 u}{\partial x^2} + \frac{1}{h^2} \frac{\partial^2 u}{\partial \lambda^2} \right] , \tag{9}$$

$$u \frac{\partial v}{\partial x} + \frac{v}{h} \frac{\partial v}{\partial \lambda} = - \frac{1}{h\rho} \frac{\partial p}{\partial \lambda} + \nu \left[\frac{\partial^2 v}{\partial x^2} + \frac{1}{h^2} \frac{\partial^2 v}{\partial \lambda^2} \right] , \tag{10}$$

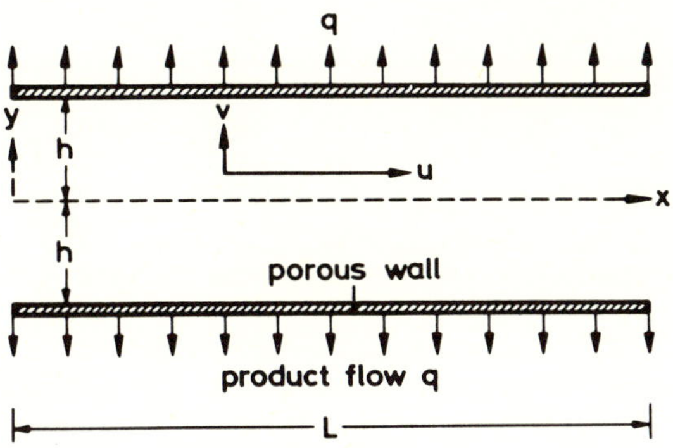

Fig. 4: Flow conditions in a two-dimensional duct.

and the continuity equation is:

$$\frac{\partial u}{\partial x} + \frac{1}{h}\frac{\partial v}{\partial \lambda} = 0, \tag{11}$$

where the dimensionless variable $\lambda = y/h$ has been introduced, and where ρ is the density.

The boundary conditions are:

$$u(x,\pm 1) = 0 \tag{12}$$

$$(\partial u/\partial \lambda)_{\lambda=0} = 0 \tag{13}$$

$$v(x,0) = 0 \tag{14}$$

$$v(x,\pm 1) = q = \text{constant}. \tag{15}$$

For a two-dimensional incompressible flow a stream function ψ exists such that:

$$u(x,\lambda) = (1/h)(\partial\psi/\partial\lambda), \tag{16}$$

$$v(x,\lambda) = -(\partial\psi/\partial x), \tag{17}$$

and the continuity equation, (11), is satisfied. For constant wall velocity q and the given boundary conditions, a suitable choice of stream function is:

$$\psi(x,\lambda) = [h\overline{u}(0) - qx]f(\lambda), \tag{18}$$

where $\overline{u}(0)$ is the mean velocity of the solution at x=0, giving for the velocity components of Eqs. (16) and (17):

$$u(x,\lambda) = [\overline{u}(0) - (qx/h)]f'(\lambda), \tag{19}$$

$$v(\lambda) = qf(\lambda). \tag{20}$$

In these equations $f(\lambda)$ is some function, yet to be determined, of the distance parameter λ. It is to be noted that under the assumption of constant wall velocity q, the y component of velocity v becomes a function of λ only. If Eqs. (19) and (20) are substituted into the equations of motion (9) and (10), there results, after some rearrangement, a differential equation of third order for $f(\lambda)$. This can be solved by means of a perturbation solution. With the first-order perturbation solution, one obtains the following expressions for the velocity components:

$$u(x,\lambda) = [\overline{u}(0) - qx/h][1.5(1-\lambda^2)][1-(R/420)(2-7\lambda^2-7\lambda^4)], \tag{21}$$

$$v(\lambda)/q = (\lambda/2)(3-\lambda^2) - (R/280)\lambda(2-3\lambda^2 + \lambda^6), \tag{22}$$

where $R = (qh/\nu)$.

The velocity "profile" in the major flow direction is found to deviate from the Poiseuille parabola by being flatter at the center of the channel and steeper in the region close to the walls. Using this solution of the Navier-Stokes equations, Sherwood et al.[4] dealt with the problem of salt diffusion in the boundary layer. As a result of the low water fluxes prevalent in the reverse osmosis equipment, Sherwood et al. neglected the terms involving R; in making this simplification an error not exceeding 0.2% for values of $R < 0.02$ is produced which greatly simplifies the analysis. Salt diffusion in the flowing stream is described by the equation:

$$\frac{\partial}{\partial X}(UC) + \Gamma \frac{\partial}{\partial \lambda}[VC - \Theta(\partial C/\partial \lambda)] = 0, \tag{23}$$

where

$$X \equiv x/h \qquad C \equiv c_s/c_s'$$

$$U \equiv u/\overline{u}(0) \qquad \Gamma \equiv q/\overline{u}(0)$$

$$V \equiv v/q \qquad \Theta \equiv D_s/qh\ .$$

The initial and boundary conditions are:

$$C(0,\lambda) = 1 \tag{24}$$

$$C(X,1) = \Theta(\partial C/\partial \lambda)_{\lambda=1} \tag{25}$$

$$(\partial C/\partial \lambda)_{\lambda=0} = 0 \tag{26}$$

Sherwood et al.[20] used a series expansion in λ and X plus a digital computer to obtain the solution $C(X,\lambda)$ to Eq. (23). Concentration profiles were obtained by varying the value(s) of ΓX and/or Θ. The results are plotted as r versus the function $\xi = \Gamma X/3\Theta^2$ and given by Johnson et al.[21]. In order to avoid the necessity of using a computer to evaluate the infinite series, approximate analytical solutions to Eq. (23) have been given by Dresner[22] for three distinct regions along the channel. Near the channel entrance, the approximate solution is:

$$r = \xi + 5[1 - \exp(-\sqrt{\xi/3})]. \tag{27}$$

In the immediate neighborhood of the channel entrance ($\xi \ll 1$) a better representation of r is:

$$r = 1.536(\xi)^{1/3}. \tag{28}$$

Far downstream from the channel entrance, the approximate solution is:

$$r = 1/3\Theta^2. \tag{29}$$

The approximate solutions agree well with the evaluated curves. The same mathematical analysis can be made for tubular membranes as was shown by Fisher, Sherwood, and Brian[23].

A further development was made by Gill et al.[24,25] by taking into account the change in q down the tube length L. In addition to the equations used by Sherwood et al. the following relation was added:

$$-q = L_p \Delta P[1 - B_2 M(x,0)], \tag{30}$$

where

$M(x,0) = m_s/m_s(0,y)$,

m_s = mass fraction of salt in brine solution,

$B_2 = \Pi_0/\Delta P$,

Π_0 = osmotic pressure of brine solution at $x = 0$.

Solving the differential equations by series expansion, Gill et al. obtained a solution for the two cases of complete (r=1) and incomplete (r < 1) intrinsic salt rejection. A comparison of these solutions with the solutions given by Sherwood et al. showed that concentration polarization is greater for the variable flux case near the channel entrance and is lower near the channel exit. The average polarization, however, for the constant flux case is very nearly equal to that for the variable flux case.

EXPERIMENTAL INVESTIGATIONS OF CONCENTRATION POLARIZATION WITH LAMINAR FLOW

There has been very little experimental investigation concerning concentration polarization with laminar flow. Strathmann [17] and Bixler and Cross[18] measured indirectly the salt concentration at the membrane surface as a function of streaming velocity u for $0 < u < 60$ cm/sec using NaCl solutions and cellulose acetate membranes in one series and tetrabutylammonium chloride and polyelectrolyte resin membranes in another series. The measurements were performed at 102 atm pressure. Over a wide range of recovery rates and bulk velocities, the theoretically predicted values were, in nearly all cases, approximately 20% lower than the experimental results. It is remarkable that the deviations between experimental and theoretical findings decrease with increasing streaming velocity. This would seem to indicate that at small streaming velocities, there is an increase in salt concentration as a consequence of water loss.

Goldsmith and Lolachi[26] measured the concentration near the membrane surface in hyperfiltration systems with laminar flow using Ag-AgCl microprobes. Using a 0.5 molar solution, cellulose acetate membranes, and an operating pressure of 102 atm, values for $c_s^m/c_s^{\prime}$ of approximately 3 were obtained. Unfortunately these measurements were not compared to theoretical calculations.

SUMMARY

The agreement between experimental and theoretical results for turbulent flow, in contrast to laminar flow, is rather good. Due to the extensive investigations into hyperfiltration systems for turbulent flow, it is now possible to quantitatively determine the concentration polarization if r, D_s, q, and $\overline{u}$ are measured. The intrinsic rejection r is calculated from measurements of r_{obs} as a function of $\overline{u}$. Hyperfiltration systems having turbulent flow have the advantage over laminar flow systems due to their ability to limit sedimentation of organic contaminants upon the membrane, thereby avoiding the necessity of extensive water pretreatment.

At the present time, unfortunately, one is unable to forthrightly consider experimental confirmation of laminar flow hyperfiltration theories. Since Strathmann has shown that laminar flow hyperfiltration may offer economic advantages, further efforts should be directed to improve the experimental evidence to test the theory.

Optimally, the objective should now be the formulation of a simple set of reliable working equations, tables, graphs, or nomographs by which a working engineer can estimate the magnitude, and therefore, the effect of concentration polarization. These aids should be applicable for various geometries, flow conditions, and solute diffusivities. The achievement of such an objective shall be clearly the result of the determined work by many investigators. Thus, standardized notation and the precise description of the operating conditions will be of increasing importance in the future. Only by this means can the work of many investigators be culminated into an overall description of hyperfiltration.

It is well known that an inverse relationship exists between q and r due to the phenomenon of concentration polarization. It is the author's contention that membrane technology has progressed to a point where polarization places a definite limit upon the performance of a membrane. Accordingly, increased attention should now be directed to reducing the boundary layer thickness by means of improved flow configurations, thereby decreasing concentration polarization effects.

ACKNOWLEDGMENT

I would like to thank Mr. P. Eschenauer, Max-Planck-Institut für Biophysik, Frankfurt am Main, for reading the proofs.

LIST OF SYMBOLS

c_s salt concentration, mol/cm^3

c_s^m salt concentration at the membrane surface, mol/cm^3

c_s' bulk concentration of salt in the brine, mol/cm^3

c_s'' product concentration of salt, mol/cm^3

C c_s/c_s'

d tube diameter, cm

D_s diffusion coefficient of salt, cm^2/sec

h half the channel height, cm

j Chilton-Colburn factor $[j \equiv (k/\overline{u})N_{Sc}^{2/3}]$

k mass transfer coefficient, cm/sec

k' semiempirical constant in the Chilton-Colburn analogy

L channel length, cm

L_p membrane constant, cm/sec-atm

m_s mass fraction of salt

N_{Sc} Schmidt number ($N_{Sc} \equiv \nu/D_s$)

P hydrostatic pressure, atm

q permeation velocity or volume flux, cm/sec

R Reynolds number

r intrinsic salt rejection

r_{obs} observed salt rejection

u velocity component, cm/sec

$\overline{u}$ mean velocity of bulk solution, cm/sec

U $u/\overline{u}(0)$

v velocity component perpendicular to the channel wall, cm/sec

V v/q

x coordinate, cm

X x/h

y coordinate, cm

Greek letters

Γ $q/\overline{u}(0)$

δ boundary layer thickness, cm

Θ D_s/qh

λ y/h

ν kinematic viscosity, cm^2/sec

ξ $\Gamma X/3\Theta^2$

Π osmotic pressure, atm

ρ density, g/cm^3

ϕ salt flux, mol/cm^2-sec

ψ stream function

REFERENCES

1. U. Merten, Ind. Eng. Chem. Fundamentals, 2, 229 (1963).
2. U. Merten, H. K. Lonsdale, and R. L. Riley, Ind. Eng. Chem. Fundamentals, 3, 210 (1964).
3. W. Banks and A. Sharples, "The Mechanism of Desalination by Reserve Osmosis and its Relation to Membrane Structure", Office of Saline Water Res. Develop. Progr. Rept. No. 143, June 1965.
4. T. K. Sherwood, P.L.T. Brian, R. E. Fisher, and L. Dresner, Ind. Eng. Chem. Fundamentals, 4, 113 (1965).
5. R. J. Raridon, L. Dresner, and K. A. Kraus, Desalination, 1, 210 (1966).
6. P.L.T. Brian in Desalination by Reverse Osmosis, U. Merten (Ed.), M.I.T. Press, Cambridge, Mass. 1966, pp. 161-202.
7. A. J. Shor, K. A. Kraus, J. S. Johnson, and W. T. Smith, Ind. Eng. Chem. Fundamentals, 7, 44 (1968).
8. T. K. Sherwood, Chem. Eng. Progr. Symp. Ser., 55, 71 (1959).
9. Chandrasekhar, Hydrodynamic and Hydromagnetic Stability, Academic Press, New York, 1954.
10. R. Gröpl and W. Pusch, Desalination, 8, 277 (1970).
11. J. D. Sheppard and D. G. Thomas, A.I.Ch.E. Journal, 17, 910 (1971).

12. D. G. Thomas, Ph.H. Hayes, W. R. Mixon, J. D. Sheppard, W. L. Griffith, and R. M. Keller, Environmental Sci. & Technology, 4, 1129 (1970).

13. D. G. Thomas, A.I.Ch.E. Journal, 11, 520 (1965).

14. D. G. Thomas, A.I.Ch.E. Journal, 11, 848 (1965).

15. D. G. Thomas, A.I.Ch.E. Journal, 12, 124 (1966).

16. J. S. Watson and D. G. Thomas, A.I.Ch.E. Journal, 13, 676 (1967).

17. H. Strathmann, "Control of Concentration Polarisation in Reverse Osmosis Desalination of Water", Office of Saline Water Res. Develop. Progr. Rept. No. 336, April 1968; and H. Strathmann and B. Keilin, Desalination, 6, 179 (1969).

18. H. J. Bixler and R. A. Cross, "Final Report on Control of Concentration Polarization in Reverse Osmosis Desalination of Water", Office of Saline Water Res. Develop. Progr. Rept. No. 469, October 1969.

19. A. S. Berman, J. Appl. Phys., 24, 1232 (1953).

20. T. K. Sherwood, P.L.T. Brian, and R. E. Fisher, MIT Desalination Research Laboratory, Rept. 295-1, August 1963.

21. J. S. Johnson, Jr., L. Dresner, and K. A. Kraus in Principles of Desalination, by K. S. Spiegler (Ed.), Academic Press, New York, 1966, pp. 345-439.

22. L. Dresner, Oak Ridge Natl. Lab. Rept. No. 3621, May 1964.

23. R. E. Fisher, T. K. Sherwood, and P.L.T. Brian, MIT Desalination Research Laboratory, Rept. 295-5, 1964.

24. W. N. Gill, Chi Tien, and D. W. Zeh, Intl. J. Heat Mass Transfer, 9, 907 (1966).

25. W. N. Gill, Chi Tien, and D. W. Zeh, Ind. Eng. Chem. Fundamentals, 4, 433 (1965).

26. H. Goldsmith and H. Lolachi, Office of Saline Water Res. Develop. Progr. Rept. No. 527, May 1970.

II. CELLULOSIC MEMBRANES

THE USE OF SOLUBILITY PARAMETERS FOR SOLVENT SELECTION IN ASYMMETRIC MEMBRANE FORMATION

Elias Klein and James K. Smith
Gulf South Research Institute
New Orleans, Louisiana

INTRODUCTION

The ability to form asymmetric membranes is the basis upon which relatively impermeable polymers have been adapted to yield reverse osmosis membranes that can provide purified water at technologically significant rates. The experiments reported by Loeb and Sourirajan[1], which were subsequently shown[2,3] to yield membranes having a relatively dense surface layer followed by a gradual transition to a supporting macroporous structure, initiated the subsequent research to explain the mechanism of formation of such structures. The preparation of porous membranes had been well known before that time. Elford[4] described the manufacture of porous cellulose nitrate; Vaughan[5] adapted the method to cellulose acetate; and Loeb attributes the leads which led to their experiments to earlier literature[6] on the formation of porous cellulose acetate membranes. But the unique property of the Loeb-Sourirajan membranes is their ability to exhibit the selectivity characteristic of the dense cellulose acetate, and yet provide water fluxes which indicate that the rate controlling layer can only be of the order of 0.25 μ, even when the gross thickness of the membranes is about 100 μ. It has been shown[3] that the structure is a composite exhibiting the characteristics of both a porous membrane and a thin dense membrane. With this fact it is possible to reconcile the diffusive transport data of cellulose acetate for both solute and water.

Once the significance of the asymmetry of the Loeb-Sourirajan membranes was appreciated, great efforts were expended to achieve the same graded structures with a variety of other polymers. Although the inherent properties of many polymers indicated that they may be useful for membrane separation processes, it is only recently that significant progress has been reported in the literature. Michaels and Baker[7] claimed the formation of asymmetric membranes from a number of different polymer types. Their membranes were analogous to the Loeb-Sourirajan structures, but the "skin" of their membranes allowed relatively high molecular weight solutes to diffuse through. Asymmetric membranes have been prepared from cellulose acetate-butyrate[8], from ethyl cellulose[9], from cellulose triacetate[10], and more recently from polyamides [11]. Thus it is evident that the properties of cellulose acetate are not unique, and that given the proper solvent choices it may be possible to prepare asymmetric membranes from any polymer.

The mechanism by which a polymer solution is converted to an asymmetric membrane structure has been studied by Kesting et al.[12,13,14], who viewed the process as a result of formation of coacervate phases, whose dispersion sizes and coalescing properties governed the relative preponderance of "skin" and "pore" phases in the final membrane. In a more recent series of articles, Frommer et al.[15,16,17] have investigated the dynamics of the membrane formation process and have shown that the activity of the quenching fluid, the type of solvent, and the conditions of precipitation may control the rate and direction of mass transport in the nascent membrane structure.

Whatever the mechanism of asymmetric structure formation is, it begins with formation of polymer solutions. But the properties of the starting solutions must be unique, for Loeb reported that not all solvents were equally useful[18]. While good solutions can be prepared from ternary cellulose acetate-plus-acetone solutions containing other good solvents, such as tetrahydrofuran, acetonitrile, or p-dioxane, such mixtures do not lead to asymmetric membrane formation. Clearly then, a good solution is not sufficient to form asymmetric membranes. Additional ingredients are required. Loeb and Sourirajan used $Mg(ClO_4)_2$. Kesting[14] postulated that this chelated with the hydroxyls of secondary cellulose acetate, but Banks and Sharples[19] used the same salt to form asymmetric membranes of cellulose triacetate, which could not chelate with the electrolyte. Manjikian[20] found that formamide could be used in lieu of the magnesium salt. Even Vaughan[5] and Elford[4] found additives necessary to form porous membranes.

With this confusing background, one brings to understand the difficulties which faced researchers in converting new polymers into asymmetric membranes. The report which follows is an outgrowth of the authors' attempts to rationalize the selection of solvents for the formation of asymmetric membranes. The polymers to which the analysis will be applied are ethyl cellulose and cellulose acetate, but the approach hopefully will be valid for other polymers of interest.

SOLUBILITY PARAMETERS

The solubility parameters first enunciated by Hildebrand and Scott[21] have been found to have limited applicability when tested on polar polymers in concentrated solutions. The concept is valid, but the assumptions of the initial theory are violated when forces other than dispersion forces are operative. For such solutions a number of approaches have been devised which use the Hildebrand concept, but establish corrections for the stronger forces encountered. Of the various studies which deal with descriptive techniques, that of Hansen is especially useful[22], since it has been extended to a large number of solvents, and has provisions for the prediction of solvent power of mixtures. Hansen assumed that the total energy of evaporation, ΔE^V, was an additive quantity

$$\Delta E^V = \Delta E_d{}^V + \Delta E_p{}^V + \Delta E^V{}_h , \qquad (1)$$

of ΔE_d, the energy required to overcome dispersion forces; ΔE_p, the energy required to overcome polar forces; and ΔE_h, the energy required to break the hydrogen bonds in the liquid. Dividing Eq. (1) by the molar volume, V, yields the parallel equation in terms of the solubility parameters:

$$\delta_T{}^2 = \delta_d{}^2 + \delta_p{}^2 + \delta_h{}^2 . \qquad (2)$$

To utilize Eq. (2) Hansen and Skaarup[23] used the hydrocarbon homomorphs to determine δ_d, assigned 5000 cal per OH group to estimate δ_h, and used the dielectric constants, refractive index, and dipole moment to estimate δ_p from an equation derived by Bottcher. The most recent compilation of the fractional solubility parameters includes data for 226 solvents.

To use this system for describing the solubility of a polymer, the δ_i values are used as coordinates. Hansen found that in many cases the two dimensional coordinates based on only the δ_p and δ_h values were adequate to describe the solubility boundary of a given polymer. More recently Bagley et al.[24] have shown the relationship between the experimentally accessible internal pressure measurements $(dP/dV)_T$ and the Hansen parameters, δ_p and δ_d, to be

$$(dE/dV)_T \ (V) = (\delta_p^2 + \delta_d^2) = \delta_v . \qquad (3)$$

In practice, the solubility of the polymer under study is determined in a series of solvents whose locations on a two or three dimensional grid are represented by the ordered pairs given by their solubility parameters. For a homogeneous polymer, the resulting diagram will delineate a region within which are found liquids which are solvents for the polymer and outside of which are liquids which are not solvents. The exact position of the boundary is a function of the concentration used in the measurement, the temperature, and the uniformity of the polymer sample.

For mixtures of solvents, the assumption is made that the cohesive energies are additive when normalized for the volume fraction of the constituent present. Thus for component 1, the energy of vaporization is given by n_1E_1, and the total energy by

$$E_T = n_1E_1 + n_2E_2 = n_1V_1\delta_1^2 + n_2V_2\delta_2^2 = V_T\delta_T^2 , \qquad (4)$$

$$\delta_T^2 = \frac{n_1V_1\delta_1^2 + n_2V_2\delta_2^2}{n_1V_1 + n_2V_2} , \qquad (5)$$

but n_1V_1/V_T is the volume fraction of component 1, f_1, and similarly for component 2, $n_2V_2/V_T = f_2$ leading to:

$$\delta_T^2 = f_1\delta_1^2 + f_2\delta_2^2 . \qquad (6)$$

With these definitions, one can now prepare a solubility diagram using either the two dimensional characterizations suggested by Hansen[25] or by Bagley et al.[24], or create a solubility space diagram using all three coordinates, as suggested by Hansen.

Since secondary cellulose acetate has been characterized very well in the literature, both in terms of its solubility and in terms of the asymmetric membranes that can be made from it, it was well suited for testing the solubility parameter concept. In Fig. 1 are shown the boundaries of solubility for cellulose acetate based on data obtained from Ott and Spurlin[26], Eastman Chemical Co.[27], Hoernschemeyer and Saltonstall[28], and Frommer et al.[29]. The values for the Hansen parameters are those published most recently[30]. To simplify the diagram, solvents have been assigned numbers and these are tabulated in the Appendix. It is clear that with very few exceptions the entire spectrum of solvents is encompassed within the area designated, and that most if not all non-solvents are excluded from the area. There is

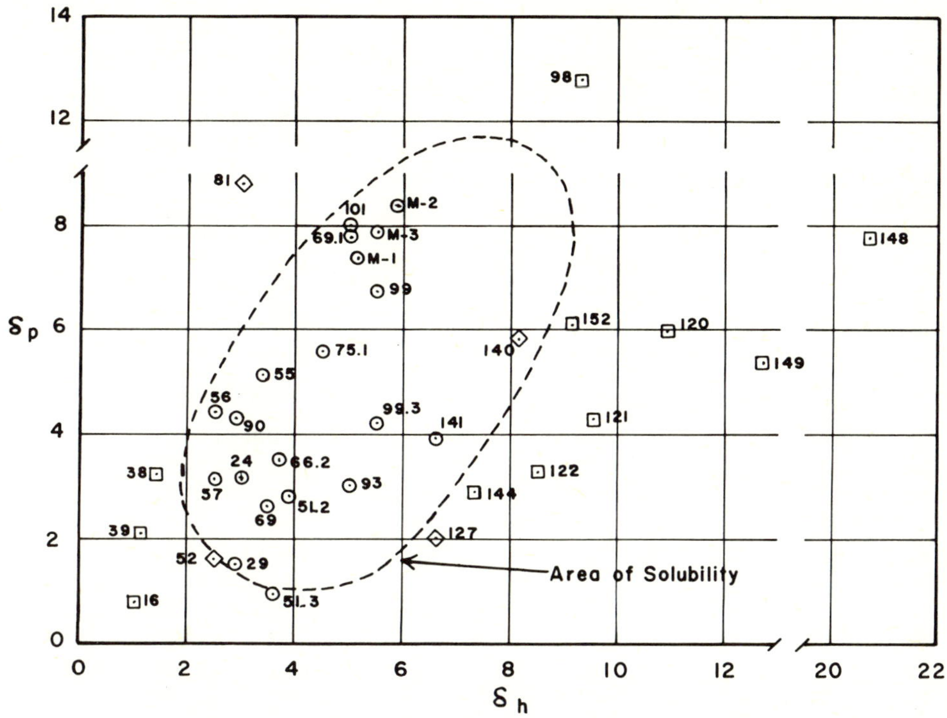

Fig. 1 Solubility Diagram of 2.5 D.S. Cellulose Acetate δ_p vs. δ_h

Legend: ○ Soluble
◇ Swells or Partially Soluble
□ Insoluble

Numbers correspond to Solvent List in Appendix.

some uncertainty at the boundaries of the diagram because of the qualitative measure of solubility employed, and because of some uncertainty in the coordinates taken for certain solvents.

The most severe test is, of course, how well the hypothesis handles the predictions of solvent mixtures. The Manjikian[31] data on membranes formed from acetone-formamide mixtures serve as a good starting point. The compositions from that paper have been plotted as points M-1 through M-3 in Fig. 1. All of the solutions fall inside the diagram, with those reported by Manjikian as leading to the best membranes falling nearest the boundary. The viscosity measurements of Lee[32] on the same solvent combination show a distinct break at the acetone/formamide ratio of 60/40

(V/V). This corresponds to the δ_h, δ_p coordinates of 6.44, 9.01, which on the diagram indicates a transition from solution to gel structure. The Frommer et al. data[29], which were established by the addition of water to cellulose acetate and solvent mixtures, are generally quite consistent with this approach, except in the case of acetic acid. Since the coordinates for water are in some doubt, the precision is less than that for the other mixtures. One can appreciate that construction of the solubility map, based on the Hansen coordinates, is the logical first step when attempting to form asymmetric membranes initially from any given polymer. Even though the exact boundaries may be in some doubt, and the exact calculations for the computation of mixtures are still under investigation, it does provide logical starting points.

To investigate ethyl cellulose, we began by constructing the diagram shown in Figure 2. The data for the boundaries were taken from several literature sources, and checked in our laboratory by examining the behavior of 10% (W/V) solutions in the solvents indicated. Note that the boundary for the cellulose ether is much more constricted than the one for cellulose acetate. Ethyl cellulose of degree of substitution 2.25 is much less soluble than the acetate in a number of solvents, but primarily those having high δ_p's; the hydrogen bonding coordinates are almost the same for both polymers, probably due to the presence of free hydroxyl groups in both polymers.

From the literature which describes the formation of asymmetric membranes from cellulose acetate, and the data in Figure 1, one can draw some working hypotheses regarding the requirements of a casting solution:

a) The casting composition should be near the solubility boundary facing the quench medium.

b) A volatile component should be such that its loss will move the composition out of the solubility area, rather than into it.

c) The solids content at the solution boundary must be high in order to cause a rapid transition from solution to gel.

d) All components of the system should be miscible with the quench medium.

With these working rules in mind, a series of solvent combinations was evaluated for their ability to allow asymmetric membrane formation with ethyl cellulose. The first of these was

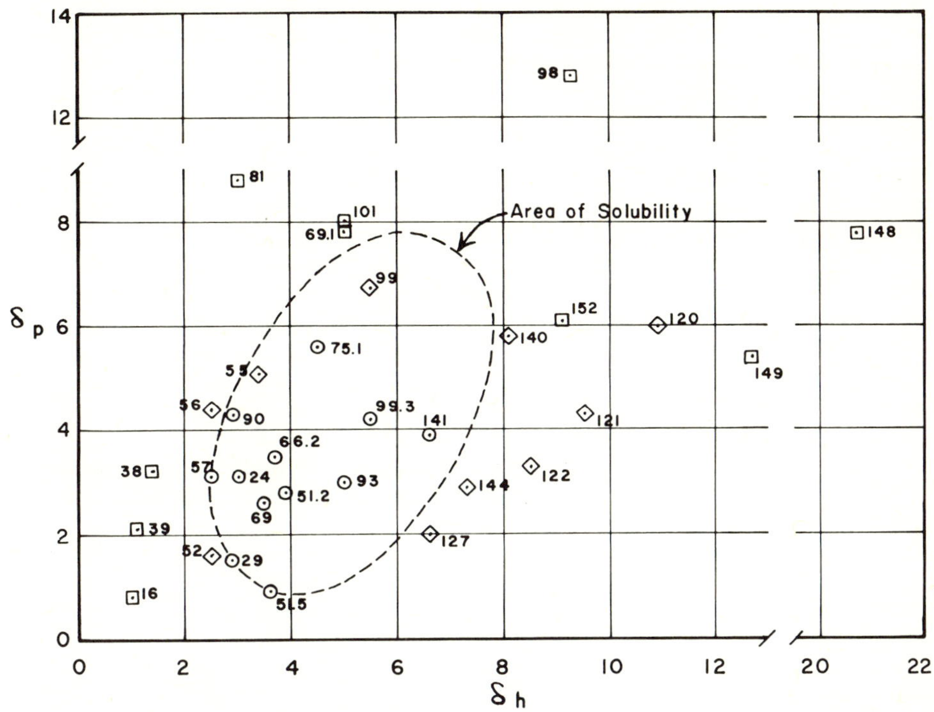

Fig. 2 Solubility Diagram of 2.25 D.S. Ethyl Cellulose δ_p vs. δ_h

Legend: ○ Soluble
◇ Swells or Partially Soluble
□ Insoluble

Numbers correspond to Solvent List in Appendix.

based on a combination of methyl acetate-formamide-methanol, which had been the principal components of a solvent for ethyl cellulose membranes reported earlier[33]. A second solvent combination consisted of tetrahydrofuran-trimethyl phosphate-methanol, and another combination consisted of pyridine-dioxane-formamide. All of these experiments used ambient temperature quenching with water, and no secondary heat treatment. The results of reverse osmosis experiments indicated all formed asymmetric membranes; the data have been summarized earlier[34,35]. Finally, to see if the hypotheses formulated were applicable to quench systems other than water, a membrane was prepared from a solvent mixture containing benzene-methylene chloride, and the quench liquid was hexane. The membrane was asymmetric but the step-wise solvent exchange process required to bring it to an aqueous state for testing weakened the structure. Rejection for NaCl was over 95%, even after the solvent exchange steps.

Having established that the solubility parameter map could be used as a guide to formulating solvent mixtures, it was of interest to see if the fine details of solvent composition could be related to the resultant membrane. From Figure 3, which shows the areas representing the compositions examined, some deductions can be made. The methyl acetate-formamide-methanol solvent was nearest the solubility boundary, and loss of the very volatile methyl acetate-methanol components would immediately move the composition outside of the solubility limits. The evaporation times required were short, and the resultant fluxes were in the range 2-4 x 10^{-4} g/cm^2-sec (4-8 gal/ft^2-day) for highly rejecting membranes. The trimethyl phosphate-tetrahydrofuran-methanol mixture was also very close to the solubility boundary, but quite a bit of the volatile THF and methanol would have to be lost to exceed the solubility boundary because of the proximity of the remaining trimethylphosphate coordinates to the boundary. The loss of THF and methanol needed to move the composition outside the boundary might, in fact, be so great that precipitation would occur before the solvents were gone due to the increased solids content. The precipitated polymers would be high in solvent content.

The pyridine-dioxane-formamide solvent combination was located initially closer to the center of solubility than either of the others, but the experiments used relatively long evaporation times, and the loss of pyridine and dioxane moved the composition toward the formamide coordinate. Thus the rate of surface layer precipitation could be intermediate between the values of the first two solvent combinations listed. Examination of the averaged flux data[34] shows that the methyl acetate solvent yielded the highest fluxes, the pyridine solvent the next, and the trimethylphosphate solvent the lowest. While generalizations are useful we are not justified in expecting more precise prediction solely from examination of the solubility diagram. It is true that high flux asymmetric membranes cannot be made without first establishing the proper casting solution, but the converse is not necessarily true; it is possible not to obtain good asymmetric membranes even from the proper casting formulation. One must certainly consider not only the composition of the casting mixture, but also the dynamics of the process which follows. Since the rate of solvent loss and quench fluid ingress is the subject of another chapter in this book[36], the reader is referred there.

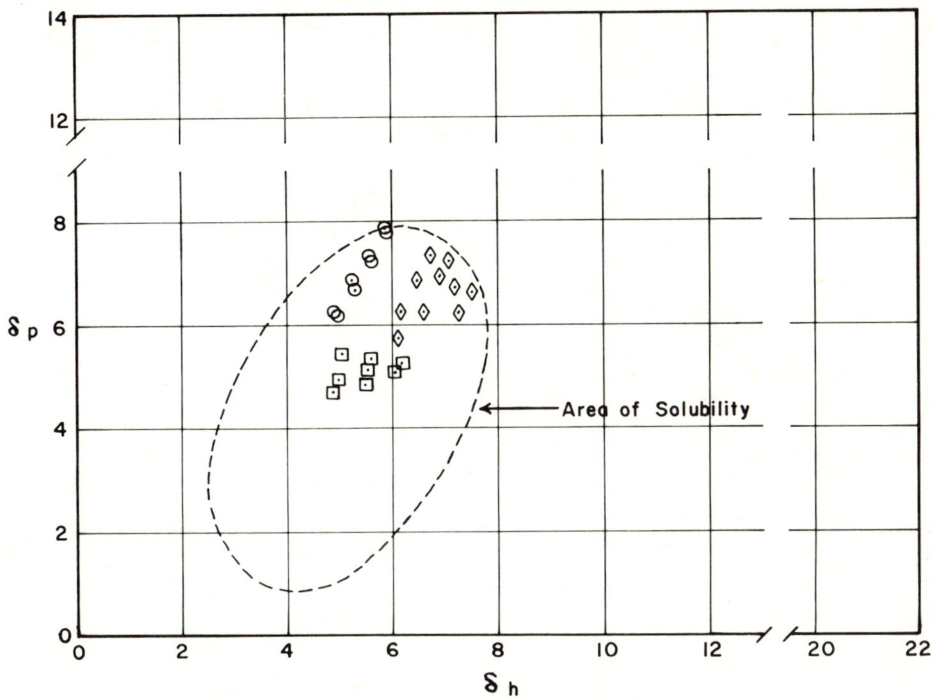

Fig. 3 Selection of Solvent Compositions for Asymmetric Ethyl Cellulose Membrane Formation

Legend: □ Pyridine-Dioxane-Formamide
○ Trimethyl Phosphate-Methanol-THF
◇ Methyl Acetate-Formamide-Methanol

To improve on the reliability of the solvent predictions, two alternate methods of structuring the solvent spaces were examined. Bagley et al.[24] separate the hydrogen bonding component from the volume dependent components of the solubility parameter. This reduces the number of solvent parameters to two without discarding any of the contributions of the dispersion components, such as was done in the use of δ_h and δ_p. In Figure 4, the solubility of ethyl cellulose is shown in terms of the δ_v and δ_h coordinates suggested by Bagley et al. To compute the δ_v coordinates, the Hansen values for δ_p and δ_d were used. The Bagley information eliminates certain dilemmas present in Figure 2, but presents new ones. For example, dioxane is an excellent solvent for ethyl cellulose, but is shown in Figure 2 near the boundary; in Figure 4, it is near the center of solubility; but acetone, a marginal solvent for ethyl cellulose, also

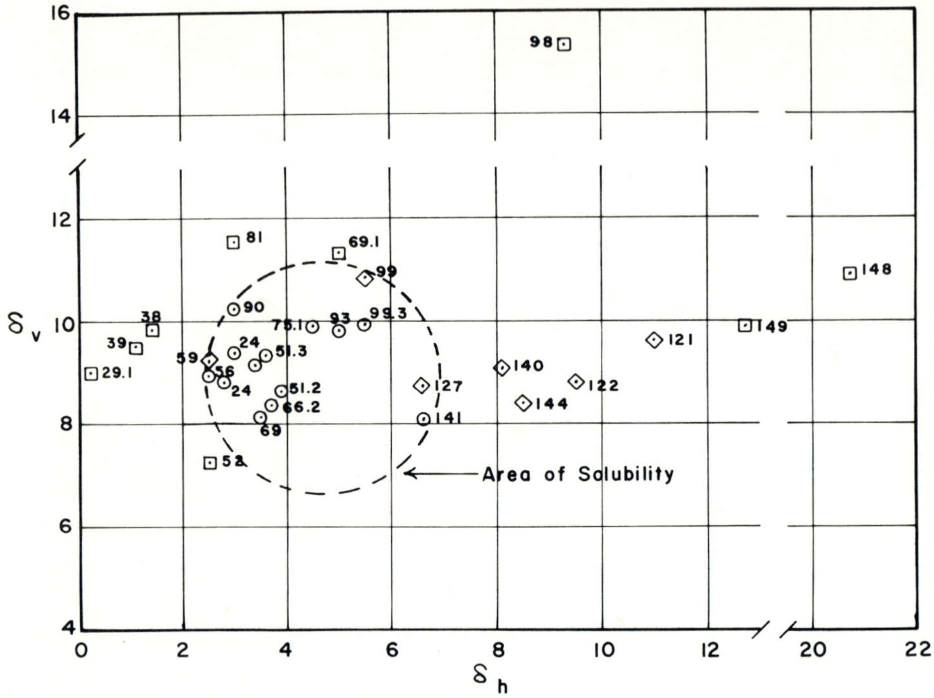

Fig. 4 Solubility Diagram of 2.25 D.S. Ethyl Cellulose δ_v vs. δ_h

Legend: ○ Soluble
◇ Swells or Partially Soluble
□ Insoluble

Numbers correspond to Solvent List in Appendix.

appears near the center of solubility in Figure 4. The greatest difficulty encountered with the use of the δ_v values comes in the handling of mixtures. The pyridine-dioxane-formamide mixtures which formed useful casting solutions are computed to lie entirely outside of the solubility diagram of Figure 4.

The methods of computing δ_v values for mixtures have not been investigated, and there remains the possibility that other combination formulae will be derived which will provide better fits for the data on mixtures; but for the present one must conclude that the combination of δ_p and δ_d into a single term does not offer any substantial benefit over the method suggested by Hansen.

The analyses of other theoretical approaches for the treatment of concentrated polymer solutions have been reviewed recently by Burrell[37]. The availability of solvent data for a large number of compounds in the recent Hansen and Beerbower[30] publication recommends their method over others currently available, at least in the solution of membrane problems.

A more exact application of the Hansen parameters requires consideration of all three of the components of the Hildebrand cohesive energy density. With the use of computers, such calculations become more feasible, despite the difficulty of visualizing three dimensional spaces. Two difficulties remain, however, even in this more complete treatment. The two dimensional representations shown in Figures 1-4 indicate that solubility in the δ_p, δ_h plane is represented by an oval at best, and not a circle. Furthermore, the magnitude of the variations of the third component now to be considered (δ_d) is much smaller than those of the other two. Thus we must make some allowance in its scale if a symmetrical model is to be used. Hansen and Skaarup[23] suggested the use of a sphere to describe the volume such that coordinates of solvents would be located inside the sphere, and coordinates of non-solvents would be outside the sphere. The center of the sphere corresponds to the polymer coordinates. They corrected for the diversity in scale by doubling the units to be used for the δ_d parameter. Thus the radius of the sphere within which a solvent will be found is given by:

$$\underline{R}^2 = (\delta_h - \delta_h^*)^2 + (\delta_p - \delta_p^*)^2 + 4(\delta_d - \delta_2^*)^2, \qquad (7)$$

where the parameters with superscripts refer to the coordinates of the polymer. To establish the values of $\underline{R}$, and the coordinates for the polymer, we can use the Flory-Huggins theory as an approximation. The magnitude of $\underline{R}$ is also the vector distance between two coordinates in the solubility space. Thus it corresponds to the "breadth of the solubility spectrum" given by:

$$\Delta_{max} = |\delta_{pol} - \delta_{solv}|_{max} ,$$

which Huggins[38] related to the Flory-Huggins parameter χ:

$$(\Delta_{max})^2 = [0.5(1 + 1/0.5)^2 - \chi_s]RT/V_s .$$

For a polymer at 30°C with a D.P. of 200, χ_s estimated as 0.2, and a molar volume for the solvent, V_s, of 100 cc/mole, the value of $\underline{R}$ is 2.5 $(cal/cc)^{1/2}$.

With a starting value of $\underline{R}$, one can then use a computer search technique to establish the value of δ_h^*, δ_p^*, and δ_d^*. For ethyl cellulose the procedure adapted was as follows:

The coordinates of the solvents and non-solvents tested were used as data for the program. A trial set of coordinates for the center was given to the computer, and the program directed the calculation of a distance between the solvent or non-solvent coordinates and the trial center. This distance was then compared to the radius value of 2.5, and the deviations printed. The trial center was then moved to a new locus, and the process repeated. When the maximum number of solvents were found within the sphere, and the maximum number of non-solvents were simultaneously excluded from the sphere, that locus was taken as the coordinates for the polymer. For ethyl cellulose this was found to be at the values of 3.4, 4.2 and 8.6, with all solvents found within ($\underline{R}$ + 0.25), and all non-solvents found outside of ($\underline{R}$ - 0.25).

The coordinates of mixtures were also examined in terms of the three dimensional space. A number of difficulties must be resolved in this instance. The computation of the coordinates of mixtures is contingent on the validity of Eq. (6); furthermore, the errors in the coordinates of the components are magnified when they are squared (Bagley et al.[24]); and finally, we know from the planar diagrams that the mixtures are at the limits of the solubility boundary so that small variations will quickly indicate them to be non-solvents.

All of these difficulties were encountered in the attempts to use the three dimensional diagrams. In Table I are listed the coordinates for a number of casting compositions comprised of methyl acetate-formamide-methanol and other solvent systems. The compositional coordinates were calculated as outlined earlier. When these values are used with the coordinates found for the polymer on the basis of single component solvents for ethyl cellulose, it is found that the mixtures are well outside the solubility boundary. To reconcile this divergence one can move the center of the sphere to higher δ_p and δ_h values; when the mixtures are now tested it is found that some will be inside the solubility boundary and others still remain outside. But the shift in the center will imply that some known solvents for the polymer with low δ_p, δ_h coordinates are not solvents. The net effect is that one can shift the center and compromise the descriptions at either extreme. Alternatively, one can increase the dimensions of $\underline{R}$; but this leads to difficulties in that the sphere of solubility will now include some marginal solvents or non-solvents, especially of the alcohol type.

Table 1

Calculated Solubility Coordinates for Solvent Mixtures

Methyl Acetate, Formamide, and Methanol

Volume Fraction						
Me Acet	Formamide	MeOH	δ_h	δ_p	δ_d	δ_v
0.662	0.149	0.189	6.66	6.27	7.69	9.92
0.582	0.181	0.236	7.19	6.73	7.70	10.23
0.616	0.210	0.174	6.88	6.94	7.74	10.40
0.591	0.130	0.279	7.24	6.21	7.65	9.85
0.721	0.159	0.120	6.15	6.26	7.71	9.93
0.545	0.163	0.292	7.50	6.63	7.68	10.14

Acetone and Formamide

Acetone	Formamide					
0.798	0.202		5.16	7.34		
0.741	0.259		5.56	7.86		
0.683	0.317		5.94	8.35		
0.600	0.400		6.44	9.01		

Trimethyl Phosphate, Methanol, and Tetrahydrofuran

TMP	MeOH	THF				
0.281	0.062	0.658	4.94	4.95	8.16	9.54
0.323	0.124	0.552	5.59	5.33	8.10	9.70
0.372	0.063	0.565	5.04	5.42	8.15	9.79

Perhaps the best solution to the problem is to question the a priori assumption of Hansen that the solubility space is spherical. Indeed, the planar diagrams indicate that in all likelihood it is not! In the future it should be possible to use a figure based on an ellipsoid of revolution to describe the solvent boundaries. With such a figure there is a wider latitude in describing the solvent space, and little of the usefulness is lost.

DISCUSSION

Apart from the pragmatic application of finding solvent mixtures for polymers of interest, how can these concepts aid in the formation of asymmetric membranes? This is perhaps the most interesting, and possibly also the most speculative aspect of the problem. In review, we know now that asymmetric membranes are formed from many polymers; each polymer has unique, but not single, casting composition requirements; the compositions alone do not govern the quality and extent of asymmetric layers; and the compositional variables for casting are not just those related to solubility, i.e., not every good solvent mixture will give asymmetry.

Both the intuitive and experimental evidence then is that the process is a dynamic one, and that the time variables interact with the equilibrium variables. Thus far we have only discussed the equilibrium variables, and the question one faces is how these can be related to the time dependent variables, such as mass transfer, heat transfer, viscosity, and nucleation rate. In this discussion only a qualitative description will be attempted, since the requisite data to test some new models are not available.

A convenient point with which to begin, is the recent publication by Githens et al.[39], who showed that the "skin" formation of membranes cast from a dioxane and methanol solution of cellulose acetate increased, and then decreased with time. Frommer and Lancet[36] have shown the rate of penetration of a front from the "skin" side of a membrane toward the porous side. Boddie[40] has calculated that as acetone evaporates from the surface of a casting solution containing acetone-formamide, there is a sharp drop in surface temperature. However, the calculations of the heat and mass transfer used by Boddie assumed no step function for the skin as precipitation occurred. It is evident that for the most widely studied system, i.e., cellulose acetate dissolved in acetone-formamide, there is a period of rapid solvent loss from the surface leading to at least a transient phase separation of the surface layer. Based on evaporation rate studies we have found that acetone accounts for all of the solvent loss. Thus one can expect that the cooling of the surface is caused by a transient heat imbalance. The planar solubility diagram of

Figure 1 predicts that a small loss of acetone will move the solubility coordinates outside of the solution range because of the combination of cooling and an increase of δ_p, δ_h. For example, a 75:25 (V:V) mixture of acetone-formamide has coordinates of δ_h, δ_p = (5.4, 7.6). If the acetone composition decreases so that the composition is 50:50 (V:V), the coordinates are (7.0, 9.7), which is well outside the solubility bounds, even at the same temperature. Since we do not know the response of the perimeter to temperature as yet, it is not possible to state how far out of the boundary the composition has moved; but certainly one can predict that the surface layer will have a "skin" for the time frame within which this imbalance of acetone concentration and temperature persists.

If sufficient time is allowed for diffusion to occur, the gradient of acetone concentration between the bulk of the casting and the acetone depleted surface layer will act to resupply the surface concentration. But what is sufficient time for this event? If the diffusion rate through the surface layer were to remain the same as it is in the viscous solution, one could assume a linear concentration gradient, and the rate of acetone movement would be proportional to the mass transfer resistance at the air/skin interface. Sletmoe[41] has shown that the rate is proportional to this boundary layer in the initial stages of drying of a solvent film. But what if the "skin" were to be less permeable to acetone diffusion than the bulk solution by one or two orders of magnitude? In this case the acetone resupply rate would be more like a dissolution rate for the polymer, and all those factors which govern dissolution would be of significance.

To test such a view of the relationship between the equilibrium variables and the time dependent variables, we try to explain some of the observations of Loeb[18]; why did the combination of acetone and dioxane not lead to asymmetric membranes? Clearly, from Figure 1 the loss of acetone moved the overall composition into a highly soluble area. Similarly, tetrahydrofuran-acetone mixtures reported by Loeb cannot lead to the formation of an insoluble transient skin. But before one concludes that for skin formation one needs only to select solvent components that lead to a transient precipitation, it is necessary to explain why combinations of acetone and glycerin do not lead to good asymmetric membranes. Certainly a mixture sufficiently high in glycerin can be formulated which, upon loss of a small fraction of acetone, will lead to precipitation.

The most apparent conclusion is that precipitation per se is not sufficient. There must be something in the character of the precipitated layer which facilitates the secondary processes that lead to pore structure, dense structure, or even combinations of the two. Frommer et al.[29], who studied the two component mixtures of cellulose acetate and several solvents, found that the movement of solvent out of the cast layer and movement of water into the same layer was a function of the type of solvent used. Furthermore, they demonstrated that the activity of the water governs the rate of diffusion into the layer, but does not markedly affect the movement of solvent out of the layer. If the chemical potential of water affects its transport into the layer, we can consider the layer as a separate phase. This leads to the immediate question of how we can best describe the properties of this phase. Is it a swollen gel? A dense polymer membrane? What are the magnitude of diffusion coefficients of solvent and water through the phase?

Unfortunately, we do not have a good characterization yet of the properties of the transient phase. Its composition is difficult to establish, and the theories which describe diffusion through such films normally assume simplifications which deal with averaged compositions. Since we are interested in the surface layer, only the transient states are of interest. Nevertheless, we can examine what the properties are of the compositions corresponding to the boundary limits of the transient state. For example, if the Manjikian formulation of 45:30 (w:w) of acetone: formamide containing 25% cellulose acetate is evaporated, the limiting composition of the surface layer will be 54% formamide and 46% cellulose acetate. Similarly, if the solvent were 45:40, the limiting surface composition would be 61% formamide and 39% polymer. Formamide is not a solvent for cellulose acetate at room temperature but it is a secondary plasticizer, e.g., it lowers the glass transition temperature of the polymer. Cellulose acetate undergoes glass transitions at 50°C and 115°C in the absence of a plasticizer, but the presence of either a solvent or a plasticizer serves to lower these values[42]. Since we are interested in the properties of the surface layer at the end of evaporation but before the quenching step, we examined compositions corresponding to the limiting values which were postulated above on the basis of the Manjikian formulation.

The glass transition temperature of a polymer is both composition and time-scale dependent. For the purposes of this discussion, we can conclude that the low frequency T_g is the value of interest since we seek to incorporate the value into mass and heat transfer considerations. The glassy structure of a polymer in which segmental motions have been frozen exhibits two properties of interest to us. First, there may be as much as a one-hundred fold decrease in transport coefficient for liquids passing through the glassy regions as opposed to those passing through the amorphous regions[43]. Secondly, the glassy structure may be the morphology upon which the "skin" structure is based. Annealing of a glassy structure below the T_g leads to densification of the structure in the same way that annealing of crystalline polymers below the melting point can increase the degree of crystallinity.

The experimental method which we used to study the effect of polymer morphology was based on the device described by Gilham[44]. A glass braid was coated with the Manjikian formulation and the braid mounted into a torsional braid analyzer. Dry nitrogen was used to sweep the acetone from the coating on the braid at ambient temperatures. The temperature of the analyzer was then lowered to -90°C, and the torsional periods and damping losses determined. As the temperature was allowed to rise from -90°C to ambient, and then programmed above room temperature, the polymer coating on the braid responded by going from a glassy state to an amorphous viscous state. Major transitions were found at -20°C and -30°C for the samples containing 54% and 61% formamide, respectively (Figures 5 and 6). Smaller transitions which may be associated with other segmental motions and perhaps the bonding of residual solvent to the polymer were found at higher temperatures. Of particular interest is the finding that as the temperature of the 54% formamide mixture approached -15°C from room temperature, the stiffness, $1/P^2$, (P = torsional period) increases sharply. If after casting the heat of evaporation of acetone is sufficient to lower the temperature of the surface film to say 0°C, then these data indicate that a stiff, glassy layer will form. This layer will remain until either more acetone moves from the interior into the layer - which will be impeded by the now reduced diffusion coefficient - or until enough heat is provided from the surface to raise the temperature above the T_g. If, however, ice water is used to quench this structure, then one can expect that the glassy form will remain, and the formamide will exchange for the water. Since water is a poorer plasticizer than formamide, one could expect the T_g to rise. The experience that 80°C annealing in water further densifies the structure is consistent with this model.

But what of the layer below the surface? The composition will be essentially unchanged from that existing at the time of casting. There should be a lowering of the temperature at the boundary nearest the "skin" side which may lead to some dense precipitation. But below this second layer, the conversion of a solution of the polymer into a porous gel structure must depend on the processes studied by Kesting et al.[12-14] and more recently by Strathmann et al.[45]. The dynamics of this process have been studied by Frommer and Lancet[36]. From the point of view of the solubility diagrams, we can say very little about the distribution of the polymer as water ingress causes phase separation to occur. One can project the direction of the diagram in which

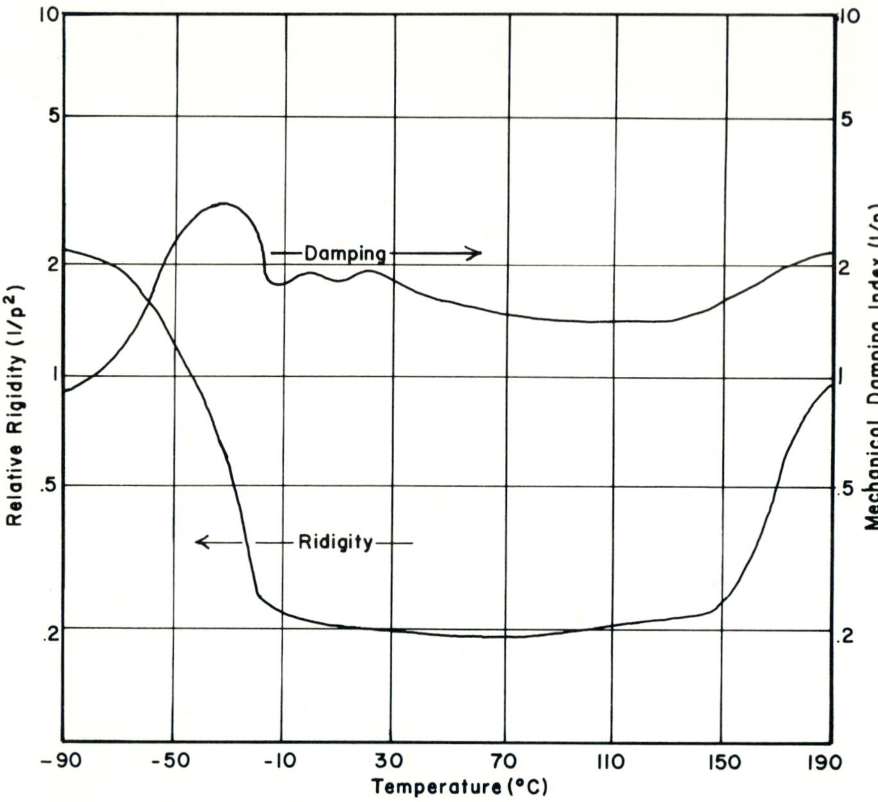

Fig. 5 Thermochemical Behavior of Cellulose Acetate Plasticized with Formamide. 5 Parts Cellulose Acetate/5 Parts Formamide.

the homogeneous composition will move prior to phase separation, but the solubility parameters tell us nothing of the composition of the resultant phases. Thus, this area of research is one in which classical physical chemistry could well lead to a clearer understanding of the pore formation conditions.

Can these conclusions be applied to those membranes which have been studied less intensively than has cellulose acetate? One can only deal in qualitative descriptions here. The polybenzimide membranes described in the patent by Richter and Hoehn [46] are formed from dimethylacetamide solutions containing $LiNO_3$. As the dimethylacetamide is evaporated the composition must certainly move toward lower solvent power. If the quenching process causes the surface layer to become glassy, the same conclusions would apply.

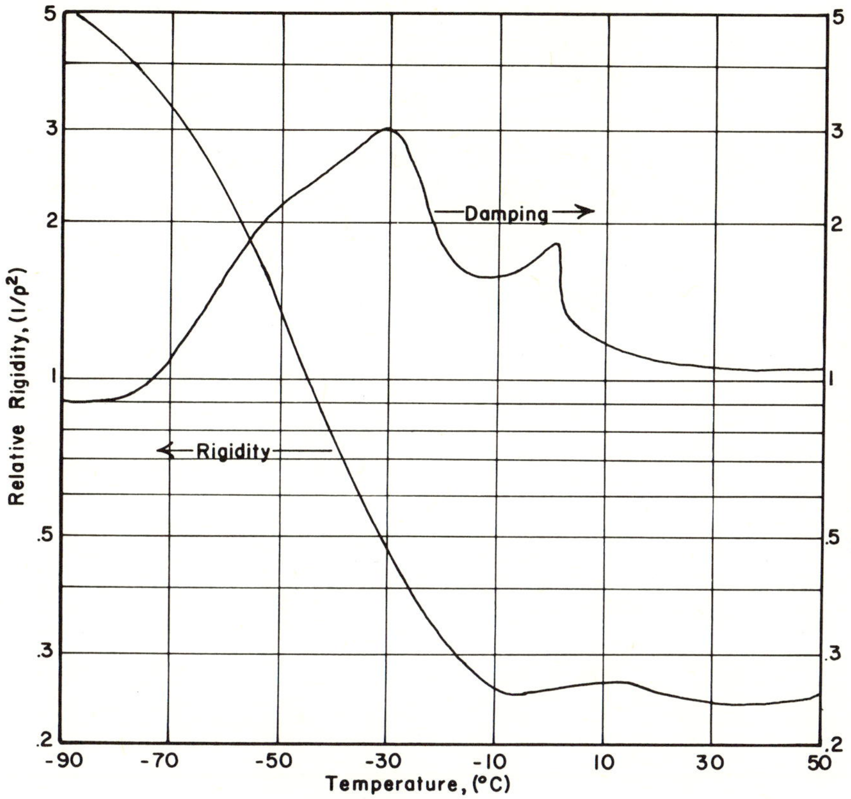

Fig. 6 Thermomechanical Behavior of Cellulose Acetate Containing Formamide. 5 Parts Cellulose Acetate/ 7 Parts Formamide.

Certainly in the case of ethyl cellulose, these hypotheses are self-consistent. The loss of solvent leading to a boundary layer which is below its T_g can be postulated for each combination investigated. For the trimethylphosphate system where relatively dense membranes were formed, it is apparent that large losses of THF are required, and that thick "skins" needed to be formed to cause salt impermeability. Even in the examples of asymmetric membrane formation from single solvents, the possibility exists that transients in mass and heat transfer can lead to adjacent layers of glassy and rubbery polymer.

The combination of solvent power as predicted from solubility parameters and the use of continuity equations for heat and mass transfer, incorporating due considerations of phase properties, is a most powerful tool. To utilize such techniques in the future for predicting, rather than explaining, membrane formation conditions will require more intensive laboratory study. Specifically, we need better models for predicting how the solubility characteristics of solvent mixtures should be calculated; we need to know the change of the solubility boundary both with polymer concentration and with temperature; and we need to find some practical relationship between glass transition temperatures and solubility parameters. (There will not be any simple relationship between the T_g and the δ values since one is a function of kinetic parameters, and the other is a thermodynamic value.) We will also need more information on the transport of solvents through glassy polymers. All of these studies will not be fruitful, obviously, since the inter-relationships may be complex. But empirical relationships do exist, and study is needed to see how they may apply to the membrane formation problem.

ACKNOWLEDGMENT

This work was supported by Contract No. 14-30-2732, with the Polymer and Biophysics Division, Office of Saline Water, United States Department of the Interior.

APPENDIX

Number	Name	Solvent Coordinates Used in the Figures δ_v	δ_d	δ_p	δ_h
13	Cyclohexane	8.2	8.2	0	0.1
16	Toluene	8.8	8.8	0.7	1.0
24	Methylene chloride	9.4	8.9	3.1	3.0
27.1	1,1-dichloroethylene	8.96	8.3	3.3	2.2
28	Ethylene dichloride	9.97	9.3	3.6	2.0
29	Chloroform	8.82	8.7	1.5	2.8
29.1	1,1-dichloroethane	9.03	8.1	4.0	0.2
32	Trichloroethylene	8.93	8.8	1.5	2.6
36	Carbon tetrachloride	8.70	8.7	0	0.3
37	1,1,1-trichloroethane	8.56	8.3	2.1	1.0
38	Tetrachloroethylene	9.84	9.3	3.2	1.4
39	Chlorobenzene	9.53	9.3	2.1	1.1
39.2	1,1,2,2-tetrachloroethane	9.53	9.2	2.5	4.6
51.2	Tetrahydrofuran	8.66	8.2	2.8	3.9
51.3	1,4-dioxane	9.34	9.3	0.9	3.6
52	Diethyl ether	7.24	7.1	1.4	2.5
53	Bis (2-chloroethyl) ether	10.20	9.2	4.4	2.8
55	Acetone	9.15	7.6	5.1	3.4

Number	Name	Solvent Coordinates Used in the Figures δ_v	δ_d	δ_p	δ_h
56	Methyl ethyl ketone	8.95	7.8	4.4	2.5
57	Cyclohexane	9.24	8.7	3.1	2.5
60	Methyl i-butyl ketone	8.08	7.5	3.0	2.0
66.2	Methyl acetate	8.37	7.6	3.5	3.7
69	Ethyl acetate	8.13	7.7	2.6	3.5
69.1	Trimethyl phosphate	11.32	8.2	7.8	5.0
75.1	Triethyl phosphate	9.93	8.2	5.6	4.5
76	Diethyl phthalate	9.80	8.6	4.7	2.2
77	Tricresyl phosphate	11.07	9.3	6.0	2.2
81	Acetonitrile	11.56	7.5	8.8	3.0
90	Pyridine	10.25	9.3	4.3	2.9
93	Aniline	9.82	9.5	2.5	5.0
98	Formamide	15.31	8.4	12.8	9.3
99	Dimethylformamide	10.82	8.5	6.7	5.5
99.1	Dimethylacetamide	9.93	8.2	5.6	5.0
99.3	Hexamethyl phosphoramide	9.93	9.0	4.2	5.5
120	Methanol	9.66	7.4	6.0	10.9
121	Ethanol	8.82	7.7	4.3	9.5
122	1-propanol	8.47	7.8	3.3	8.5
127	Cyclohexanol	8.73	8.5	2.0	6.6
140	Formic acid	9.09	7.0	5.8	8.1
141	Acetic acid	8.10	7.1	3.9	6.6
148	Water	10.9	7.6	7.8	20.7
149	Ethylene glycol	9.90	8.3	5.4	12.7
150	Glycerol	10.35	8.5	5.9	14.3
150.1	Propylene glycol	9.40	8.2	4.6	11.4
152	Triethylene glycol	9.90	7.8	6.1	9.1
144	Phenol	-	8.8	2.9	7.3
101	DMSO	-	9.0	8.0	5.0
66	Ethylene carbonate	-	9.5	10.6	2.5

REFERENCES

1. S. Loeb and S. Sourirajan, Adv. Chem. Ser., 38, 117 (1962).

2. R. L. Riley, J. O. Gardner, and U. Merten, Science, 143 801 (1964).

3. R. L. Riley, U. Merten, and J. O. Gardner, Desalination, 1, 30 (1966).

4. W. J. Elford, Trans. Faraday Soc., 33, 1094 (1937).

5. M. F. Vaughan, Nature, 183, 43 (1959).

6. A. Dobry, Bull. Soc. Chim. France, 5^{e}Serie III, 312 (1936).

7. A. S. Michaels and R. W. Baker, U.S. Patent No. 3,526,588 (1970).

8. United States Department of the Interior, Office of Saline Water, 1970-1971 Saline Water Conversion Report, pp. 304-5.

9. ibid, p. 306.

10. T. E. Davis, G. W. Skiens, and O. Cotton, ACS Polymer Preprints, 12, No. 2, p. 378 (1971).

11. See, for example, the papers by Applegate and Antonson and by McKinney in this volume.

12. R. E. Kesting and A. Menefee, Kolloid Z. Z. Polym., 230, 341 (1969).

13. R. E. Kesting, M. K. Barsh and A. Vincent, J. Appl. Polymer Sci., 9, 1873-93, (1965).

14. R. E. Kesting, J. Appl. Polymer Sci., 9, 663 (1965).

15. R. Bloch and M. A. Frommer, Desalination, 7, 259 (1970).

16. M. A. Frommer, R. Matz and U. Rosenthal, Ind. Eng. Chem. Prod. Res. Devel., 10, 193 (1971).

17. R. Bloch and M. A. Frommer, United States Department of the Interior, Office of Saline Water, 1969-70 Saline Water Conversion Report, p. 29.

18. S. Loeb, in Desalination by Reverse Osmosis, U. Merten (Ed.) M.I.T. Press, Cambridge, Mass., 1966, p. 69.

19. W. Banks and A. Sharples, J. Appl. Chem., 16, 28, 94 (1966).

20. S. Manjikian, S. Loeb and J. W. McCutchan, Proceedings of the First International Desalination Symposium, Washington, D.C., 1965.

21. J. Hildebrand and R. Scott, Solubility of Non-Electrolytes, 3rd Ed., Reinhold, New York, 1949.

22. C. M. Hansen, Ind. Eng. Chem. Prod. Res. Devel., 8, 1 (1969).

23. C. M. Hansen and K. Skaarup, J. Paint Tech., 39, 511 (1967), or Ref. 25.

24. E. B. Bagley, T. P. Nelson and J. M. Scigliano, J. Paint Tech., 43, 35 (1971).

25. C. M. Hansen, Doctoral Dissertation, Technical University of Denmark, Danish Technical Press, Copenhagen, 1967.

26. E. Ott and H. Spurlin, Cellulose and Cellulose Derivatives, Part II, Interscience, New York, 1954.

27. Eastman Chemical Co., Technical Bulletins.

28. D. L. Hoernschemeyer and C. W. Saltonstall, Jr., Envirogenics Co., private communication.

29. M. A. Frommer, I. Feiner, O. Kedem and R. Bloch, Desalination, 7, 393 (1970).

30. C. M. Hansen and A. Beerbower, "Solubility Parameters", Encycl. Chem. Tech., Suppl. Vol., 1971.

31. S. Manjikian, Ind. Eng. Chem. Prod. Res. Devel., 6, 1 (1967).

32. L. A. Lee, United States Department of the Interior, Office of Saline Water, 1970-71 Saline Water Conversion Report, p. 381.

33. C. W. Saltonstall, F. C. Burnett, W. S. Higley, W. M. King and A. L. Vincent, United States Department of the Interior, Office of Saline Water Research and Development Progress Report No. 232, Jan. 1967.

34. E. Klein and J. K. Smith, ACS Polymer Preprints, 12, 268 (1971).

35. J. K. Smith and E. Klein, United States Department of the Interior, Office of Saline Water Research and Development Progress Report No. 630, 1971.

36. See the paper by M. A. Frommer and D. Lancet in this volume.

37. H. Burrell, Preprints, Division of Organic Coating and Plastics Chemistry, ACS, Vol. 31, 367 (1971).

38. M. L. Huggins, Ann. N. Y. Acad. Sci., 43, 1 (1942).

39. G. J. Githens, P. A. Hitchcock, D. C. Sammon and G. E. Wakley, Desalination, 8, 369 (1970).

40. W. W. Boddie, Doctoral Dissertation, Dartmouth University 1970.

41. G. M. Sletmoe, J. Paint Technology, 38, 641 (1966).

42. J. H. Daane and R. E. Barker, Polymer Letters, 2, 343 (1964).

43. G. S. Park, Chapt. 5 in Diffusion in Polymers, J. Crank and G. S. Park, (Eds.), Academic Press, New York 1968.

44. J. K. Gilham, Appl. Polymer Symposia, 2, 59-75 (1966).

45. H. Strathmann, P. Scheible, and R. W. Baker, J. Appl. Polymer Sci., 15, 811 (1971).

46. J. W. Richter and H. H. Hoehn, U.S. Patent No. 3,567,632 (1971).

THE MECHANISM OF MEMBRANE FORMATION: MEMBRANE STRUCTURES AND THEIR RELATION TO PREPARATION CONDITIONS*

Moshe A. Frommer and Doron Lancet
Hydronautics, Incorporated
Laurel, Maryland 20810

INTRODUCTION

The application of membrane techniques for filtration, concentration, and separation purposes is continuously increasing. An outstanding example is desalination by reverse osmosis. Nevertheless, most of the procedures developed for manufacturing membranes were formulated empirically, and the understanding of the factors governing membrane structure and its characteristics is still poor. One of the most severe problems associated with the practical application of high pressure filtration techniques is the elimination of membrane defects. The formation of voids and large cavities in cellulose acetate (CA) desalination membranes has been observed by many investigators[1-3] but except for empirical procedures for eliminating such defects [1,3] no study has been made of the mechanism of their formation. In this article we attempt to elucidate factors governing the formation of voids and cavities in CA membranes. We shall also show that simple and quick optical microscope experiments can serve as useful guides for outlining procedures for eliminating the formation of such defects (large finger-like cavities) in tubular desalination membranes.

---

*This work was carried out at the Polymer Interfaces Section of the National Bureau of Standards as part of the Research Associate Program of the NBS. Part of this work has been presented at the 162nd National Meeting of the American Chemical Society, Washington, D.C.

Understanding of the structures of cellulose acetate desalination membranes is important for understanding the mechanism of desalination by reverse osmosis. It is now well established that the high filtration rates achievable with these membranes are due to their asymmetric structure.

It has furthermore been shown[4-6] that many of these membranes are "skinned", i.e., their salt selective layer appears as a very thin discrete dense uniform polymeric layer comprising the uppermost portion of the membrane. These observations have been extrapolated and have led to a widely accepted two-layer model for desalination membranes[6-8] in which the entire "upper" layer, even though relatively thick, has been identified as the salt-selective layer[9].

In this paper we shall present our observations on the structure of CA membranes cast from Manjikian modified Loeb-Sourirajan solution (25:30:45 weight ratio of CA: formamide: acetone) which indicate that these membranes are composed of two thick layers and that the salt-selective layer is only a portion of the upper layer of the membrane. It is furthermore shown that the thickness of this upper layer (most of which is not salt-selective) increases with the increase in the acetone-evaporation period prior to leaching.

EXPERIMENTAL

Materials

The cellulose acetate (CA) was Eastman Chemical Co.* type E-398-10. Acetone, dioxane, acetic acid, dimethylformamide (DMF) and dimethylsufoxide (DMSO) were of analytical grade purity. Formamide and triethylphosphate (TEP) were of Practical grade. Binary polymer solutions contained 20 percent or 7 percent (by wt.) of CA. The ternary solution consisted of 25:30:45 (weight ratio) CA: formamide: acetone. This Manjikian modified Loeb-Sourirajan solution (10) is abbreviated in this paper as the modified Loeb solution.

Membrane Preparation

A thin (0.3 mm) film of CA solution was cast at ambient temperature on a glass plate and precipitated in an ice-water bath. Films of binary solutions were immersed in water immediately after their casting, whereas the modified Loeb solution

*Company names are given for identification purposes only and do not imply an endorsement by the National Bureau of Standards.

was allowed to evaporate at ambient atmosphere for a predetermined period prior to its precipitation in the water bath. All membranes were annealed in a water bath at 80°C for 30 minutes, and stored in water.

Observations

Optical microscope studies of the processes taking place during polymer precipitation were performed employing the technique developed by Matz[11]. In this technique a drop of the casting solution is placed on a microscope slide and covered with a cover slide to form a thin circular film. Drops of water are placed in contact with the periphery of the cover slide. These drops are immediately drawn by capillary forces in between the microscope slides to form a circular water-polymer solution interface. By utilizing dark field microscopy, it is possible then to observe and follow the existence and formation of light scattering zones such as interfaces between homogeneous phases and turbid areas.

Scanning electron micrographs were taken of cross sections of membranes (initially stored in water) which were allowed to dry at ambient conditions after immersion in a 66:30:4 water: glycerol: Triton X-100* solution. Cross sections were obtained from fresh edges prepared by fracturing the membranes at liquid nitrogen temperatures[12]. These membranes were mounted obliquely on the specimen mount (so that the cross section was exposed to the incident electron beam) and coated with gold. The axis about which the plane of the cross section was tilted relative to the incident electron beam runs horizontally (left to right) in all the micrographs. The scales indicated refer to sample magnification along that direction. The actual dimensions of the cross sections in the vertical direction are larger by 10-30 percent than indicated by the scale, depending on the tilting.

Rate of Acetone Evaporation

The rate of acetone evaporation was determined, at ambient conditions, from the weight loss of a glass slide coated with a thin (∿0.3mm) film of the modified Loeb solution. (The boiling point of formamide is 210°C as compared to 56.5°C for acetone. The evaporation of formamide in our experiments was, therefore, neglected.) The CA content in the cast film was then determined by complete evaporation of the acetone and the formamide, achieved by evacuation at 80°C overnight. From the measured weight of the

*A solution of isooctylphenoxy polyethoxyethanol, produced by Rohm and Haas Co.

dry CA and the known composition of the casting solution, it was possible to compute the net weight of the cast solution before any evaporation took place.

RESULTS AND DISCUSSION

Factors Responsible For The Formation of Voids And Imperfections In CA Membranes Cast From Binary Solutions

Structures of CA Membranes Cast From Binary Solutions. Scanning electron micrographs (SEMS) of cross sections of membranes cast from six different binary solutions are shown in Figure 1. Since the preparation procedure was identical in all cases it is obvious that differences in the structures of the resultant membranes must be attributed to the nature of the casting solvent. The cross sections of the membranes cast from acetic acid or from TEP (Type A membranes) appear uniform compared to those cast from DMF, DMSO, or dioxane (Type B membranes) which contain cavities almost as large as their thickness (200 μm). Type A membranes are transparent and Type B membranes are white and opaque. This is in accordance with the observations shown in Figure 2 that in highly magnified cross sections of Type A membranes no distinguishable pores of the order of 1 μm in size can be observed, whereas Type B membranes contain many pores larger than 0.4-0.7 μm (the range of the visible spectrum) which scatter light. The acetone-cast membranes contain many micron size pores and smaller, as shown in Figure 2d, and therefore seem to belong to Type B membranes. However, since this system (acetone-cast membrane) displays many unique properties it will be discussed separately.

It may appear from Figure 1 that the difference between Type A and Type B membranes results from a different extent of shrinkage which the cast polymer solution undergoes during its precipitation in water. Type A membranes (cast from acetic acid or from TEP) may appear to be uniform because they shrink considerably during phase separation (the thicknesses of the SEM cross sections of the TEP- and acetic acid-cast membranes are 60-80 μm as compared to the casting thickness of 300 μm) and conversely the large cavities contained in Type B membranes (cast from dioxane, DMF, or DMSO) appear to be associated with a small decrease in thickness during precipitation.

However, the thicknesses of the dried cross sections used for scanning electron microscopy are not representative of the thicknesses of the wet membranes. Actually the thickness of the wet membranes (excluding the acetone-cast membrane) are not very different as indicated by their total water content (68-76 percent)

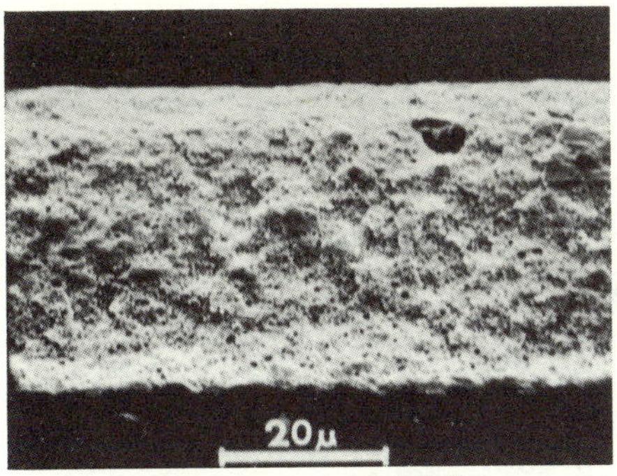

SOLVENT: ACETONE

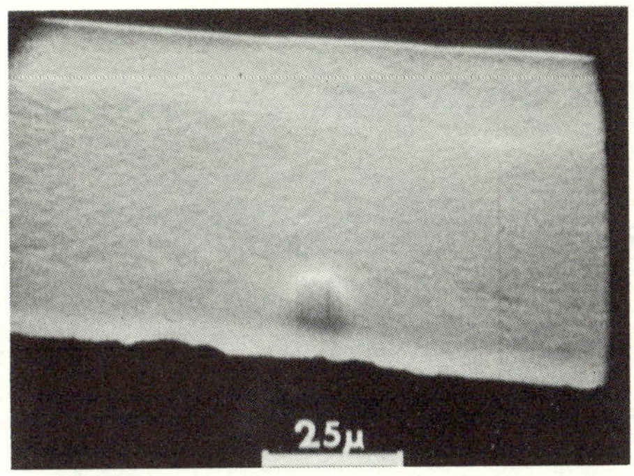

T. E. P.

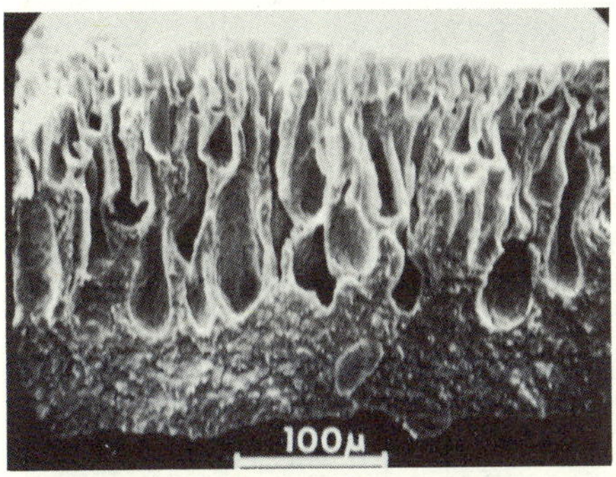

DIOXANE

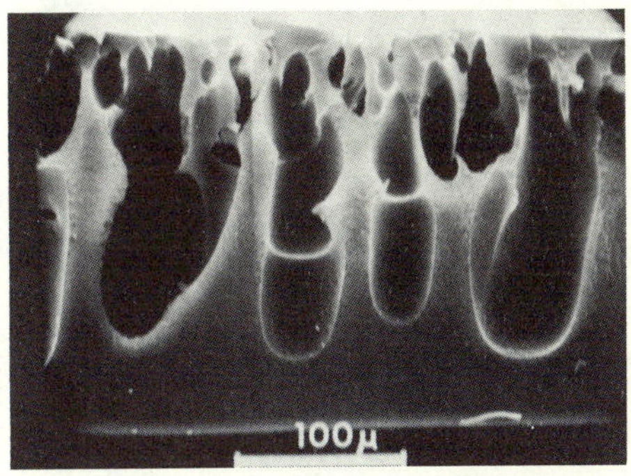

D. M. F.

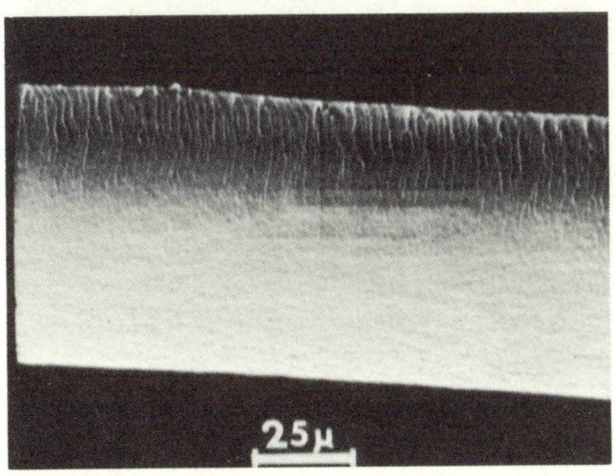

ACETIC ACID

D. M. S. O.

Fig. 1 - Scanning Electron Micrographs of Cross Sections of Membranes Cast From 20% C.A. in Six Different Binary Solutions. (The upper surface of the cross-sections corresponds to the air-solution interface of the cast solution.)

TYPE - A

TYPE - B

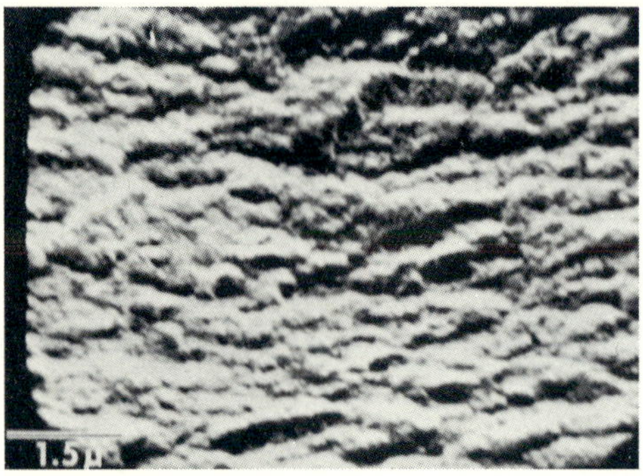

a. ACETIC ACID

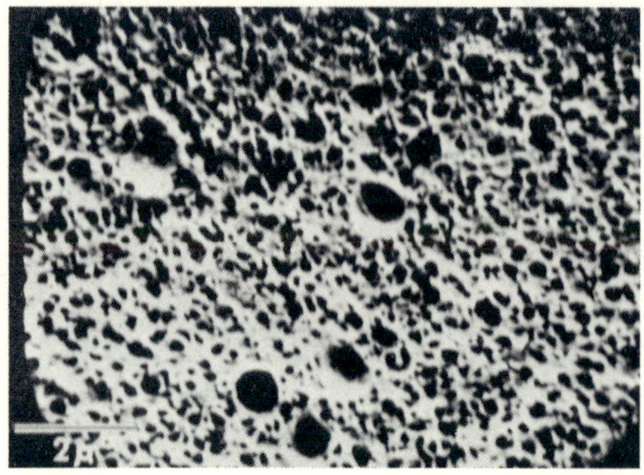

c. D.M.S.O.

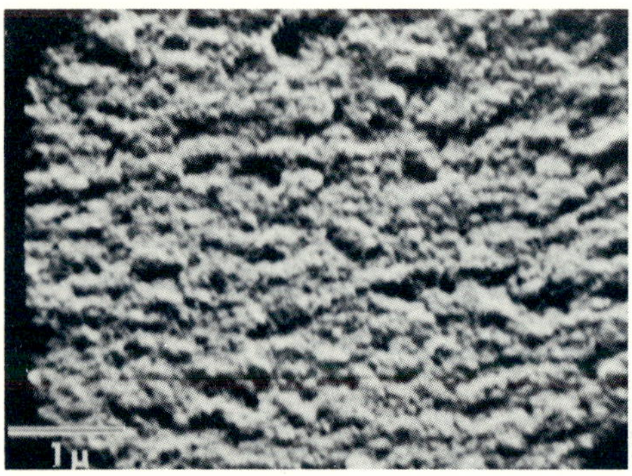

b. T.E.P., 80°C ANNEALED

d. ACETONE

Fig. 2 - High Magnification Scanning Electron Micrographs of Cross Sections of C.A. Membranes Cast From Acetic Acid or T.E.P. (Representing Type A membranes) or From D.M.S.O. or Acetone (Representing Type B membranes).

given in Table 1. The large differences in the thicknesses of the SEM cross sections result primarily from different extent of shrinkage which the wet membranes undergo during their drying and evacuation. Both Type A and Type B membranes (in five out of the six cases) contain essentially the same relative amounts of polymer and water. However, they differ in the wet state in the way the water is distributed in the membrane.

Table 1

Total Water Contents and Shrinking During Drying* of Wet CA Membranes Cast From Six Different Binary Solutions

Casting Solvent	Water Content, % by Weight	Wet Thickness* / Dry Thickness
Acetone	37	1.20
Dioxane	68	1.10
Acetic Acid	71	1.69
TEP	71	1.5 - 1.7
DMF	76	1.00
DMSO	76	1.03

* The thicknesses of the dry membranes were measured after immersion in a 66:30:4 water: glycerol: Triton X-100 solution followed by "drying" at ambient conditions. Additional shrinking of the cross sections used in electron microscopy might have taken place during evacuation and gold coating of the samples.

Relations Between Structure of Membranes and Rates of Their Precipitation. To clarify the reasons for the large differences in membrane structures we have studied the rate and the nature of precipitation of cellulose acetate from the various casting solutions employed. Typical pictures of precipitation structures observed through an optical microscope 2.0 minutes after drops of the various casting solutions were contacted with water are shown in Figure 3. Let us discuss first the differences in the rate of precipitation of CA from the various solutions, and later the fact that in most cases finger-like structures appear, which are similar to the shapes of the cavities shown in Figure 1 for Type B membranes.

The rate of penetration of the precipitation front from the periphery of a circular droplet towards its center is shown in Figure 4 for each of the six different binary solutions. While it is realized that quantitatively the rates of precipitation in the case of a thin cast film might be different from those given in Figures 3 and 4 due to differences in interfacial conditions, relative reactant quantities, and temperature of precipitation (the optical microscope experiments were conducted at room tempera-

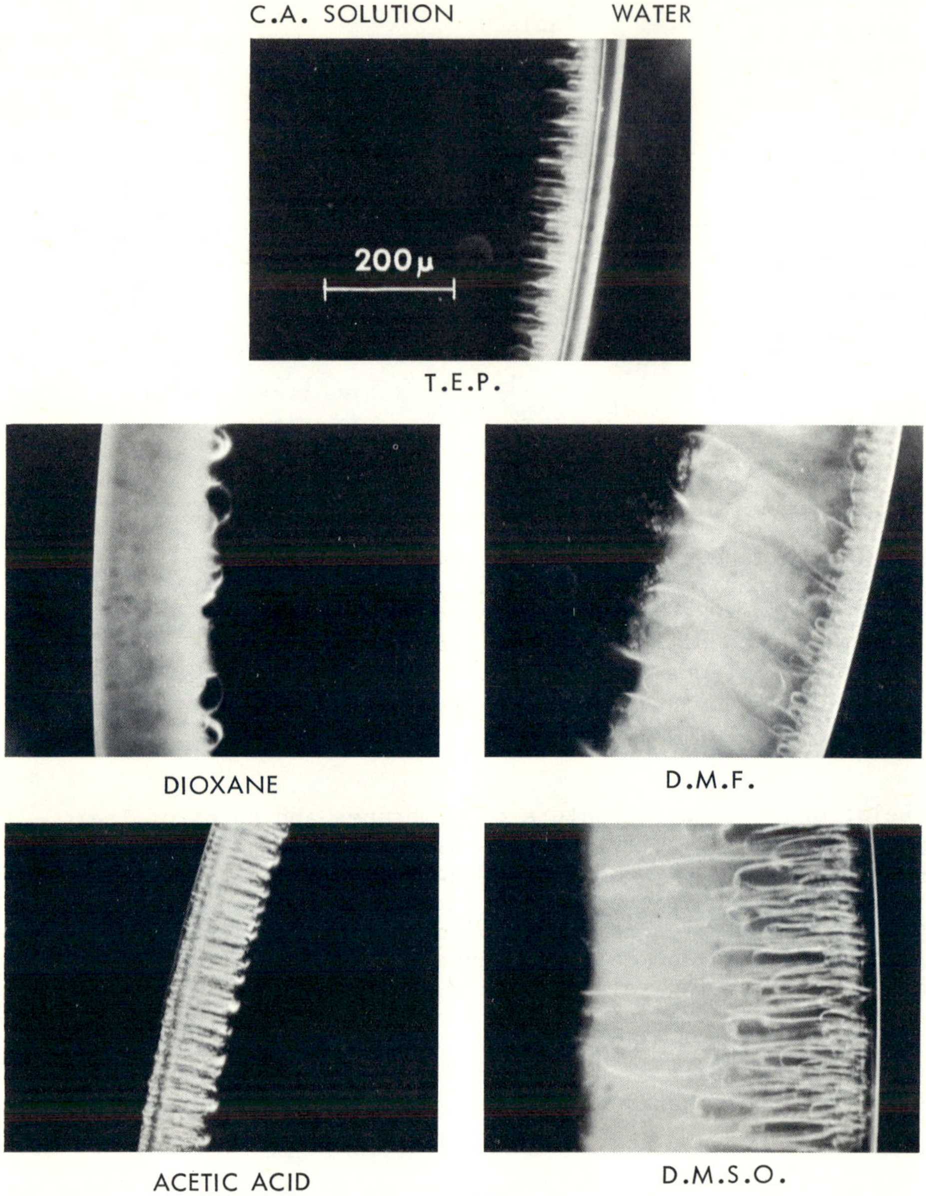

Fig. 3 - Photomicrographs Showing The Influence of Solvent on The Rate and The Nature of Precipitation of Cellulose Acetate From a 20% Polymer Solution. (Water was introduced in between the microscope slides 2-3 seconds after formation of the thin circular film of polymer solution. Pictures were taken 2.0 minutes after contact with water.)

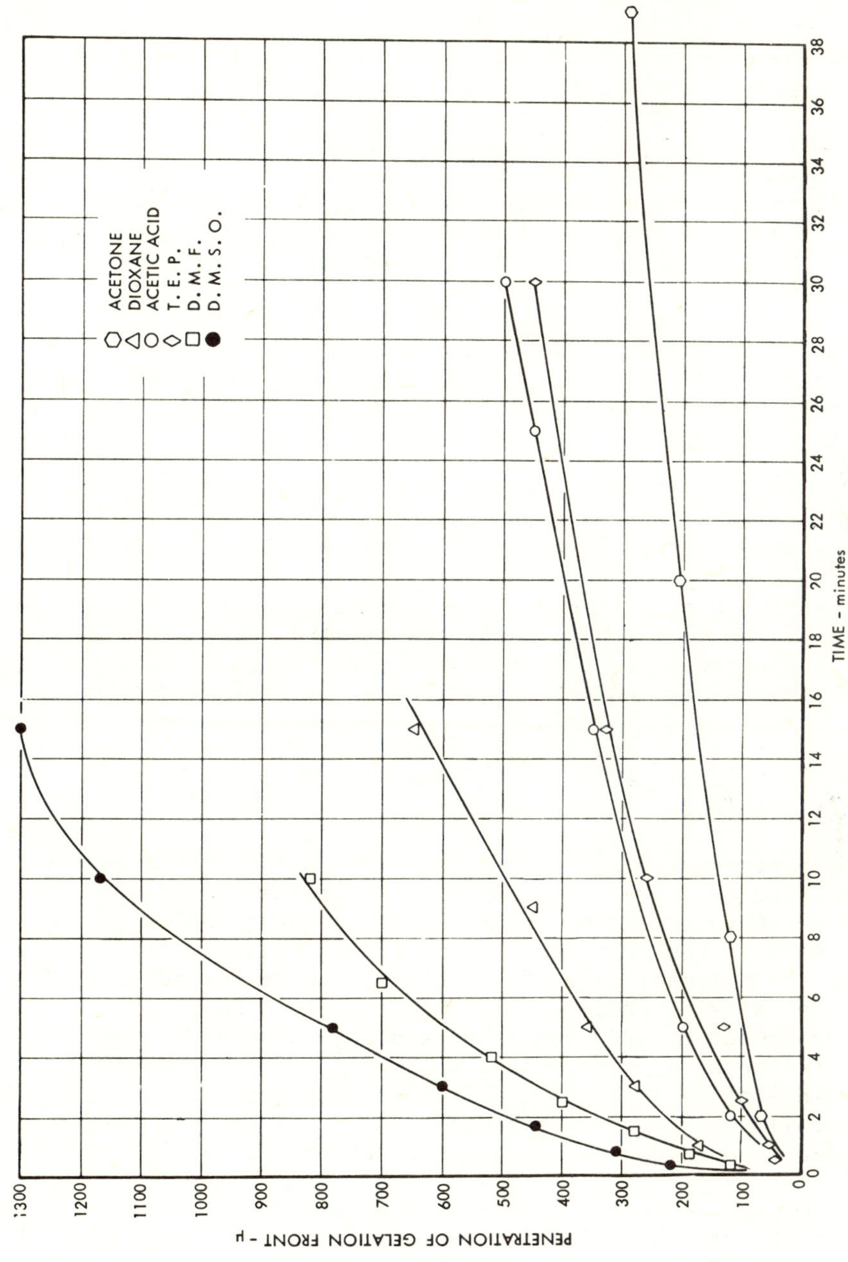

Fig. 4 - Rate of Precipitation of C.A. From Six 20% Binary Solutions. (Water was introduced in between the microscope slides 2-3 seconds after the formation of the thin circular drop of polymer solution.)

ture), it is reasonable to assume that the relative rates are similar. The results suggest that the formation of large cavities in Type B membranes (represented by the DMF or DMSO systems) is related to the high rate of their precipitation from the casting solution. In contrast, the formation of uniform Type A membranes (TEP or acetic acid) is associated with slow rates of precipitation.

Since the structures of membranes and their rates of precipitation are so significantly determined by the nature of the casting solvent it is of interest to correlate these structures with properties of the CA-water-solvent system which are affected by the nature of the solvent. It has been shown[13-15] that other factors being constant, a change in the casting solvent induces changes in: (a) the rate of flow of solvent out of the polymer solution into the leaching bath, (b) the rate of penetration of water into the polymer solution, and (c) the concentration of water required for precipitating cellulose acetate from the cast solution.

The rates of flow of solvents out of a cast polymer solution into the leaching bath were measured during membrane formation with apparatus described in Reference 14. The results expressed in terms of the slope of C_t/C_∞ (the concentration, C_t, of solvent at time, t, expressed as a fraction of the final concentration of the solvent, C_∞, in the leaching bath) against the square root of time, are given in the first column of Table 2. Rates of penetration of water into a cast polymer solution may be estimated from the rate of flow of water through a porous CA membrane into a 2:3 water:solvent mixture[12] and are given in Table 2. The concentration of water at the precipitation point[12] is given in the third column of Table 2.

The rate of penetration of the precipitation front in the optical microscope experiments is expected to increase with: (a) increase in solvent outflow (b) increase in water inflow and, (c) decrease in the concentration of water necessary for precipitating the polymer. A useful arbitrary scale for anticipating relative rates of precipitation may therefore be the product of the slope of C_t/C_∞ versus $\sqrt{t}$ and J_{H2O} divided by the water concentration at precipitation, C. This arbitrary scale, given in column 4 of Table 2, is qualitatively consistent with the experimentally measured rates of penetration of the precipitation front taken from Figure 4 and given in column 5.

Let us now examine whether the relative rates of precipitation of the CA from the various casting solutions can be predicted from the tendencies of the pure solvent (with no polymer) and the non-solvent (water) to dissolve in one another. The parameter describing the tendency of the water and of the solvent to mutually

Table 2

Factors Governing The Rate of Precipitation of Cellulose Acetate From Various Binary Solutions, and Experimentally Observed Precipitation Rates

	J_S Measure of Solvent Outflow, Slope of C_t/C_∞ vs. $\sqrt{t}$, $sec^{-1/2}$	J_{H_2O} Estimated Rate of Water Inflow, (microliters)/cm^2-sec	C Approximate Water Concentration at Precipitation Point, percent	Arbitrary Scale for Relative Rate of Precipitation, $100 J_S J_{H_2O}/C$	Average Experimental Rate of Precipitation (μm/min.)*
Acetone	0.10-0.13	None	28	0	10
Dioxane	0.12	0.06	30	2.4	36
Acetic Acid	0.12	>0.06	31	>2.3	24
T.E.P.	0.07	0.07	10	5	24
D.M.F.	0.13	0.16	12	17	67
D.M.S.O.	0.12	0.27	15	22	90

* The figures given are average rates of penetration of the precipitation front between 1-10 minutes after contact with water.

dissolve in one another is the free energy of mixing. There is, however, very little data concerning free energies of dissolution. Another parameter which may be indicative of the tendency of mutual mixing is the heat of mixing. The heats of solution for five of the binary systems (H_2O in acetone, dioxane, acetic acid, DMF, and DMSO) are summarized in Figure 5, in units which seem appropriate for our purpose, namely calories per gram of solution as a measure of ΔH_{mix}, and percent water as a measure of concentration. The order of decreasing tendency of mixing with water indicated by this graph is DMSO > DMF > acetone > dioxane > acetic acid. Excluding acetone, this is identical to the order of decreasing rates of precipitation (column 5 in Table 2).

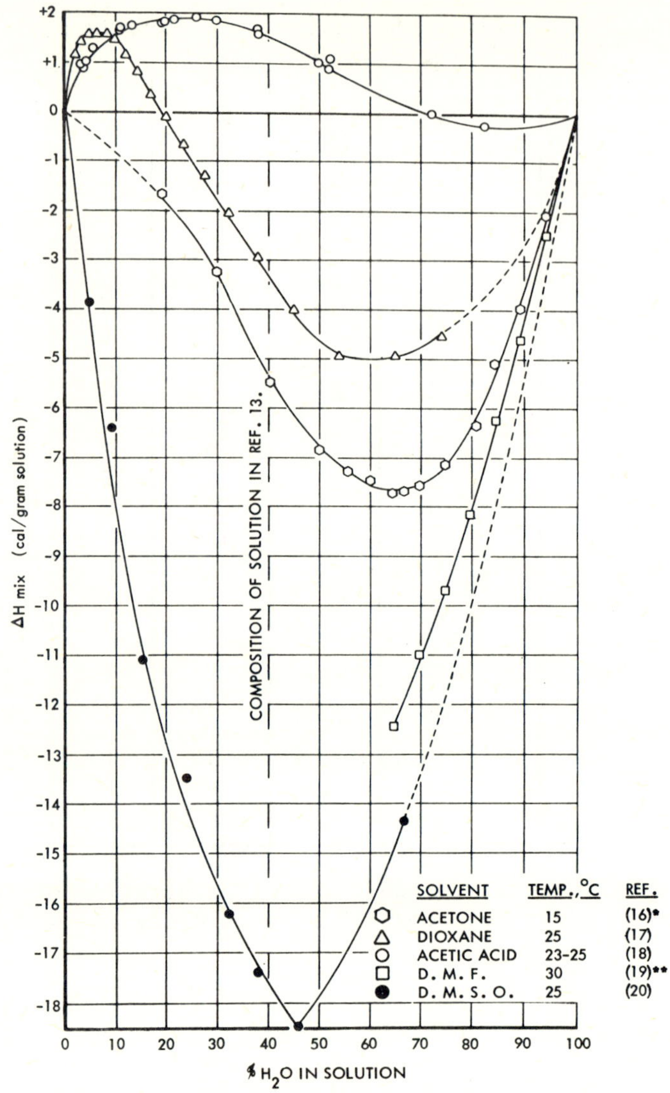

Fig. 5 - Dependence of the Heats of Mixing of Various Binary Water-Solvent Solutions on Composition.

Effects of Solvent Evaporation on Membrane Structure. The acetone-cast membranes and the CA-acetone-water system appear to contradict almost every "rule" presented until now. Thus, although the heat of mixing of acetone with water is higher than that of dioxane-water and of acetic acid-water, the rate of CA precipitation from an acetone solution is the slowest one. Despite this slow rate of precipitation, acetone-cast membranes display some of the characteristics of Type B membranes containing many micron-size pores and occasionally also a few large voids. These discrepancies may partly be explained by two unique characteristics of the acetone-CA-water system, namely the high volatility of acetone (compared to the volatility of the other five solvents used in this study), and the exceptionally low permeability of CA membranes to water in the presence of acetone. A good measure for the relative volatility of liquids might be their boiling point. The boiling point of acetone (56°C) is considerably lower than that of the other solvents used in this study (101°C, 118°C, 215°C, 153°C, and 189°C for dioxane, acetic acid, TEP, DMF, and DMSO, respectively). The volatility of acetone under ambient conditions is so high that about 25 percent of the acetone evaporates from the cast film (used for preparing the membrane) within 30 seconds. Because of this, even apparently "immediately" leached membranes (namely membranes which were immersed in the water bath within a few seconds from their casting) contain a much more concentrated solution at their interface with the air. On this basis alone we may expect that the upper layer, when precipitated in water, will be denser and consequently less permeable than interfacial layers precipitated from more dilute solutions. Furthermore, it has been shown[13] that even porous CA membranes separating water from an acetone solution, while being highly permeable to acetone, are only slightly permeable to water (Table 2), indicating that apparently there is no water flow into the cast solution through the interfacial precipitate. The slow rate of CA precipitation from an acetone solution may be attributed therefore to an exceptionally low water inflow. Because of the high volatility of acetone, and because of its capability to flow out of the cast solution through the interfacial water-impermeable precipitate, the acetone-cast film shrinks considerably prior to phase separation. As a result, starting from the same initial cast thicknesses, the thickness (as estimated from its water content, which is 37 percent by weight) of acetone-cast membranes are considerably lower than those of membranes cast from the other five binary solutions.

The reason why the relation between the rate of precipitation and membrane structure, found for the five systems, is not applicable to the acetone system is not yet fully understood.

Shrinking During Precipitation and Void Formation. We would now like to discuss why in all the droplet precipitation experiments shown in Figure 3 (including the TEP and acetic acid systems which yield uniform Type A membranes), finger-like protrusions appeared while, as shown in Figure 1, finger-like cavities exist only in Type B membranes. As will be explained below, this difference in behavior appears to result from the shrinking of the polymeric phase during its precipitation from the casting solution.

In the precipitation of a flat cast film part of the shrinking of the polymeric phase accompanying phase separation can be compensated for by a uniform reduction in its thickness. Thus the volume fraction of the polymer in a sheet membrane is 20-40 percent and is higher than its ∿15 percent volume fraction in the casting solution. However, in the optical microscope experiment the droplet is contained between two glass slides which inhibit or at least disturb the movement of an interfacial precipitate adhering to the glass. Under these conditions the volume reduction associated with polymer precipitation can only be compensated for by the formation of large (finger-like) cavities.

If the extent of shrinking is large, cavities will form in sheet membranes even when the tendency of water to mix with the solvent is low (i.e., when the rate of precipitation may be expected to be slow). This can be demonstrated by the fact, shown in Figure 6, that membranes cast from 7 percent by weight CA in TEP contain very large cavities, although membranes cast from 20 percent by weight polymer in the same solvent appear smooth and uniform (Figure 1).

Fig. 6 - Scanning Electron Micrograph of a Cross Section of C.A. Membrane Cast From 7% Solution in T.E.P. (This membrane was annealed at 95°C.)

The Structures of Desalination Membranes Cast From Modified Loeb-Sourirajan Solution

It has been shown above that solvent evaporation prior to immersion of the cast polymeric film into water may affect considerably the structure of membranes cast from binary solution. In this section we shall describe the influence of solvent evaporation on the structure of actual desalination membranes cast from the well known ternary Manjikian modified Loeb-Sourirajan solution[10].

Influence of Air Exposure Period on Membrane Structure. Shown in Figure 7 are scanning electron micrographs of cross sections of CA membranes cast from modified Loeb solution and dried 5-10 sec. (a), one min. (b), and 5 min. (c), prior to their immersion in the ice water bath. These pictures suggest the existence of two distinguishable layers in the samples dried for 5-10 sec and for one min. The thickness of the upper and what appears to be also a denser layer, increases with the increase in evaporation period, and varies between about 10 percent of the thickness for the membrane obtained by 5-10 sec. evaporation to practically the total thickness of the membrane evaporated for 5 min. The two thick layer structures and the thickening of the upper layer with an increase in the evaporation period were reproducibly observed in many membranes of each type.

The types of precipitated layers, and an increase in the thickness of the "outer" layer (corresponding to the "upper" layer in actual desalination membranes) with an increase in air exposure time prior to contact with water, could also be observed in optical microscope studies (using dark field illumination) of the precipitation of CA from a modified Loeb solution, as shown in Figure 8. In this case the "outer" precipitated layer is transparent (and, therefore, appears dark in the micrographs), whereas the inner precipitated portion contains many voids and therefore scatters light and appears bright. It may be noticed that in the immediately leached drop there exists only one type of precipitate - the bright one. It is important to point out that the outermost white band which appears in the 0.5, 1.0 and 1.5 minutes evaporated drops shown in Figure 8 is not a third type of precipitate but a consequence of light scattering associated with the curvature of the water-solution interface.

A typical graph describing the rate of acetone evaporation at ambient conditions, similar to those employed in preparing actual membranes, from a 25:30:45 (w/w) CA: formamide: acetone casting solution is given in Figure 9. The rate is very high immediately after film casting with very little change occurring after 5 minutes. (It may be noted that the amount of acetone

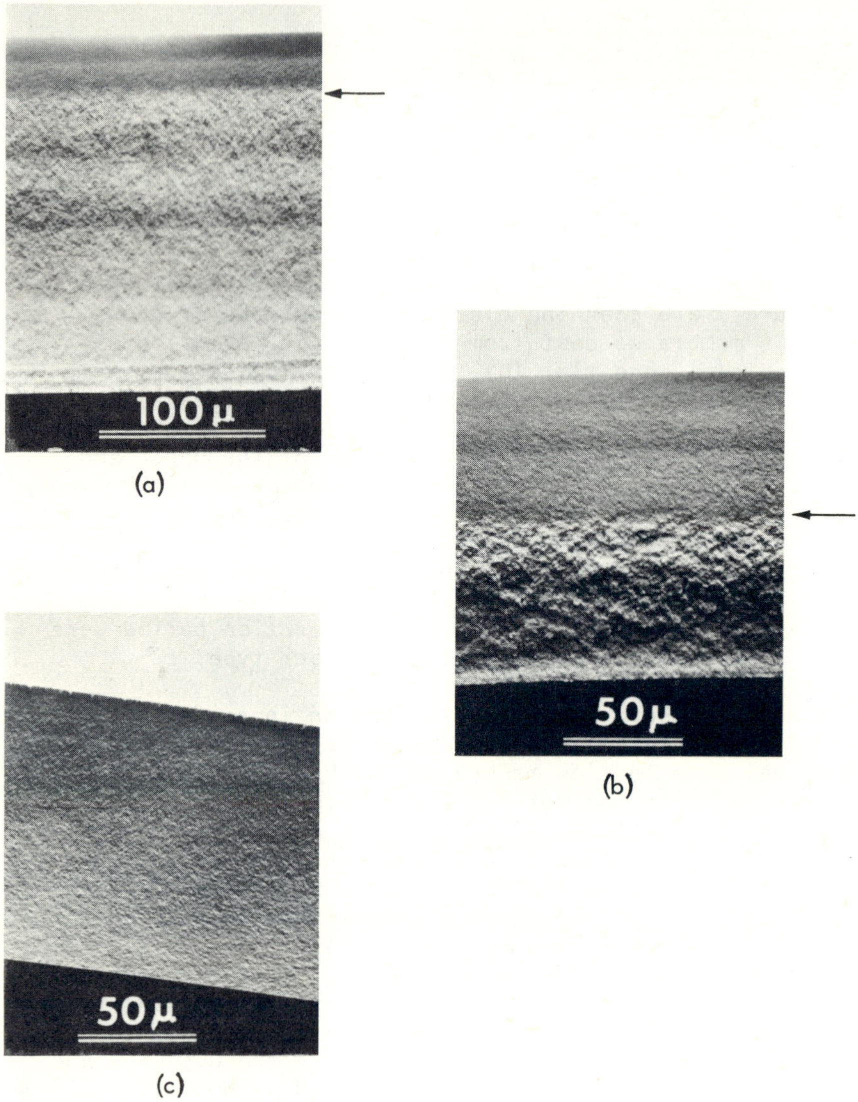

Fig. 7 - Scanning Electron Micrographs of Cross Sections of C.A. Membranes Cast From Modified Loeb Solution and Dried 5-10 Sec. (a), 1 Min. (b), and 5 Min. (c), Prior to Immersion in Water. The upper surface on the cross sections corresponds to the air-solution interface of the cast solution. The arrow designates the boundary line between the upper and the lower layers in (a) and (b) (see text).

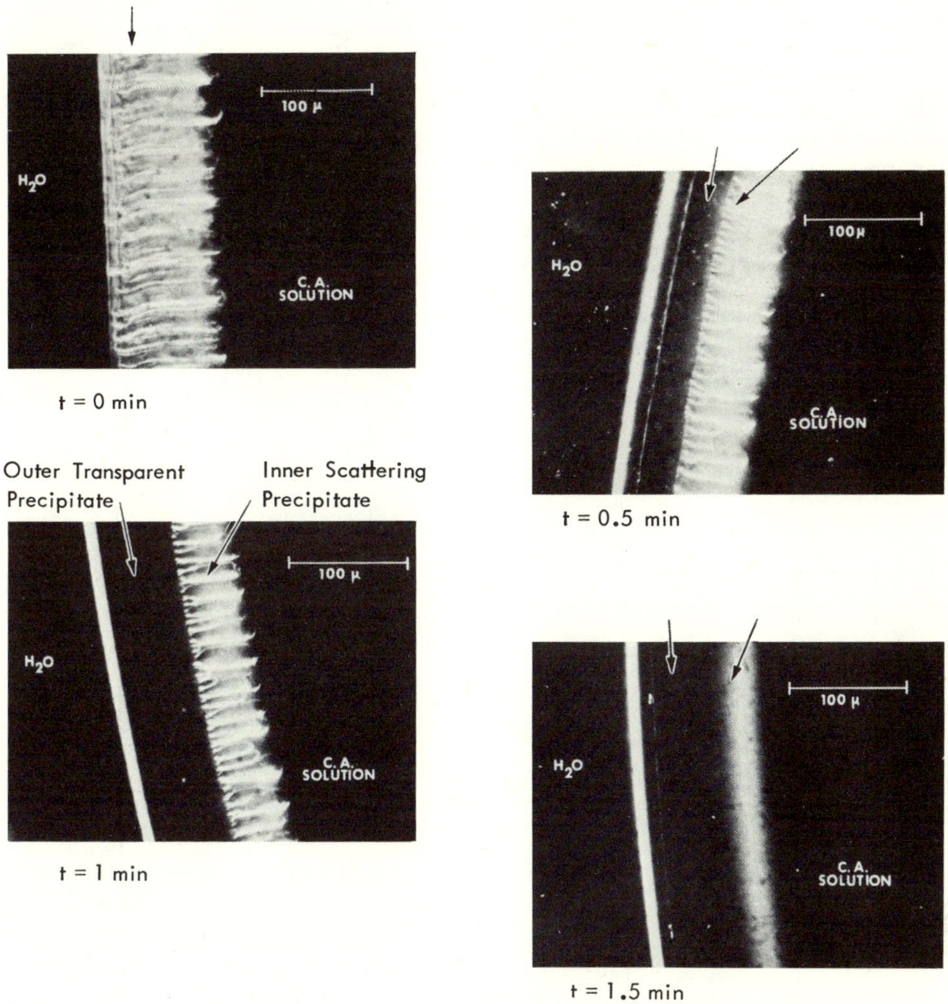

Fig. 8 - Photomicrographs Showing the Influence of Air Exposure Period on the Precipitation of C.A. From a Modified Loeb Solution. t = Period Between Formation of the Thin Circular Drop and Its Contact With Water. (Pictures were taken one minute after contact with water.)

evaporated from the film is higher than the weight loss given in Figure 9 due to absorption of water from the atmosphere).

Comparison of the rate of acetone evaporation to the rate of thickening upon evaporation of the upper layer in actual membranes (Figure 7) suggests that the formation of this layer results from acetone evaporation. The rate of thickening of the outer transparent precipitate in the optical microscope drop precipitation experiments (Figure 8) is consistent with the rate of acetone evaporation shown in Figure 9 in that it is very high in the first 30 sec. and levels off in the next few minutes. This suggests that the formation of this transparent layer results also from acetone evaporation. Let us discuss now two questions:

(1) Why and how can evaporation of acetone lead to the formation of two types of precipitated layers?

(2) How is the thickness of the upper acetone-evaporated layer related to the thickness of the salt-selective layer commonly called the "skin"?

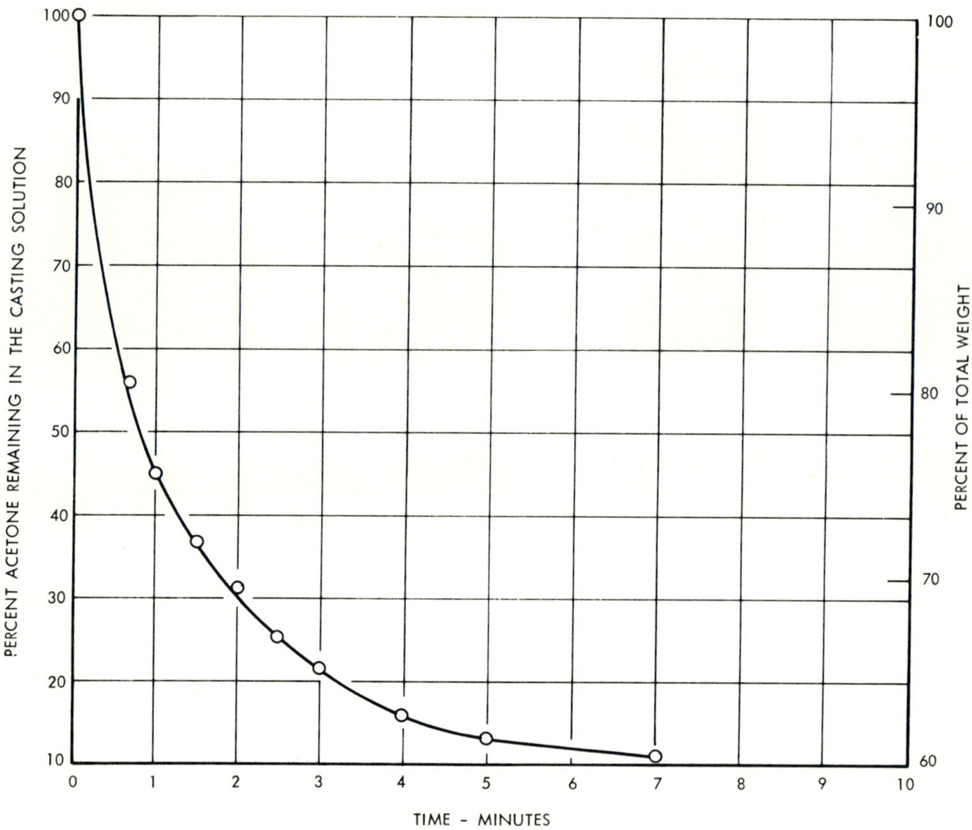

Fig. 9 - Rate of Acetone Evaporation From a Modified Loeb Solution (25:30:45 CA: Formamide: Acetone, W/W).

Acetone Evaporation and the Formation of Two "Thick" Precipitated Layers. The composition of the modified Loeb solution is such that a slight decrease in the acetone content is associated with a considerable increase in viscosity. This can clearly be seen in the phase diagram of the CA-acetone-formamide system shown in Figure 10. It requires only about 20 percent decrease in the acetone content to induce a sol-gel transition within the casting solution. It is apparent therefore, from the data shown in Figure 9 that the thickness of the upper solidified-gel layer covering the fluid casting solution will increase rapidly with an increase in the acetone evaporation period, and will occupy the total thickness of a ∿ 0.3 mm cast film in less than 5 minutes.

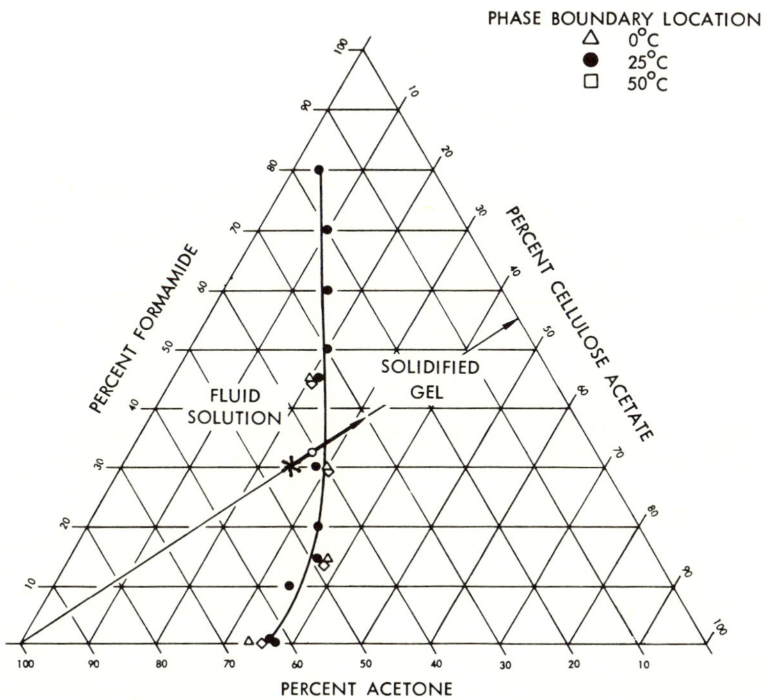

Fig. 10 - Phase Diagram for CA-Formamide-Acetone at 23-40°C. (Taken from Ref. 21.) The star (*) designates the composition of Manjikian modified Loeb-Sourirajan solution, and the arrow indicates the direction of composition change induced by acetone evaporation.

The processes taking place upon immersion of a solidified film in water are distinctly different from those occurring during immersion of a freshly cast solution in the same bath. These differences in the nature of phase separation of a solidified gel and a fluid solution most probably are larger than the small differences in precipitation properties originating from composition differences across the solution or across the gel. Since the borderline between the upper gel layer and the lower solution layer is quite sharp, the immersion of a cast film containing these two layers yields a two layered membrane, as observed experimentally. Our observations are in accordance with those of Trudelle and Nicolas[22] who found that a membrane cast from a Manjikian modified Loeb-Sourirajan solution and evaporated 30 seconds prior to immersion consists of two thick layers, the upper one being 16 percent (19 μm) of the total (120 μm) thickness of the membrane. The water contents reported by Trudelle and Nicolas for the "lower" and "upper" layer in their 30 sec.-evaporated membranes are 62 percent (w/w) and 38 percent (w/w), respectively. These values are comparable to the total water contents found by us for an "immediately" leached and a 25-minute-evaporated membrane, namely 63 percent (w/w) and 43 percent (w/w), respectively. Having shown that the "upper" layer in our membranes is denser than the "lower" one (in that it contains lower amounts of water) let us investigate which portions of the "upper" layer are responsible for the salt-rejection characteristics of the membrane.

The "Upper" Layer and the Salt Selective Portion of the Membrane. In our membranes the selectivity to salt disappeared after abrading with a fine emory paper only a thin (∿5 μm) layer from the upper surface of either the one-minute or the five-minute evaporated membrane. The disappearance of selectivity to salt was accompanied with a considerable increase in the rate of volume flow. Since the thicknesses of the upper layers (which were already gelled when immersed in water) in these one-minute and five-minute acetone-evaporated membranes are ∿60 μm (Figure 7b) and more than 100 μm (Figure 7c), respectively, it is obvious that the salt-selective portion of the membrane is only a part, the uppermost portion, of this upper "thick" acetone-evaporated layer. This observation also is in accordance with the observations of Trudelle and Nicolas[22] who found for a 30 sec. acetone-evaporated membrane a continuous increase in the water content and a corresponding decrease in salt rejection capability from the surface of the membrane inwards. This increase in the water content was very sharp in the surface region.

Hollow cellulose acetate fibers spun from a solution in acetone and formamide have also been found to be composed of two "thick" layers[9], similar to those described above. In the case of the hollow fibers, the entire denser ∿8 μm layer out of an approximate 200 μm total thickness was identified as a salt-selective skin. Such a thickness is larger than the common estimates given for the thickness of the "skin", which are generally lower than 3 μm[6]. Whether the entire denser 8 μm layer is salt-selective or not remains an open question.

A three layer structure consisting of an upper "active layer", a "sub-layer" and a "porous layer" has recently been reported by Gittens et al.[5] for CA membranes cast from a 15:35:50 (w/w) CA: dioxane: methanol solution. While the membranes of Gittens et al. are similar to ours in that their active layer is located at the uppermost portion of the cast membrane and its thickness is smaller than 5 μm, the mechanism of formation of the various layers in their membranes may well be different from the mechanism of formation of the two "thick" layers in our membranes. The dioxane-methanol casting solution contains two volatile components, the non-solvent (methanol) being more volatile than the solvent (dioxane). Since in our casting solution the good solvent (acetone) is much more volatile than the second liquid component (formamide), the structures of membranes prepared under identical conditions, and the dependence of the structure on the preparation variables (e.g., evaporation time prior to leaching) could well be different from those in the CA-dioxane-methanol system.

<u>Elimination of Defects in Tubular Desalination Membranes.</u> A comparison of the structures observed in the optical microscope precipitation experiments (Figure 8) to the structure of CA membranes cast from the same solution (Figure 7) reveals that the large finger-like protrusions shown in Figure 8 are not evident in sheet membranes prepared from the same modified Loeb solution. It was shown earlier that finger-like protrusions were formed during optical microscope precipitation experiments of CA from TEP or acetic acid (Figure 3), while sheet membranes cast from these solutions appear uniform (Figure 1). We would like to discuss now an additional observation, namely that voids and large cavities similar in shape to the finger-like protrusions shown in Figure 8 are commonly found in tubular (internal flow) CA membranes [1,2,3] prepared from the modified Loeb casting solution. The question which arises therefore is: what is the difference in the precipitation conditions between a sheet membrane, a drop of casting solution, and a tubular membrane which induces large cavity formation in the latter two cases but not in the first one?

It was shown earlier that if the possibility of free shrinking of the polymeric phase accompanying phase separation is eliminated, voids will be occluded in the precipitate. While the sheet membrane is free to shrink through reduction of its thickness, the interfacial precipitates in the drop-precipitation experiment and in the tubular membrane are immobilized. In the optical microscope experiment, the precipitated layer formed at the periphery of the drop immediately after its contact with water adheres to the two glass slides between which the drop is contained, and therefore, the possibility of free shrinking is eliminated. In the tubular membrane situation the resistance of the solidified internal surface to stretching inhibits free shrinking of the cast tubular film and thereby induces cavity formation.

If these explanations are correct then the simple and quick optical microscope drop-precipitation experiments may be used for outlining procedures for eliminating defects in tubular membranes. It can be seen in Figure 8 that if the acetone-evaporation period is sufficiently long "fingers" do not grow. Indeed it has recently been shown by Gollan[3] that void formation could be eliminated in "hydrocast" tubular CA membranes, if the length of the forming air bubble (namely the length of the air exposure period) was high enough. The mechanism by which an increased acetone evaporation period can eliminate "finger" formation in tubular CA desalination membranes seems to be related to the thickening of the upper layer in these membranes associated with increased evaporation time. As can be seen in Figure 8, the finger-like protrusions grow only inside the transparent precipitate, namely in the solution layer which has not yet gelled. Increasing the acetone evaporation period reduces the thickness of the solution layer in which these fingers can grow and thereby reduces the extent of their growth.

It was shown earlier that even if a cast polymeric film is allowed to shrink freely during its precipitation, finger-like protrusions may form in the sheet membrane (see Figure 1) and that generally void formation is associated with a high rate of precipitation. The higher the rate of precipitation of the polymer from the casting solution, the greater is the probability of void formation. We may therefore expect that by lowering the rate of precipitation of drops or of tubular films of modified Loeb solution the formation of finger-like cavities in these precipitates may be eliminated. Lowering the rate of precipitation of CA from the same casting solution may be achieved by substituting salt solution for water as a precipitation medium. Indeed, as can be seen in Figure 11, large voids disappeared when a drop of a CA-acetone-formamide solution was precipitated in salt solutions containing $\geq$ 2.5 percent NaCl. A similar observation was reported recently by Gollan[3] for hydrocast tubular membranes precipitated in NaCl solutions.

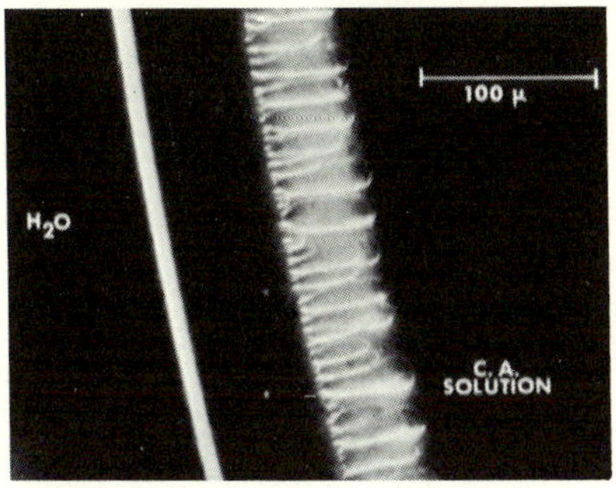

0 Conc.

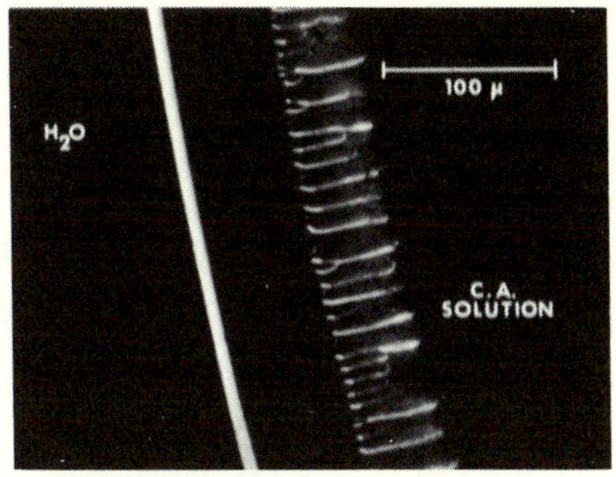

1.25% Conc. of NaCl
in Leaching Bath.

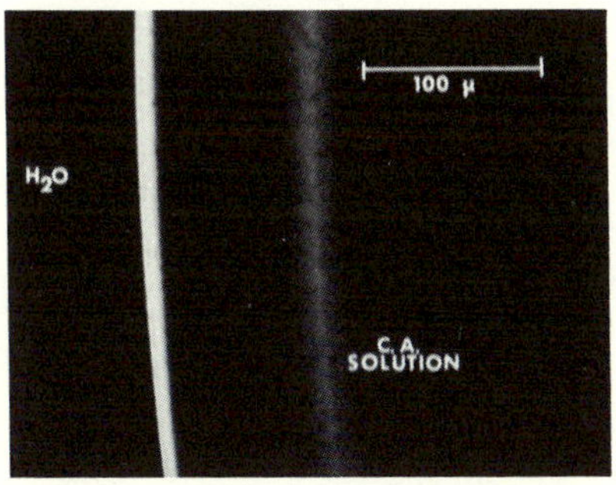

2.5%

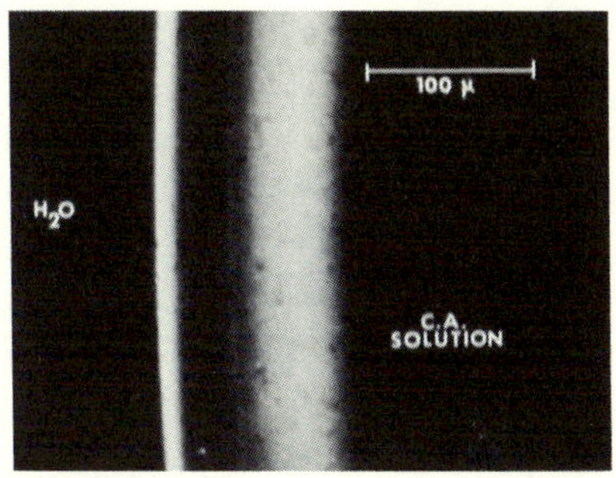

5%

Fig. 11 - Photomicrographs Showing the Influence of the Composition of the Leaching Bath on the Precipitation of C.A. From a Modified Loeb-Sourirajan Solution. (Water was introduced in between the microscope slides 1 minute after formation of the drop. Pictures were taken 1 minute after contact with water.)

CONCLUSIONS

The factors responsible for the formation of voids and imperfections in CA membranes and the structure of desalination membranes cast from modified Loeb-Sourirajan solution have been studied. It has been suggested that:

1. Tubular membranes are more susceptible to occlude voids in their matrix than sheet membranes. This is because tubular

membranes are not as free as sheet membranes to shrink during their precipitation.

2. The formation of finger-like cavities in CA membranes is generally associated with a high rate of precipitation. The rate of polymer precipitation upon immersion in the water bath will increase with an increase in (a) the rate of solvent flow out of the cast polymer solution, and (b) the rate of water flow into it. The rate of precipitation also increases with a decrease in the concentration of non-solvent (water) required for precipitating the polymer from the casting solution. These observations suggest that the rate of CA precipitation from the same casting solution can be lowered by immersing the cast film in an aqueous salt solution rather than in pure water. "Salt-gelling" can be used for eliminating "finger" formation in tubular desalination membranes.

3. Manjikian modified Loeb-Sourirajan membranes (cast from a CA-acetone-formamide solution) are apparently composed of two "thick" layers of comparable thickness. The thickness of the "upper" and denser layer varies from $\sim$15 µm for 5-10 secs. acetone-evaporated membranes to practically the total thickness of the membrane (> 100 µm) for a 5 min. "dried" one. The salt-selective portion of the membrane is only part (the uppermost portion) of this upper layer, and can be removed by a slight abrasion of the membrane surface.

4. It is suggested that the formation of membranes consisting of two "thick" layers results from the existence of two layers in the cast CA solution prior to its immersion in water, an upper, solidified gel supported by a layer of fluid solution. These two layers are formed in the CA-acetone-formamide solution due to a sol-gel transition induced by acetone evaporation. Finger-like cavities can grow only in the fluid lower layer of the cast solution. Since the thickness of this fluid layer decreases sharply with an increase in the acetone-evaporation period prior to leaching, the formation of cavities in tubular desalination membranes can be diminished by increasing this "drying" period.

ACKNOWLEDGMENT

The authors wish to thank the Office of Saline Water, U.S. Department of the Interior, for Grant No. 14-30-2529, supporting this work. The authors acknowledge gratefully the cooperation and the many helpful and stimulating discussions with Dr. Stromberg, Dr. Khoury, Dr. L. Smith, Dr. Morrissey and other members of the Polymer Division of the National Bureau of Standards.

REFERENCES

1. W. M. King, P. A. Cantor, L. W. Schoellenbach, and C. R. Cannon, Applied Polymer Symposia, 13, 17 (1970).

2. Burns and Roe Construction Corp., Report to Office of Saline Water, February, 1971.

3. A. Gollan, Hydronautics, Inc., private communication. See also the paper by A. Gollan and M. P. Tulin in this volume.

4. H. K. Lonsdale, in Desalination by Reverse Osmosis, U. Merten (Ed.), The MIT Press, Cambridge, Massachusetts (1966).

5. G. J. Gittens, P. A. Hitchcock, D. C. Sammon, and G. E. Wakley, Desalination, 8, 369 (1970).

6. R. E. Kesting, Synthetic Polymeric Membranes, McGraw Hill New York, 1971.

7. H. Yasuda and C. E. Lamaze, Applied Polymer Symposia, 13, 157 (1970).

8. R. Hays and O. Kedem, Office of Saline Water Research and Development Progress Report No. 401 (1969).

9. T. A. Orofino, Office of Saline Water Research and Development Progress Report No. 549, May 1970.

10. S. Sourirajan, Reverse Osmosis, pp. 58-63, 112-116, Logos Press, London, 1970.

11. R. Matz, Hydronautics, Inc., private communication.

12. R. E. Kesting, M. Engdahl and W. Stone, Jr., J. Macromol. Sci., A3, 157 (1969).

13. M. A. Frommer, I. Feiner, O. Kedem, and R. Bloch, Desalination, 7, 393 (1970).

14. M. A. Frommer, R. Matz, and U. Rosenthal, Ind. Eng. Chem. Prod. Res. Devel., 10, 193 (1971).

15. U. Rosenthal, J. Nechushtan, O. Kedem, D. Lancet and M. A. Frommer, Desalination, 9, 193 (1971).

16. International Crit. Tables, Vol. V, p. 157. *It should be noted that the heats of mixing cited from the same author by J. Timmermans (The Physico Chemical Constants of Binary Systems in Concentrated Solutions, Vol. 4 Interscience, New York, 1960, p. 47) are 3.5-4 times larger than those given in the graph.

17. J. R. Rex and R. J. Sullivan, J. Phys. Chem., 62, 188 (1958).

18. A. N. Campbell and J.M.T.M. Gieskes, Canad. J. Chem., 43, 1004 (1965).

19. DuPont Co. product information, **Heats of mixing inconsistent with this graph were reported by B. E. Geller, Russian J. Phys. Chem., 35, 542 (1961).

20. J.M.G. Cowie and P. M. Toporowski, Canad. J. Chem., 39, 2240 (1961).

21. Saline Water Conversion Report for 1969-1970, Office of Saline Water, U.S. Department of the Interior, p. 449.

22. Y. Trudelle and L. Nicolas, Desalination, 4, 198 (1968).

INFLUENCE OF PROCESSING VARIABLES ON THE TRANSPORT PROPERTIES OF CELLUOSE ACETATE MEMBRANES. ULTIMATE ELONGATION, BIREFRINGENCE, AND TRANSPORT CHARACTERISTICS

Eberhard A. Meinecke and Devendra V. Mehta*

Institute of Polymer Science
The University of Akron
Akron, Ohio 44304

INTRODUCTION

Two approaches are being taken to improve the desalination characteristics of cellulose acetate (CA) membranes: the first involves the improvement of present membrane formulations[1], and the other, the development of tubular membranes[1,2]. The tubular approach shows a strong potential for commercialization, mainly because the tubular geometry offers large surface area per unit volume. The surface area available for desalination can be further increased by the use of small diameter hollow fibers, since a large number of these can be packed per unit volume.

The two major problems in the tubular approach are: (1) successful extrusion of hollow fibers having the desired modified membrane structure, characteristic of all high flux CA membranes and (2) knowledge of the effect of orientation of the extruded fibers on the desalination characteristics.

EXPERIMENTAL

Fabrication of Cellulose Acetate Membranes (CAM)

Membranes were fabricated from a solution of 25% by weight cellulose acetate (Eastman Chemical Products, Inc. E398-3) in spectro quality acetone. Casting was performed at room temperature in a dust-free laboratory hood on a polished plate-glass

*Present address: Miles Laboratories, Ames Company Division, Elkhart, Ind. 46514

surface with the specific edge of a multiple-clearance "Gardner" applicator. After a one-hour air evaporation period the membrane was immersed in a distilled water bath at room temperature. Membrane thicknesses were measured on a "Federal" thickness gauge. An average value of at least three readings per specimen was used.

Dumbbell shaped specimens having a neck width of 1/2 inch and an overall length of 5 1/2 inches were used for ultimate elongation and birefringence measurements. Specimens were cut from membranes having a casting thickness of 15 mils.

Ultimate Elongation Measurements

A conventional motor-driven stretching unit was employed for stretching the specimens. Stretching was performed in a constant temperature glycerine-water bath (96% by volume glycerine) maintained within ± 0.1°C. The ultimate elongation, λ_B, and the corresponding time-to-break (t_B) values were calculated from constant stretch rate (R_s) data.

The ultimate elongation of CA was determined at stretching rates ranging from 1.13×10^{-3} to 62.9×10^{-3} in./sec. and at stretching temperatures ranging from 30° to 170°C.

Birefringence Measurements

The same stretching apparatus and procedure was used except that the specimen was not stretched to its ultimate elongation but to a specific elongation ratio λ.

The specimen, after being stretched to a specific λ at some rate and temperature, was immediately removed from the glycerine-water bath and then quenched in a room temperature water bath.

The experimental birefringence set-up was arranged according to the method described by Treloar[3]. A commercially available Babinet compensator (Gaertner Scientific Co.) and a polarized beam of white light and one of sodium light were consecutively used.

Birefringence was determined at two elongation ratios, $\lambda = 1.5$ and 2.0. Stretching rates from 0.62×10^{-3} to 63.4×10^{-3} in./sec. and temperatures from 130° to 160°C were employed for each of the two elongation ratios.

Desalination Measurements

The range of membrane casting thickness, t_c, was from 1.5 to 15 mils. The membranes were always stored in distilled water at room temperature. All membranes were checked for pinholes by a copper sulfate-potassium ferrocyanide test[4].

Each test specimen was about 3 inches square and was cut from a stock of known wet thickness (t_w). Specimens of membranes stretched to an elongation ratio λ were obtained from membranes that were stretched in a glycerine-water ($\sim$ 96% by vol. glycerine) or a water bath at a specific stretching temperature, T_s, and rate of stretching, R_s.

Desalination testing was by reverse osmosis. A conventional high pressure reverse osmosis desalination test set-up was used. All tests were performed at 27°C and 102 atm (1500 psig), using a 1% by weight sodium chloride feed solution containing 0.02 gram Biebrich Scarlet dye per liter of feed solution.

Salt contents of feed and product water were conductometrically determined at 27°C, using a Beckman dip cell.

Desalination characteristics of unstretched and stretched low flux cellulose acetate membranes (LFCAM) were studied as a function of the membrane thickness, elongation ratio, the rate of stretching, and the temperature of stretching.

RESULTS AND DISCUSSION

Ultimate Elongation of Cellulose Acetate

Ultimate elongation (λ_B) of low flux CA membranes, obtained over a 56-fold variation in stretching rate, are double-logarithmically plotted in Fig. 1 as a function of time-to-break (t_B) for several values of the stretching temperature, T_s.

A low value of $\lambda_B \approx 1.06$ was obtained for specimens stretched at room temperature. This low value of λ_B corresponds to the high modulus of CA and is in good agreement with the values of λ_B = 1.05 - 1.10 reported in the literature[5]. With the rate of stretching increasing from $R_s = 1.2 \times 10^{-3}$ in./sec. to 58×10^{-3} in./sec., the temperature at which CA exhibited initial elastic behavior showed a corresponding increase from T_s = 90° to 150°C as expected [6].

It was observed that at bath temperatures of $T \geq 170°C$, the glycerine acted as a solvent for CA. It was therefore difficult to obtain stretching data above 160°C. At these temperatures, the specimens were highly swollen. This was followed by an increase in length before stretching was commenced. The turbulence in the stirred bath contributed further to the elongation of the softened specimens. The unstretched specimen length had therefore to be measured after this undesirable increase in length had taken place.

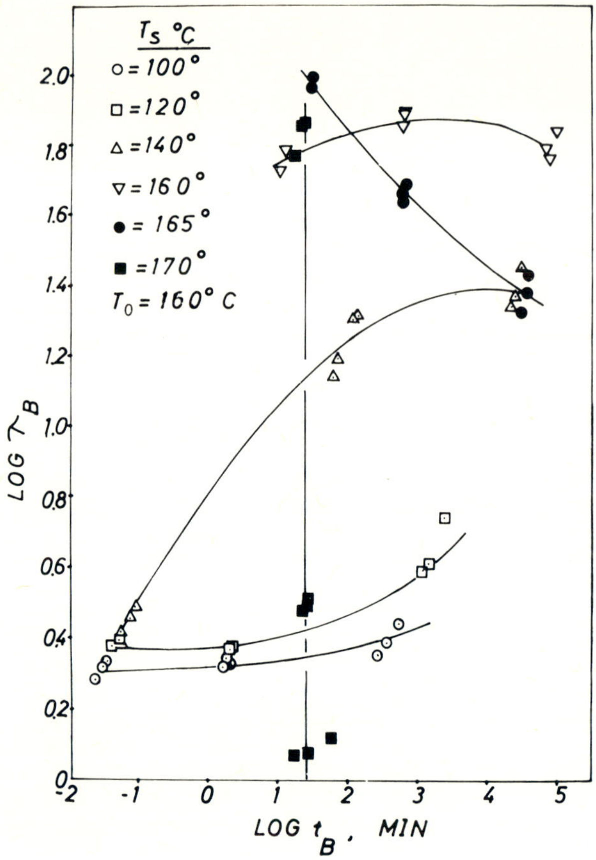

Fig. 1 - Ultimate Elongation of Cellulose Acetate as a Function of Time-to-Break for Several Stretching Temperatures. T_O is the reference temperature for the time-temperature shift.

As no detailed viscoelastic studies of CA have been made to date, preliminary time-temperature shifts were performed according to the method of Williams-Landell-Ferry, WLF[6,7]. A reference temperature T_O = 160°C was used for calculating the shift factor, a_T, values.

The WLF method of reduced variables was applicable only to a narrow temperature range of $140° \leq T \leq 160°C$. This lack of applicability of the WLF method is believed to be partly due to the multiple secondary transitions of CA[5]. In the region where experimental data were obtained, cellulose acetate is, however, in its transition from glass-like to rubbery behavior. The method of reduced variables is generally applicable to data obtained in the transition, plateau, and terminal zones of the time scale only when the same monomeric friction coefficient of polymer segments is involved, i.e., the structure of the material does not change. The multiple transitions in CA, however, indicate structural changes.

In order to successfully apply a time-temperature superposition, each of the isothermal curves in $\ln \lambda_B$ versus $\ln t_B$ (Fig. 1) was horizontally displaced, either to the right ($T_s > 160°C$) or left ($T_s < 160°C$) of the reference curve at $T_o = 160°C$, until a continuous curve was obtained for each pair of adjacent curves. The master curve obtained by such a horizontal displacement method is shown in Fig. 2.

The experimental shift factors, $\ln a_{T_{EXP}}$, are plotted as a function of temperature in Fig. 3. The shift factors as calculated from the WLF equation are also shown in Fig. 3. It is evident that the WLF equation applies to a very narrow temperature range only, as mentioned earlier. The deviations for $T_s >$ 160°C are believed to be due to the onset of dissolution of CA in the glycerine-water mixture, as discussed earlier. In the temperature region below 140°C, $\ln a_{T_{EXP}}$ does not change much with tem-

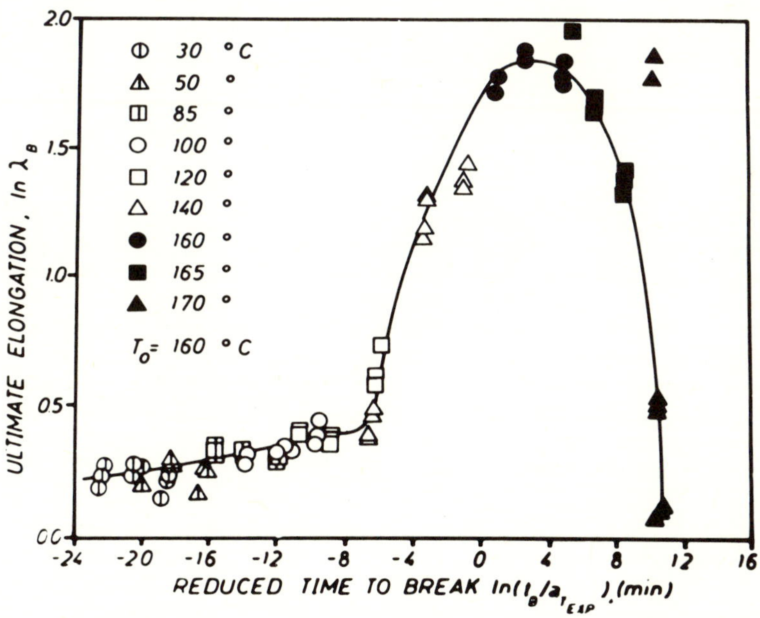

Fig. 2 - Master Curve of Ultimate Elongation of Cellulose Acetate Using Experimental Shift Factors, $a_{T_{EXP}}$

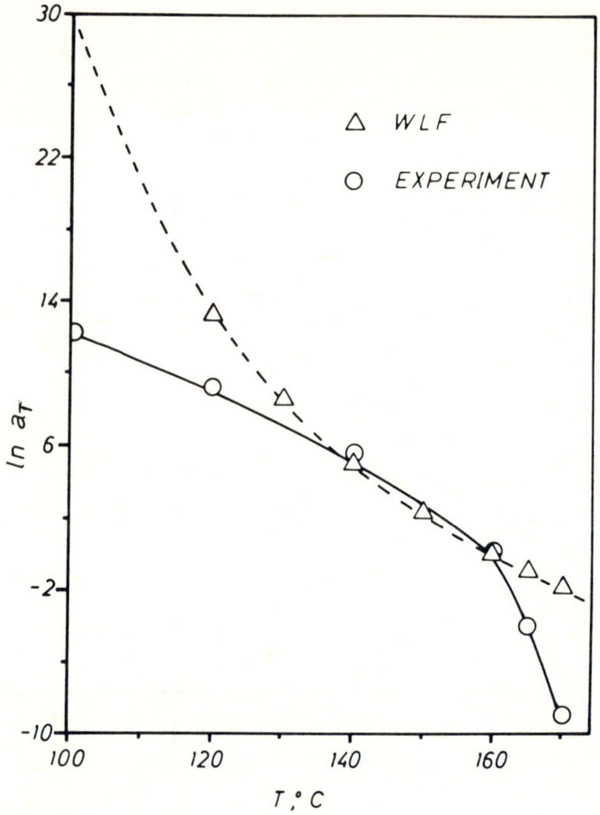

Fig. 3 - Temperature Dependence of Shift Factors of Cellulose Acetate

perature. Similar results have been reported for other glassy polymers[8,9]. Rusch and Beck, for example, reported data on poly(methyl methacrylate) below its glass transition temperature. This material is known to exhibit other minor transitions[3] just as cellulose acetate does. With the help of their data they could show that the WLF equation can be extended to a range below the glass transition temperature if the temperature T is replaced by a parameter related to the frozen free volume in the glass.

Similar agreement could be obtained for cellulose acetate. The flagged data in Fig. 2 show that the time-temperature equivalence can be extended to the glassy zone. The values of ln a_T used for evaluating ln (t_B/a_T) were obtained by extrapolation of the shift factor vs. temperature curve (Fig. 3). These shift factors yield good superposition in the region below T_g.

Birefringence of Cellulose Acetate

Birefringence measurements were employed in order to determine the effect that time-temperature dependent stretching had on the relative degree of molecular orientation in CAM. Based on the method of reduced variables, applicable to most time- and temperature-dependent viscoelastic properties of polymers it is conceivable that isothermal birefringence data obtained as a function of λ and R_s can also be transformed into a master curve independent of time and temperature. One of the requirements for the applicability of the time-temperature superposition[6] to birefringence data is that the shift factors be the same as those used for the superposition of ultimate properties of CA (Fig. 3).

In order to construct master curves for the birefringence of CA, the data were obtained at two elongation ratios, λ = 1.5 and 2.0, as a function of temperature and rate of stretching. The birefringence data at λ = 2.0 are plotted logarithmically as a function of R_s for several values of T_s in Fig. 4. The scatter in the birefringence data is mainly attributed to membrane thickness variations (about ± 0.1 mils). Most birefringence measurements are performed on specimens of about 125 mils (∿ 0.5 mm) thickness, while the specimens used in these studies were about 1.5 mils. Since birefringence is inversely proportional to the thickness[3], slight variations in thickness will greatly affect the results.

The master curve of birefringence of CA using the experimental shift factors (Fig. 3) is shown in Fig. 5. The use of experimental shift factors shows an improvement in fitting the experimental

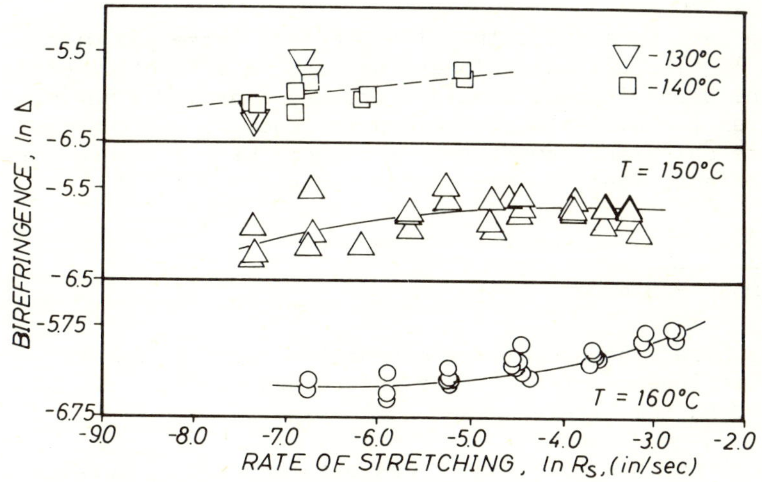

Fig. 4 - Effect of Rate of Stretching on the Birefringence of Cellulose Acetate (λ = 2.0)

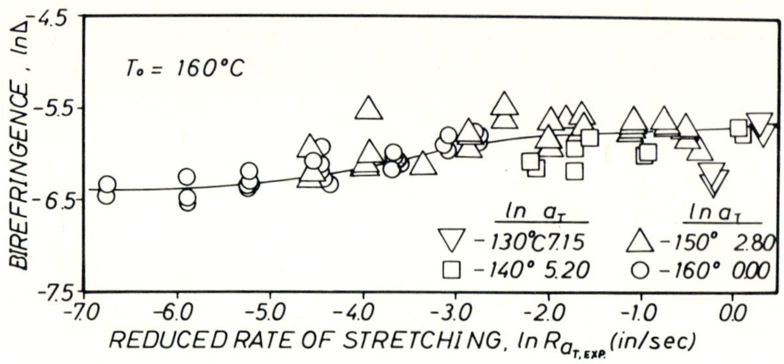

Fig. 5 - Master Curve of Birefringence of Cellulose Acetate (λ = 2.0), Using Experimental Shift Factors

results as compared to those when WLF shift factors are used. Examination of Fig. 5 indicates that for an approximately 1160-fold change in R_s, the corresponding increase in birefringence is only of the order of 144 percent, thus reinforcing the fact that the stiff CA molecules[5] cannot undergo substantial anisotropic deformations even at temperatures above the glass transition temperature [reference temperature T_o (= 160°C) > T_g].

In order to generalize the relative changes in birefringence occurring at various λ, R_s, and T_s, an attempt was made to develop an empirical method by which such changes could be expressed as a function of a reduced degree of elongation, independent of R_s and T_s.

At constant R_s and T_s, the birefringence increased with increasing λ. The ultimate elongation is also a function of R_s and T_s. The birefringence should therefore approach a maximum value, Δ_m, as the ultimate elongation is reached. At constant rate of stretching and temperature a reduced birefringence (or degree of orientation), ϕ, may be defined as:

$$\phi = \left[\frac{\Delta_\lambda - \Delta_o}{\Delta_m - \Delta_o}\right]_{R_s,\,T_s}, \qquad (1)$$

with Δ_λ the birefringence at any elongation ratio λ, Δ_o the initial birefringence of CA in the unstrained state, i.e., at λ = 1.0, and Δ_m the birefringence at $\lambda = \lambda_B$.

In deriving this equation it was assumed that Δ_m obtained just before failure was not affected by the onset of crystalline order occurring with elongation ratios, i.e., the value of Δ_m can be obtained by extrapolating a Δ vs. λ curve to λ_B.

At this point, it is reasonable to define a rate-independent reduced elongation, ε_r:

$$\varepsilon_r = \left(\frac{\varepsilon}{\varepsilon_B}\right)_{T_s} = \left(\frac{\lambda - 1}{\lambda_B - 1}\right)_{T_s} . \qquad (2)$$

The elongation, $\varepsilon(= \Delta L/L_o)$, is employed instead of λ because ϕ must be zero for $\varepsilon \to 0$, and if $\lambda_r(= \lambda/\lambda_B)$ is used, then $\phi \to 0$ when $\lambda_r = 1/\lambda_B$. Furthermore, as λ_B is a function of R_s and T, the value of $1/\lambda_B$, representing the unstrained state, will not have a unique value. On the other hand, ε_r will always be zero in the unstrained state.

The results of ϕ as a function of ε_r are plotted in Fig. 6. Results of birefringence at T = 130°, 140°, and 150°C at varying rates of stretching, and λ = 1.5 and 2.0 were obtained from the curves of ln Δ vs. ln R_s (Fig. 4). The corresponding ϕ values were calculated from equation (3), below. The values of ε_B at each R_s and the three temperatures were obtained from the curves of ln λ_B vs. ln t_B (Fig. 1).

A slightly better superposition was achieved when the vertical shift T/T_o (T_o = 160°C, reference temperature) was used for the reduced birefringence at temperature T:

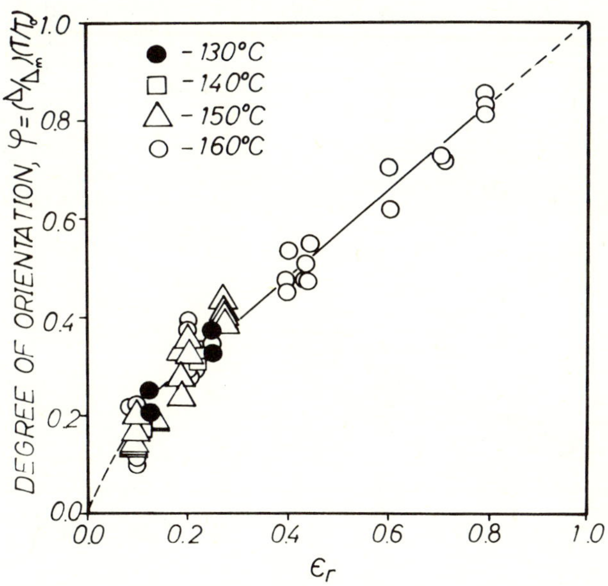

Fig. 6 - Master Curve of Reduced Birefringence of Cellulose Acetate

$$\phi = \frac{\Delta_\lambda}{\Delta_m} \frac{T}{T_o} . \tag{3}$$

A plot of ϕ as a function of ε_r, therefore, yields a master curve. For any set of λ, R_s, and T_s values, the birefringence Δ can be determined from the master curve of ϕ.

Desalination Characteristics of LFCAM

Transport Characteristics of Unstretched LFCAM. The flux of desalinized water, Q, in gal/ft^2-day (gf^{2}d) is graphically presented as a function of reciprocal membrane thickness in Fig. 7. The observed inverse relationship between Q and t_w at constant conditions of test temperature, applied pressure, and feed concentration confirms the applicability of Fick's law to the transport characteristics of unstretched LFCAM[1]. The scatter in the Q values for thinner membranes at $t_w \leq 0.33$ mils, is attributed to variations in membrane thickness ($\pm$ 0.05 mils). Variations in thickness would result in the formation of regions of non-uniformities which could fail under the high test pressure (1500 psi) and set up salt concentration polarization on the high pressure side of the membrane. The corresponding results of percent salt rejection (%SR) were observed to be independent of the membrane thickness at $t_w > 0.33$ mils.

The symmetric nature of the LFCAM[1] was confirmed by obtaining desalination data using the "bad-side" of LFCAM. The "bad side" is the membrane surface which is in contact with the casting surface.

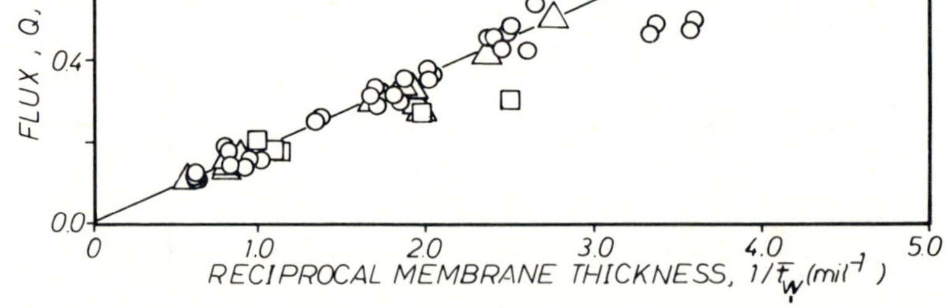

Fig. 7 - Desalinized Flux of Unstretched Low Flux Cellulose Acetate Membranes vs. The Average Reciprocal Thickness. Treatment temperature (T_t) was 23°C.

These results are included in Fig. 7. The Q and %SR values are identical to those obtained for the corresponding "good-side" membranes, thereby indicating symmetry of the membrane structure.

As stretched specimens were obtained by stretching in the presence of a plasticizing medium (glycerine-water bath), the effect of such a plasticizing treatment on the membrane characteristics of room-temperature plasticized, unstretched LFCAM were also studied. The results of this series are included in Fig. 7. The Q and corresponding %SR characteristics are identical to those obtained for the unplasticized membranes, indicating that room temperature plasticization of LFCAM does not have any measurable effect on the membrane transport characteristics.

Transport Characteristics of Stretched LFCAM. A series of desalination tests was performed using LFCAM stretched to an elongation ratio of $\lambda \approx 1.4$ at a stretching rate of 0.015 in./sec. Stretching was performed in a water bath at 100°C. The flux as a function of reciprocal stretched membrane thickness ($1/t_s$) is shown in Fig. 8. Also included are the results obtained on unstretched LFCAM heat treated in a water bath at 100°C for either 15 or 30 minutes. An inverse relationship between Q and t can be observed. This indicates that under conditions of constant pressure, temperature, and feed concentration Fick's law is applicable not only to the transport characteristics of unstretched LFCAM but also to those of stretched and heat treated membranes as well. The scatter at t_s below about 0.25 mils, corresponding to an unstretched thickness, t_w, of about 0.3 mils, is believed to be due to thickness variations, as discussed earlier.

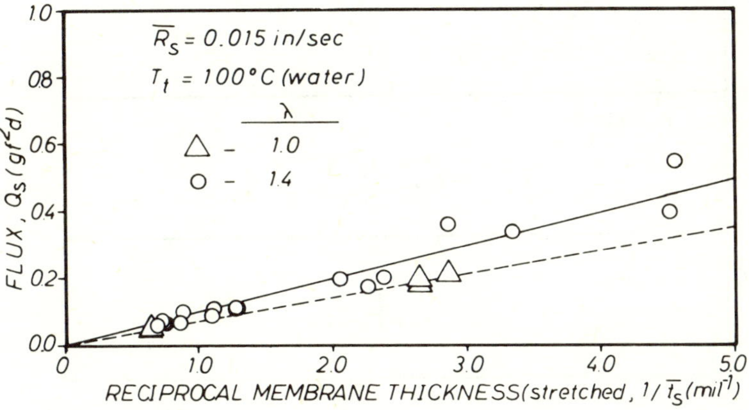

Fig. 8 - Desalinized Flux of Unstretched (λ = 1.0) and Stretched (λ = 1.4) Low Flux Cellulose Acetate Membranes at Treatment (T_t) or Stretching (T_s) Temperature of 100°C.

Comparison of desalination performance of unstretched and stretched membranes indicates that lower degrees of stretching ($\lambda \simeq 1.4$) improve the flux characteristics with essentially no detrimental effect on salt rejection. The salt rejection characteristics are essentially unaffected by the stretching and heating process.

In order to apply the time-temperature superposition using the experimental shift factors[10,11] to the transport properties of stretched LFCAM, a series of tests were performed on membranes of constant casting thickness, t_c = 15 mils, stretched to an elongation ratio of 2.0. Stretching was performed in a glycerine-water bath at varying rates and temperatures of stretching. The flux Q is plotted logarithmically as a function of stretch rate, R_s, in Fig. 9. The results were normalized to a constant thickness of t_s = 1.5 mils.

Each of the isotherms of lnQ - lnR_s in Fig. 9 was shifted to a reference temperature curve at T_o = 160°C. The shift factors used were the experimental shift factors, ln $a_{T_{EXP}}$, determined from the ultimate elongation results[10,11]. The master curve is presented in Fig. 10. Examination of this master curve indicates that the principle of time-temperature superposition is also applicable to the transport characteristics (desalinized flux) of LFCAM. For an almost 84-fold change in R_s the flux decreased by

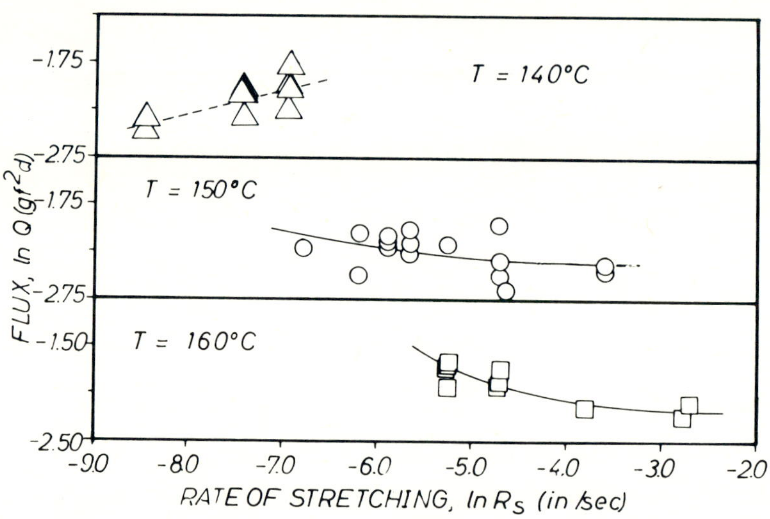

Fig. 9 - Desalinized Flux of Stretched (λ = 2.0) Low Flux Cellulose Acetate Membranes as a Function of Rate of Stretching.

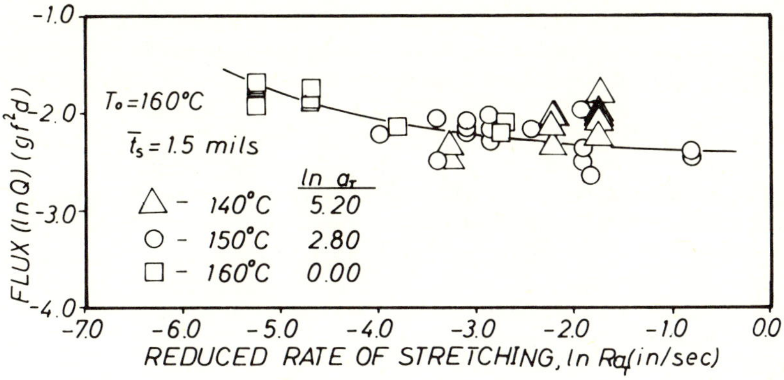

Fig. 10 - Master Curve of Desalinized Flux of Low Flux Cellulose Acetate Membranes as a Function of Rate of Stretching

about 49%. At constant λ this decrease in flux with increasing R_s is attributed to the resultant increase in the degree of molecular orientation occurring at the higher test rates. This increasing molecular alignment of CA molecules or segments thereof acts as a barrier to transport, decreasing the water permeability of the membrane.

The corresponding results of %SR as a function of R_s are plotted in Fig. 11. The reduction in salt rejection at low rates, i.e., R_s less than about 8×10^{-3} in./sec., is believed to be reflecting the possible membrane changes occurring at the high test temperature[11]. From a viscoelasticity viewpoint, low rates of

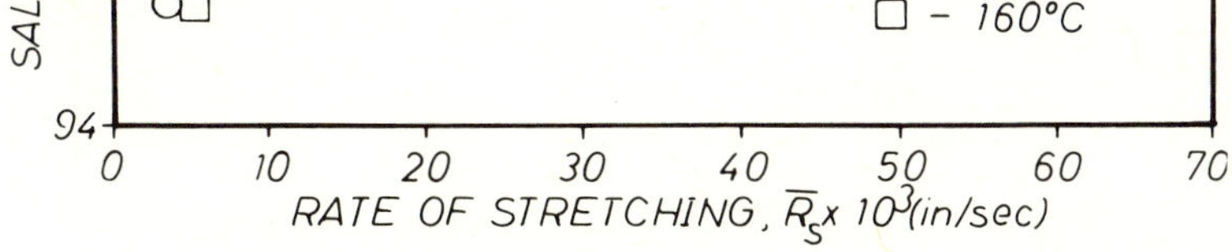

Fig. 11 - Salt Rejection of Stretched (λ = 2.0) Low Flux Cellulose Acetate Membranes as a Function of Rate of Stretching

stretching are equivalent to high temperature operation. The change in salt rejection at low R_s can also be due to a membrane thinning effect occurring at the thinner regions existing in the membrane surface due to thickness variations.

The corresponding relative degree of orientation expressed in terms of reduced birefringence[10,11], ϕ, were calculated as follows: (i) the ultimate elongation data at specific R_s and T_s values were obtained from the master curve (Fig. 2), (ii) values of the ratio of elongation [$\varepsilon(=\Delta L/L_o)$] to elongation at break (ε_B), i.e., reduced elongation, ε_r, were calculated, and (iii) the corresponding reduced birefringence, ϕ, values were then obtained from the master curve of ϕ vs. ε_r (Fig. 6).

These results of ϕ as a function of reduced rate of stretching, ln Ra_T, are logarithmically plotted in Fig. 12. A master curve is obtained, as expected. The reduced rates of stretching correspond to those utilized in the study of the transport properties (Figs. 10 and 11). These results show that ϕ decreases with decreasing rates of stretching. The decrease in birefringence or molecular orientation (ϕ) shows also a corresponding increase in the flux Q, indicating that transport properties are dependent on the molecular structure of the polymer membrane.

Only a limited amount of experimentation was possible at higher values of λ because, even at medium rates ($R_s \simeq 13.2 \times 10^{-3}$ in./sec.) and high temperature (T_s = 160°C), membranes which were stretched to λ greater than about 2.7 exhibited a considerable loss of lateral strength.

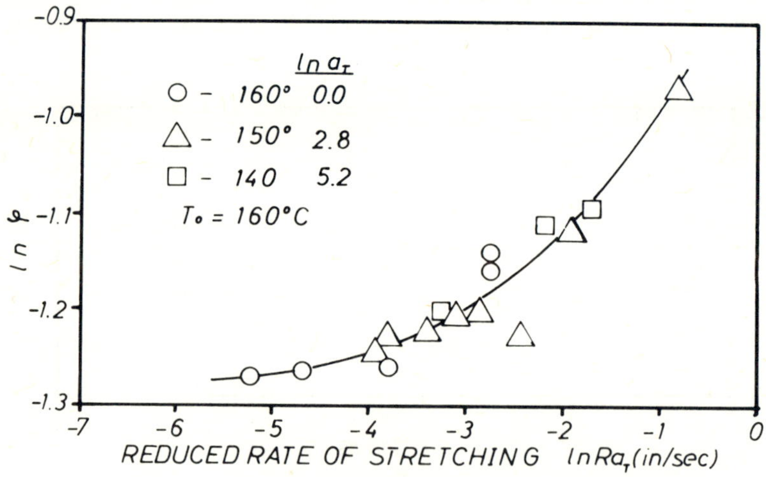

Fig. 12 - Master Curve of Reduced Birefringence of Low Flux Cellulose Acetate Membranes Stretched to λ = 2.0

The results of water flux and salt rejection as functions of the degree of stretching (or elongation ratio) are shown in Figs. 13 and 14, respectively. Stretching was performed in a water bath at 100°C or in a glycerine-water bath at 160°C.

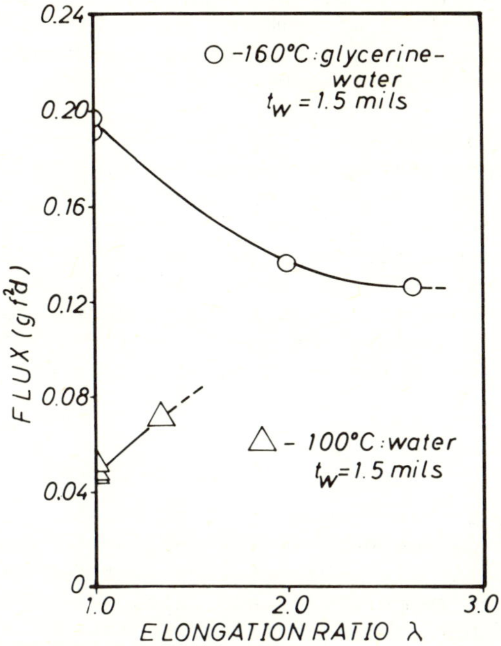

Fig. 13 - Effect of Elongation Ratio on Desalinized Flux of Low Flux Cellulose Acetate Membranes

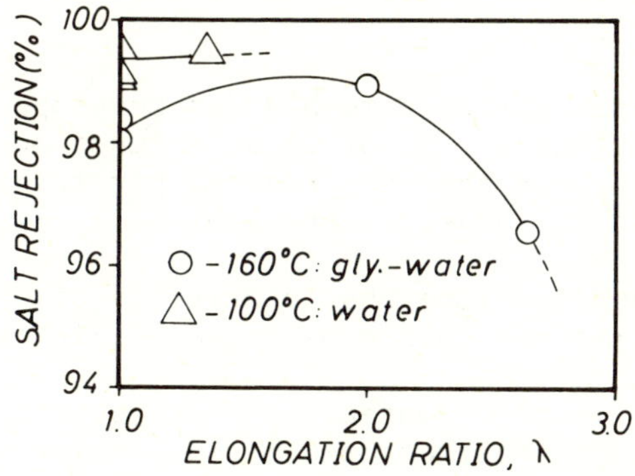

Fig. 14 - Effect of Elongation Ratio on Salt Rejection of Low Flux Cellulose Acetate Membranes

It is of interest to note that the water selectivity characteristics of LFCAM are a strong function of the chemical nature of the treatment medium. Stretching of LFCAM in water improves the transport characteristics as indicated by the increase in Q with λ. The salt rejection remained essentially constant at about 99.3%. This improvement in the performance with λ is believed to be partly due to the shrinkage of membrane pores at the high treatment temperature (100°C)[10,11], and also due to favorable molecular alignment resulting from stretching.

In the case of membranes stretched in glycerine-water at T_s = 160°C, the flux decreases with increasing λ. The %SR increases with λ up to about λ = 1.75, followed by a rapid decrease. This behavior is attributed to a number of factors, such as increased CA-plasticizer interactions occurring at the high test temperatures (160°C); excessive molecular motion of CA, as the polymeric membrane is above its glass transition temperature; and the approach of membrane dissolution condition ($T \rightarrow 170°C$). The effect of stretching adds to the existing complexities.

CONCLUSIONS

The applicability of time-temperature superposition to the ultimate elongation of CA has been established. The appropriate shift factors required for the superposition of these results over the entire range of temperatures and rates were obtained by appropriate amounts of horizontal shifting, as the WLF equation applies to a very narrow temperature range only.

The principle of time-temperature superposition is also shown to be applicable to birefringence of 2.5-acetyl cellulose acetate, oriented at varying conditions of temperature and stretching rate. A master curve is obtainable for each specific extent of elongation, i.e., for each λ value.

An empirical reduced variable method was developed for expressing reduced birefringence, ϕ, of CA as a function of reduced elongation, ε_r. It is shown that the reduced birefringence can be represented by a master curve, independent of λ, R_s, and T_s.

The transport (desalination) characteristics of stretched LFCAM, like their unstretched counterparts, obey Fick's law.

Transport properties of oriented LFCAM are strongly dependent on the chemical nature of the treatment media. Membranes stretched in water at elevated temperature showed an improvement in both the flux and salt rejection over that observed for heat treated, unstretched membranes. Membranes stretched in glycerine-water showed a deterioration in the flux characteristics with both increasing rates of stretching and elongation. A corresponding increase in the relative degree of orientation (birefringence) was observed.

The principle of time-temperature superposition is shown to be applicable to the transport properties of 2.5-acetyl cellulose acetate membranes.

ACKNOWLEDGMENT

This research was supported by the U. S. Department of the Interior, Office of Saline Water, under Grant No. 14-01-001-1456.

LIST OF SYMBOLS

a_T shift factor

$a_{T_{EXP}}$ experimental shift factor

L specimen length, in.

L_f final (stretched) length of specimen, in.

L_o original (unstretched) length of specimen, in.

Q flux of desalinized water, gal/ft^2-day

R_a reduced rate of stretching, in./sec.

R_s rate of stretching, in./sec.

%SR percent salt rejection

T_g glass transition temperature, °C

T_o reference temperature for time-temperature shifts, °C

T_s temperature of stretching, °C

T_t membrane annealing temperature, °C

t_B time-to-break, min.

t_c membrane casting thickness, mils

t_s stretched membrane thickness, mils

t_w wet membrane thickness, mils

λ elongation ratio, L_f/L_o, in./in.

λ_B elongation ratio-at-break, in./in.

λ_r reduced elongation ratio, $= \lambda/\lambda_B$

ϕ reduced birefringence (degree of orientation)

Δ_m birefringence at λ_B

Δ_o birefringence at $\lambda = 1.0$ (unstretched state)

Δ_λ birefringence at λ

ε elongation, $(L_f-L_o)/L_o = \Delta L/L_o = \lambda - 1$, in./in.

ε_B elongation-at-break, in./in.

ε_r reduced elongation, $= \varepsilon/\varepsilon_B$

REFERENCES

1. U. Merten (Ed.), Desalination by Reverse Osmosis, Chaps. 3 and 4, The M.I.T. Press, Cambridge, Mass. (1966).

2. S. Loeb, Desalination, 1, 1 (1966).

3. L. R. G. Treloar, The Physics of Rubber Elasticity, The Clarendon Press, Oxford, England (1958), Chap. 10.

4. Office of Saline Water Research and Development Progress Report No. 434 (1968).

5. U. Merten (Ed.), Desalination by Reverse Osmosis, The M.I.T. Press, Cambridge, Mass. (1966), Chap. 4.

6. J. D. Ferry, Viscoelastic Properties of Polymers, Chaps. 11 and 12, John Wiley & Sons, Inc., New York, N. Y. (1970).

7. M. L. Miller, The Structure of Polymers, Reinhold Publishing Corp., New York, N. Y. (1966), Chap. 6.

8. N. G. McCrum and G. A. Pogany, J. Macromol. Sci.-Phys., B4(1), 109 (1970).

9. K. C. Rusch and R. H. Beck, Jr., J. Macromol. Sci.-Phys., B3(3), 365 (1969).

10. E. A. Meinecke and D. V. Mehta, Preprints for 162nd National Meeting of the American Chemical Society, Washington, D. C., Sept. 1971.

11. D. V. Mehta, Ph.D. Thesis, The University of Akron (1971).

CELLULOSE ACETATE BLEND MEMBRANES

W. M. King, D. L. Hoernschemeyer, and C. W. Saltonstall, Jr.

Envirogenics Company
El Monte, California 91734

INTRODUCTION

Dense films of cellulose diacetate have been shown to be intrinsically capable of very high salt rejection in reverse osmosis, albeit at uneconomical water fluxes[1,2,3]. The water and salt permeabilities obtained in these measurements indicate that membranes of cellulose diacetate should reject at least 99.8% of the salt in a 3.5% sodium chloride solution at 102 atm (1500 psi)[2]. Both Reid and Breton[1] and Lonsdale et al.[2] also showed that the intrinsic permeabilities to both salt and water decrease as the acetyl substitution of the cellulose diacetate is increased. Since the salt permeability decreases much more rapidly with increasing substitution than the water permeability, cellulose triacetate should provide a higher salt rejection than cellulose diacetate.

Although they provide economically attractive fluxes, neither ultrathin films nor asymmetric membranes prepared from cellulose diacetate exhibit salt rejections as large as expected from the intrinsic permeability of relatively thick dense films. For example, asymmetric membranes of cellulose diacetate generally reach only 99% salt rejection, instead of 99.8%, in operation with 3.5% sodium chloride at 1500 psi. This rejection falls short of the 99.5% required for a 30 to 50% recovery of potable water from seawater in a single pass. While cellulose triacetate would be expected to provide higher salt rejection, the fabrication of asymmetric membranes from this ester has not produced economically attractive fluxes together with the necessary salt rejection.

High inherent viscosity and limited solubility in acceptable solvents appear to be the major impediments in the fabrication of asymmetric membranes from cellulose triacetate. The relatively wide solubility range of cellulose diacetate leads to the hypothesis that esters having degrees of substitution intermediate between cellulose diacetate and cellulose triacetate might provide favorable processing characteristics together with the salt rejection required for the single-pass desalination of seawater. Unfortunately, cellulose acetates of intermediate degrees of substitution are not commercially available.

In an effort to answer the needs of seawater desalination, an exploratory program was undertaken in these laboratories in which asymmetric membranes were fabricated from blends of cellulose diacetate and cellulose triacetate to circumvent the lack of commercial esters in the intermediate substitution range. Membranes made from blends with average substitutions between about 41.5 and 42.5% acetyl provided salt rejections between 99.0% and 99.6% when tested with 3.5% sodium chloride at 1500 psi. Surprisingly, their fluxes were all about the same as those obtainable with similar cellulose diacetate membranes.

It is particularly interesting that the flux of the "blend" membranes did not decrease with increasing acetyl content over the 41.5 to 42.5% acetyl range as predicted by the measurements on dense films. Indeed, later work showed that blend membranes of very high flux can be fabricated for use in the desalination of brackish waters. Although not subject to precise measurement, it appears that the active layer thickness of the blend membranes was considerably less than that of the concurrently developed cellulose diacetate membranes.

A substantial effort was required to develop the processing parameters for the reproducible fabrication of blend membranes in large quantities. The difficulties overcome in this effort and the properties of the resultant membranes are described in more detail below.

MEMBRANE FABRICATION

Casting Solution Formulation

Asymmetric membranes of cellulose diacetate have usually been made from solutions of the polymer in a mixture of solvent, nonsolvent, and "swelling agent." The solvent has usually been acetone, although such materials as acetic acid, dioxane, and methyl ethyl ketone have been used with less success. Swelling agents are materials which have the effect of producing a microgel structure in the membrane, increasing its water content, and greatly increasing the membrane's water permeability. That is, when added to the casting solution, swelling agents are responsible for transforming the structure of the membrane from the dense form with a very low permeability to a quite permeable microcellular form. A great variety of compounds, such as inorganic electrolytes, inorganic and organic acids and amides have been used successfully as swelling agents[4]. The role of the nonsolvent is usually to maintain a high salt rejection, while permitting a large degree of membrane swelling and a high water flux[5]. Water or an alcohol is commonly used, with the choice partially determined by the nature of the swelling agent, e.g., to enhance its solubility.

The same considerations and techniques have been used to fabricate membranes from blends of cellulose diacetate and triacetate.* Because of the limited solubility of cellulose triacetate, it is necessary to include solvents other than acetone. Water-miscible solvents for cellulose triacetate (actual degree of substitution 2.83) include dimethyl sulfoxide, acetic acid, meta-cresol, N-methyl-2-pyrrolidone, 1,4-dioxane, and sulfolane (tetrahydrothiophene-1,1-dioxide); and possible cosolvents (materials which swell the polymer) include acetone, dimethyl formamide, pyridine, benzaldehyde, acetic anhydride, and cyclohexanone. However, most single or binary mixtures selected from this list produce very poor quality membranes. In fact, the only combination which has yielded high quality blend membranes is dioxane and acetone. A more detailed description of the influence of casting solution composition is given in a later section.

Casting Conditions

An evaporation period of 2 to 4 min was used in the fabrication of the first asymmetric cellulose acetate membranes, which were usually cast onto glass plates at 0° to -10°C (Ref. 6). More recent work has shown that the lengthy evaporation period is not necessary to the formation of an asymmetric membrane, provided that the casting solution is formulated to compensate for the solvent formerly lost during the long evaporation period. Similarly, the requirement for low temperature casting was found to be associated with the heat capacity of the glass plate which interfered with gelation of the membrane if it was not already cold prior to immersion in ice water. Consequently, the drying time can be reduced to a few seconds or less at room temperature

*Unless noted otherwise, cellulose diacetate refers to Eastman Chemical Co. Type E-398-3 and cellulose triacetate refers to Eastman Type A-432-130B.

when the membranes are cast from a properly formulated solution onto a lightweight low heat capacity support such as Mylar.

Most of the sheet-form membranes used in the study of blend membranes were cast at 10 mils wet thickness onto an aluminized Mylar belt moving at 244 cm/min into a 1°C gelation bath located 20 cm from the casting knife. After they had floated free of the Mylar support, they were heat treated by immersion in hot water.

Experience has shown that there is little or no change in the osmotic properties of membranes prepared with drying times of 1 to 10 sec. Thus, the 5 sec drying time used in sheet-form casting is an adequate precursor to the conditions encountered in tubular membrane extrusion where very little evaporation can take place within the enclosed air space formed by the casting solution between the extrusion nozzle and the gelation bath. Tubular blend membranes were extruded into Dacron sleeves at 366 cm/min from an extrusion nozzle located immediately above the gelation bath. They were heat treated by immersion in hot water before insertion into the perforated stainless steel tubes used to support the membrane during reverse osmosis testing.

Defect Formation and Prevention

Any imperfection which disrupts the integrity of the active layer is detrimental to the performance of a reverse osmosis membrane because it allows the feed solution to pass unchanged into the product. Defects may allow the salt rejection to decrease drastically before any change is noted in the water flux. A 1% leak will, for example, double the apparent salt permeability of a high retention (99%) membrane. Consequently, it was necessary to insure the absence of certain defects which occurred in blend membrane in order to reliably realize their intrinsic capabilities, particularly for use in the single-pass desalination of seawater.

Wavemarks. Wavemarks are defects in the active layer region of a membrane resulting from irregular wetting of the still liquid membrane as it enters the gelation bath. Normally, the water in the gelation bath advances over the membrane in a continuous manner[7]. If, however, it momentarily pauses in its advance, a wavemark is formed along the line where the water failed to wet the film in a continuous manner. When viewed microscopically, wavemarks appear as lines across the surface of the membrane. Under the scanning electron microscope a wavemark appears as a slight prominence running across the membrane surface as shown in the elevated cross-sectional view in Figure 1. At the peak of the prominence the active layer bends quite sharply, giving rise to the possibility of the active layer cracking under pressure. Whether the lack of selectivity of wavemarks is due to bending and

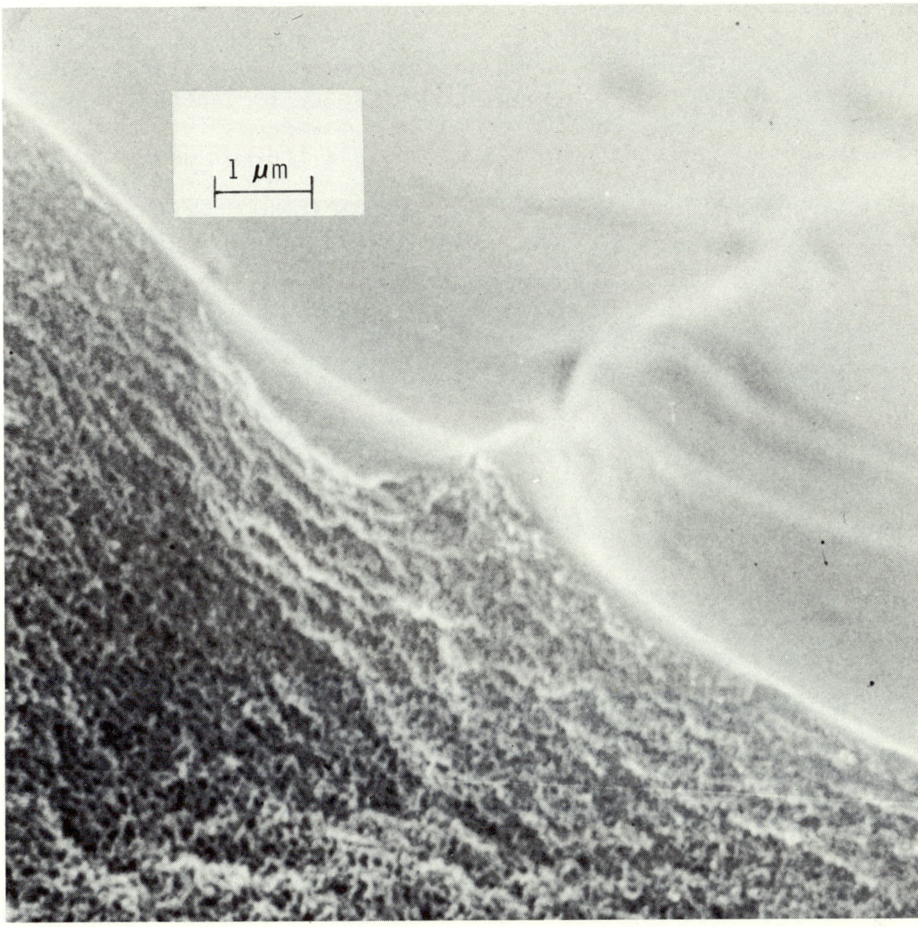

Fig. 1 - Elevated Cross-sectional View of a Wavemark by Scanning Electron Microscope.

cracking during reverse osmosis or to tearing of the nascent active layer as the wavemark is formed, it appears certain that some type of active layer discontinuity is responsible since even large dye molecules (e.g., Niagara Sky Blue 6B) penetrate wavemarks at pressures as low as 100 psi.

Higher casting speeds aid in preventing wavemarks since the advancing wave front of water tends to stabilize as the rate at which the membrane enters the gelation bath increases. Similarly, membranes prepared from highly viscous casting solutions, which are less susceptible to deformation, are less prone to wavemark formation. Tubular membranes are more subject to wavemark formation than sheet-form membranes because of the tendency of the gelation-bath meniscus to fluctuate from positive to negative in the narrow confines of the tube. Apparently, a wavemark is formed each time the meniscus shifts from one mode to the other. This is illustrated in Figure 2 by dye penetration in lines transverse to the casting direction in a tubular membrane. In general, membranes prepared at more than 60 cm/min from casting solutions having a viscosity greater than 300 poises are free of wavemarks.

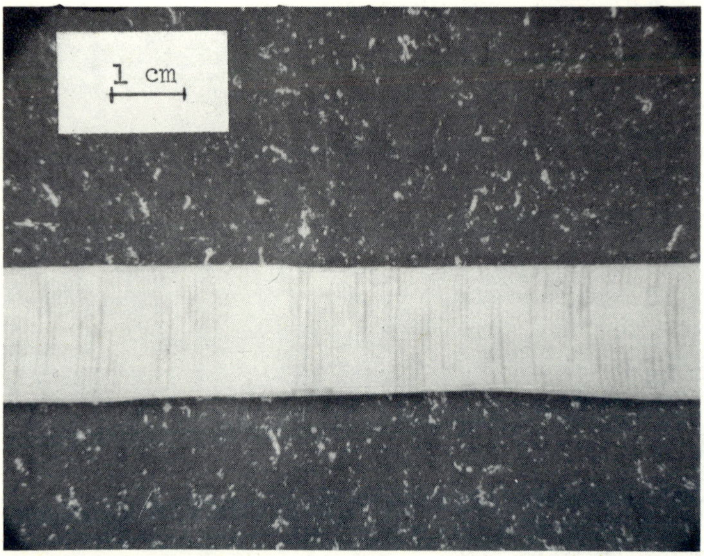

Fig. 2 - Wavemarked Section of Membrane After Exposure to Pressurized Niagara Sky Blue 6B Dye.

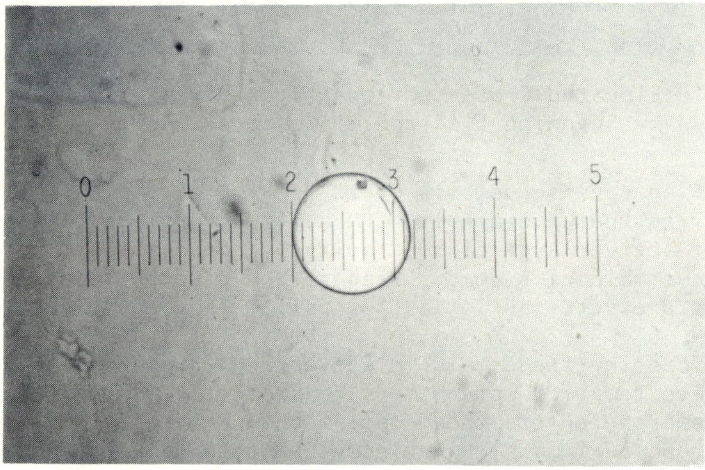

Fig. 3 - Optical Photomicrograph of a Water-Filled, Substratum-Related Void in a Cellulose Acetate Membrane. Each div = 10 μm

Voids. Voids, which are defined as bubbles or other areas in the membrane where there is no polymer, may also result in salt leaks and their formation is often more difficult to prevent than wavemarks. Two types of voids are commonly found in blend membranes[7]. One type is related to the casting surface (substratum) while the other type is related to the casting solution composition.

Only occasional substratum-related voids are found in blend membranes cast on smooth surfaces such as glass or aluminized Mylar. They are found in large numbers, however, in membranes cast on nylon parchment or on certain papers. These voids range from 20 to over 150 micrometers (μm) in diameter (Figure 3) and are located at the membrane-substratum interface (Figure 4). Substratum-related voids appear to be due primarily to incomplete wetting of the substrate by the casting solution. Irregular wetting can come about as a result of surface irregularities (e.g., depressions in a paper surface or holes in a cloth surface) in

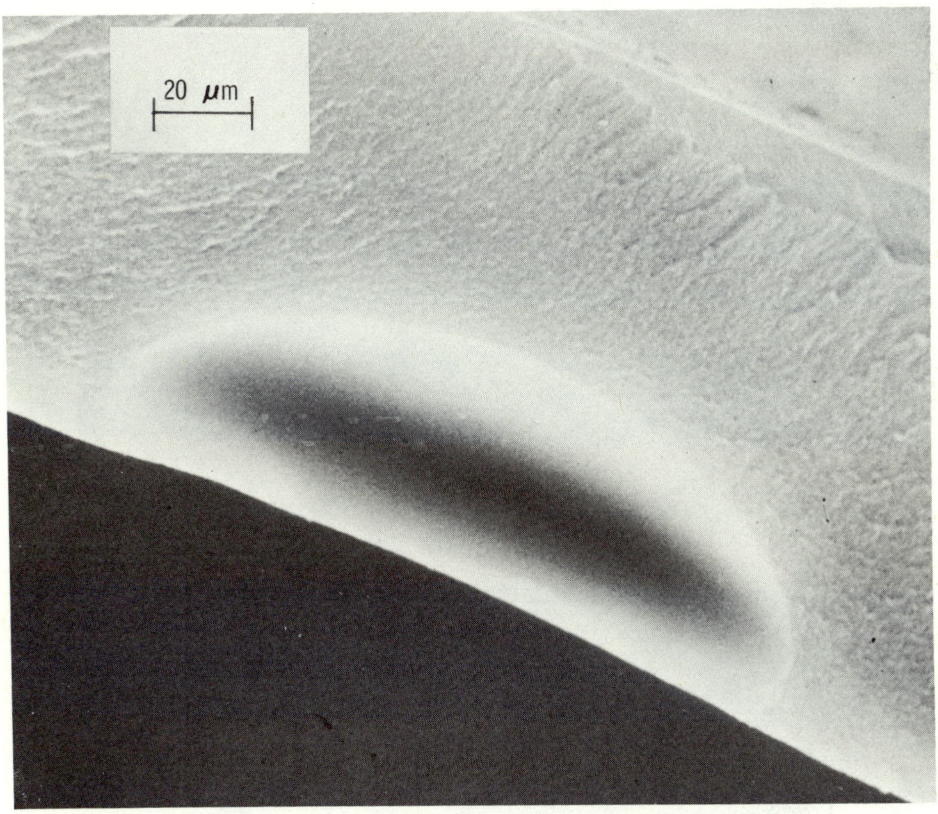

Fig. 4 - Scanning Electron Photomicrograph of Cross-Sectioned Substratum-Related Void.

the support material or by poor wetting of the support material by the casting solution. In either case, the result is the entrapment of small air pockets between the casting solution and the support. Although they are numerous, especially in membranes cast on porous materials, they rarely rupture since they are located so deep in the membrane.

Voids related to the casting solution are generally present in much smaller quantities than substrate-related voids, but they are much more harmful because they are nearer the active layer surface. These voids are conical, and if the apex lies near the active layer they often rupture upon pressurization as shown by dye tests. The maximum void diameter, which varies between 20 and 80 μm, is dependent upon the point of origin since the void generally grows until its base is near the bottom of the membrane (Figure 5). An optical photomicrograph in top view with the focus

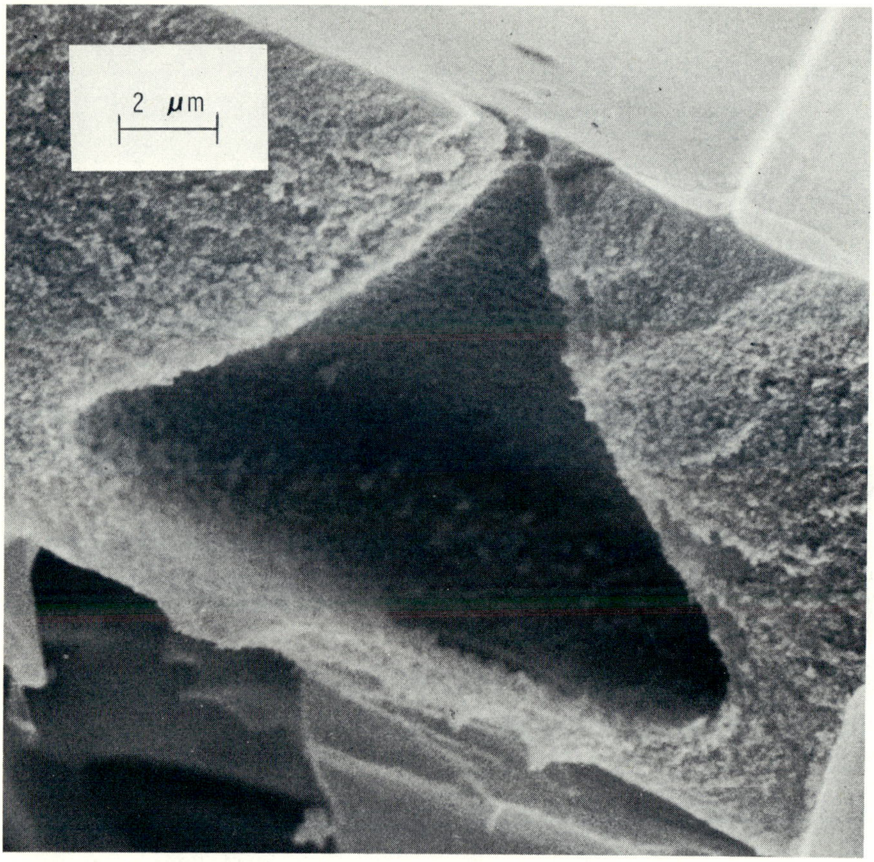

Fig. 5 - Scanning Electron Photomicrograph of a Cross-Sectioned Conical Void in a Tubular Blend Membrane.

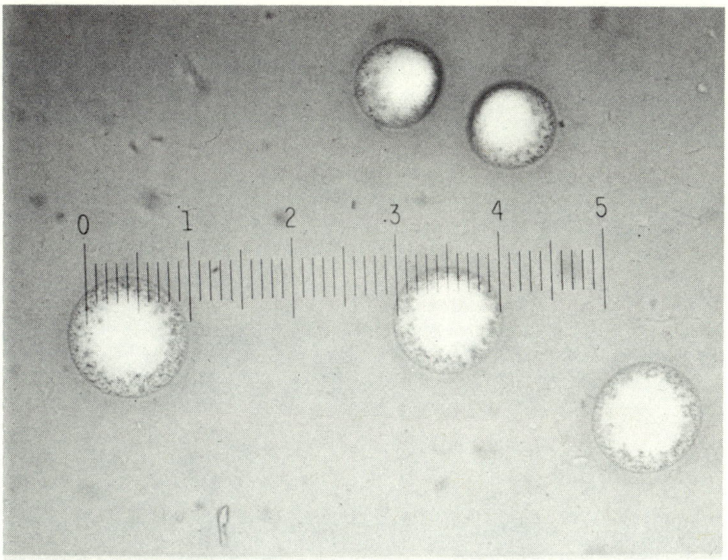

Fig. 6 - Optical Photomicrograph of a Conical Void.
Each div = 10 μm

on the maximum diameter of the void is shown in Figure 6. The apparent structure is probably due to the rough interior texture of the void since scanning electron photomicrographs made at a higher power show the void walls to be extremely porous (Figure 7). It is quite possible that the "intrusion cells" observed by Frommer, who studied the gelation process under the microscope, are closely related to the conical voids found in asymmetric membranes[8]. He noted that small, balloon-like protrusions appear in the polymer system near the interface shortly after contact with water. These extend into the membrane and grow in length and thickness as gelation proceeds, developing a conical shape with the apex near the active layer and the rounded base extending into the membrane.

In summary, it now appears that the substratum-related voids are relatively harmless but the conical voids must either be eliminated or reduced in size and removed from the vicinity of the active layer.

Void Prevention. When the casting-solution-related voids were first observed, they were thought to be due to the presence of foreign material in the casting solution. This theory was soon dispelled when rigorous filtration of the casting solution failed to change the void frequency or size. The casting solutions were next centrifuged to remove any microscopic trapped air bubbles and finally degassed to remove dissolved air, all without any effect on the propensity toward void formation.

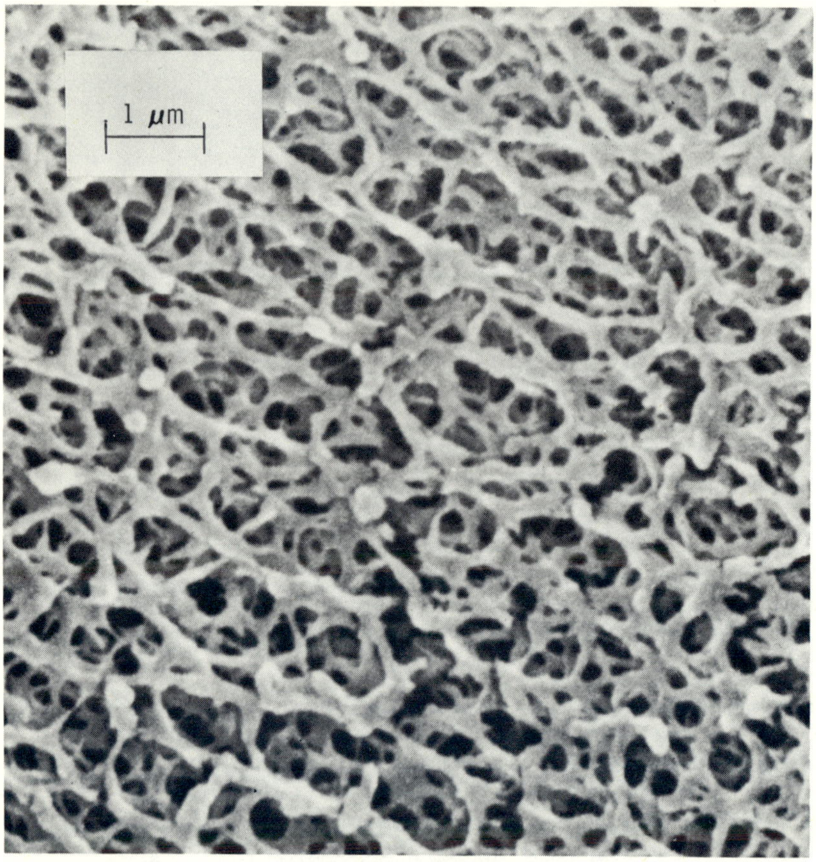

Fig. 7 - Scanning Electron Photomicrograph of the Interior Wall of a Conical Void in a Blend Membrane.

Having shown that filtration and rigorous removal of air from the casting solution had no effect on conically-shaped void formation, the search for other fabrication parameters which might affect void formation was extended to include drying time, casting speed, gelation temperature, atmospheric humidity, and membrane thickness.

Void frequency increases very rapidly with increasing gelation bath temperature[9]. In general, the void concentration in blend membranes gelled at 15°C is about ten times that of membranes gelled at 5°C. The response of a typical blend membrane to increased gelation bath temperature is shown in Figure 8. Gelation at temperatures higher than the customary 1°C have been used to increase the void frequency and thereby increase the precision of the void concentration data. Both void frequency and size were

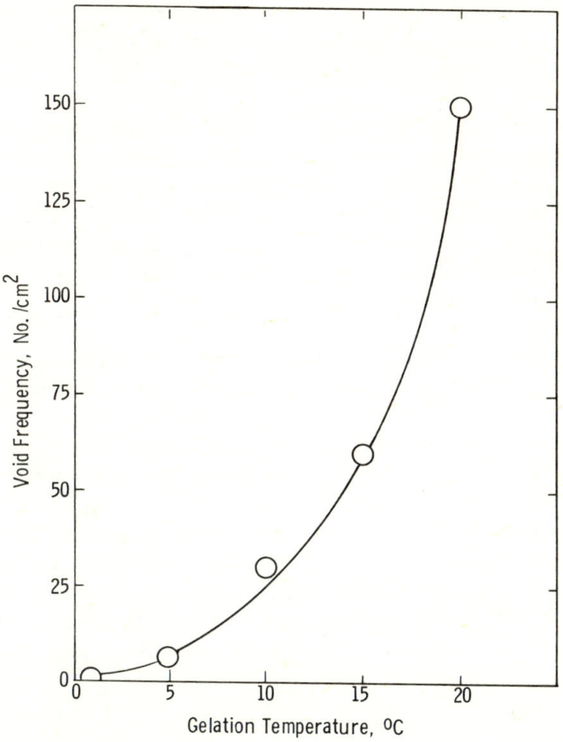

Fig. 8 - Dependence of Void Frequency Upon Gel Bath Temperature.

the same in membranes cast at 0% and 35% relative humidity. Variations in casting thickness of up to 75 μm from the standard 250 μm had little effect upon void frequency or size. Casting speed at constant drying time, however, was found to have a marked effect on void formation. Between 180 and 380 cm/min the void frequency was constant, but at 75 cm/min the voids were 3 to 4 times more numerous, which indicates that casting speeds of 180 cm (6 ft)/min or faster are desirable for void as well as wavemark prevention. There was no change in void size over the speed range from 75 to 380 cm/min. Drying times of 3, 4, 6, and 8 sec had no significant effect on either void size or frequency.

Three formulations were repeatedly cast over a period of eight weeks to determine the effect of casting solution age upon void formation. There was less than 10% variation in void size and frequency after one week of aging, but after three weeks there was a 25% increase in void frequency, and a 50 to 100% increase in void size. After eight weeks there was little additional change in void frequency or void size. These results indicate that casting solutions of this type should be less than one week old at the time of casting in order to obtain the minimum number and size of voids.

The ratio of dioxane to acetone in the casting solution was found to have a pronounced effect on the propensity toward void formation in blend membranes[9]. A marked increase in void frequency was observed as the dioxane-to-acetone ratio was increased. As shown in Figure 9, formulations with 90 parts of solvent per 20 parts of polymer also gave fewer membrane voids than did solutions with 80 parts of solvent.

Since the frequency of the voids can be altered by changes in the casting solution formulation, both the solvent system and the swelling agent system were modified in an effort to eliminate the voids without sacrificing the excellent osmotic properties of the blend membranes. Before proceeding with variations in the nonsolvent portion of the casting formulation, several alternative water-miscible organic liquids were evaluated as possible replacements for acetone and/or dioxane in the casting solutions. Dimethylacetamide and cyclohexanone substituted in turn for acetone did not produce complete solutions. In each case the solutions were clear but they were filled with undissolved but highly

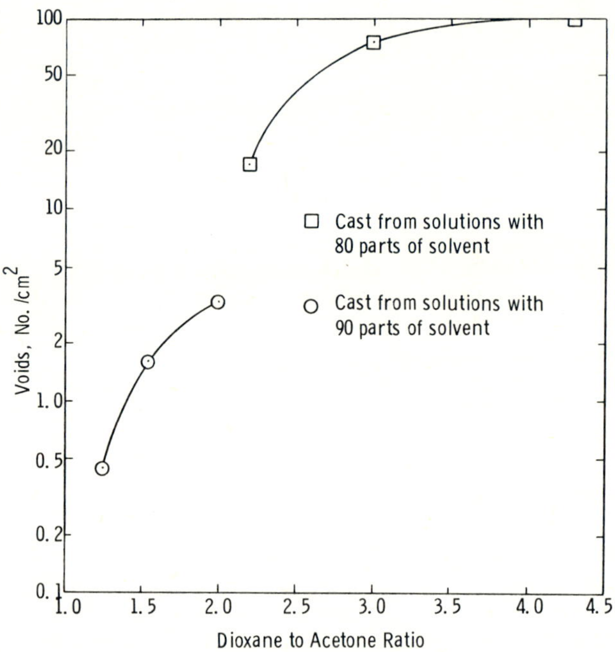

Fig. 9 - Influence of Dioxane-to-Acetone Ratio Upon Void Frequency.

swollen cellulose triacetate particles. When N-methyl-2-pyrrolidone, which is a solvent for both cellulose diacetate and cellulose triacetate, was used as the sole solvent or as a replacement for dioxane, the solutions had very high viscosities and produced membranes filled with voids. Tetrahydrofuran, methyl ethyl ketone, methyl acetate, and pyridine were examined in various combinations with both dioxane and acetone as cosolvents for cellulose triacetate. Only dioxane-methyl ethyl ketone was capable of dissolving the cellulose triacetate. Membranes prepared from casting solutions based on dioxane with methyl ethyl ketone had osmotic properties comparable to those prepared from dioxane with acetone but they contained the usual concentrations of voids. Therefore, the emphasis of the study was shifted to the nonsolvent portion of the casting solution formulation.

Higher alcohols, i.e., those up to and including all the butanols, were examined as possible substitutions for methanol. They were all effective in reducing the void frequency, but unfortunately reduced the flux of the membranes without a concomitant increase in salt rejection.

Acetic acid, which is an effective swelling agent as well as a solvent for both cellulose diacetate and cellulose triacetate was studied as a possible addition to the standard formulation containing methanol and maleic acid. Ten parts of acetic acid was effective in suppressing void formation in all of the membranes prepared from six casting solutions containing 20 parts of polymer and varying ratios of dioxane to acetone, whereas some of the membranes prepared from the casting solutions containing no acetic acid contained nearly 200 voids per cm^2 when gelled at 15°C. Most of the membranes suffered a 50 to 100% reduction in flux, however, and one formulation was selected for further study with lesser amounts of acetic acid in an attempt to maintain a high flux and still prevent void formation. The membranes were cast at 10 mils wet thickness on aluminized Mylar, dried 5 sec, gelled at 1° $\pm$ 0.5°C, and annealed for 5 min at 90°C. As in previous void studies, a second series of membranes gelled at 15°C, to deliberately enhance void formation, was prepared for a void count. A microscopic examination of both sets of membranes indicated that while 4 parts of acetic acid was sufficient to suppress void formation in membranes gelled at 1°C, 8 parts was required to prevent void formation in membranes gelled at 15°C. The use of sufficient acetic acid to effectively prevent void formation at 15°C was again accompanied by a considerable reduction in flux as shown in Table 1. The performance of these membranes in reverse osmosis tests is also presented in Table 1 in terms of water flux and salt rejection. Rejection is defined as 1-(concentration of feed solution)/(concentration of product solution).

TABLE 1

VOID REDUCTION IN METHANOL-MALEIC ACID MEMBRANES

Test Conditions: 1500 psi, 3.5% NaCl

Acetic Acid*	Void Frequency/cm^2		Water Flux (gfd)**	Salt Rejection (%)
	1°C Gelation	15°C Gelation		
0	5	11	15.3	99.4
1	1	8	14.9	99.4
2	1	9	14.8	99.2
3	1	8	13.4	99.1
4	0	6	13.8	99.3
5	0	10	13.7	99.0
6	0	1	13.5	98.8
7	0	1	13.0	98.9
8	0	0	11.4	99.1
9	0	0	11.2	99.1
10	0	0	11.2	99.1

*Parts by weight in addition to 10 parts of E-398-3 cellulose diacetate, 10 parts A-432-130B cellulose triacetate, 60 parts dioxane, 30 parts acetone, 9 parts methanol, and 6 parts maleic acid.

**gfd = gallons/ft^2-day. To convert to ml/cm^2-sec, divide by 2.12×10^4.

During the exploration of the formulations described above, several membranes were prepared from casting solutions containing only polymer, dioxane, acetone, and varying amounts of acetic acid. The excellent homogeneity of the casting solutions and the exceptional salt rejection of the membranes prepared from them, especially at the 9 part level (4.6 gfd, 99.8%), prompted a further investigation of this system. Both the solvent level and ratio were varied while the acetic acid was held constant at 9 parts per 10 parts each of cellulose diacetate and cellulose triacetate. The membranes were cast on aluminized Mylar, gelled at 1°C, and annealed at 90°C. Two of the formulations produced void-free membranes, one of which had a reasonable flux with outstanding salt rejection as shown by the results summarized in Table 2.

As an additional check on the resistance of the last formulation in Table 2 to void formation, a membrane was gelled at 15°C, which normally increases the void frequency by approximately

TABLE 2

ACETIC ACID AS A SWELLING AGENT

Test Conditions: 1500 psi, 3.5% NaCl

Parts by Weight*			Flux	Salt
Dioxane	Acetone	Voids/cm^2	(gfd)	Rej.(%)
55	25	3	6.8	99.7
60	20	3	8.6	99.4
60	30	0	8.6	99.0
65	25	0	8.9	99.6

*In addition to 10 parts of E-398-3 cellulose diacetate, 10 parts A-432-130B cellulose triacetate, and 9 parts acetic acid.

ten-fold. A microscopic examination of the membrane revealed only one conically-shaped void in a 16 cm^2 area. Since this formulation had ample viscosity (410 poises) to prevent the formation of wavemarks it was selected for further evaluation in tubular form.

The membranes were extruded into Dacron liners and inserted into 9/16-in. stainless steel tubes for testing. They failed to achieve over 99% salt rejection and a high pressure dye test revealed the presence of a few ruptured voids. Therefore, a new casting solution was prepared with 12 parts instead of 9 parts of acetic acid to further improve the void resistance of the membranes. Tubular membranes prepared from this formulation are free of voids when dye checked at pressures from 500 to 1500 psi and, as shown in the section on membrane performance, are suitable for the single-pass desalination of seawater.

MEMBRANE PROPERTIES

Intrinsic Permeability Coefficients

In the course of developing asymmetric blend membranes, the intrinsic permeation characteristics of blends of cellulose diacetate and cellulose triacetate were examined to determine if the transport properties differ from those of homopolymer acetates of the same degree of substitution. In the absence of coupling between water and salt transport in a membrane, the water flux (J_w) and salt flux (J_s) through a section of polymer of thickness X are related to the intrinsic water permeability (θ_w) and salt permeability (θ_s) by the following two equations:

$$J_w = \frac{\theta_w}{X} \frac{(\Delta P - \Delta\pi)\overline{V}_w}{RT}, \qquad J_s = \frac{\theta_s}{X}(C_s^f - C_s^p)$$

Here, ΔP is the difference in hydrostatic pressure of the feed solution and permeate; $\Delta\pi$ is the difference in the osmotic pressure of the feed solution, of concentration C_s^f, and permeate, of concentration C_s^p; $\overline{V}_w$ is the molar volume of water; and RT is the product of the gas constant and the temperature. The water permeability was measured by high pressure (1500 psi) reverse osmosis tests with sodium chloride solution. The presence of salt in the feed solution permits the detection of leaky films, but is not useful as an indicator of intrinsic salt permeability because even the most minute defect will significantly increase the measured salt permeation. Intrinsic water permeabilities were only calculated from measured water fluxes when the measured salt permeation was less than two percent. The inherent salt permeability was determined from the equilibrium electrical conductivity of films immersed in 0.1 M sodium chloride solution. These quantities are related by $\theta_s \cong D_s^o \, k_{sp}/k_{sp}^o$, where D_s^o is the diffusion coefficient of salt in water, k_{sp} is the specific conductivity of the film, and k_{sp}^o is the specific conductivity of the aqueous salt solution[10].

Dense films were cast onto glass plates from 10% solutions of the polymer in dioxane and acetone, dried for 24 hr, and immersed in water until the film spontaneously released from the glass plate. Some of the films were annealed by heating in water.

From this study of dense films it was learned that the magnitude of the intrinsic permeability coefficients of blends of cellulose diacetate and triacetate is the same as that of homopolymer cellulose acetates of the same degree of substitution[3]. This equivalence in water and salt transport was later corroborated with asymmetric membranes made from blends of cellulose diacetate and triacetate, with an average ds of 2.63, and from homopolymer cellulose acetate with a ds of 2.61. Both sets of membranes gave an initial flux of 12.5 gfd and a salt rejection of 99.6% when tested with 3.5% sodium chloride at 1500 psi[11].

The dense films exhibited very low water fluxes and very high intrinsic salt rejections. A considerably more permeable film was prepared by casting a typical membrane casting solution (i.e., one containing dioxane, acetone, methanol, and maleic acid), drying for only 4 hours before immersion in water, and eliminating the hot water annealing. The osmotic coefficients of three films prepared from 1:1 mixtures of cellulose diacetate and triacetate

TABLE 3

WATER AND SALT PERMEABILITY OF BLEND FILMS

Film Type*	Water Permeability, θ_w (g/cm-sec)	Salt Permeability, θ_s (cm^2/sec)		Salt Permeation (%)	
		Before Testing	After Testing	Measured**	Calculated
A	6.4×10^{-8}	1.8×10^{-12}	1.8×10^{-12}	-	0.04
B	1.4×10^{-7}	2.5×10^{-11}	2.5×10^{-11}	-	0.25
C	8.3×10^{-7}	3.4×10^{-9}	1.6×10^{-9}	3.9	4.2

*Films A and B were cast from acetone-dioxane solutions while the casting solution for C also contained methanol and maleic acid.

**Salt permeation measured at 1500 psi with 0.1 M NaCl.

are listed in Table 3. These data show that the salt permeability increases by a factor of approximately 1000 as the water permeability increases by a factor of 10, and that the actual salt permeation is closely approximated by the value calculated from intrinsic salt permeability (θ_s) data obtained from electrical conductivity measurements. (To calculate the salt permeation for film C in Table 3, the θ_s value used was an average of that before and after reverse osmosis testing.) But more importantly, the data for film C can be used to estimate the thickness of the active layer in asymmetric blend membranes.

An estimate of active layer thickness can be obtained by comparing the reverse osmosis properties of film C and asymmetric membranes tested under identical conditions. If film C were tested with 1% sodium chloride at 800 psi, it would be expected to give a salt permeation of 8.5%.* Experience indicates that an asymmetric blend membrane with a salt permeability of 8.5% will have a flux of approximately 45 gfd. Substituting the value of 45 gfd for J_w, the values of θ_w for film C from Table 3, and pressures of 800 psi and 115 psi for ΔP and $\Delta\pi$ respectively, in

*This figure is obtained by correcting for the change in effective pressure (ΔP-$\Delta\pi$); 4% x (1500-65)/(800-115) = 8.5%. Salt permeation is defined as 1-(salt rejection).

the transport equation for water flux, yields a value of 1340 Å for the thickness of the active layer in such a blend membrane.*

Effect of Casting Solution Composition on Reverse Osmosis Properties

The development of suitable casting solution compositions may be viewed as an attempt to simultaneously accomplish two tasks: minimize active layer thickness while maintaining high salt rejection, and obtain a strong porous gel structure whose water permeability is very much larger than that of the active layer. Efforts to achieve these goals largely consisted of studying the influence of variations of the casting formulations based on dioxane and acetone because no other solvent system has been found to give membranes of comparable quality. Dioxane and acetone are used in conjunction for two reasons. First, although dioxane is capable of dissolving both polymers, it gives solutions with high viscosities (> 1000 poises) when the polymer concentration is in the useful range of 17 to 20%. Second, the use of dioxane alone results in membranes of extremely high permeability and low inherent salt rejection. The partial substitution of acetone rapidly increases the salt rejection and decreases the solution viscosity. The influence of the dioxane-to-acetone ratio upon the formation of voids in asymmetric membranes was discussed previously; its effect upon the reverse osmosis properties of membrane[12] annealed at the same temperature is shown in Table 4.

TABLE 4

INFLUENCE OF DIOXANE AND ACETONE PROPORTIONS ON MEMBRANE PERMEATION

Dioxane-to-Acetone Ratio	Flux (gfd)	Salt Permeation (%)
4.33	75	44
3.00	52	15
2.20	44	11
2.00	37	5.2
1.57	30	3.5
1.25	26	3.2

Membranes tested with 1% sodium chloride at 800 psi.

---

* Note that 1 ml/cm^2-sec is equal to 2.12 x 10^4 gfd.

These membranes were prepared from solutions containing 85 parts of acetone plus dioxane, 9 parts of methanol, 6 parts of maleic acid, and 10 parts of each polymer. It is seen that as the proportion of dioxane decreases the membrane permeability decreases quite markedly. The performance of membranes with the higher permeability can be made equivalent to the performance of membranes of lower permeability by annealing at a higher temperature.

Methanol and maleic acid are added to the casting solution to improve the performance (membranes cast from dioxane-acetone solutions have a flux of 15 gfd with 97% rejection of sodium chloride at 800 psi when annealed at only 65°C) and to decrease the number and size of the voids. The detailed effects of methanol and maleic acid concentration upon membrane osmotic properties are strongly dependent upon the dioxane-to-acetone ratio, but, generally, small concentrations of either increase the flux at constant salt rejection. At larger concentrations (5 to 10% for methanol and 2 to 5% for maleic acid), the water and salt permeability increase in a manner very similar to that achieved by using lower annealing temperatures. At larger levels of maleic acid the flux continues to increase, but with rapidly decreasing salt rejection. By contrast, as the methanol concentration is increased from 10% to 20% (at which point the polymer begins to precipitate) there is no significant change in the reverse osmosis properties.

The general response of the salt permeation and the flux of blend membranes[12] is shown in Figure 10. This figure summarizes the performance of membranes fabricated from approximately 100 casting formulations with slightly different amounts of dioxane, acetone, methanol, and maleic acid. The reproducibility of a single type of membrane is better than indicated by the width of the curved region; for example, five castings from five different solutions of the same composition gave 38.2 ± 3.4 gfd and 5.9 ± 2.0% salt permeation. If one of the higher permeability membranes is annealed at a higher temperature, its salt permeability and water flux diminish along a line parallel to the boundary envelopes shown in the figure.

Polymer mixtures varying in molecular weight and the ratio of diacetate to triacetate were examined because of the possible influence of these two parameters on flux stability. It was thought that the use of the higher molecular weight diacetate or a higher proportion of the stronger cellulose triacetate might reduce the rate of creep, and therefore reduce the rate of flux decline.

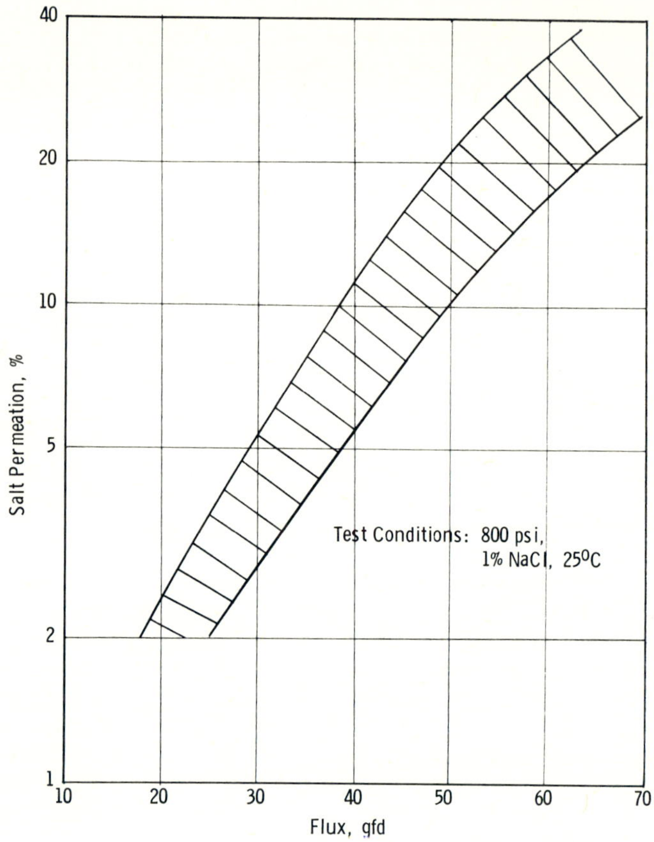

Fig. 10 - Reverse Osmosis Performance of Blend Membranes.

The osmotic response to varying the molecular weight of the cellulose diacetate in blend membranes[12] is shown in Table 5.* To compare the osmotic quality of these membranes it is necessary to plot the performance data on a response curve of flux versus salt permeation (e.g., of the type in Figure 10). When this is done, it is found that E-398-3 gives membranes with the best reverse osmosis performance. Furthermore, the smallest number of voids was obtained by using E-398-3 cellulose acetate.

The Effect of Test Conditions on Reverse Osmosis Performance

The osmotic response of membranes to pressure, salt concentration, and salt type was examined with brackish-water type membranes prepared from 1:1 mixtures of cellulose diacetate

*Except for the type of cellulose diacetate used, all casting solutions had the same composition.

TABLE 5

EFFECT OF MOLECULAR WEIGHT UPON MEMBRANE PROPERTIES

Cellulose Diacetate	Degree of Acetylation	Solution Viscosity (Poises)	Flux (gfd)	Salt Rejection (%)	Void Frequency (No./cm^2)	Void Diameter (μm)
E-398-3	2.41	150	50	91	<0.05	-
E-398-10	2.41	193	44	86	3	30
E-400-25	2.48	261	27	93	3	10
E-394-45	2.37	325	50	83	∿1	20
E-394-60	2.41	365	40	86	∿2	30

All casting solutions contained dioxane, acetone, methanol, maleic acid, and equal quantities of A-432-130B cellulose triacetate and one of the cellulose diacetates.

All membranes were annealed at 60°C for 5 min. Testing was at 800 psi with 1% sodium chloride.

and cellulose triacetate. It was found that the sodium sulfate permeability was approximately one-tenth the sodium chloride permeability for membranes with a sodium chloride rejection between 85 and 95%. To analyze the results of changing salt concentration or hydrostatic pressure it is useful to relate the measured performance to that predicted by the two transport equations described earlier. When membranes were tested at 800 psi and the sodium chloride concentration was changed from 1% to 3.5%, the flux went from 20 gfd to 10 gfd while the salt rejection changed from 99.2% to 98.53%. This result is in reasonable agreement with the calculated properties at the higher concentration - a flux of 11.4 gfd and a salt rejection of 98.6%.

When the hydrostatic pressure is increased, the flux generally increases more slowly than the effective pressure ($\Delta P - \Delta\pi$) because of increased compaction of the membrane. The osmotic response to changes in hydrostatic pressure is listed in Table 6. In calculating the expected behavior at the higher pressure, it was assumed that the hydraulic permeability (flux divided by the effective pressure) remained constant; in fact, it decreased by approximately 6% for the first two membranes, and by 4% for the last. This pressure effect is not instantaneous, and the hydraulic permeability is expected to continue to diminish with time; that is, larger hydrostatic pressures are expected to increase the rate of flux decline.

TABLE 6

PRESSURE DEPENDENCE OF REVERSE OSMOSIS PROPERTIES

		Reverse Osmosis Characteristics				
		Measured			Calculated	
Pressure (psi)	Sodium Chloride Conc. (%)	Flux (gfd)	Hydraulic Permeability (gfd/atm)	Salt Rej. (%)	Flux (gfd)	Salt Rej. (%)
400	0.5	25	1.08	82	-	-
600	0.5	37	1.01	93.6	39	92.7
600	0.3	44	1.14	91.6	-	-
800	0.3	56	1.08	93.0	62	94.1
600	0.3	30	0.78	96.0	-	-
800	0.3	39	0.75	96.9	42	97.1

Membrane Stability

Membrane stability is one of the most important factors which determine the economic suitability of a membrane for demineralizing various types of saline water. The two major stability considerations are the chemical and biological stability of the polymer, and the mechanical stability of the membrane, i.e., its stability with regard to creep and consequent decline in flux.

Chemical and Biological Stability. Because cellulose is a natural polymer, bacteria capable of depolymerizing it (and to a lesser extent, its acetyl ester) are ubiquitous. Some bacteria can also deacetylate cellulose acetate. The kinds of active bacteria and the stability of the membrane towards conditions favoring biological attack have been examined[13]. Following that work, another program was designed to discover a suitable method for protecting the membrane from biological degradation. Twenty-five materials were examined and five of these exhibited the required potency in killing bacteria, showed no deleterious effect upon the membrane properties, and were sufficiently nontoxic to permit safe handling by personnel[14]. One of the most effective and innocuous of the materials was formaldehyde. It was shown that a 0.2% aqueous solution of formaldehyde would completely kill the bacteria which degrade cellulose acetate. Blend membranes were stored in 0.2% formaldehyde for 32 weeks at 24°C with no change in their osmotic performance.

Another important factor is the pH of the feed solution. It is common practice to acidify feed waters to a pH of approximately 5 in order to prevent the deposition of calcium carbonate on the membrane. To examine the stability of membranes in acidic solutions, random specimens taken from one large roll of blend membrane were stored at room temperature in dilute sulfuric acid at various concentrations and periodically tested. It was discovered that 130 days of storage in a solution with a pH of 2.0 had no effect upon flux or salt rejection, and only a very slight change (about 10%) in the permeation characterisitics occurred after 450 days storage at a pH of 2.0. Storage of membranes for one year at 25°C in a solution of pH 3.0 produced no change in the reverse osmosis performance[15].

Mechanical Stability. A universal characteristic of all plasticized polymers is their susceptibility to continuous plastic flow when subjected to constant loading. This behavior is particularly noticeable with desalination membranes because the polymer is plasticized by water and because the membrane is largely a porous cellular structure, with a compressive strength far less than that of solid polymer. As a result of plastic creep, the flux continuously diminishes with time. There are three mechanisms commonly postulated to be responsible for this diminishing flux: (1) the hydraulic permeability of the porous portion of the membrane is reduced, (2) the density of the active layer region is increased with a consequent reduction in its water and salt permeability, or (3) the thickness of the active layer region is increased due to accretion from material at the interface between the active layer and the porous substructure. With the first model of flux decline, the rate of salt transport (J_s) should be unchanged, so that the percent salt permeation (proportional to J_s/J_w) should increase. The second mechanism would be expected to give a smaller percent salt permeation because any alteration in a homogeneous film of cellulose acetate which causes a decrease in the intrinsic water permeability results in a much larger decrease in the intrinsic salt permeability. According to the third model there should be no change in salt rejection because J_s/J_w is independent of active layer thickness.

Examples of all three types of changes in salt permeation will be seen in Table 9 (vide infra). Of the eleven sets of membranes tested, the salt rejection of eight decreased significantly after 200 hrs of operation. It might be concluded from this observation and from the three postulated mechanisms, that flux decline is largely caused by diminished hydraulic permeability in the porous portion of the membrane. However, the two cases of constant salt rejection and one case of increasing salt rejection suggest that significant changes in the active layer portion can also occur. The importance of physical changes in the active layer region, caused by hydrostatic compression, are scored by many other cases of constant salt rejection, such as those shown in Table 8.

It is generally found that the logarithm of flux is linearly related to the logarithm of time and the slope of the log flux vs. log time curve is taken to be a measure of the rate of flux decline. More explicitly, the flux-decline parameter, m, is defined by the equation $\log J(t) = \log J_1 - m \log t$, where $J(t)$ is the flux at time t, and J_1 is the calculated flux at 1 hr. The relationship between the projected change in flux after one year and the magnitude of the flux-decline parameter is shown in Figure 11. A flux-decline parameter of 0.03, corresponding to a 24% decrease in flux after one year, is generally considered to be a useful goal.

An important characteristic of most desalination membranes is the relationship between their initial flux (or their hydraulic permeability) and the rate at which their flux will decrease. Typical relationships between J_1 and the flux-decline parameter, m, are shown in Figure 12 for two different types of cellulose

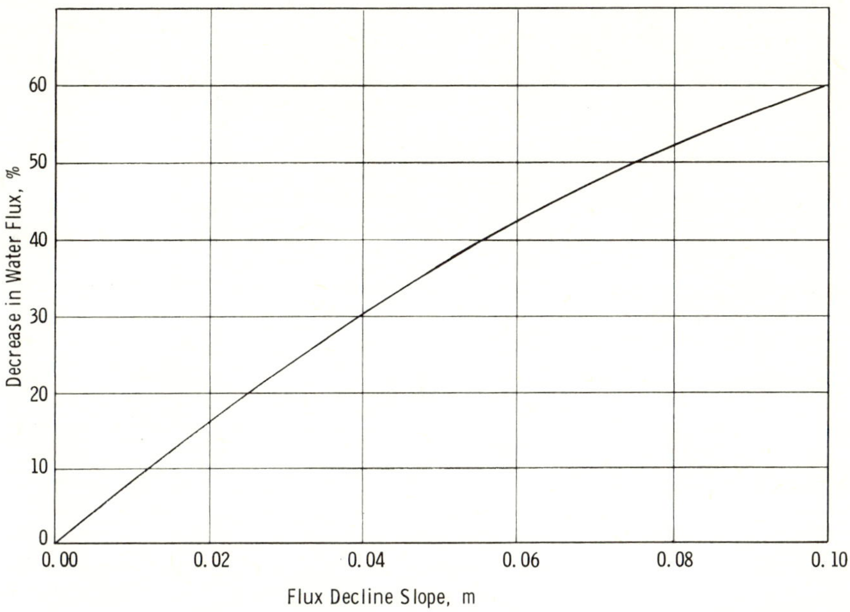

Fig. 11 - Relationship of Flux-Decline Slope to Water Flux After One Year.

acetate membranes annealed at various temperatures. (Type A is a cellulose diacetate membrane and Type B is a blend membrane.) The effect of the relationships shown in Figure 12 is such that the water productivity increases very slowly as the initial flux of the membrane increases. For example, if the initial flux of Type B membrane was increased from 35 gfd to 50 gfd by reducing the annealing temperature, the projected average first-year flux would only increase from 30 gfd to 33.5 gfd. Furthermore, for membranes with m values of 0.05 or larger, the large diminution in flux is generally accompanied by a decreasing salt rejection. For all these reasons it is clear that is important to increase the flux stability of practical membranes.

The effect of the cellulose diacetate-to-triacetate ratio upon the stability of high retention membranes at 1500 psi (with 3.5% sodium chloride) is shown in Table 7. The first three membranes[11] were fabricated by casting at 10 mils and drying for approximately 40 sec at 0°C before gelation in water at 1°C. It is seen that an equal proportion of diacetate and triacetate produced the most stable high retention membrane. The lower and higher proportions yielded membranes of diminished stability although their initial flux and salt permeation were essentially

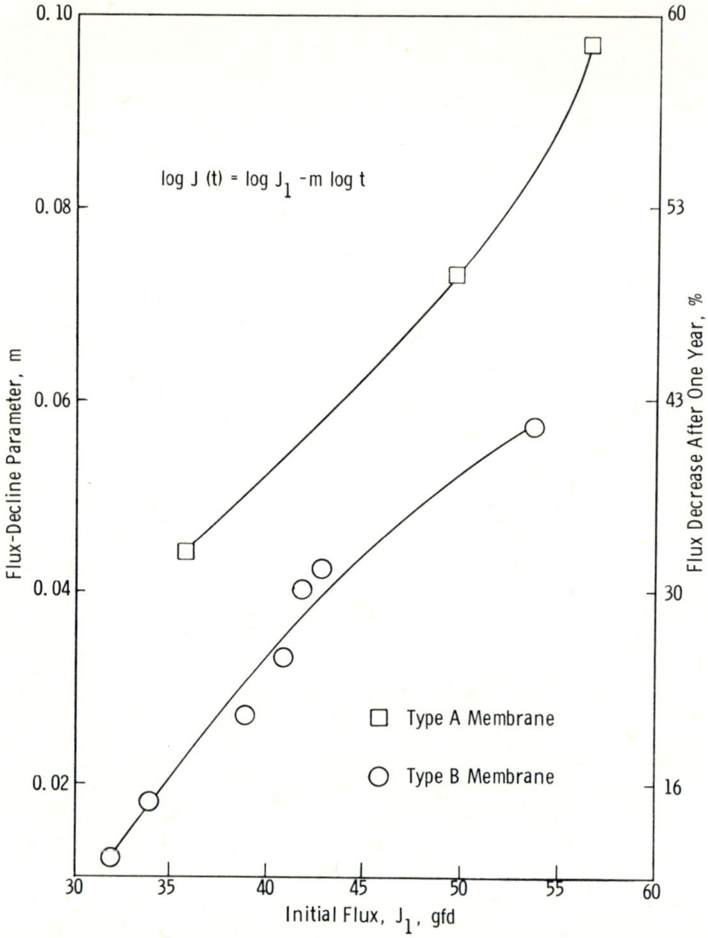

Fig. 12 - Effect of Initial Flux on Rate of Flux Decline

TABLE 7

FLUX STABILITY OF HIGH RETENTION MEMBRANES

Proportions of CA & CTA	J_1 (gfd)	m	Salt Rejection (%)
12:8	10.8	0.041	99.50
10:10	10.9	0.032	99.65
8:12	10.8	0.055	99.60
10:10	12.3	0.019	99.60

Tested for 10 days with 3.5% sodium chloride at 1500 psi.

TABLE 8

REVERSE OSMOSIS STABILITY OF TUBULAR MEMBRANES

Pressure (psi)	Feed Solution	J_1 (gfd)	Flux-Decline Parameter (m)	Average First Year Flux (gfd)	Salt Rejection(%) 12 hrs	Salt Rejection(%) 320 hrs
1500	Seawater	13.5	0.050 ± 0.001	9.1	99.20	99.30
1500	Seawater	11.7	0.042 ± 0.002	8.3	99.30	99.30
400	0.5% NaCl	24.2	0.010 ± 0.006	22	90.5	90.5
600	0.5% NaCl	38	0.041 ± 0.007	27	93.5	93.5

the same. The fourth membrane[16] was prepared from the same casting formulation as the others, i.e., 55g of dioxane, 35g of acetone, 9g of methanol, 3g of maleic acid, and 20g of cellulose acetate blend, but was dried for a period of 60 to 90 sec at a temperature of 23°C. This particular membrane is characterized not only by an improved initial flux but also by a considerably smaller value of flux-decline parameter.

Tubular membranes were examined both at 1500 psi with sea-water and at lower pressures with more dilute concentrations of sodium chloride. These membranes were extruded into Dacron sleeves (of 9/16-in. diameter), gelled in ice water, annealed in hot water, and subsequently supported in perforated stainless steel tubing[7]. The flux stability and osmotic characteristics of these tubular membranes are listed in Table 8. It is apparent that these tubular high retention membranes, as well as the flat-sheet high retention membranes described in Table 7, are adequate to the task of demineralizing seawater in a single pass operation. The performance of the blend membrane with 5000 ppm sodium chloride at 400 psi would be adequate for demineralizing most brackish waters and the low operating pressure should permit the use of inexpensive plastic supporting tubes.

Blend membranes made from eleven different casting formulations were examined for their long-term suitability for brackish water desalination at 800 psi. To compare the behavior of these membranes it is most useful to compare the values of their projected average first-year fluxes; this parameter simultaneously takes into account the magnitude of the initial flux and the flux-decline parameter. The performance of eleven types of membranes[12] is shown in Table 9. It is seen that increasing the polymer concentration in a constant casting solvent mixture results in a slight decrease in the projected average first-year flux. Substituting a higher molecular weight cellulose diacetate for E-398-3 resulted in significantly improved water productivity but with a concomitant decrease in salt rejection. Increasing the proportion of cellulose diacetate to cellulose triacetate had very little effect upon flux stability. This behavior differs from that observed with different ratios of diacetate to triacetate in high retention membranes (Table 7). This difference may have resulted from the fact that the same casting composition was used for the high retention membranes, whereas the proportion of methanol in the casting solutions for the brackish-water membranes of Table 9 was increased with larger diacetate proportions. Membranes made from higher proportions of cellulose triacetate were not examined in these tests because of the inferior initial osmotic performance of the membranes. One formulation (the tenth) is decidedly superior to the others, giving 93% salt rejection and a projected average first-year flux of 33 gfd, which is about 30% larger than most of the other membranes examined.

TABLE 9

FLUX STABILITY OF BLEND MEMBRANES

Casting Solution Composition						Reverse Osmosis Properties				
									Salt Rejection (%)	
CA	CTA	Dioxane	Acetone	Methanol	Maleic Acid	$\bar{J}$	J_1	m	At 2 hrs	At 200 hrs
9	9	60	30	9	6	26	39	0.044	93.0	92.3
10	10	60	30	9	6	24	40	0.061	94.6	91.3
11	11	60	30	9	6	22	39	0.067	94.0	90.0
12	12	60	30	9	6	22	32	0.043	95.7	93.9
11	11	60	30	9	6[a]	22	30	0.040	94.7	92.3
11	11	60	30	9	6[b]	21	28	0.036	93.4	91.6
9[c]	9	60	30	9	6	32	48	0.052	90.0	88.0
15	7	60	30	12	6	29	35	0.024	94.8	94.4
19	5	60	30	15	6	26	36	0.043	92.8	91.4
9	9	60	20	9	3	33	43	0.035	92.2	93.4
10	10	60	20	15	3	26	38	0.047	95.4	93.2

All membranes were tested for 200 hrs with 1% sodium chloride at 800 psi. CA and CTA represent E-398-3 and A-432-130B cellulose acetates, respectively. J_1 is the calculated flux at 1 hr (in gfd), m is the flux-decline parameter, and $\bar{J}$ is the projected average first-year flux.

[a]Plus 4g of acetic acid.

[b]Gelled in 6% acetic acid at 1°C.

[c]E-394-45 instead of E-398-3.

CONCLUSIONS

Asymmetric reverse osmosis membranes made from a blend of cellulose diacetate and cellulose triacetate exhibit desalination properties superior to conventional cellulose diacetate membranes. They are capable of demineralizing seawater in a single pass, with initial fluxes of 11 to 13 gfd and salt rejections of 99.3 to 99.6% at a pressure of 1500 psi. Projected fluxes after one year of operation are 8 to 10 gfd. For the purification of various brackish waters, blend membranes give higher fluxes, at any particular salt rejection, than cellulose diacetate membranes. At 800 psi with a 1% sodium chloride feed water, brackish-water blend membranes will give fluxes as high as 45 gfd at 90% salt rejection. In addition to their higher fluxes, the brackish-water blend membranes also show superior flux stability in comparison with conventional cellulose diacetate membranes with the same salt rejection.

ACKNOWLEDGMENT

The work reported in this paper was supported by the Office of Saline Water under Contracts 14-01-0001-1732, 1767, 2182, 2205, 2224, and 14-30-2664. The authors also wish to recognize the contributions of M. L. O'Hair, O. S. Schaeffler, L. W. Schoellenbach, A. J. Secchi, and A. L. Vincent.

REFERENCES

1. C. E. Reid and E. J. Breton, J. Appl. Pol. Sci., 1, 133 (1959).

2. H. K. Lonsdale, U. Merten and R. L. Riley, J. Appl. Pol. Sci. 9, 1341 (1965).

3. P. A. Cantor, et al., in "Development of Improved Cellulose Ester Reverse Osmosis Membranes," Office of Saline Water Research and Development Report No. 434, U. S. Government Printing Office, Washington, D.C. 20402.

4. S. Loeb in Desalination by Reverse Osmosis, U. Merten, (Ed.), The M.I.T. Press, Cambridge, Mass. 1966, Chapter 3.

5. B. Keilin in "The Mechanism of Desalination by Reverse Osmosis," Office of Saline Water Research and Development Progress Report No. 117, U. S. Government Printing Office, Washington, D.C. 20402.

6. S. Loeb and S. Sourirajan, Advan. Chem. Ser. 38, 117 (1962).

7. W. M. King and M. L. O'Hair in "Seawater Tubular Reverse Osmosis Membranes," Office of Saline Water Research and Development Report No. 695, U. S. Government Printing Office, Washington, D.C. 20402.

8. M. A. Frommer, R. Matz, and R. Bloch in "Saline Water Conversion Report for 1970-1971," U. S. Government Printing Office, Washington, D.C. 20402.

9. W. M. King, et al., in "High Retention Tubular Membranes for Reverse Osmosis," Office of Saline Water Research and Development Report No. 682, U. S. Government Printing Office, Washington, D.C. 20402.

10. C. W. Saltonstall, Jr., W. M. King and D. L. Hoernschemeyer, Desalination 4, 309 (1968).

11. P. A. Cantor, et al., in "Development and Scale-up of New and Improved Cellulose Ester Membranes," Aerojet-General Corporation Report No. 4868-01-F, a final report to the Office of Saline Water, U. S. Department of the Interior, Contract No. 14-01-0001-1767, May 1969.

12. D. L. Hoernschemeyer, et al., in "Research and Development of New and Improved Cellulose Ester Membranes," Envirogenics Co. Report No. 1319-02-F, a final report to the Office of Saline Water, U. S. Department of the Interior, Contract No. 14-01-0001-2205, April 1971.

13. P. A. Cantor, et al., in "Biological Degradation of Cellulose Acetate Reverse-Osmosis Membranes," Office of Saline Water Research and Development Report No. 340, U. S. Government Printing Office, Washington, D.C. 20402.

14. R. W. Lawrence, et al., in "Development of Storage Techniques for Reverse-Osmosis Membranes," Office of Saline Water Research and Development Report No. 673, U. S. Government Printing Office, Washington, D.C. 20402.

15. A. J. Secchi, unpublished results, Envirogenics Co.

16. C. W. Saltonstall, Jr., et al., "Development of Stable, High-Flux Brackish-Water Membranes," Aerojet-General Corporation Report No. 1350-F, a final report to the Office of Saline Water, U. S. Department of the Interior, Contract No. 14-0001-2182, June 1970.

STABILIZATION OF CELLULOSIC DESALINATION MEMBRANES BY CROSSLINKING

D. L. Hoernschemeyer, R. W. Lawrence,
C. W. Saltonstall, Jr., and O. S. Schaeffler
Envirogenics Company
El Monte, California 91734

INTRODUCTION

C. E. Reid and coworkers[1] discovered that, among many polymers, cellulose acetate exhibits an exceptionally selective permeability toward water in salt solutions under conditions of reverse osmosis. Subsequently, Loeb and Sourirajan[2] developed a method of fabrication for water-swollen cellulose acetate membranes with greatly increased flux rates but which retain much of the selectivity ("salt rejection") of ordinary films. The practicality of the so-called asymmetric membranes of this type is now established by their use in commercial reverse osmosis desalination equipment. Further increases in water flux, however, are highly desirable in order to reduce the cost of the reverse osmosis process. The preparation of membranes with both increased flux and improved flux stability was the object of the work described herein.

Figure 1 shows the asymmetric structure of a cellulose acetate membrane in cross section. This microphotograph, taken with a scanning electron microscope, shows an extremely thin layer (in the upper part of the photograph) surmounting a relatively thick (50 to 100 μm) substructure with pores about 0.3 μm in diameter. The high flux rates of water in reverse osmosis result from the thinness of the top layer (ca. 0.05 μm) and the selectivity depends on the absence of pores in this thin "active layer." Water (and salt) permeate freely through the supporting porous substructure.

Fig. 1 - Cross section of Asymmetric Cellulose Acetate Membrane.

Water is believed to permeate the active layer by a pressure-driven solution-diffusion mechanism while hydrated solute ions are relatively efficiently excluded by their large size and low solubility in cellulose acetate. More highly charged ions are more efficiently rejected than those of lower charge (e.g., $SO_4^{=} > Cl^-$, $Ca^{++} > Na^+$) [3,4]. The overall permeabilities of both water and ions are believed to be controlled by the amount of water dissolved in the active layer of the cellulose acetate (cellulose acetate sorbs about 12% by weight of water at 100% relative humidity). For example, Yasuda et al. have shown that for both salt and water $\log D = \log D_o - K[(1/H) - 1]$, where D is the diffusion coefficient in the polymer, D_o is the diffusion coefficient in aqueous solution, H is the weight fraction of water in the polymer, and K is a constant[5,6]. The amount of water in the active layer is, in turn, controlled by annealing the swollen membrane in water at temperatures up to 95°C, lower annealing temperatures yielding membranes with a higher water flux but lower salt rejection.

Water is an excellent plasticizer for cellulose acetate (reducing the modulus and tensile strength by 50% or more), and hence the membrane is subject to creep under the pressures (400 to 1500 psi) employed in reverse osmosis. The creep process is accompanied by a gradual decrease in water permeability with time, probably due to an increase in thickness and perhaps loss of sorbed water in the active layer. Water flux, J, at time t (in hours) is well predicted by the expression: $J = J_1 t^{-m}$, where J_1 is the flux after 1 hour and the exponent is called the flux-decline parameter. In any series of membranes, differing only in the annealing temperature employed, m becomes larger (the rate of flux decline becomes greater) as the annealing temperature is reduced and the initial flux is increased. This was shown in Figure 12 of the previous paper by King et al. Thus, cellulose acetate membranes of very high flux, although retaining useful salt rejection properties (especially for $SO_4^{=}$, Ca^{++}, etc.), exhibit unstable water flux rates and their high fluxes cannot be utilized to improve the cost effectiveness of the reverse osmosis process.

Numerous workers have postulated that the flux-decline parameter of high-flux cellulose acetate membranes could be greatly reduced by crosslinking the polymer chains in the membranes. Crosslinking should minimize creep and accretion to the active layer. Although attractive in principle, treatment of membranes with crosslinking reagents which act through formation of bonds between unsubstituted hydroxyl groups in the cellulose acetate has not proved successful in our hands. With many such reagents (diisocyanates, diacid chlorides, etc.) the membranes must be dried before treatment, but more importantly, the relatively large organic molecules must permeate the active layer. Permeation of the active layer requires sufficient solubility of the reagent in cellulose acetate that the osmotic characteristics of the membrane are seriously damaged. It was reasoned, therefore, that the crosslinking entity should preferably be attached to the cellulose acetate chains so that after establishing the osmotic properties of the membranes, only low molecular weight initiators need permeate the active layer. This has been accomplished through the attachment of methacrylyl groups to cellulose acetate, and high flux membranes made from the resulting cellulose acetate methacrylate have been crosslinked by free-radical initiation. These crosslinked membranes, described below, have shown the expected stabilization of flux.

SYNTHESIS AND MEMBRANE FABRICATION

In order to obtain a methacrylyl substitution large enough to provide satisfactory crosslinking, it was necessary to choose cellulose acetates with more unsubstituted hydroxyl groups (lower acetyl degree of substitution) than those generally employed in the fabrication of membranes. The two cellulose acetates chosen on this basis were an experimental material with an acetyl degree of substitution of 2.1 (Eastman Chemical Co. Type E-360-60) and a commercial material with an acetyl degree of substitution of 2.3 (Eastman Chemical Co. Type E-383-40).* These esters were methacrylated by a method typified by the following procedure. One kilogram of vacuum-dried cellulose acetate is dissolved in 10 liters of a 9:1 (by volume) solution of acetone and triethylamine. After raising the temperature to 35°C, a solution of 0.5 kg of freshly distilled methacrylyl chloride and 0.5g of p-dimethoxybenzene in 1 liter of acetone is added slowly so as to maintain the reaction temperature between 35 and 40°C. When the addition is completed, stirring is continued for 24 hours without heating, then the reaction mixture is poured into 90 liters of deionized water with rapid stirring. The precipitated polymer is washed with water and 2-propanol and air dried. Purification is conducted by redissolving the polymer in acetone (10g per 100 ml) then reprecipitating in water followed by washing with water and 2-propanol. This procedure yields a product with a methacrylyl substitution of 0.2 to 0.4 (by bromination) and with the same acetyl content as the starting material.

Asymmetric membranes were prepared from cellulose acetate methacrylate casting solutions containing polymer, acetone, and additives by the technique widely used in preparing cellulose acetate membranes. The casting solution was coated 0.010 inches thick on a glass plate (at -10°C) or an aluminized polyester film (at 25°C), gelled immediately in water at 0 to 1°C, then annealed in hot water for 3 minutes. With the polymer made from Type E-360-60 cellulose acetate, a casting solution containing (parts by weight) 10 of polymer, 35 of acetone, and 14 of maleic acid was used. The casting solution used with the polymer made from Type E-383-40 cellulose acetate contained (parts by weight) 10 of polymer, 22.7 of acetone, 12.7 of 2-propanol, and 9 of water. These casting solutions were developed to yield membranes with initial reverse osmosis properties (uncrosslinked) of 35 to 40 gal/ft^2-day (gfd)** and 90% salt rejection*** when tested with

*The degree of substitution is the average number of substituents (out of a maximum of three) per anhydroglucose unit.

**Divide gal/ft^2-day by 2.12 x 10^4 to convert to cm^3/cm^2-sec.

***Salt rejection (%) = $100 \left[1 - \frac{\text{conc. of salt in permeate}}{\text{conc. of salt in feed}}\right]$.

1% sodium chloride solution at 800 psi. The flux and salt rejection were varied by adjustment of the annealing temperature.

The membranes were tested at 25 ± 1°C in reverse osmosis cells of a type previously described at feed flow rates of 5 gal/hr unless otherwise stated[7]. In general, the volume of permeate was less than 10% of the feed volume and was returned continuously to the feed reservoir.

CROSSLINKING AND DESALINATION PROPERTIES

Three requirements were considered in the search for a crosslinking initiator system. The first requirement was that the process function in the presence of water. This condition was chosen because treatment with an organic solution of a free-radical initiator would require prior drying of the wet membrane. Moderate conditions of pH and temperature were desired so that the membrane would not be chemically damaged. In particular, a pH between 3 and 7 was desired because the maximum hydrolytic stability of cellulose acetate occurs at a pH of 5 to 6. The third requirement was that the initiators should easily permeate the active layer of the membrane without damaging its structure. This suggested the use of inorganic redox systems. Two water soluble inorganic species commonly used in redox initiator systems are hydrogen peroxide and a persulfate salt. These, together with sulfite or bisulfite, were chosen for investigation.

The general features of the crosslinking procedure used consisted of immersing the membrane into a stirred bath at a temperature of 65 to 85°C, adding the reagents, and after a period of 10 to 60 min, quenching the membrane in cool water. Under these conditions and with potassium persulfate and sodium bisulfite in concentrations of approximately 40 millimolar, the pH changes from 5 to 2 after a period of about 10 min. Because of the low terminal pH of the solution, the use of buffers was briefly investigated. It was found that sodium acetate or disodium hydrogen phosphate were effective in maintaining a higher pH during the course of the reaction, but with some reduction in the extent of crosslinking obtained.* However, since there was no evidence of ester hydrolysis at terminal pH values as low as 2, most of the reactions were carried out without buffers.

Representative results obtained by crosslinking cellulose acetate methacrylate (CAM) membranes with the sodium bisulfite/potassium persulfate system are given in Table 1. Three major

*The extent of crosslinking was measured by the fraction of the membrane insoluble in acetone after a period of two hours.

TABLE 1

PERSULFATE/BISULFITE CROSSLINKED MEMBRANES

Methacrylate Substitution ds_m	Reagent Concentration (m$\underline{M}$)		Degree of Insolubility (%)	Reverse Osmosis Properties After Crosslinking	
	$K_2S_2O_8$	$NaHSO_3$		Flux (gfd)	Salt Rejection (%)
0.38	37	44	97	35	87
0.32	37	44	85	33	84
0.22	37	44	75	28	85
0.22	37	87	92	42	93
0.22	75	87	95	48	90

Membranes crosslinked at 65°C for 15 min. Polymer derived from cellulose 2.3-acetate. Reverse osmosis properties measured with 1% NaCl at 800 psi and 25°C.

---

factors which can effect the extent of crosslinking are the concentration of methacrylate groups in the polymer (i.e., the methacrylate substitution, ds_m), the membrane permeability (related to the annealing temperature), and the concentration of reagents. It is seen that the ester with the largest amount of methacrylate crosslinked to the greatest extent, giving an insoluble fraction of 97%. Even at the lowest ds_m, however, an increase in the initiator concentration resulted in 95% insolubility. The reverse osmosis properties of the crosslinked membranes were equivalent or superior to those of the membranes before crosslinking.

Although a high degree of crosslinking was obtained with the bisulfite/persulfate system, it exhibited a shortcoming not shown by the data in Table 1. For some purposes it is desirable to crosslink the membrane by exposure of only one side of the membrane to the crosslinking solution. This is the case with tubular membranes, where it is more convenient to circulate crosslinking solution through the tube, thereby exposing only the active surface of the membrane to the crosslinking solution. As might be expected, this resulted in a smaller degree of crosslinking with the persulfate/bisulfite system; 73% insolubility was obtained by crosslinking from the active layer side, in contrast to 97% insolubility when crosslinked by immersion in the redox solution. For this reason a superior redox system was sought.

It was believed that the substitution of hydrogen peroxide for potassium persulfate would result in a higher degree of crosslinking because of the much greater permeability of the membrane to hydrogen peroxide (or the OH and HO_2 radicals) than to the SO_4^- radical ion. For example, the permeability to hydrogen peroxide would be expected to be comparable to that to water, or approximately four orders of magnitude larger than the permeability to an ionic solute such as persulfate. Although greater crosslinking was anticipated with the hydrogen peroxide/bisulfite system, the first two entries in Table 2 show that this expectation was not realized. This situation was readily corrected, however, by using metallic salts as crosslinking promoters. Although little crosslinking was obtained with the unpromoted reagents, the addition of small amounts of iron or copper salts led to a large increase in the extent of crosslinking. Nearly complete insolubilization was obtained with iron or copper concentrations as low as 2 millimolar. Moreover, with the iron-catalyzed reaction it was possible to obtain an acetone insoluble fraction of 90% by treating only the active layer side of the membrane.

An important feature of crosslinking is the influence exerted by the membrane's initial state on the extent of crosslinking achieved. It would be expected that the "crosslinkability" would increase as the permeability of the membrane to reagents or free-radical species increases. Indeed, it is found that the extent

TABLE 2

THE EFFECT OF PROMOTERS WITH THE HYDROGEN PEROXIDE SYSTEM

Reagent Concentration (m$\underline{M}$)			Promoter Concentration (m$\underline{M}$)	Degree of Insolubility (%)
$NaHSO_3$	Na_2SO_3	H_2O_2		
-	200	100	none	30
223	-	202	none	7
223	-	202	20 ($FeSO_4$)	100
112	112	143	20 ($FeSO_4$)	100
112	112	136	2.0 ($FeSO_4$)	98
112	112	136	0.20 ($FeSO_4$)	66
112	112	136	2.0 ($CuSO_4$)	96

Polymer derived from cellulose 2.3-acetate. Membrane annealed at 75°C for 5 min. The reaction temperature for crosslinking was 74 $\pm$ 4°C.

of crosslinking increases when the permeability is increased by using lower annealing temperatures. This behavior is shown in Fig. 2. The ordinate in Fig. 2 labeled Relative Salt Permeability refers to relative values of the rate of salt transport through the membrane, defined as [flux (gfd) x (100% - % salt rejection)].

The permeability can also be altered by changing the casting solution composition and the effect of such changes upon "crosslinkability" was examined. Some results of this study are listed in Table 3. Although the insoluble fraction decreased with increasing annealing temperature, there was no correlation between either flux or salt rejection and the insoluble fraction. Each type of membrane, when annealed at 78°C, gave an insoluble fraction of 62 $\pm$ 3%. The mechanism by which the annealing temperature

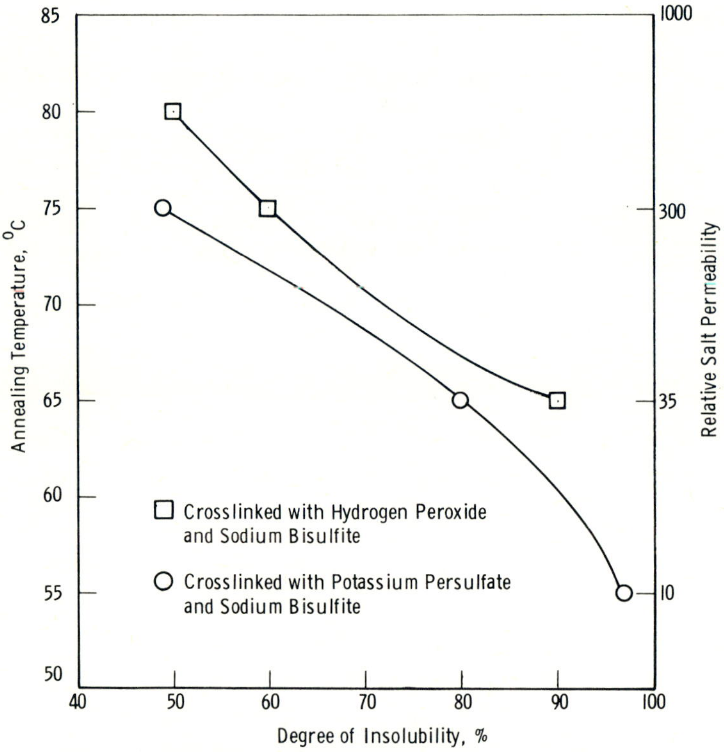

Fig. 2 - The Relationship Between Membrane Annealing Temperature, Relative Salt Permeability, and Extent of Crosslinking

TABLE 3

"CROSSLINKABILITY" AND INITIAL MEMBRANE PROPERTIES

Type of Casting Solution	Annealing Temp. (°C)	Flux (gfd)	Salt Rejection (%)	Acetone Insoluble Fraction (%)
1	75	38	80	70
2	65	38	91.4	90
1	81	24	91.5	56
3	75	22	96.1	73

Osmotic properties of uncrosslinked membranes were measured with 1% NaCl at 800 psi. The casting solutions all contained 10g of CAM, 25g of acetone, and 14g of isopropyl alcohol; plus 11g of water in solution No. 1, 9g of water in No. 2, and 10g of water in No. 3.

---

controls the "crosslinkability" is not known. One possibility is that the polymer chain mobility is determined by the annealing temperature and the chain mobility in turn controls the probability that neighboring methacrylyl radicals will come within reaction distance and subsequently join to give a crosslink.

Although the goal of the crosslinking work was the attainment of better flux stability, an auxiliary benefit was obtained. It was found that the reverse-osmosis properties of membranes which were highly crosslinked were generally superior to those of the uncrosslinked membrane. The changes in osmotic properties of membranes crosslinked to an acetone insoluble fraction of at least 90% are shown in Table 4 (XL denotes crosslinking).

Of primary importance, crosslinking led to a great improvement in flux stability of the membrane. Three sets of membranes exhibited either very small or no decrease in flux over a ten-day interval of testing at 800 psi, as shown by the data in Table 5.

TABLE 4

THE EFFECT OF CROSSLINKING ON REVERSE-OSMOSIS PROPERTIES

Feed Solution	Pressure (psi)	Water Flux (gfd) Before XL	Water Flux (gfd) After XL	Salt Rejection (%) Before XL	Salt Rejection (%) After XL
1% NaCl	800	26	34	94	93
0.1% NaCl	300	11	16	89	96
0.3% Na_2SO_4	300	19	26	92	99.5

TABLE 5

FLUX STABILITY OF MEMBRANES CROSSLINKED WITH HYDROGEN PEROXIDE/SULFITE SYSTEM

Degree of Insolubility (%)	Initial Flux, J_1 (gfd)	Flux-Decline Parameter*, m	Salt Rejection (%) 1 Hr	Final
none	40	0.046 ± 0.010	93	95
59	36	0.000 ± .002	97	97
90	37	0.006 ± .003	97	97
91	49	0.008 ± .008	96	95

Polymer derived from cellulose 2.3-acetate. Specimens tested for 10 days with 0.1% NaCl at 800 psi and 25°C. Crosslinking solutions as in 5th line of Table 2.

*Uncertainties listed are standard deviations.

TABLE 6

CELLULOSE 2.1-ACETATE METHACRYLATE MEMBRANES

Membrane*	Flux-Decline Parameter, m	Initial** Flux (gfd)	Salt Rejection(%) NaCl	Na_2SO_4
80	0.042	50	86	-
80-XL	0.022	67	77	-
83	0.037	35	90	-
83-XL	0.019	52	93	-
80-XL	-	96	-	98
83-XL	-	73	-	99

*The number is the annealing temperature, XL denotes the same type of membrane which had been crosslinked (to an acetone insoluble fraction of 90 to 95%) with sodium bisulfite and potassium persulfate. The polymer had an acetyl ds of 2.1 and a methacrylyl ds of 0.03.

**All reverse osmosis tests were made at 800 psi and 25°C. The flux-decline data were obtained with a flow rate of 5 gal/hr but the tests with Na_2SO_4 solution used a flow rate of 14 gal/hr. The feed solutions were 1% in NaCl or 0.3% in Na_2SO_4.

The uncrosslinked membrane exhibited a flux-decline parameter of 0.05, which would correspond to a loss of about 35% of the initial flux during the first year of operation. In contrast, the flux of the crosslinked membranes with the highest flux-decline parameter would be predicted to be at least 90% of its original value after one year.

Membranes of still higher flux were prepared from cellulose acetate methacrylate derived from the cellulose 2.1-acetate. From the results given in Table 6, it is seen that crosslinking reduced the flux-decline parameter by 50% and increased the flux by 34 to 48%. The result of the combined effect of these changes induced by crosslinking is to increase the projected flux after one year by 63% (for the type 80 membrane) and 83% (for the type 83 membrane). The osmotic properties obtained with sodium sulfate are not significantly better than those exhibited by certain cellulose acetate membranes, but the flux stability should be vastly improved. Cellulose acetate membranes with fluxes of 75 to 85 gfd have flux-decline parameters of 0.05 to 0.065, in marked contrast to values of only 0.02 for these crosslinked cellulose acetate methacrylate membranes. Furthermore, it is anticipated that improvements in the crosslinking procedure will result in flux stabilities comparable to those listed in Table 5, i.e., m values of 0.01 or less.

An exploratory investigation was made to discover how crosslinking altered the permeability characteristics of CAM membranes. If hydroxyl or hydroperoxide radicals initiate crosslinking of contiguous methacrylyl groups, one might expect to find hydroxyl groups bonded to the crosslinks. This possibility was first checked by examining the infrared spectrum of the membranes (by the multiple internal reflection technique), but no difference in the spectrum due to crosslinking was discernible. However, the sensitivity is poor because the hydroxyl increment is likely to be small compared to the initial hydroxyl ds of 0.7.

Significant changes due to crosslinking were observed in the dielectric behavior of the membranes. The capacitance, dissipation factor, and A.C. resistance of wet membranes were measured between 100 and 10^4 Hz. Over this frequency interval there was no difference in the dissipation factor of the crosslinked and uncrosslinked membranes. On the other hand, the conductivity (at 1000 Hz) increased by a factor of 3 to 4 with crosslinking and the capacitance increased roughly in proportion to the extent of crosslinking. Capacitance ratios (crosslinked to uncrosslinked) at 5000 Hz were approximately 1.6, 1.8, and 3.0 for membranes with acetone insoluble fractions of 56%, 73%, and 94%, respectively. These changes in capacitance and resistance are consistent with the model of hydroxylation (or the introduction of some other very polar group) during crosslinking.

During earlier studies of the mechanism of flux decline, all evidence pointed to changes in the active layer, rather than in the porous substructure, as responsible for flux decline. Particularly telling was the observation that the compressive creep rates of crosslinked and uncrosslinked membranes were identical[8]; after eight days at 800 psi both membranes diminished in thickness by 63%. This occurred in spite of the great difference in flux stability of the two membranes.

Tensile properties of wet membranes were measured to give a more complete characterization of the effect of crosslinking on mechanical properties. The elastic modulus increased slightly but the yield stress was unaffected by crosslinking. Similarly, the rate of stress relaxation (for specimens stressed to 75% of their yield stress) was unchanged. The elongation at break, however, decreased from 13% to 8% as a result of crosslinking. This behavior is similar to that of crosslinked elastomers, where the ultimate elongation monotonically decreases with extent of crosslinking, while the tensile strength goes through a maximum[9].

CONCLUSIONS

Incorporation of crosslinked cellulose acetate methacrylate membranes in reverse osmosis equipment should serve to improve the efficiency of the process in desalination of brackish water due to their high flux and excellent flux stability. Adaptation of the membranes for very high salt rejection (99.5%), currently under study, would be similarly advantageous in the single-pass desalination of seawater.

ACKNOWLEDGMENT

The work reported in this paper was supported by the Office of Saline Water of the U.S. Department of the Interior, Contract No. 14-01-0001-2205.

REFERENCES

1. C. E. Reid and E. J. Breton, J. Appl. Polymer Sci., 1, 133 (1959).

2. S. Loeb and S. Sourirajan, Advan. Chem. Ser., 38, 117 (1962).

3. S. Sourirajan, Ind. Eng. Chem. Fundamentals, 2, 51 (1963); 3, 206 (1964).

4. T. D. Hodgson, Desalination, 8, 99 (1970).

5. H. Yasuda, C. E. Lamaze, and L. D. Ikenberry, Makromol. Chem. 118, 19 (1968).

6. H. Yasuda, C. E. Lamaze, and A. Peterlin, to be published.

7. R. E. Kesting, M. K. Barsh, and A. L. Vincent, J. Appl. Polymer Sci., 9, 1873 (1965).

8. Office of Saline Water Research and Development Progress Report No. 556.

9. L. E. Nielsen, Mechanical Properties of Polymers, Reinhold Publ. Co., New York 1962, p. 114.

THE PREPARATION AND PROPERTIES OF STYRENE GRAFTED CELLULOSE ACETATE MEMBRANES FOR DESALINATION

F. Kimura-Yeh, H. B. Hopfenberg, and V. Stannett
Department of Chemical Engineering
North Carolina State University
Raleigh, North Carolina

INTRODUCTION

Membranes formed from secondary cellulose acetate have been the most successful for the reverse osmosis method of desalination. However, the high pressures used for seawater desalination lead to the gradual compaction of the expanded Loeb-Sourirajan type membrane structures with a consequent decrease of the product flux rate and apparent salt rejection. It seemed of interest, therefore, to study modifications of cellulose acetate which would reduce the rate of compressive creep of the modified membrane and, in turn, stabilize the flux-rejection characteristics of the resulting membranes.

The effect of styrene grafting to cellulose acetate was studied as a possible technique for mechanically stabilizing cellulose acetate membranes[1,2]. Styrene was chosen because of its inherently low creep and insensitivity to water. Graft copolymerization is an attractive means for modifying base polymers because grafting frequently results in the superposition of properties relating to backbone and pendant chains. Radiation was chosen as the cleanest and most versatile method of grafting available and because the laboratory has had extensive prior experience in this field. The ideal result from this study would be synthesis of membranes exhibiting the time-dependent mechanical properties of polystyrene and the transport properties of cellulose acetate.

It is important to recognize the considerable array of system parameters which affect the characteristics of a graft copolymer (e.g., percent graft, molecular weight of the pendant side chains) and in turn the ultimate properties of the bulk graft copolymer. For a given base polymer and a given monomer, one can conveniently vary the total radiation dose, the dose rate, and swelling agent type and concentration, the type and concentration of chain transfer agent, and the addition of small quantities of difunctional monomer. It is important, therefore, to recognize that a systematic study of the effect of grafting on ultimate properties must take into account these variables. For instance, grafting in the presence of chain transfer agents can result in the formation of extremely short pendant chains. The properties of short chain grafts are remarkedly different from their long chain analogs. The effect of the molecular weight of the grafted side chains on flux, rejection, and compaction properties was explored in detail. The use of purified graft copolymer as a coupling agent for the blending of the parent homopolymers was also studied for the case of styrene-cellulose acetate.

The work extended into some novel areas of film grafting. The properties resulting from direct grafting to dense cellulose acetate were compared and contrasted with the effects of grafting to asymmetric membranes. The properties of membranes cast from purified graft copolymer have also been evaluated in a similar manner.

The cografting of styrene with the "quaternizable" monomer, 2-vinyl-pyridine, to cellulose acetate was also studied in an attempt to improve the water transport properties by introducing more hydrophilic moieties into the grafted side chains.

LONG CHAIN STYRENE GRAFTING TO CELLULOSE ACETATE

Grafting to Dense Cellulose Acetate Films

Grafting Procedure and Kinetics. Styrene-grafted cellulose acetate dense films were prepared in styrene-pyridine mixtures using the mutual irradiation technique. Irradiation was carried out with γ-rays at room temperature in vacuo. The percent graft as a function of dose is shown in Fig. 1.

The grafting process approaches a limiting value of percent graft with increasing dose for the 90:10 styrene-pyridine monomer solution. Grafting with the 80:20 mixture does not approach a limiting value although the percent graft changes much more slowly with dose at doses above 2 Mrads. Both styrene-pyridine systems require a continuing supply (via diffusive transport) of monomer during the course of the radiation induced grafting. The results suggest quite strongly that the overall kinetics of the styrene grafting is rate determined by diffusion of monomer to the growing

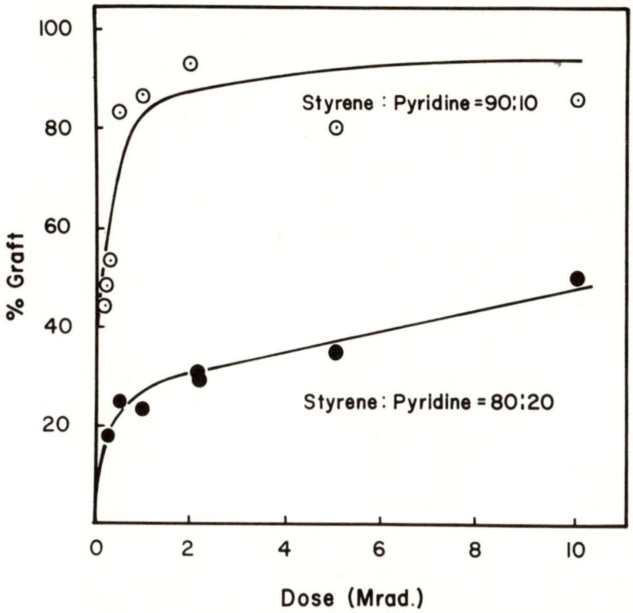

Fig. 1 - The Grafting of Styrene onto Gulf General Atomic Co. Dense Cellulose Acetate Film.

graft chain. Similar results were obtained by many workers; see, e.g., the results of Stannett et al.[3] for the grafting of styrene on wool and the results of Chapiro[4] for the grafting of styrene onto Teflon.

The increased percent graft for the 90:10 system relative to the 80:20 system is a manifestation of low rates of termination in the film matrix swollen with 90:10 monomer solution in comparison with the termination rates likely in the more plasticized "80:20 swollen film." It follows, therefore, that the morphology and ultimately the transport and creep properties of the resulting graft will be a strong function of the ratio of styrene to pyridine in the monomer solution. Undoubtedly, extremely long pendant polystyrene chains will result from grafting with 90:10 styrene-pyridine.

The weight increases are shown as a function of dose in Fig. 2 for two dose rates. The percent graft increased with total dose during the initial stage of irradiation and then saturated at a certain dose. The maximum weight increases were 90 and 30% at dose rates of 0.16 Mrad/hr and 1.05 Mrad/hr, respectively. At the higher dose rate radicals may be produced too rapidly to be "fed" effectively by the slow diffusing monomer. The weight

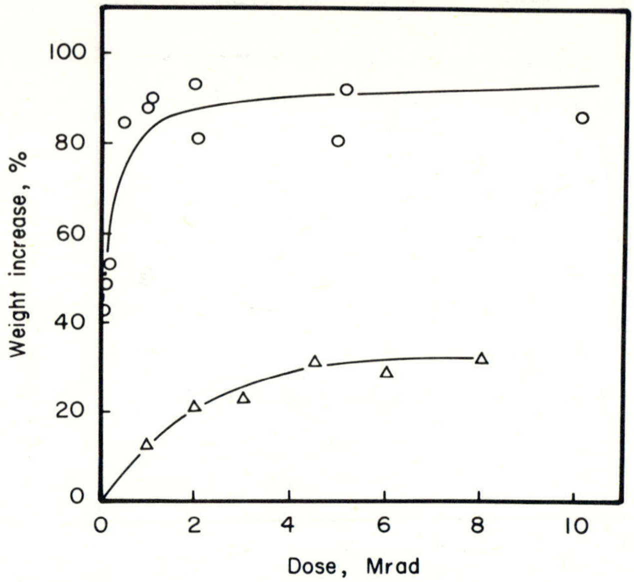

Fig. 2 - Percent Graft as a Function of Total Dose for Styrene Grafting to Dense Cellulose Acetate Films. (o) Dose rate: 0.16 Mrad/hr; (Δ) Dose rate: 1.05 Mrad/hr.

increase also decreased with increasing thickness of film. The dose rate effect and the thickness effect are consistent with a diffusion controlled system involving surface grafting rather exclusively.

Some additional experiments were performed to help explain more fully the diffusion controlled nature of the grafting system. The effect of film treatment and "regrafting" on the weight increase of grafted films obtained at a dose before the weight increase was saturated are shown in Table 1. When the irradiated glass tube, sealed in vacuo, was left at room temperature for 3 days after irradiation, post-polymerization occurred and the weight increased considerably. The weight actually increased further after the above sample was again irradiated. These values obtained after treatment and regrafting agree with the maximum weight increases observed in this grafting system, well within the experimental error. An additional 10% grafting, however, was obtained by a combination of post-effect and re-irradiating. This is quite consistent with, for example, the effects of lower dose rates.

The same treatments were carried out using the grafted films which were obtained at a dose after the weight increase was satu-

TABLE 1

EFFECTS OF TREATMENT AND REGRAFTING ON OVERALL WEIGHT INCREASE

Sample No.	Overall Dose (Mrad)	Treatment After Grafting	Dose for Reirradiation (Mrad)	Over-all Weight Increase (%)
1	1.0	None	-	60.8
2	3.5	None	-	86.9
3	1.0	Irradiated ampoule left for 3 days at room temperature	-	89.0
4	1.0	Irradiated ampoule left for 3 days at room temperature	2.5	99.0

Composition of monomer solution: styrene:pyridine = 90:10
Dose rate: 0.16 Mrad/hr

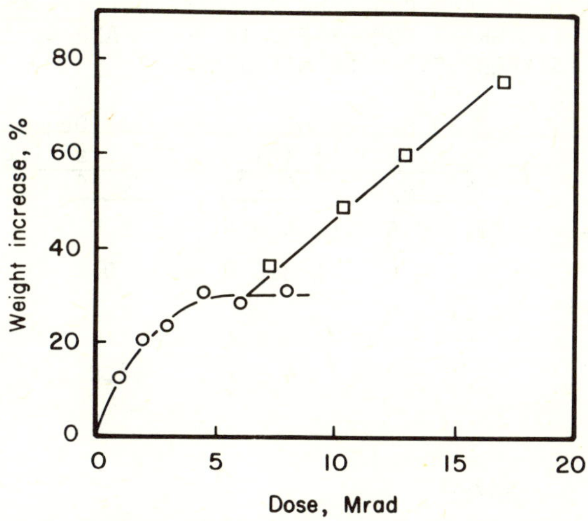

Fig. 3 - The Effect of Recasting on the Grafting Behavior of Styrene-Cellulose Acetate Graft Copolymer Films. (o) Initial Grafting Behavior; (□) Regrafting Consequent to Dissolution and Recasting.

rated. The weight increases were essentially the same before and after treatment and regrafting.

Regrafting of styrene to recast film, on the other hand, occurred smoothly and the weight increased linearly with dose, as can be seen in Fig. 3.

The degrees of swelling of cellulose acetate film, grafted film, and recast film with styrene:pyridine (90:10 and 80:20) mixtures at room temperature were measured. The results are shown in Table 2. The grafted films which were obtained after the weight increase was saturated were only slightly swollen with the monomer solutions. However, the degree of swelling of the recast film was extremely high compared with that of grafted film. These results can help explain why the regrafting to grafted film in this grafting system did not occur. It is clear that the grafting imposes some restraint on the swelling of the film. That this is not chemical in nature can be seen from the fact that the recast film swells considerably. The reduction in swelling could be caused by the restraining influence of some crystallites in the original film. This would explain both the limited swelling and the limited grafting. After solution and recasting such restraints would disappear, resulting in increased swelling and unrestricted grafting.

TABLE 2

DEGREE OF SWELLING AND GRAFTING OF CELLULOSE ACETATE FILMS AND THEIR GRAFT COPOLYMERS IN 90:10 AND 80:20 STYRENE:PYRIDINE MIXTURES

Sample	Degree of Swelling (%)		Degree of Grafting (%)*	
	90:10	80:20	90:10	80:20
Cellulose acetate	18.5±0.5	52.4±0.4	25.0	45.2
Styrene (28.1%) grafted cellulose acetate	1.2±0.2	4.3±0.8	∿0	∿0
Recast film from 28.1% grafted film	39.6	111.6	68.0	-

*7 Megarads at 1.0 Mrad/hr

Apparently with very low dose rates there is sufficient time for the slow solution or rearrangement of the crystallites to take place since higher degrees of grafting are obtained in spite of the same initial swelling. Alternatively, the grafting could be highly asymmetric through the thickness of the film and this, coupled with the considerations outlined above, would also explain the results. This explanation is attractive in that the low modulus ungrafted core could be compressed by the pressure in reverse osmosis testing, causing the rapid loss of flux which actually was found to occur.

Sorption and Diffusion of Water Vapor in Cellulose Acetate and Styrene-Grafted Cellulose Acetate. The data describing the effects of styrene grafting on the equilibrium sorption of water and diffusion of water in cellulose acetate are presented graphically in Fig. 4. Unsteady-state, water-vapor sorption experi-

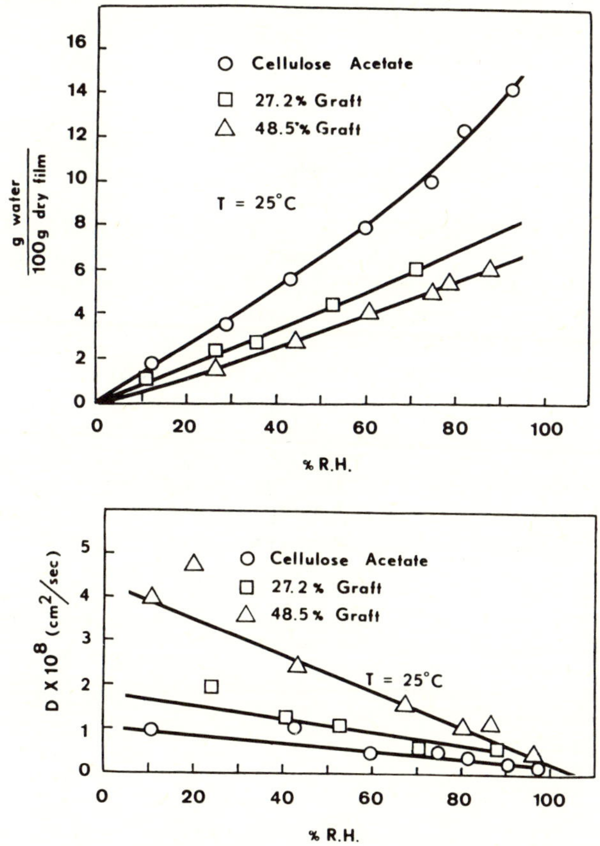

Fig. 4 - Equilibrium Water Sorption and Diffusion Coefficients for Water as a Function of Relative Humidity in Cellulose Acetate and Cellulose Acetate-Styrene Graft Copolymers ments revealed that the equilibrium water content was decreased as a consequence of grafting, and water sorption, in all cases, deviated positively from Henry's law. Desorption studies made it clear that hysteresis is found both in the cellulose acetate and graft copolymers. The hysteresis is commonly observed with cellulosic materials and has been discussed in detail by Stamm[5] and by Urquhart and Eckersoll[6]. The water sorptions at equilibrium have also been calculated on the basis of the cellulose acetate content and these values at 50% relative humidity are given in Table 3. It can be seen that except for one case, there is about a 13% reduction in water content due to grafting even when based on the cellulose acetate content only. The marked hysteresis of the sorption isotherms suggests some subtle disordering of hydrogen bonds consequent to sorption in cellulose acetate, rather than the more dramatic disturbance of crystalline order as suggested by Kesting and Stannett[7] for the system styrene-grafted cotton. The grafting process itself also must lead to some disordering of the hydrogen-bonded areas. This disordering will be manifested by increased sorption.

Opposing this effect is the reduction in swelling of the graft copolymers by the presence of domains of polystyrene. This latter effect apparently overcomes the dissipation of the small amount of order except at lower degree of grafting under high swelling conditions. In any case, comparatively small reductions in the water sorption are brought about by the styrene grafting process amounting to a maximum of about 45% in the series studied. The water diffusivity, as measured by the rate of sorption, was increased as percent graft was increased and was concentration dependent, decreasing with increasing water concentration. These

TABLE 3

THE EFFECT OF STYRENE GRAFTING ON WATER SORPTION IN CELLULOSE ACETATE FILM (T = 25°C, RELATIVE HUMIDITY = 50%)

Graft (%)	Monomer Solution Used In Grafting	Water Based on Total Weight of Dry Film (%)	Water Based on Cellulose Acetate Content in Dry Film (%)
0	-	4.5	4.5
27.2	90:10	3.0	3.8
44.0	90:10	2.6	3.8
48.5	90:10	2.4	3.6
25.7	80:20	3.8	4.8
36.7	80:20	2.8	3.8

TABLE 4

THE EFFECT OF STYRENE GRAFTING ON THE MEMBRANE CONSTANT FOR WATER TRANSPORT IN GRAFTED CELLULOSE ACETATE FILM (T = 25°C, RELATIVE HUMIDITY = 50%)

Graft (%)	Monomer Solution Used for Grafting Styrene: Pyridine	Water Diffusivity $D_1 x 10^8$ (cm^2/sec)	Equilibrium Water Content C_1 (g/g)	Film Density (g/cc)	C_1' (g/cc)	Membrane Constant $D_1 C_1' x 10^{10}$ (g/cm-sec)
0	-	2.5	0.044	1.30	0.0572	14.3
27.2	90:10	1.1	0.020	1.25	0.0363	3.99
44.0	90:10	1.5	0.026	1.23	0.0320	4.80
48.5	90:10	2.8	0.024	1.22	0.0293	7.20
25.7	80:20	2.2	0.038	1.25	0.0473	10.4
36.7	80:20	8.3	0.026	1.10	0.0286	23.7

data are consistent with a model involving aggregation of the pendant styrene chains into domains that dissolve virtually none of the sorbed water. The domains do, however, sufficiently disrupt the cellulose acetate matrix to increase the water diffusion coefficient as the percent graft is increased.

The water diffusivities at 25°C at a relative humidity of 50% are shown in Table 4. The actual values of the diffusion constants of the grafts are highly dependent on the conditions of grafting used and the amounts grafted.

The effect of grafting on the water diffusivity in the graft prepared with the 80:20 monomer mixture was much more dramatic than the similar effect resulting from grafting with the 90:10 mixture. Swelling increases as the pyridine content of the monomer solution increases; therefore, it is possible that grafting occurring in the presence of the 80:20 (styrene-pyridine) solution effectively locks the membrane in a more expanded configuration than similar graft contents obtained with 90:10 monomer solution. This expanded structure leads to larger diffusion constants at similar levels of grafting as shown by the data in Table 4.

The overall effect of grafting on the water transport properties may be estimated by comparing the "membrane constants" defined as the product of water diffusion coefficient and the water solubility. The membrane constant has the dimensions of g/cm-sec. The membrane constants at 25°C at a relative humidity of 50% are compared in Table 4.

The membrane constant increases with styrene content for grafts synthesized with 90:10 (styrene-pyridine) mixtures. Since the water sorption is decreased with grafting, this is a manifestation of the increased water diffusivity. The membrane constant for the graft synthesized with the 80:20 mixture is much higher in comparison with the 90:10 series suggesting, again, that grafts synthesized from the 80:20 mixtures contain more free volume. With 36.7% grafting, for example, the membrane constant is much larger than for even the conditioned unheated cellulose acetate film. The most important result of these water-transport studies is that although grafting markedly retards tensile creep, it apparently does not greatly reduce the specific water permeability of the resulting copolymer; in fact, it can lead to actually increased water permeability.

Earlier work by Stannett et al.[8] with films cast from pure graft copolymers of cellulose acetate and styrene showed the diffusion constants to increase somewhat and the solubilities to decrease with grafting. With those films, however, the water vapor permeability decreased somewhat with grafting in contrast to some of the above results. This coupled with the difference between the 80:20 and 90:10 series of grafted films dramatize the effect of the structure of the grafted films and their method of formation on resultant properties.

<u>Other Water Transport Properties of the Grafted Films</u>. Water transport properties of the film recast from styrene graft were compared with styrene grafted films. The styrene grafted film was prepared in a styrene:pyridine = 90:10 mixture at a dose after the grafting yield was saturated. Recast film was cast from this graft.

The reverse osmosis properties for 5,000 ppm salt solution at 800 psi are shown in Table 5. The flux for the film grafted into the plateau region of the grafting-dose curve showed essentially zero flux, even after extraction of a few percent of homopolymer with cyclohexane. This remarkable result is fully in keeping with the low swelling of such films in the styrene-pyridine grafting mixtures presented in Table 2. On dissolving and recasting these films, however, essentially normal fluxes were obtained. They were not enhanced, however, by extensive extraction and removing small quantities of homopolystyrene from the film.

The actual water sorption figures at 50% relative humidity are listed in Table 5. It is interesting that these values based on cellulose acetate only are all similar to cellulose acetate itself except for the unextracted original grafted film. This again shows the remarkable effect of direct grafting in "filling" or otherwise changing the swelling capacity of the film.

TABLE 5

WATER SORPTION[a], MEMBRANE CONSTANTS FOR WATER TRANSPORT, AND REVERSE OSMOSIS PROPERTIES[b] OF STYRENE-GRAFTED CELLULOSE ACETATE FILMS

Membrane Description	% Graft	Water Based on Weight of Dry Film (%)	Water Based on Cellulose Acetate (%)	Water Diffusivity $D_1 \times 10^8$ (cm^2/sec)	Membrane Constant $D_1 C_1' \times 10^{10}$ (g/cm-sec)	Flux (gfd-mil)	Density[d] (g/cm^3)
Cellulose Acetate	0	4.5	4.5	2.5	14.3	0.035	1.285
Styrene-Grafted Cellulose Acetate without Extraction	43.2	2.3	3.3	3.4	9.2	0.006	1.223 (1.214)
Extracted with Cyclohexane	40.2	3.4	4.8	2.15	9.0	∿0	1.227 (1.218)
Recast Film	40.2	3.1	4.3	1.8	6.5	0.017	1.202 (1.218)
Recast Film Extracted with Cyclohexane	28.2[c]	3.9	4.8	1.6	7.3	0.016	1.214 (1.224)

[a]Water sorption and membrane constant, at 25°C and R.H. 50%.

[b]Reverse osmosis properties were measured at 5,000 ppm NaCl and 800 psi.

[c]Weight loss by extraction, 8.6%; % graft calculated using this value, 28.2%.

[d]Calculated densities are in parentheses.

The integral diffusion constants are in excellent agreement with those previously reported[8] and decrease somewhat with increasing humidity also in agreement with previous data[1]. It is curious that the diffusion constants for water are higher in the unextracted original grafted films than in cellulose acetate itself. The membrane constants for the recast films are about one-half that of cellulose acetate, in good agreement with the reverse osmosis flux values* of Table 5. The original films, however, have somewhat higher membrane constants than the recast films which is not in keeping with the very low flux values found with these films as can be seen in Table 5. Apparently at the high water contents and pressures used in reverse osmosis the non-recast films show highly restricted water transport.

In conclusion it can be said that these experiments fully confirm the remarkable barrier properties of the original "plateau-grafted" films. However, their reverse osmosis fluxes are correspondingly very low.

The densities of the grafted films are also included in Table 5. The density decreased with grafting of styrene and the densities of grafted films were found to be somewhat less than the calculated values using the composition of grafts. The density of recast film, however, was less than that of grafted film as anticipated. This is presumably due to the prevention of crystallization and the reduction in packing of the cellulose acetate by grafted polystyrene. The un-recast films, on the other hand, were somewhat closer packed, i.e., had reduced free volumes compared with the recast films. This was reflected in the densities being higher than calculated and by the water transport properties.

Tensile Creep Behavior of Styrene-Grafted Cellulose Acetate. Tensile creep data are presented in Fig. 5. Clearly styrene grafting retards creep, and the retardation is increased with increasing grafting. At low percent graft (i.e., 3%) the shape of the creep-log time curve is markedly different from the nearly linear behavior observed for ungrafted cellulose acetate and the grafts with styrene contents in excess of 20%.

Apparently little is gained in retardation of creep by grafting in excess of ca. 30%. Conversely, low levels of grafting merely give rise to an induction period followed by somewhat accelerated creep (in comparison with ungrafted cellulose acetate). This acceleration may be a manifestation of the destruction of the micro-domains of polystyrene-rich regions which serve as

*These fluxes are expressed in gal/ft^2-day (gfd), normalized to a film thickness of 1 mil. Divide gfd by 2.12x10^4 to convert to cm^3/cm^2-sec.

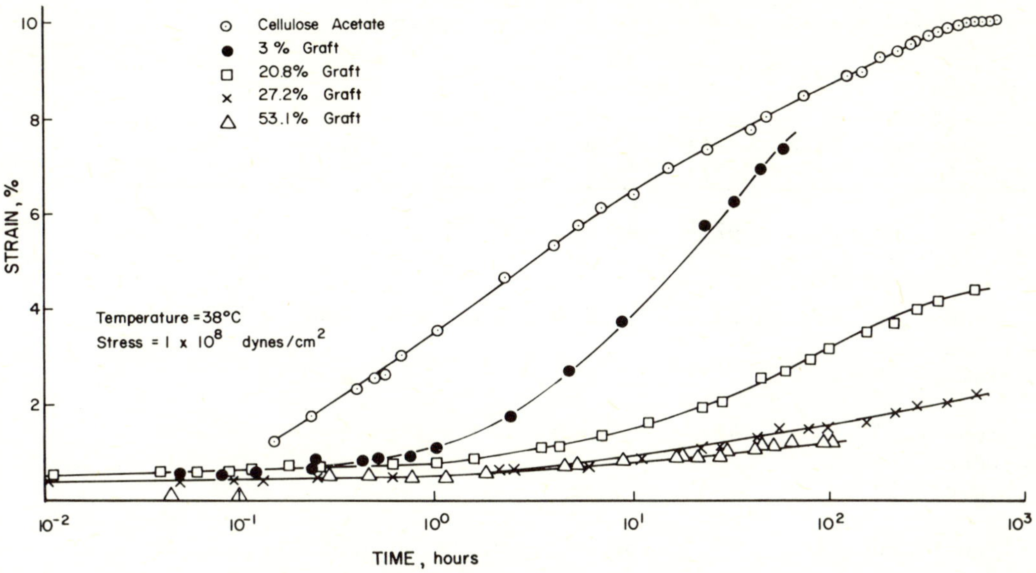

Fig. 5 - Time-Dependent Tensile Creep of Wet Styrene-Cellulose Acetate Grafts Prepared from 90 Styrene:10 Pyridine Solution.

secondary valence crosslinks. The presence of domain morphology in grafts has been suggested by several independent sources[8,9]. Presumably, at high levels of grafting the domains are more stable and the creep is retarded over a much longer time scale.

These tensile creep data were interpreted to suggest that styrene grafting would be an effective tool for improving the compaction resistance of cellulose acetate.

Preparation and Properties of Purified Graft Copolymers of Styrene and Cellulose Acetate[2]

Pure graft copolymer was separated from the graft by alternative elution with an acetone-water (70:30) mixture and benzene. The grafting conditions and molecular weight of the side chain are shown in Table 6.

Dense membranes cast from purified graft copolymer were subjected to tensile creep tests as well as reverse osmosis tests. The tensile creep behavior was correlated with percent graft as well as molecular weight of the pendant polystyrene chains. These data are presented graphically in Figs. 6 and 7.

TABLE 6

CHARACTERIZATION AND PROPERTIES OF PURIFIED STYRENE-CELLULOSE ACETATE GRAFT DENSE MEMBRANES

Volume Composition of Grafting Solution		Dose (Mrad)	Graft (%)	Molecular wt. of Side Chain	Compaction Slope*	Flux* (gfd-mil)	Salt Rej. (%)
Styrene Pyridine	80 20	10	92.4	183,000	-	0.01	-
Styrene Toluene Pyridine	40 40 20	10	63.0	162,000	-0.032	0.060	98.2
Styrene Toluene Pyridine	20 60 20	10	44.7	138,000	-	-	-
Styrene Toluene Pyridine CCl_4	20 60 20 1	10	47.5	66,000	-0.012	0.0071	98.5
Styrene Toluene Pyridine CCl_4	20 60 20 2.0	10	74.0	55,000	-0.030	0.066	90.0
Styrene Toluene Pyridine CCl_4	40 40 20 1.5	11.6	113.6	100,000	-0.060	0.052	93.5
Pure cellulose acetate dense film			-	-	-0.063	0.079	96

*Reverse osmosis test conditions: ΔP = 1500 psi, feed concentration = 35,000 PPM NaCl, T = 25°C.

The relationship between tensile creep rate and percent graft for these pure grafts followed the same pattern as did the creep behavior of grafted dense membranes. Creep rate dropped monotonically with percent graft. The data of Fig. 7 illustrate

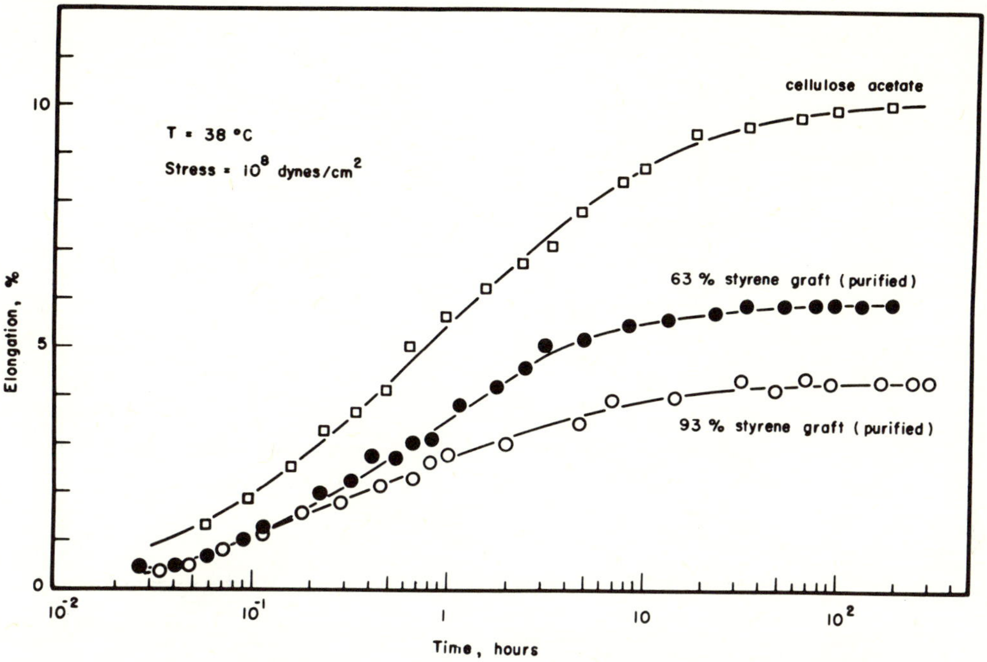

Fig. 6 - The Effect of Percent Graft on the Tensile Creep Behavior of Purified Styrene-Cellulose Acetate Graft Copolymer Films.

the effect of pendant chain length on the tensile creep behavior of the purified grafts. The shorter chain length graft (molecular weight, 66,000) was more effective in reducing the rate of tensile creep than the graft characterized by a longer polystyrene chain, molecular weight, 138,000. The molecular weight of the polystyrene chains was determined viscometrically consequent to hydrolysis of the cellulosic backbone.

The preparative conditions and the transport properties of dense membranes prepared from purified graft copolymer are summarized in Table 6. Solution casting of polymer membranes results in some anisotropy due to stresses which are created during the unidirectional drying of the membranes. The term "dense membrane" is used to distinguish the structure from that of a Loeb-Sourirajan type membrane whose structure is highly anisotropic. The low level of anisotropy in these "dense" membranes is apparently sufficient to give rise to membrane compaction manifested by quite measurable flux decline.

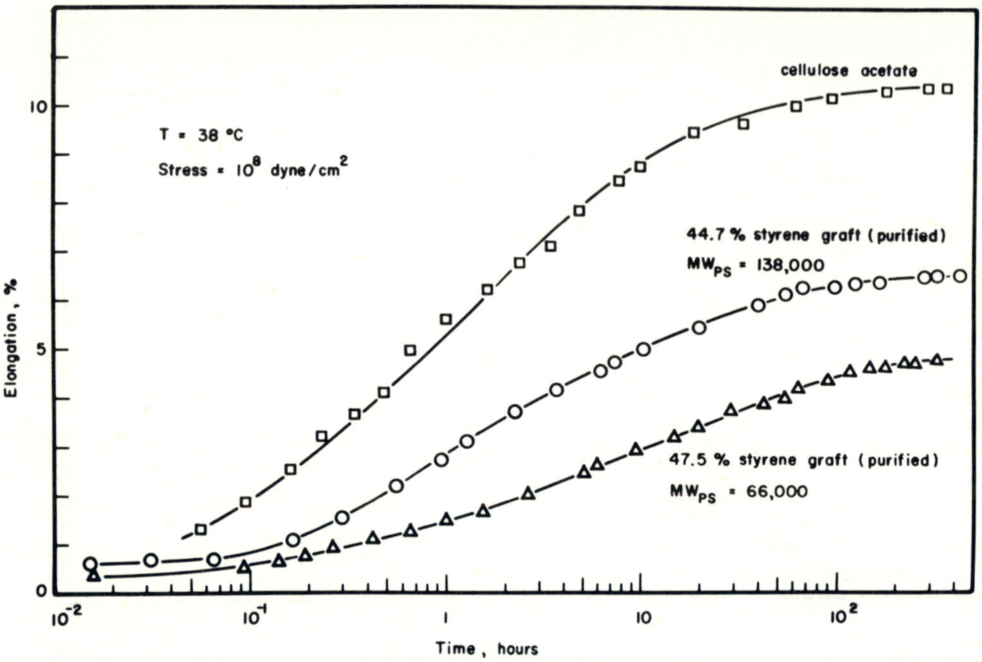

Fig. 7 - The Effect of Pendant Chain Length on The Tensile Creep Behavior of Purified Styrene-Cellulose Acetate Graft Copolymer Films.

All grafted membranes studied in this series exhibited salt rejection in excess of 90%. All grafted membranes exhibited improved compaction resistance in comparison with the cellulose acetate control. The product flux was generally between 70 and 85% of the flux observed through the ungrafted cellulose acetate control.

The flux decline data (compaction slopes), defined as the slope of a log-log plot of product flux vs. time, presented in Table 6 are consistent with the tensile creep data. Shorter chain length grafts appear more effective in reducing flux decline than grafts of longer chain length. It has been similarly observed that shorter chain length grafts creep more slowly than their longer chain analogs.

Preparation and Properties of Polyblend Membranes

One area of the styrene grafting program was the preparation of dense membranes of blends of cellulose acetate and polystyrene using their graft copolymers as "compatibilizing" agents. It had been shown in previous studies that clear films could be obtained in this way provided that the molecular weights of both the side chains and backbone were sufficiently large[8].

A number of purified graft copolymers were used in preparing these polyblend membranes; see, for example, the caption to Fig. 8. In general the films were tough and slightly opalescent. The tensile creep behavior of three typical polyblend and two pure graft copolymer membranes of similar overall composition is shown in Fig. 8. It can be seen that the long time tensile creep rate of the blends is quite comparable to that of pure graft copolymers of similar composition and greatly reduced from that of pure cellulose acetate. The reverse osmosis properties of the blends in general showed higher flux and lower salt rejection than the corresponding pure grafts. This reflects, probably, the

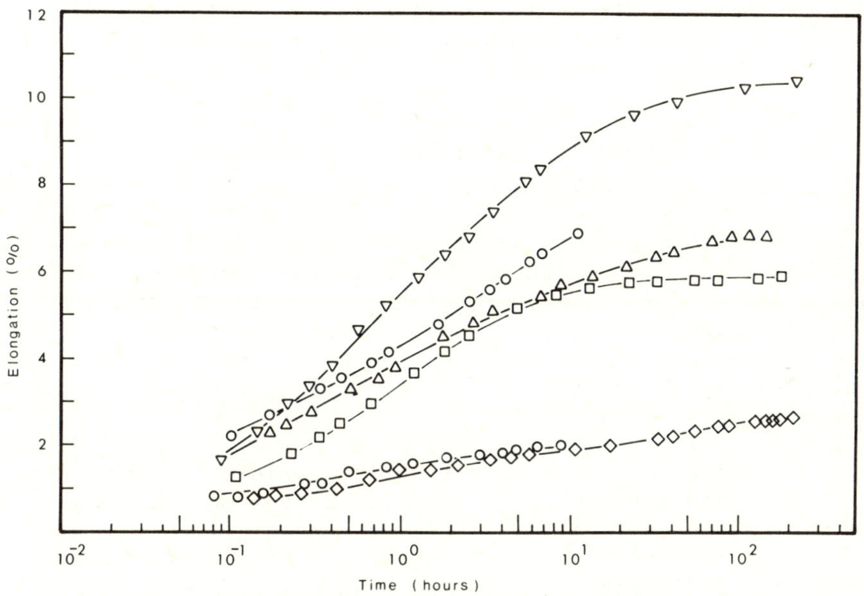

Fig. 8 - Tensile Creep Properties of Blend Membranes and Comparable Pure Grafts (38°C, 10^8 Dynes/cm^2).
▽ Pure CA - 100% CA
○ CA: PS: 63% Graft 2:1:2 - 64% CA
△ CA: PS: 113% Graft 2:1:2 - 59.0% CA
□ Pure Graft 63% - 61.3% CA
○ CA: PS: 113% Graft 2:2:1 - 49.4% CA
◇ Pure Graft 113% - 47.0% CA

slight heterogeneity of the blends as manifested by their opalescence. It is believed however that comparable properties to the pure graft could be obtained with the blends by judicious choice of components. Since these would only need, for example, about 20% addition of pure graft to virgin cellulose acetate, considerable practical and economic advantages could be achieved if such "grafted" membranes become of technical significance.

Direct Grafting to Asymmetric Cellulose Acetate Membranes

The grafting of styrene to asymmetric membranes was carried out in an aqueous system in the presence of sodium lauryl sulfate (SLS) in order to prepare grafted membranes without damage to the sensitive dense layer of the membrane. The degree of grafting was estimated by cold alkaline hydrolysis assuming this was conducted to the complete saponification of the cellulose acetate. The results are presented in Table 7. In every case much lower fluxes were found than with the original membranes but with significant salt rejection. It is interesting that grafting can be accomplished to asymmetric membranes without damage but with increased densification of the film.

PREPARATION AND PROPERTIES OF SHORT CHAIN STYRENE-GRAFTED CELLULOSE ACETATE MEMBRANES

In an attempt to prepare grafts with much shorter side chains, preparations were made with an abnormally high ratio of the chain transfer agent (carbon tetrachloride) to monomer[10]. The grafting recipes used are included in the figurehead of Fig. 9. The response of percent graft to total dose is presented graphically in Fig. 9 for the recipes including excess chain transfer agent.

The system prepared with more than 15% pyridine in the liquid phase did not exhibit the usually observed diffusion controlled behavior manifested by leveling in the percent graft-versus-dose curves. Presumably the higher concentrations of swelling agent (pyridine) are sufficient to accommodate an adequate monomer supply to the active growth sites and diffusion does not limit the grafting kinetics.

It was shown recently[11] that when the grafting proceeded at a constant or an accelerating rate, e.g., with butadiene grafted to polyoxymethylene, permeabilities of the resulting graft were higher than the permeability of the base polymer. It would appear that the grafting reaction must proceed isotropically in the film with swelling to obtain good fluxes.

TABLE 7

GRAFTING TO ASYMMETRIC MEMBRANES IN AQUEOUS SOLUTION UNDER NITROGEN GAS

Heat Treatment	Monomer	Concentration of Monomer (%)	Additive	Dose Rate (Mrad/hr)	Dose (Mrad)	Weight Increase[a] (%)	Flux (gfd)	Salt Rejection (%)
-	None	-	-	-	-	-	10.0	-
80°C, 30 mins.	None	-	-	-	-	-	9.5	94.0
-	Styrene	1.5	0.1% SLS	0.12	0.24	43.5	1.5	-
-	Styrene	1.5	0.1% SLS	0.12	0.24	91.9	0.43	96.4
-	Styrene	1.5	0.1% SLS	0.12	0.24	64.3	1.83	85.6
-	Styrene	1.5	0.1% SLS	0.12	0.24	87.7	1.49	76.0
80°C, 30 mins.	Styrene	1.5	0.1% SLS	0.12	0.24	19.1	2.6	90.1
"	Styrene	1.5	0.1% SLS	0.12	0.24	21.4	1.6	88.4
"	Styrene	1.5	0.1% SLS	0.12	0.36	43.2	1.3	94.4

[a]Estimated from saponification of grafted film.

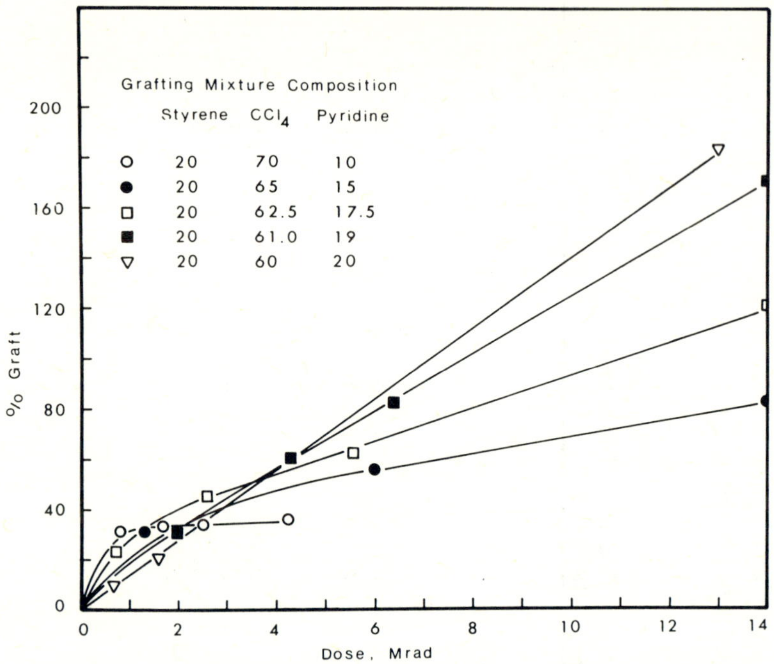

Fig. 9 - Grafting Yield as a Function of Dose for Styrene Grafting to CA in the Presence of CCl_4 and Pyridine.

The pendant chains formed in the presence of swamping excesses of chain transfer agent could, in principle, be too short to associate and form discrete regions or domains rich in polystyrene. Intrinsic viscosity measurements indicated that the molecular weight of the side chains is ca. 4,000 which is nearly a factor of 100 less than the molecular weight of the side chains in the long chain grafts. It is clear that the primary result of the grafting would be the formation of many short chains pendant from the cellulose acetate backbone. Property modifications should accrue from this internal plasticization or destructuring of the cellulose acetate.

The water sorptive capacity of the grafted membrane was found to be slightly higher (on the basis of cellulose acetate) than the sorption equilibria for the virgin cellulose acetate. These results are contrary to most reported conventional grafting studies involving hydrophobic monomers and cellulose acetate. This seeming anomaly is a direct consequence of the very short grafted side chains produced in the presence of excess chain transfer agent (CCl_4).

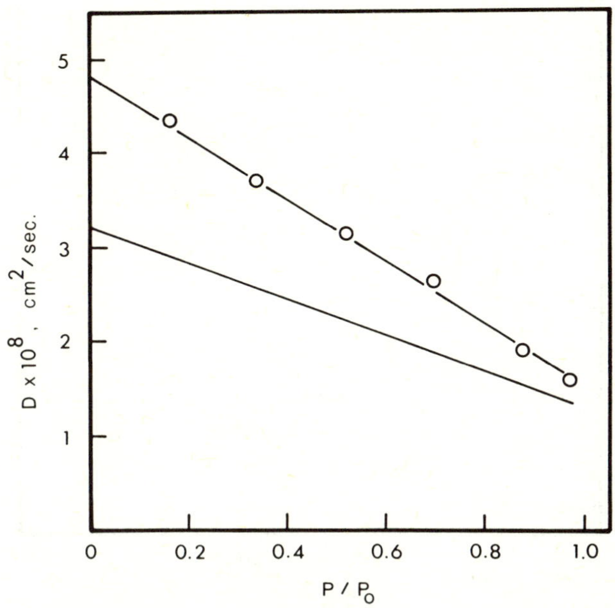

Fig. 10 - Concentration Dependence of Integral Diffusion Coefficients at 25°C for Water Vapor in Short Chain Graft CA. CA (—), and 33.8% Styrene-Grafted CA Prepared in Pyridine:Styrene:CCl_4 20:20:60 Mixture (o) (Short Chain Graft).

Diffusion coefficients for water in the "short-chain" grafted film are presented in Fig. 10. The water diffusivities are distinctly higher in the grafted membrane than in the unmodified cellulose acetate. These data viewed in concert with the grafting kinetics and the sorption equilibria suggest a rather consistent model for the physico-chemical effect of "short-chain" grafting. The predominant effect of grafting is destructuring of the cellulose acetate. These effects are distinctly different than the property changes effected by "long-chain" grafting which tend to form domains which act predominantly as "secondary valence crosslinks."

The long chain domains are rather water insensitive, resulting in decreased water sorption and a structuring rather than destructuring of the surrounding cellulose acetate. This "structuring" or free volume filling results in the observed leveling of percent graft versus total dose. The "structured," low free volume surface essentially prevents diffusion of monomer.

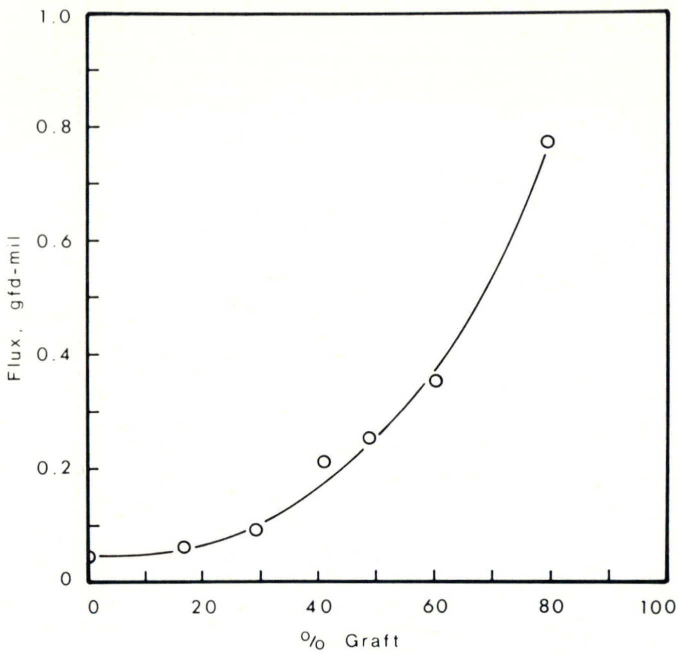

Fig. 11 - Product Flux as a Function of Percent Graft for Unannealed, Dense, Short Chain Graft Membranes.

The reverse osmosis properties of unannealed films are presented in Figs. 11 and 12. The "short-chain" grafting yielded a markedly more open or destructured membrane resulting in a monotonic increase in flux with increasing percent graft and a companion decrease in rejection at high grafting levels. The results indicating an increase in product flux with grafting suggested that swelling of the membrane per se during the grafting procedure may play a significant part in the modification of membrane properties. Annealing at 80°C in water for one hour increased the salt rejection and decreased the product flux.

The excellent salt rejection of virgin cellulose acetate was retained past 40% graft, although the product flux increased significantly for membranes in this same range of percent graft. The tensile creep behavior of these grafts was highly stabilized consequent to grafting although the fluxes were improved and the excellent salt rejection was unaffected. The data of Fig. 13 indicate that the ultimate elongation was less than 3% after 500 hours of creep under a constant load of 10^8 dynes/cm^2 although the cellulose acetate exhibited accelerated creep leading to failure at 150 hours at an ultimate elongation in excess of 10%. It is clear that even the short chain grafts can form water-insensitive domains stabilizing these wet creep performances.

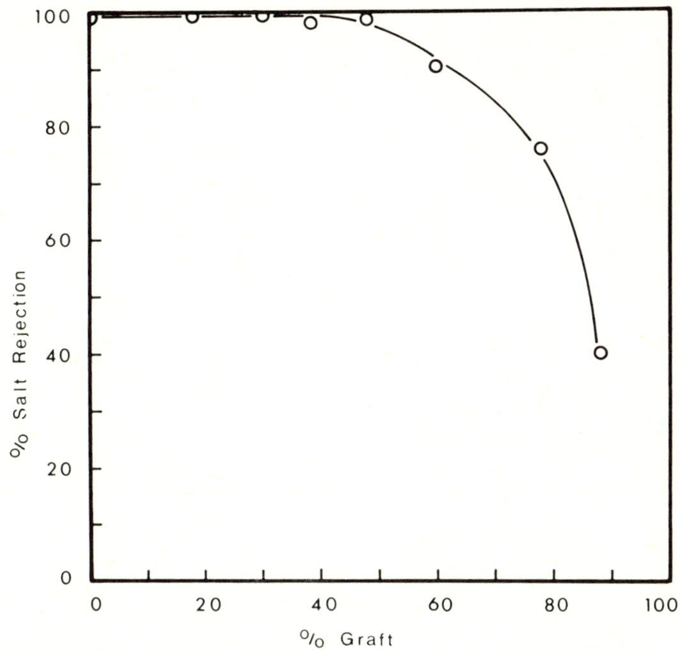

Fig. 12 - Salt Rejection as a Function of Percent Graft for Unannealed, Dense, Short Chain Graft Membranes.

These consistent results relating to short chain grafts are quite encouraging because high fluxes and rejections were coupled with reduced creep. For example, a recast film cast from an acetone solution of a 28.2% graft showed a flux of 0.06 gfd-mil and 98% salt rejection at 5,000 ppm NaCl and 800 psi. These values are similar to those for cellulose acetate itself. However, graft copolymer containing a similar amount of grafted polystyrene prepared in the presence of large excesses of CCl_4 (chain transfer agent) was cast from a 20% solution in pure acetone. The membrane when allowed to dry at room temperature for 15 seconds, submerged in water at 0°C, and subsequently annealled at 80°C for 30 minutes provided a flux of 10.4 gfd and a salt rejection of 93% under similar conditions.

Very recently, it was shown that at least part of the improved flux was due to tenaciously held, occluded pyridine hydrochloride, presumably formed from the hydrochloric acid produced by the radiolysis of the carbon tetrachloride. This lead to some experiments with grafting comonomer mixtures of styrene and 2-vinylpyridine. These are described in the following section.

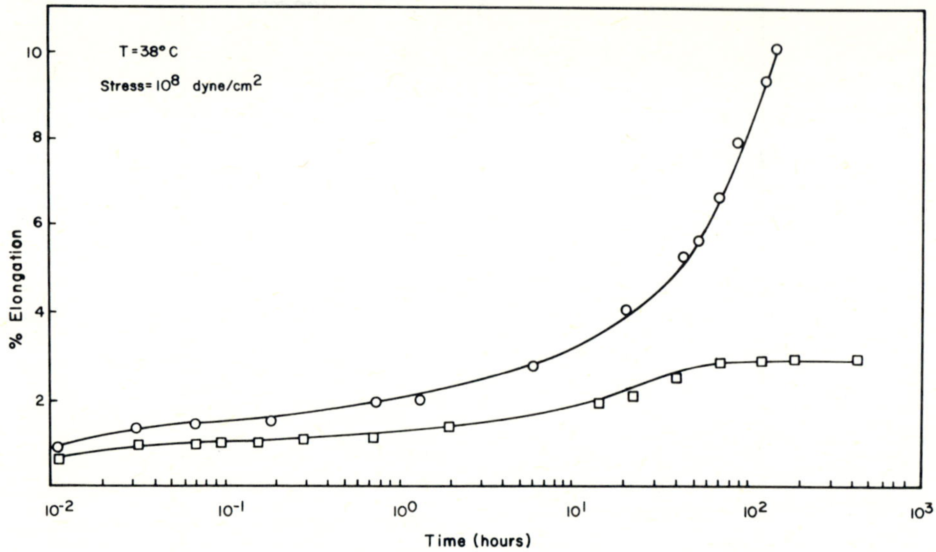

Fig. 13 - Tensile Creep of Styrene Grafted Cellulose Acetate Film Prepared from 20 Styrene, 20 Pyridine, 60 CCl_4 (38.1% Graft) (□) as Compared to Pure Cellulose Acetate Film (o).

COGRAFTING OF STYRENE AND 2-VINYLPYRIDINE TO CELLULOSE ACETATE

It was shown earlier that the grafting of styrene improved the compaction resistance of cellulose acetate. The mutual grafting of 2-vinylpyridine to dense cellulose acetate produced, after quaternizing, grafted films with extremely high normalized fluxes, significant but lower salt rejection, and essentially unaltered compaction properties[12]. For example, a 34% 2-vinylpyridine-grafted cellulose acetate membrane showed a flux of 2.4 gfd-mil, a salt rejection of 80%, and a compaction slope of -0.03 when tested at 1500 psi with 35,000 ppm NaCl. The results of the independent styrene grafting and 2-vinylpyridine grafting studies plus the apparently beneficial effect of occluded pyridine hydrochloride suggested that the cografting of these dissimilar monomers might result in synergistic effects.

The effect of 2-vinylpyridine concentration in the monomer mixture of styrene and 2-vinylpyridine on the grafting and reverse osmosis properties of cografted membranes is shown in Table 8. The data indicate that 2-vinylpyridine retards the overall rate of grafting. Conversely, the transport kinetics of the resulting grafted and quaternized membranes indicate that the

TABLE 8

REVERSE OSMOSIS PROPERTIES OF STYRENE 2-VINYLPYRIDINE-GRAFTED CA AT 800 PSI WITH 5,000 PPM NaCl

Composition of Monomer Solution	Dose (Mrad)	Weight Increase (%)	Quaternized	Flux (gfd-mil)	Salt Rej. (%)
Styrene: 2-Vinylpyridine: Trichloroethylene	7.5	42.2	CH_3Br	0.16	98.4
= 15:5:80	7.5	42.0	HCl	0.08	94.4
10:10:80	7.5	33.4	CH_3Br	0.36	98.6
"	7.5	36.5	HCl	0.36	98.6
"	12.0	49.6	-	0.016	-
"	12.0	52.9	-	-	-
"	12.0	49.4	CH_3Br	0.30	87.0
"	12.0	51.1	HCl	0.49	87.0
"	12.0	51.1	HCl	0.43[a]	87.6[a]
"	2.2	27.5	CH_3Br	0.24[b]	99.4[b]
"	6.0	42.8	CH_3Br	0.40[b]	99.0[b]
2-Vinylpyridine: Trichloroethylene = 20:80	3	19.8	CH_3Br	0.20	98.0
"	7.2	30.9	CH_3Br	0.49	90.0
"	13.5	41.8	CH_3Br	-	-

[a]Dried and retested.

[b]Cellulose acetate, Eastman Chemical Co. E-398-10.

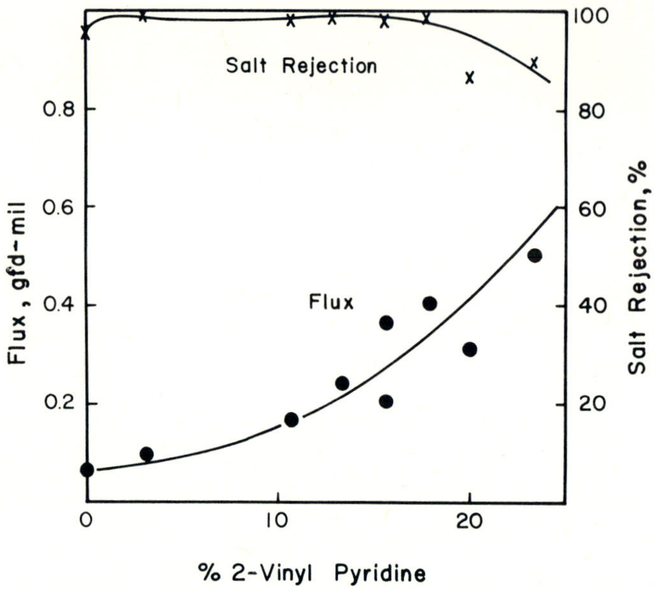

Fig. 14 - Reverse Osmosis Properties of Styrene-2-Vinylpyridine-Grafted Cellulose Acetate Membrane as a Function of 2-Vinylpyridine Content. (2-vinylpyridine content calculated from data of Table 8 according to the calculational procedure of Burlant and Green[13].)

2-vinylpyridine concentration in the monomer solution markedly increases the flux of the cografted membranes. The flux and salt rejection of cografted and quaternized membranes are plotted versus the calculated values of the 2-vinylpyridine content in the membranes in Fig. 14. The flux increased monotonically with the quaternized 2-vinylpyridine content; the salt rejection, however, was greater than 98% with up to 20% 2-vinylpyridine in the membrane. Quaternization, although increasing the flux by more than an order of magnitude, did not significantly reduce the salt rejection. It is apparent, therefore, that the rather novel technique of cografting of hydrophobic monomers and hydrophilic amine type monomers coupled with quaternization affords the possibility of producing a membrane composition with the excellent salt rejection of cellulose acetate but with fluxes improved by an order of magnitude.

Similar studies, still in progress, were made with copolymer grafts of methyl methacrylate and 2-vinylpyridine. The results were found to be very similar to the styrene graft although in general the methyl methacrylate series gave superior fluxes and salt rejection compared with styrene.

ACKNOWLEDGMENT

We would like to express our thanks to our coworkers, Dr. P. T. Rigney, Mr. D. S. Wang, Mr. J. H. Hyun, and Mr. J. Bentvelzen, and to the Office of Saline Water, U.S. Department of the Interior, for their generous support of this work.

REFERENCES

1. H. B. Hopfenberg, F. Kimura, P. T. Rigney and V. Stannett, J. Polymer Sci., Part C, No. 28, 243 (1969).

2. H. B. Hopfenberg, V. Stannett, F. Kimura and P. T. Rigney, J. Appl. Poly. Sci., Applied Polymer Symposia, No. 13, 139 (1970).

3. V. Stannett, K. Araki, J. A. Gervasi and S. W. McLeskey, J. Polymer Sci., A3, 3763 (1965).

4. A. Chapiro, Radiation Chemistry of Polymer Systems, Interscience, New York, (1962), pp. 676-680.

5. A. J. Stamm, Wood and Cellulose Science, Ronald Press, New York, 1964.

6. A. R. Urquhart and N. Eckersoll, J. Text. Inst., 23, T163 (1932).

7. R. E. Kesting and V. Stannett, Makromol. Chem., 65, 248 (1963).

8. J. D. Wellons, J. L. Williams, and V. Stannett, J. Poly. Sci. A-1, 5, 1341 (1967).

9. I. Yashitake, J. Soc. Text. Cell. Ind. (Japan), 19, 7 (1968).

10. H. B. Hopfenberg, F. Kimura-Yeh, J. M. Bentvelzen and V. Stannett, to be published.

11. J. L. Williams and V. Stannett, J. Applied Polymer Sci., 14 1949 (1970).

12. H. B. Hopfenberg, F. Kimura-Yeh, and V. Stannett, to be published.

13. W. J. Burlant and D. H. Green, J. Polymer Sci. 31, 227 (1958).

CONTINUOUS CASTING OF ASYMMETRIC TUBULAR REVERSE OSMOSIS MEMBRANES

John L. Richardson, Gilbert Segovia, Wilfred H. Bachle, and H. Andre Parker-Jones

Liquid Process Products
Philco-Ford Corporation
Newport Beach, California 92663

INTRODUCTION

The advantages of tubular reverse osmosis module configurations for desalination primarily result from the large (typically 0.2 to 1 inch in diameter) characteristic fluid flow cross section and its uniformity in the flow direction. These inherent characteristics of the tubular module yield, with proper design, advantages of minimal fouling tendency, higher fractional recovery operation, and ease of physical in situ cleaning of the membrane surface in comparison with the performance of module geometries characterized by much smaller fluid flow cross sections and flow paths that are both exceedingly tortuous and contain many flow regions of stagnation and relative quiescence. From a cost-to-manufacture standpoint, full exploitation of the tubular module's operational advantages will accrue as the cost for casting uniform, high performance tubular membrane is reduced to a level as close to the cost of polymer itself as is possible. Since, in the case of cellulose acetate, approximately 1/64 pound of polymer is required to produce one square foot of membrane 8 mils thick[1], then the cost of polymer is slightly greater than 1 cent per square foot of membrane or less than 1/10 of a cent per gallon/day of permeate flow capacity. The differential between this base cost, to which must be added the comparable cost of the other casting solution ingredients, and the cost of tubular membranes cast in the usual batch manner[2] is significant.

Continuous tubular membrane casting offers the potential of minimizing the labor content in the casting process while at the same time maximizing the tendency for uniformity in wall thickness and freedom from imperfections. In view of the lack of pertinent data[3,4], one object of this investigation was to develop a sufficient understanding of those major factors in the casting process which have a significant influence on the permselective properties of the resultant membrane and which are pertinent to a continuous casting process. The ultimate objective was the development of a process for continuous casting which demonstrated the potential, in a small scale configuration, for the direct extrusion of tubular reverse osmosis membranes. For this development, a systematic study was made of the pertinent variables in the casting process using a standard cellulose acetate-solvent-swelling agent system. The practical embodiment of the continuous casting process was designed to produce nominal 1/4 inch tubular membranes in lengths sufficiently long for use in large scale modules. No attempt was made in this investigation to effect scale-up, either in the rate at which casting was carried out or in the diameter of the extruded tube.

EXPERIMENTAL

Due to both the good desalination properties of the tubular membranes obtained from the Manjikian-Loeb formulation[1,5] and the convenience of being able to cast at ambient temperature, this system (cellulose acetate - Eastman* E398-10: 25 percent by weight; formamide - Eastman Practical Grade: 30 percent by weight; and acetone - Baker Reagent Grade: 45 percent by weight) was used in the batch and continuous casting experiments as well as in the viscosity determinations. Care was exercised in the preparation and handling of the well-mixed casting solutions in order to minimize the presence of particulate matter and the loss of solvent.

Precision batch casting of three foot lengths of tubular membrane was carried out in the manner originally described by Loeb[2] but modified so as to increase both the yield and the desalination performance reproducibility of the resulting membranes[6]. The modifications which lead to the most significant improvements in reproducibility were the use of a variable speed motor drive unit for the uniform insertion of the casting tube into the gelation bath (as opposed to a gravity-fall technique) and the use of a gelation bath designed to provide a steady uniform upward flow of water chilled to 0.6 to 1.0°C. By the use of an external refrigeration system for thermal conditioning of the gelation medium and appropriate controls, temperature surveys of the gelation bath during operation showed constancy in temperature to within $\pm$ 0.2°C.

Precision heat treating[6] was also used in order to ensure that the results from one experiment to the next could be compared without uncertainties associated with the thermal annealing process.

*Eastman Chemical Products, Inc., Kingsport, Tenn.

The principal characteristics of the heat treating process which resulted in reproducibility were the use of high flow velocities and insulated lines in order that the longitudinal temperature drop through the test cells being annealed was kept to within 0.1°C. Frequently recalibrated thermocouples whose junctions were located in the flow stream immediately before and after the test cell insured that the heat treatment temperature was well established. In addition, the pressure of the heat treating fluid (de-ionized water) and the time of heat treatment (sufficient to ensure that the "tightness" of the resulting membranes was not dependent upon the duration of the heat treating process[5]) were carefully controlled.

The desalination performance testing of the tubular membranes was carried out in carefully controlled recirculating reverse osmosis test facilities[5] characterized by low pressure pulsations (total less than 5 psi, usually $\pm$ 1 psi), corrosion resistant materials (typically passivated 316 stainless steel and PVC), constancy of flow stream temperature over the full duration of the experiment (typically to within 0.1°C), and the use of both electrolytic conductance and permeate flow rate instrumentation capable of accuracies of $\leq$ 2 percent in the salt reduction, S, and $\leq$ 3 percent in the permeate flux, J (see the Nomenclature section for the definition of symbols). Data from the test facilities was reduced to the water and salt permeation coefficients A and B by taking into account the concentration polarization phenomenon which occurs in the test cell under fully developed turbulent tube flow conditions [5,7] (smooth interior wall assumed). Thus the derived permeation coefficients are independent of the flow rate in the test cell during the experiment. A measure of the reproducibility of the desalination performance data obtained as a result of using these facilities and instrumentation can be obtained by considering the data for membranes numbered 7 to 10 in Table 1. The reproducibility in product flux is shown to be less than 1.5 percent while the reproducibility in salt reduction is less than 5.5 percent. In addition, the absolute value of performance is such that the data fall above the least squares fit to the previously obtained J, S permeation data (Ref. [5], p. 50) or to the right of the previously obtained B, A permeation coefficient data (Ref. [2], p. 51). (See Figs. 1 and 2, below, for this comparison.)

The absolute viscosity and its dependence on shear rate and temperature is required for those polymer solutions which are to be extruded in the continuous casting system in order that the dependence of frictional pressure drop on flow rate may be determined. Polymer solution pressure just upstream of the extrusion annulus can then be used as an operating variable in the continuous casting process. Since no reliable data of this type has been reported for the non-Newtonian cellulose acetate solutions used in the fabrication of asymmetric membranes for reverse osmosis desalination, it was necessary to make the required viscometric measurements of the standard casting solution (cellulose acetate-acetone-formamide). The measurements were made with a Ferranti-Shirley viscometer - a cone and plate instrument. The geometry of the surfaces containing the fluid in the viscometer was such that the shear rate and the shear stress were uniform through the fluid. The methods for reducing the measured data to shear rate, viscosity, and shear stress results have been reported[8,9]. Experimentally, it has been found necessary to keep the casting solution in the viscometer in contact at all times with a saturated acetone atmosphere in order to prevent the rapid loss of acetone from the casting solution during the course of the measurements. If an acetone atmosphere was not maintained in contact with the casting solution, the measurements tended to change with time even though the operating parameters of the viscometer were maintained constant.

RESULTS AND DISCUSSION

Effect of Selected Casting Variables on Tubular Membrane Performance

Casting Surface Characteristics. Tubular membranes were cast in brass tubing, precision bore glass tubing, and 304 and 316 stainless steel tubing with both white pickle (acid cleaned) surface finish and bright anneal surface finish. The results obtained have shown that as long as the casting surface is free of gross defects and has a uniform surface finish of $\leq$ 40 rms μinch, good casting results are obtained. No essential difference was observed in membranes cast in seamless tubing compared to those cast in welded tubing. A bright anneal finished stainless steel tube will usually have a 30 to 50 rms μinch finish on the ID and a 15 to 20 rms μinch finish on the OD. A white pickle finished stainless steel tube will normally have a surface finish greater than 60 rms μinch and is thus unsatisfactory as a casting tube.

The Characteristic Times - τ_E, τ_G, τ_{HT}, τ_{WS}. A number of tubular membranes were cast and tested in order to determine how decreasing the evapoartion time, the gelation time, and the heat treating time affected desalination performance. The data are given in Table 1 and are shown compared to the previously obtained [5] performance curves in Figs. 1 and 2. By collating the data in the table with the plotted results, the following conclusions are obtained:

(1) Decreasing τ_E (60 sec to 5 sec) significantly improves S. As shown in Figs. 3 and 4, there is an increase in S of

TABLE 1

EFFECT OF FABRICATION VARIABLES ON TUBULAR MEMBRANE PERFORMANCE - BATCH CASTING RESULTS

Standard 1/2" Dia. Tubular Membranes Heat Treated at 20 psig and Pretreated 1 Hr with Distilled Water at 920 psig. Test Conditions: 25°, 800 psig, 4.8 gpm brine flow, $N_{Re} \simeq 50,000$. Casting bob O.D. = 0.335", Casting tube I.D. = 0.365", δ = 0.0015", X_G = 10".

Memb. No.	τ_E (sec)	τ_G (min)	T_{HT} (°C)	τ_{HT} (min)	C_{B_o} (ppm NaCl)	J (gpd/ft^2)	S	R (%)	A ($\frac{10^{-5}\ cm}{sec\text{-}atm}$)	B ($\frac{10^{-5}\ cm}{sec}$)
Effect of Heat Treating Time on Membrane Performance (τ_{HT}):										
1	16	60	∿88	30	10,650	30.0	20.5	95.11	3.08	6.69
2	16	60	∿88	15	10,650	32.1	18.1	94.46	3.30	8.15
3	16	60	∿88	25	10,650	28.7	19.8	94.94	2.94	6.68
4	16	60	∿88	10	10,650	31.2	18.6	94.61	3.21	7.71
5	16	60	87.7	15	11,260	28.8	21.5	95.35	2.99	6.12
6	16	60	87.7	30	11,260	26.4	25.6	96.10	2.75	4.71
7*	12	60	87.7	30	10,650	23.1	34.1	97.07	2.37	3.09
8	12	60	87.7	30	10,650	23.4	33.8	97.04	2.41	3.17
9	12	60	88.1	30	10,730	22.7	33.0	96.97	2.34	3.14
10	12	60	88.1	30	10,730	22.8	31.3	96.80	2.35	3.34
Effect of Gelation Time (τ_G):										
11	12	30	87.9	30	10,840	22.7	32.5	96.92	2.34	3.20
12	12	30	87.9	30	10,840	22.0	32.8	96.95	2.27	3.07
13	12	15	88.1	30	10,200	22.9	35.8	97.20	2.34	2.92
14	12	5	88.2	30	10,120	22.2	35.9	97.21	2.26	2.83
15	12	5	88.1	30	10,120	21.9	34.4	97.09	2.23	2.91
16	12	15	88.3	5	11,330	25.6	26.0	96.15	2.66	4.50
17	12	15	88.3	30	11,330	21.1	33.6	97.03	2.19	2.88

*Membranes 7 through 32: machine cast. Different batch of casting solution used for membranes 7 to 10 than for membranes 20 to 32.

TABLE 1 (Continued)

Memb. No.	τ_E (sec)	τ_G (min)	T_{HT} (°C)	τ_{HT} (min)	C_{B_o} (ppm NaCl)	J (gpd/ft^2)	S	R (%)	A ($\frac{10^{-5}\ cm}{sec\text{-}atm}$)	B ($\frac{10^{-5}\ cm}{sec}$)
Experimental Membrane: Overlapped Longitudinal Edges of Casting Surface										
18	12	60	88	30	10,580	20.1	27.6	96.38	2.06	3.37
Experimental Membrane: Butted Longitudinal Edges of Casting Surface										
19	12	60	88	30	10,580	22.5	24.4	95.91	2.31	4.27
Effect of Evaporation Time on Membrane Performance (τ_E)**:										
20	60	60	88.1	30	10,800	18.7	18.3	94.54	1.92	4.85
21*	60	60	88.1	30	10,800	19.4	13.8	92.76	1.98	6.76
22	30	60	88.1	30	11,180	19.5	24.8	95.96	2.02	3.67
23	30	60	88.1	30	11,180	19.3	25.5	96.07	2.00	3.54
24	45	60	88.1	30	10,050	18.7	19.7	94.91	1.90	4.50
25	45	60	88.1	30	10,050	19.8	16.7	94.00	2.00	5.64
26	15	60	88.2	30	10,500	19.2	31.7	96.85	1.97	2.80
27	15	60	88.2	30	10,500	19.7	30.1	96.67	2.02	3.03
28	5	60	88.2	30	10,350	20.3	37.9	97.36	2.07	2.45
29	5	60	88.2	30	10,350	20.5	37.5	97.33	2.09	2.50
31	2	60	88.2	30	10,730	21.6	37.2	97.31	2.22	2.65
32	2	60	88.2	30	10,730	21.6	38.0	97.37	2.22	2.59
Effect of Length on Membrane Performance (L = 20 ft, no flow in gelation bath):										
30	12	60	88.0	30	10,650	21.6	35.8	97.21	2.22	2.76

*Pinholes detected in membrane on removal from support tube.

**As τ_E varies from 60 to 2 sec, the shear rate parameter, q, varies from 11.1 to 333 sec^{-1}.

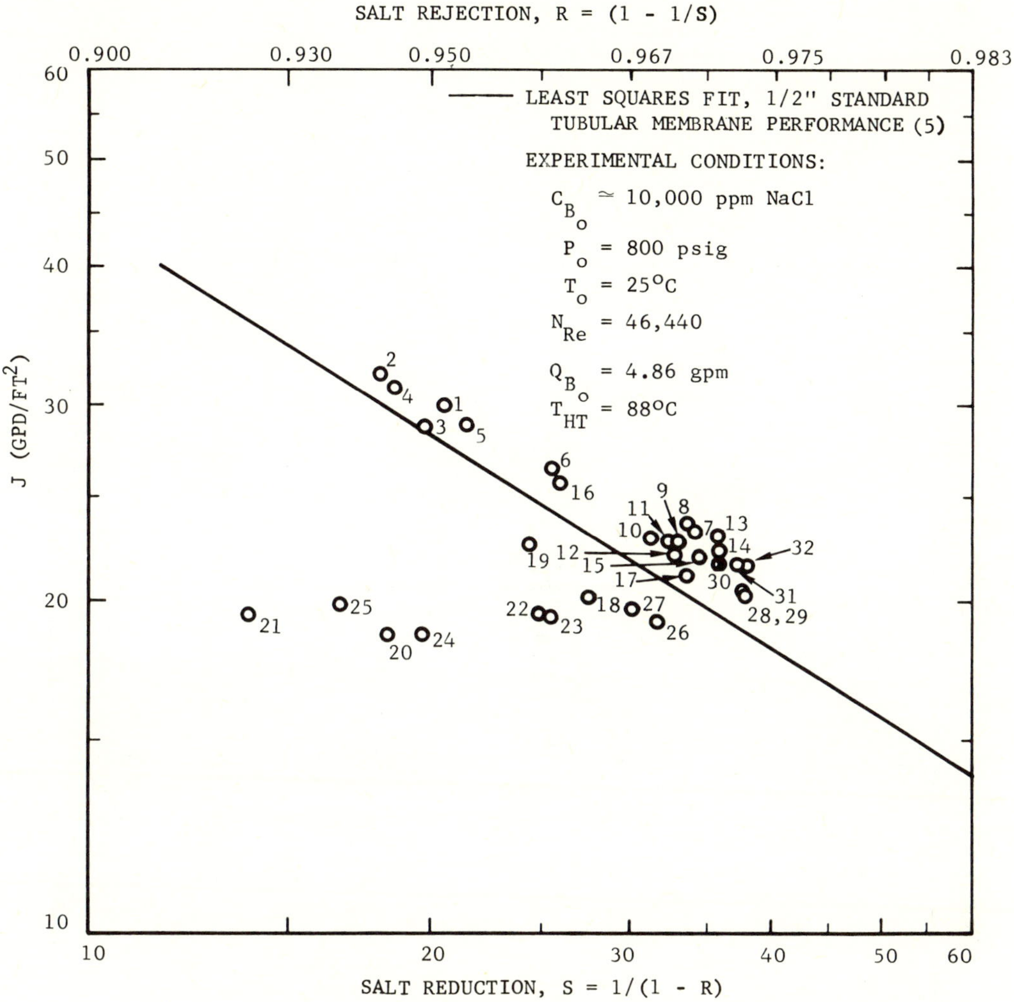

Fig. 1 - Effect on Performance of Changes in Various Fabrication Variables

approximately 20% as τ_E is reduced from 12 to 5 sec, while J increases by about 5% over the same interval. No additional improvement in S is obtained by decreasing τ_E to 2 sec; however, J is slightly increased when τ_E is decreased to 2 sec. When $\tau_E < 1$ sec, gross defects occur in the membrane structure.

(2) Decreasing τ_G (60 minutes to 5 minutes) has little, if any, effect on the performance of the resulting membranes. Gelling was observed to occur in times of < 1 minute.

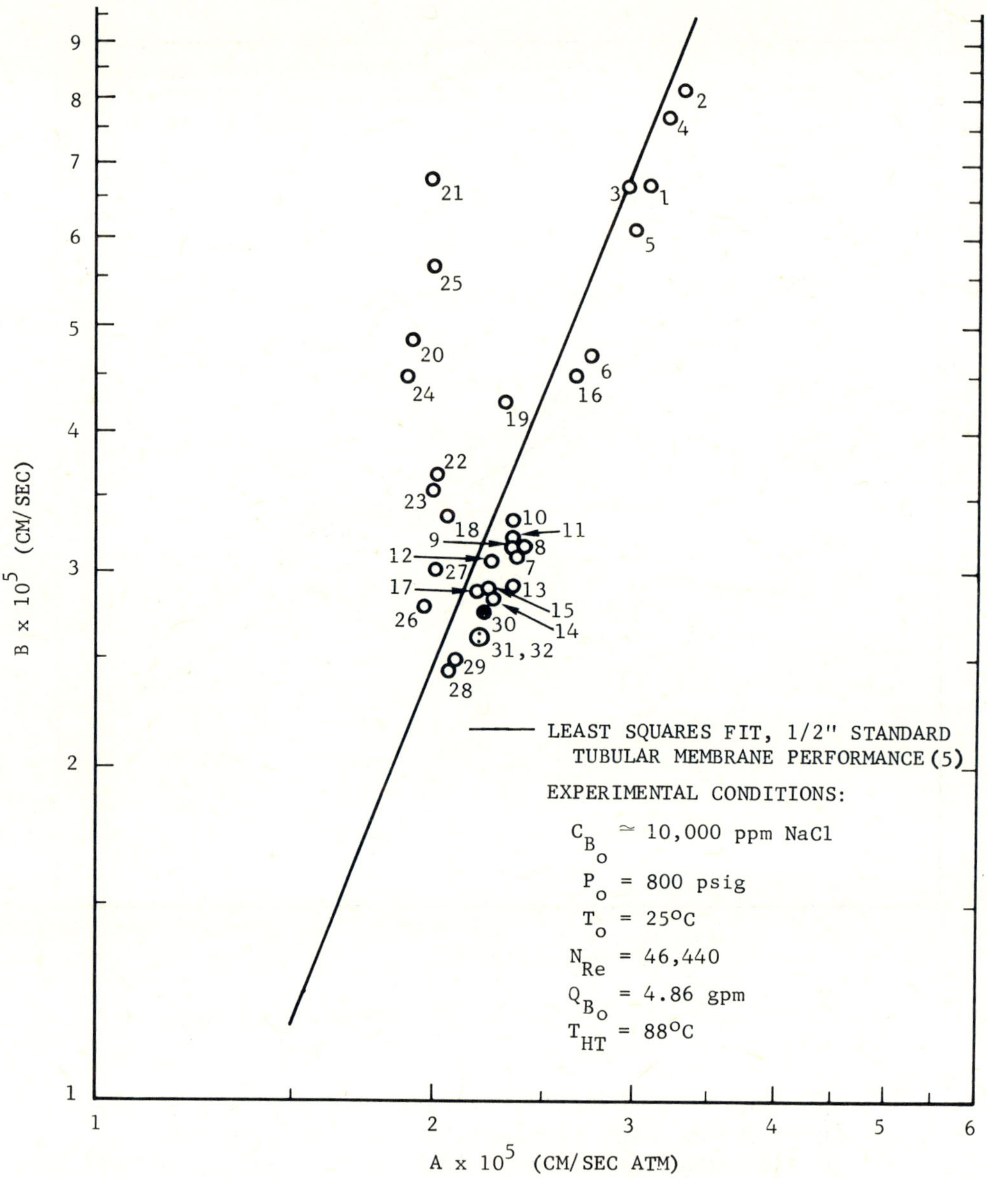

Fig. 2 - Effect on the Permeation Coefficients of Changes in Various Fabrication Variables

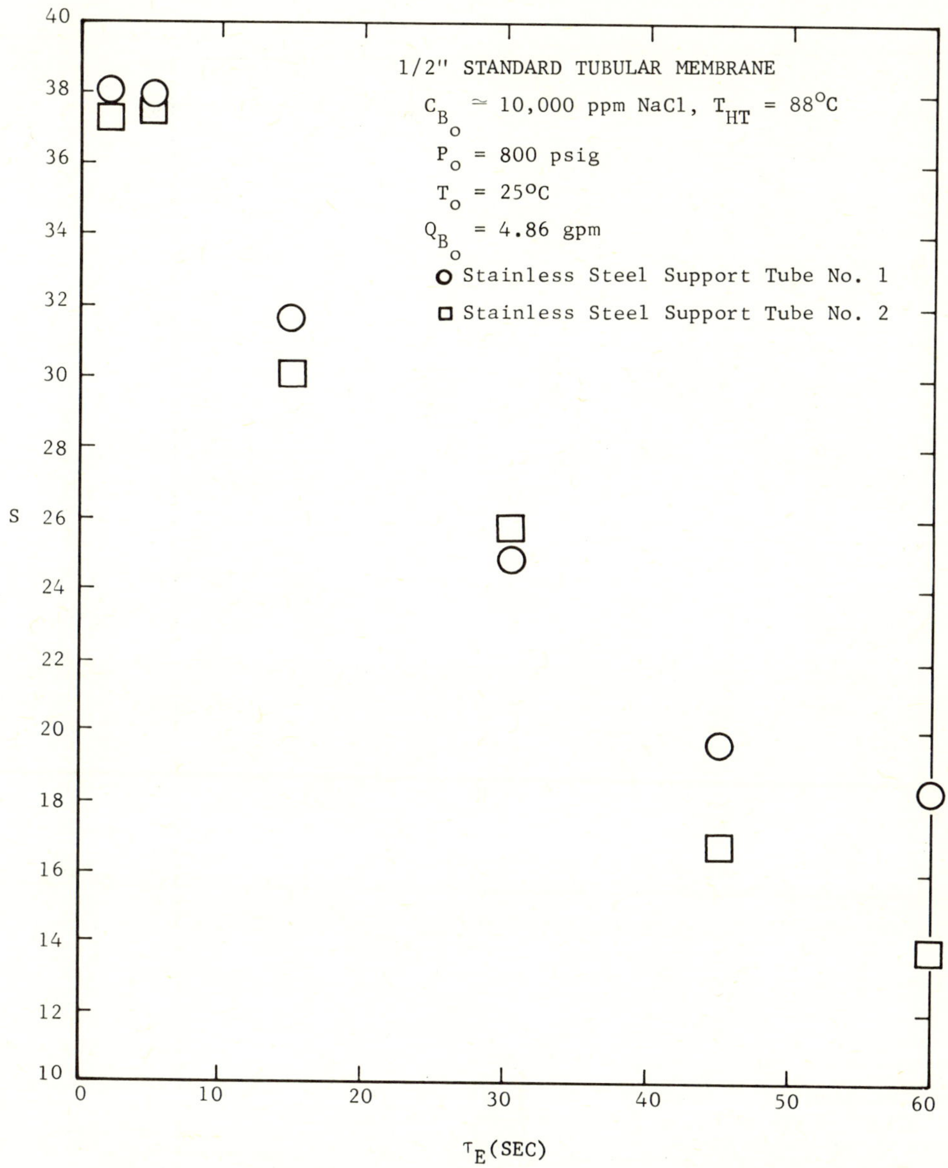

Fig. 3 - Effect of Evaporation Time on Membrane Salt Reduction

(3) Decreasing τ_{HT} (30 minutes to 5 minutes) has the effect of shifting performance along the standard performance curve to levels characteristic of more open membranes.

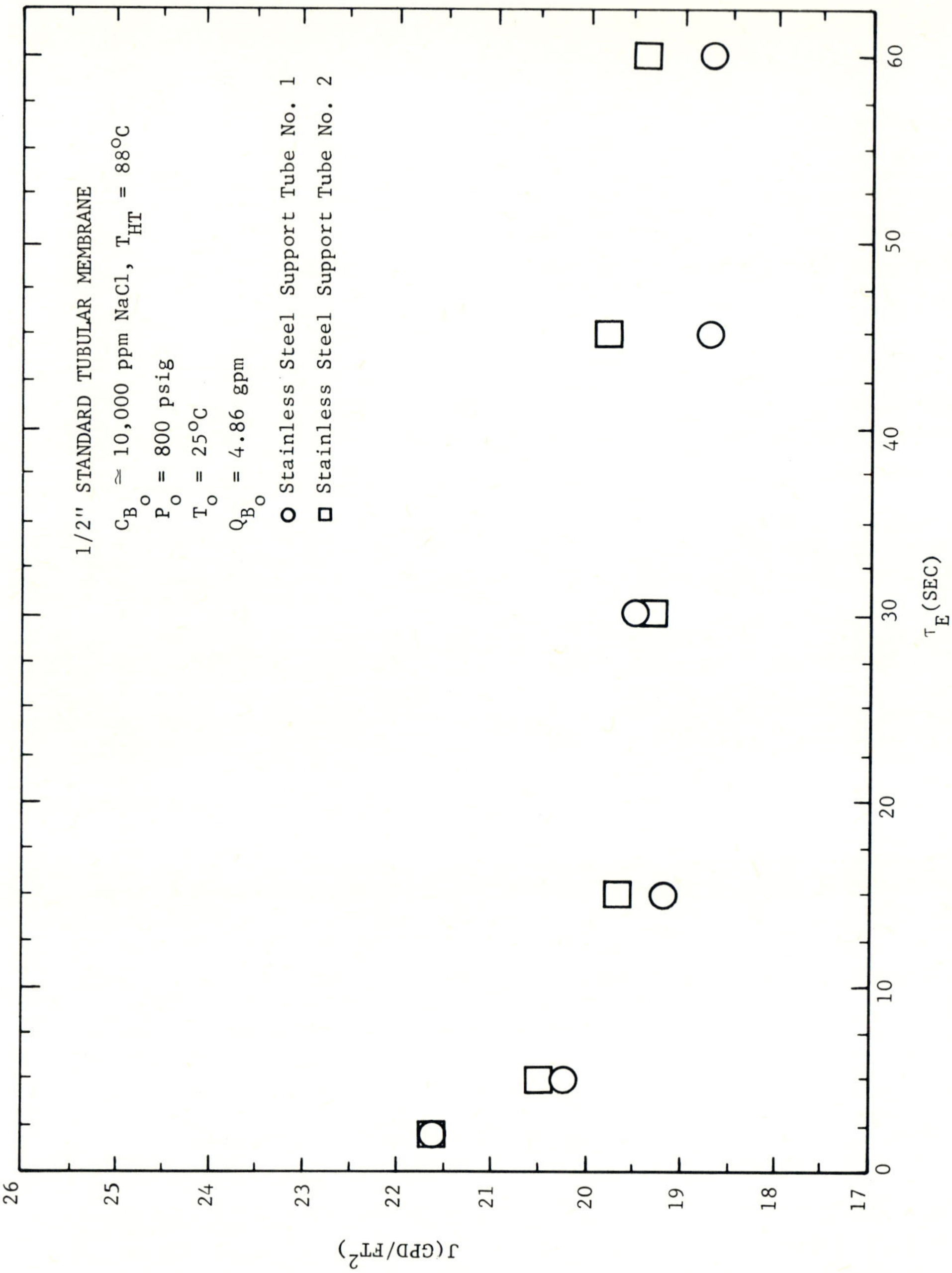

Fig. 4 - Effect of Evaporation Time on Membrane Product Flux

A summary of these results is shown schematically in Fig. 5. The standard casting conditions on which this comparison is based were[5]:

$$\tau_E = 12 \text{ to } 28 \text{ seconds}, \tau_G = 1 \text{ hour, and } \tau_{HT} = 1/2 \text{ hour.}$$

The results have shown that a continuous casting process based on much shorter characteristic times will lead to both significantly improved desalination performance as well as production rates.

The wall separation time, τ_{WS}, or the time for gelation to be sufficiently complete so that the membrane can be removed from the casting tube without damage to the membrane structure was found to be between 20 and 30 seconds for a gelation bath temperature of $\sim$ 1°C and a casting temperature of $\sim$ 23°C. A continuous caster design based on τ_E = 2 to 5 seconds, τ_{WS} > 30 seconds, and τ_G > 1 minute will lead to the fastest production rates consistent with maximum desalination characteristics of the resulting membranes.

No significant effect on membrane performance was obtained when the gelation bath flow rate was reduced to zero after the bath had previously been conditioned to a relatively constant temperature. Evidently the natural convection mixing occurring inside the short (3 feet) membrane tubes in the gelation bath is adequate for the removal of the formamide and the residual acetone not removed during the evaporation phase of the fabrication process.

<u>Continuity of the Casting Surface</u>. Tubular membranes were batch cast in standard support tubes (L $\sim$ 36") whose interior surfaces were covered with plastic film strips. In one experiment, the plastic strip (high density polyethylene film about 4 mil thick) was carefully trimmed along its longitudinal edges so that these edges would butt up against each other while the tape fully lined the inside surface of the casting tube. In another experiment, the longitudinal edges were allowed to overlap each other. Membranes were cast in these tubes, gelled, and removed from the casting tube by withdrawing the entire support film - membrane ensemble. The membranes were then removed from the support films and inserted into standard perforated stainless steel support tubes and tested to high pressures. The thickness of these trial membranes varied between 4 and 12 mils in the circumferential direction. A beading was observed along the length of both sides of the membrane tube cast on the butted casting surface. Similarly, a step, or abrupt change in membrane tube wall thickness in the circumferential direction, was observed to exist along the full length of the outside surface of the membrane tube cast on the overlapped casting surface. Both membranes, in the standard perforated support tubes, were capable of containing saline feed solution at 1500 psig without gross leakage. Desalination performance at 800 psig was only

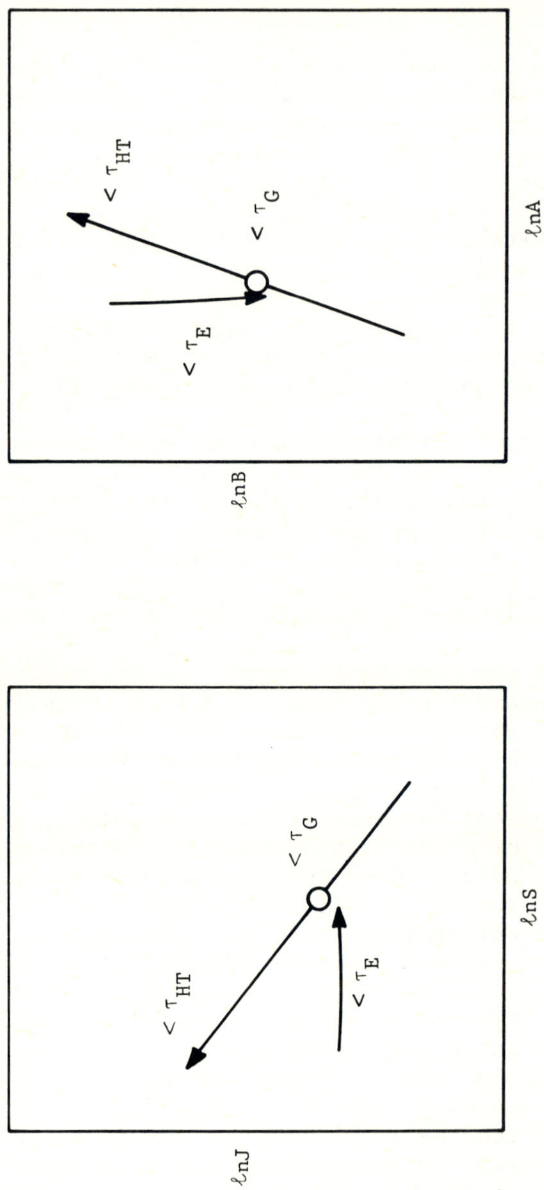

Fig. 5 - Schematic Representation of the Effect on Membrane Performance Due to Decreased Evaporation, Gelation, and Heat Treating Times slightly less than that obtained for membranes prepared in the usual manner (see Nos. 18 and 19 in Table 1).

Effect of Evaporation Time on the Compaction Slope of The Resulting Membranes. Compaction slopes for times up to 123 hours were measured for three different 1/2 inch tubular membranes - two made with τ_E = 5 seconds and one made with τ_E = 20 seconds. The compaction measurements were carried out at quite high feed flow rates (N_{Re} = 88,620). The compaction slope data themselves may be summarized as follows (data reduced as in Ref. [5]):

Standard Cellulose Acetate Tubular Membrane, Heat Treated at 20 psig and 88°C. Test Conditions: 0.50% $MgCl_2$ feed, 25°C, 800 psig, 9.3 gpm brine flow.

Membrane No.	τ_E (sec)	Data	b
102	5	J(t)	-0.0325
		S(t)	0.236
105	5	J(t)	-0.0415
		S(t)	0.208
106	20	J(t)	-0.0372
		S(t)	0.101

The results show that even under the high flow rate conditions of these tests, τ_E had little, if any, effect on the slope of either the J(t) or S(t) results. The values of b obtained from these experiments are similar to those reported earlier (Ref. [5], p. 101) for P_o = 800 psig conditions.

The Performance of Long Tubular Membranes. Methods were developed for the machine batch casting of 20 foot long tubular membranes of arbitrary diameter (0.15 to 1.0 inch). The machine caster used previously for casting 3 foot membrane tubes was used in conjunction with an extended gelation bath for the casting-gelation steps in the fabrication. Minor adjustments in the thickness of the annular gap between the casting bob and the support tube as well as in τ_E were necessary in order to produce a 20 foot long tubular membrane free of imperfections and uniform in thickness ($\delta \sim 10$ mils) along its entire length. The long tubes were stored in a special bath. The heat treating and testing facilities were modified so as to be capable of accommodating the 20 foot tubes.

The desalination performance level of the 20 foot tube was found to be just as good as that obtained previously for the best 2 foot tubes (see, for example, data for membrane number 30 in Figs. 1 and 2).

Casting Solution Viscosity and Its Dependence on Shear Rate

Experimental Viscosity Data. The absolute viscosity of the casting solution and its dependence on shear rate was measured for two lots of E398-10 cellulose acetate with only minor difference in the degree of acetyl substitution.

The reduced experimental results and the conditions of the experiment for the first lot are given in Table 2. The data are plotted in Fig. 6. Lohmander[10] has shown that the viscosity-shear rate data of dilute solutions of cellulose esters can be well approximated by the hyperbolic form,

$$\mu = \mu_\infty + \frac{a_1}{q + a_2} . \qquad (1)$$

TABLE 2

DEPENDENCE OF CASTING SOLUTION VISCOSITY ON SHEAR RATE

Standard Cellulose Acetate Casting Solution
CA Lot AC 628, Degree of Substitution: 2.43 ± 0.02

Ferranti-Shirley Viscometer:
Cone Angle, 0° 20' 50'' Cone Radius, 1.995 cm. ($1.7 < q < 32$ sec^{-1});
Cone Angle, 0° 22' 00'' Cone Radius, 1.000 cm. ($32 \leq q \leq 240$ sec^{-1}).

q (sec^{-1})	μ* (poise)	μ** (poise)	τ*** (dyne/cm^2)
1.70	460	460.0	782
3.50	313	312.7	1,100
7.00	255	256.6	1,790
16.5	230	225.6	3,800
32.0	208	212.3	6,660
65.0	200	199.5	13,000
131.5	183	181.5	24,100
163.8	173	173.6	28,300
240	155	155.3	37,200

*Experimentally measured at 25°C.

**Calculated from: $\mu = 209.6 + \dfrac{319.4}{q - 0.4265} - 0.2319q$.

***Calculated from experimentally measured μ ($\tau = \mu q$).

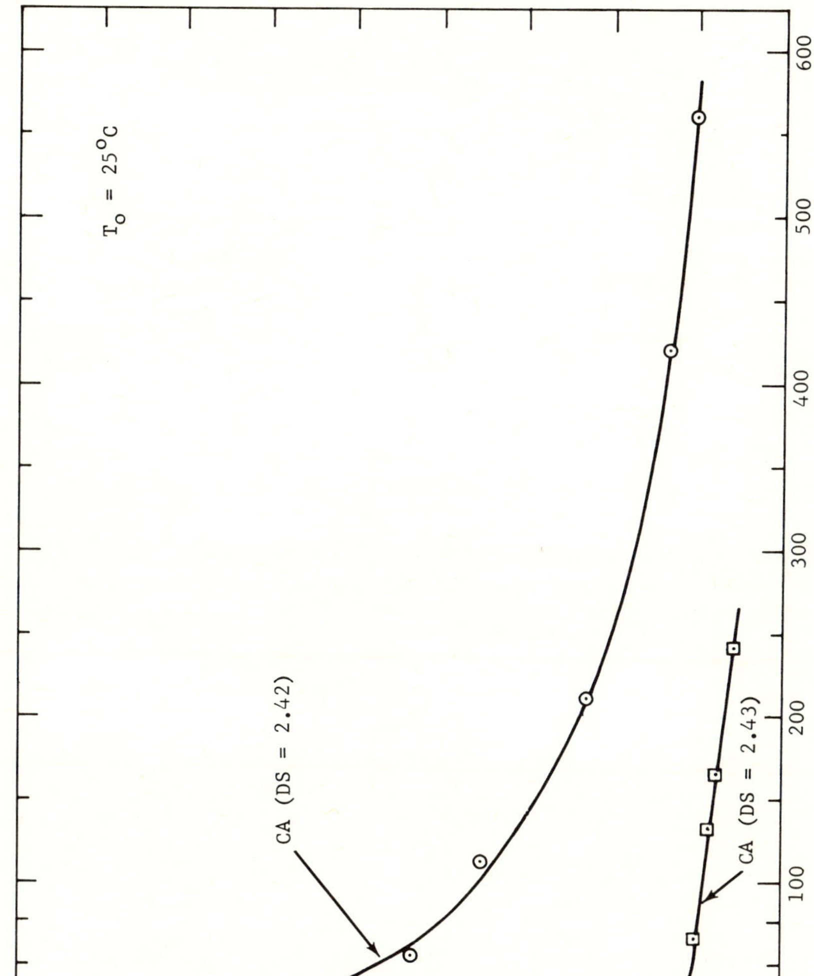

Fig. 6 - Viscoelastic Properties of Casting Solutions (Two Lots of Cellulose Acetate)

We tried to fit the casting solution viscosity data with an expression of this form using a nonlinear least squares curve fitting computer program[11]. Equation 1 was found to provide an unsatisfactory fit in the higher q range ($q > 40\ sec^{-1}$). We have found that an excellent fit over the entire range of shear rates can be obtained if the following modification of Equation 1 is used:

$$\mu = a_3 + \frac{a_4}{q + a_5} + a_6 q \ . \qquad (2)$$

This four-parameter equation can fit both the hyperbolic variation at lower shear rates and the linear fall-off in viscosity at the higher shear rates. The constants obtained for the standard cellulose acetate casting solution at 25°C using the nonlinear curve fitting calculation are (not valid for $q < 1.7\ sec^{-1}$):

$$a_3 = 209.6 \text{ poise}$$
$$a_4 = 319.4 \text{ poise/sec}$$
$$a_5 = -0.4265 \text{ sec}^{-1}$$
$$a_6 = -0.2319 \text{ poise} \ .$$

The shear stress is simply obtained, for a non-Newtonian fluid whose viscosity and shear rate have been measured in a cone-and-plate instrument with a cone angle of < 3°, by the relation (Ref. [9], pp. 6-7),

$$\tau = \mu q \ . \qquad (3)$$

Values for the shear stress are also given in Table 2.

Viscosity-shear rate data for the second lot of cellulose acetate (CA) in the standard casting solution were obtained with a cone-and-plate device at 25°C and 15°C. The experimental measurements were carried out under conditions which precluded the loss of volatile solvent from the solution while it was being sheared. The results are given in Table 3 and are shown in Figs. 6 and 7. It is apparent that in the case of the CA solution viscosity data at 25°C the two different lots of cellulose acetate polymer, even though obtained from the same supplier* and characterized as having essentially the same total degree of acetyl substitution, exhibited viscosity-shear rate results which implied significant differences between the two polymers (probably molecular weight and molecular weight distribution). The viscosity of Lot AC 1400 was significantly greater than that of Lot AC 628 and

*Eastman Chemical Products, Inc., Kingsport, Tenn.

TABLE 3

DEPENDENCE OF CASTING SOLUTION VISCOSITY ON SHEAR RATE - CELLULOSE ACETATE/FORMAMIDE/ACETONE SYSTEM

Standard Cellulose Acetate Casting Solution
CA Lot AC 1400, Degree of Substitution: 2.42 ± 0.02

Ferranti-Shirley Viscometer:
Cone Angle, 0° 25' 39''. Cone Radius, 2.000 cm

$T_o = 15°C$		$T_o = 25°C$	
μ (poise)	q (sec^{-1})	μ (poise)	q (sec^{-1})
1130	7.0	925	7.0
960	14.0	740	14.0
825	28.1	640	28.1
675	56.1	535	56.1
520	112	450	112
400	210	330	210
300	350	235	420
210	560	200	560
		170	700

decreased less rapidly with increasing shear rate in the low shear rate range while not exhibiting the linear decrease with shear rate in the higher shear rate range which was characteristic of the solution containing CA from Lot AC 628.

Instead of fitting the data from the second lot of CA to a modified hyperbolic model, a power law model (Ostwald-de Waele model) was used to fit the lower shear rate range of the 25°C data for the CA solution[8]. According to this model,

$$\mu = m\, q^{n-1} \; . \tag{4}$$

Newtonian fluids are thus characterized by $n = 1$ and pseudoplastic fluids are characterized by $n < 1$. Using the values, $\mu = 840$ poise at $q = 10\ sec^{-1}$ and $\mu = 470$ poise at $q = 100\ sec^{-1}$ (approximations to the CA viscosity data at 25°C, cf. Fig. 7), a linear extrapolation* back to $q = 1\ sec^{-1}$ yields $\mu = 1500$ poise. These values were used to find the m and n values which permit equation (4) to approximate the viscosity-shear rate dependence of the solution made from the second lot of CA in the lower shear rate range (at least $7\ sec^{-1} \leq q \leq 100\ sec^{-1}$). Thus,

$$n = 0.749 \quad (s \equiv 1/n = 1.335)$$

and $m = 3.14\ lb_f\ sec^{0.749}/ft^2$.

*On a log μ versus log q plot.

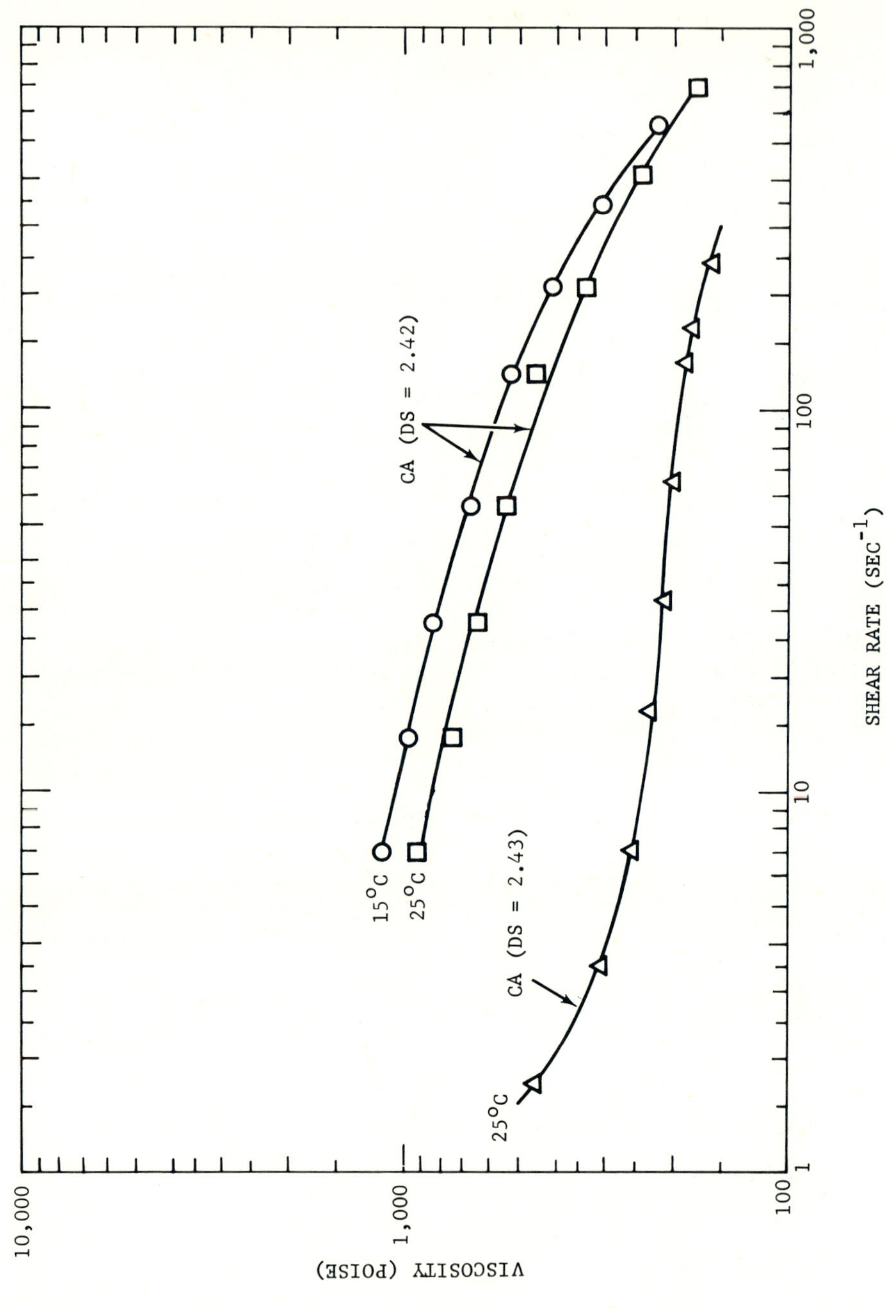

Fig. 7 - Viscoelastic Properties of Casting Solutions (Two Lots of Cellulose Acetate)

Estimation of the Frictional Pressure Drop Through the Extrusion Annulus. Since a continuous tubular membrane caster must involve extrusion of a non-Newtonian fluid through a thin annulus, pressure drop calculations used in the design of the polymer pumping system must properly take these aspects of the problem into account. Studies made of non-Newtonian flow in annuli are either based on a shear stress-shear rate model which is inappropriate for the casting solution[12] (since it may be neither a Bingham plastic nor a power law fluid), or have limited utility since they are based on flow rate predictions for Newtonian fluids in thin plane slits[13]. As a rough first approximation, we have used the latter approach (Ref. [13], pp. 258, 266-268, 271, 296-299) to estimate the pressure drop through an annular orifice of length L with a fluid whose viscosity is independent of shear rate (thin slit approximation):

$$\Delta P = [12\ \mu\ L/(R_o - R_i)^2\ g_c]\ \dot{L}\ , \tag{5}$$

where $\dot{L}$ is the lineal production rate of extrudate. The results of calculations of $\dot{L}$ as a function of ΔP for two different viscosities which bracket the range of interest are given in Fig. 8. The annular geometry used in these calculations was essentially that appropriate for extrusion of a nominal 1/4-inch diameter tubular membrane.

Assuming the validity of the power law fit to the CA viscosity-shear rate data, frictional pressure drop calculations were made for the extrusion annulus used in batch casting nominal 1/4-inch tubular membranes. Thus, for $D_o = 0.171$ inch, $D_i = 0.159$ inch, and $L = 1.54$ inch (land length), non-Newtonian power law pressure drop calculations were made using the equations developed by Fredrickson and Bird[12]. These results were compared to those obtained from the following expression valid for power law fluid flow through a very thin annular slit (Ref. [12], Eq. 35):

$$\Delta P = \frac{2mL}{R_o}\left[\frac{(1 + \alpha)(s + 2)\ \dot{L}}{R_o(1 - \alpha)^{s+1}}\right]^{1/s}\ , \tag{6}$$

where $s \equiv 1/n$ and $\alpha \equiv R_i/R_o$. If this equation is compared with that for the frictional pressure drop incurred by a Newtonian fluid ($n = 1$, $\mu = m$) flowing through a thin plane slit [Eq. (5)], then

$$\Delta P \propto \dot{L}\ ,$$

while in the case of power law fluid flow through a very thin annular slit,

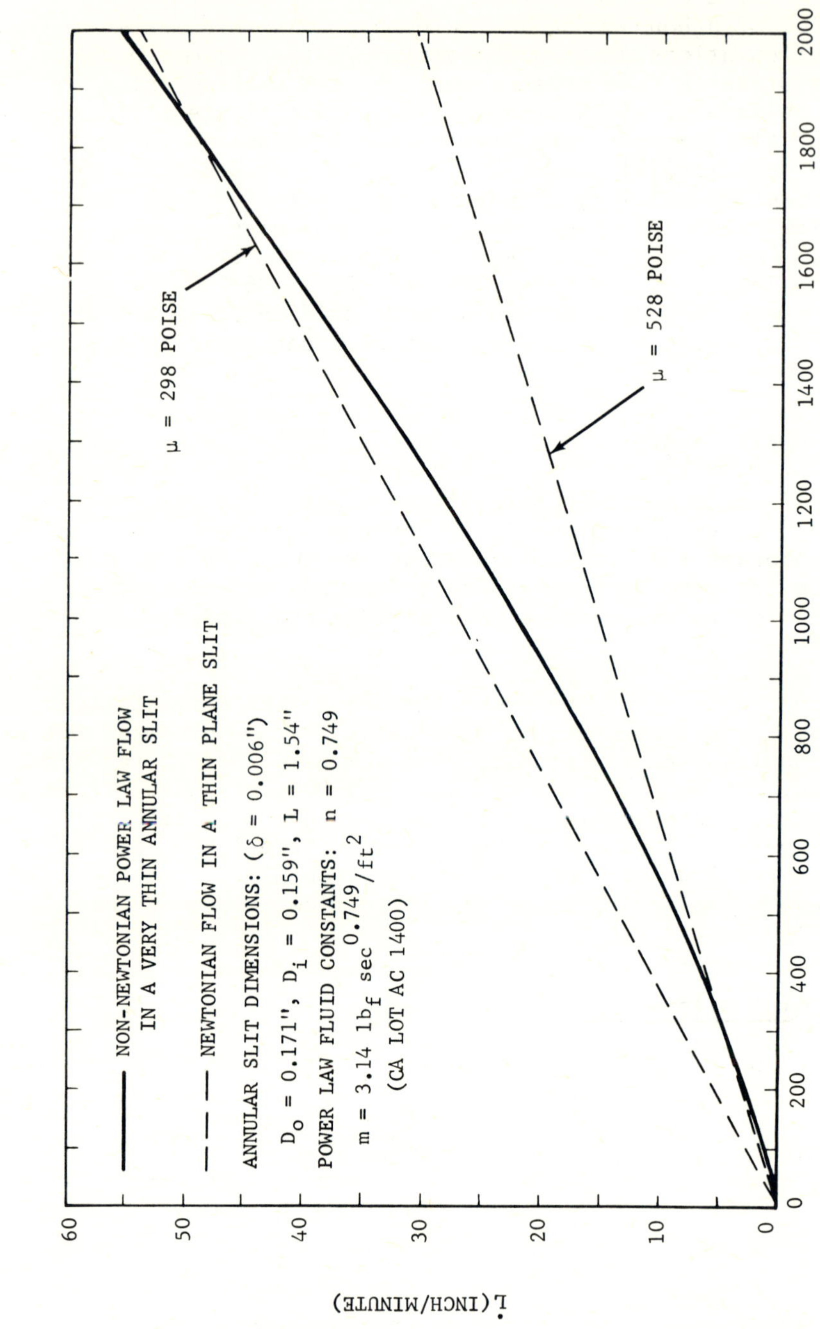

Fig. 8 - The Variation in the Static Pressure Drop Along a Thin Annulus With the Linear Extrudate Production Rate for Two Different Models of the Viscous Properties of the Solution Being Extruded

$$\Delta P \propto (\dot{L})^n ,$$

The comparison between the results of the more exact calculation and the results from Eq. (6) for the range of independent variables of interest in the design of the continuous casting process showed that Eq. (6) underestimates the actual pressure drop by approximately three percent - an insignificant difference in the case of the continuous casting device design and operating considerations dependent upon these calculated results.

In Fig. 8, a complete set of calculated results are given for the power law flow of a CA casting solution (the second CA lot AC 1400) at 25°C through an extrusion annulus characteristic of that used for batch casting nominal 1/4 inch tubular membranes [using Eq. (6)]. These results are compared with those derived from the Newtonian fluid/thin plane slit calculation [Eq. (5)] for solutions with two different shear rate-independent viscosities. It is apparent that there is considerable divergence between the two different estimates of the driving force (supply pressure imposed on the polymer casting solution upstream of the extrusion annulus) required for a given level of extrudate production rate.

In the case of the batch and continuous casting/extrusion annulus geometries, a reasonable approximation to the shear rate in the solution averaged across the annular slit and at the annulus outlet is (see Ref. [8], page 4 for a schematic representation of the velocity profile):

$$q \simeq \dot{L}/\delta = X_G/\tau_E\ \delta , \tag{7}$$

$$\text{since } \dot{L} = X_G/\tau_E . \tag{8}$$

In the final design selected for the extrusion annulus of the casting apparatus, D_o = 0.187 inch, D_i = 0.146 inch, L = 1.10 inch, and δ = 0.017 inch (when the forming channel is lined with 0.003 inch thick Teflon tapes, longitudinally butted), and thus the results given in Fig. 8 for the power law flow case represent a conservative upper limit to the pressure drops actually required for extrudate production rates in the 3 to 10 inch/minute range actually employed in the operation of the caster (when $\dot{L}$ = 1/2 ft/min and δ = 0.017 inch, q = 5.9 sec^{-1}).

The Continuous Casting Apparatus - Design and Performance

A continuous casting apparatus appropriate for the fabrication of tubular membranes was designed, built, and operated and is based on both our current understanding of the various rate processes which take place during the tubular membrane fabrication process (Fig. 9) as well as the results reported in the preceding sections.

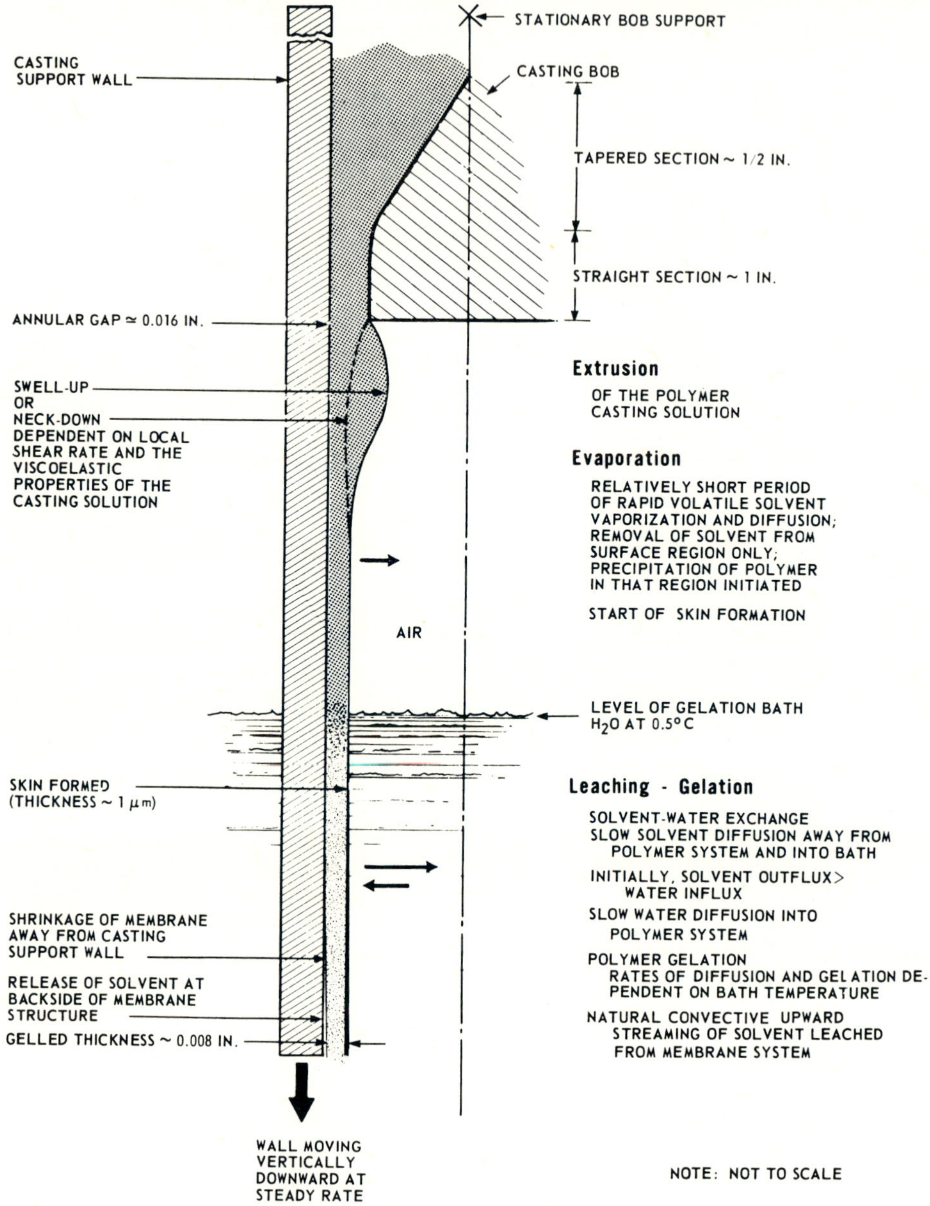

Fig. 9 - Schematic Representation of Casting Process - Cellulose Acetate-Acetone-Formamide Casting Solution, Cylindrical Geometry

A schematic representation of the experimental continuous tubular membrane casting device is shown in Fig. 10. The design is based on the principle of two plastic tapes moving at the same speed and passing through a cylindrical forming channel in which the casting solution is extruded around a bob assembly and onto the moving tape surfaces. The extruded cast solution passes through a relatively short evaporation zone (air filled) and then into a chilled water leaching and gelation bath. After sufficient residence in the gelation bath the polymer tube shrinks away from the moving tapes and is continuously removed for storage.

The design of the continuous caster which incorporated this principal of operation was developed after a series of trials and modifications[14].

The dimensions of the extrusion annulus were based on experience gained through the development of the batch machine caster. The diameter of the cast polymer tube produced from the end of the device is nominally 1/4 inch (I.D. ≃ 0.15 inch). This smaller diameter was selected for the design of the device since scale-up of the process will be mechanically more straightforward than scale-down and because less polymer solution and a smaller gelation tank are needed when the smaller diameter is used. The tapes are made of Teflon and are 0.003 to 0.004 inch thick and carefully slit to the desired width.

The use of longitudinally butted moving Teflon tapes led to the continuous extrusion of the first reasonably good salt rejecting membranes. The results obtained for five different membranes are given in Table 4. Salt rejections as high as $\sim$ 87% were obtained for the caster operated in the vacuum mode in which solvent is withdrawn from the top of the gelation zone by vacuum being applied to a tube passing through the casting bob with its end immersed in the gelation bath inside the cast tube. Both "thin" regions ($\leq \sim$ 6 mils for membrane tubes supported in desalination testing by flexible braided sleeves) and "steps," or abrupt changes in wall thickness, are thought to lead to poor salt rejection characteristics.

The injector head, bob system, tape inlet guides, and gelation medium injection system was also operated so as to permit replacement of the gelation medium. The main features of the injector design are shown schematically in Figs. 11 and 12. The point of polymer solution injection onto the moving tape surface is located at a position below the level at which the tapes are fully formed into a tubular configuration. This prevents migration of the polymer solution behind the moving tapes. The tapes are guided into this configuration by contours in the injection head which are of such a shape that the tapes do not have to yield in order to make

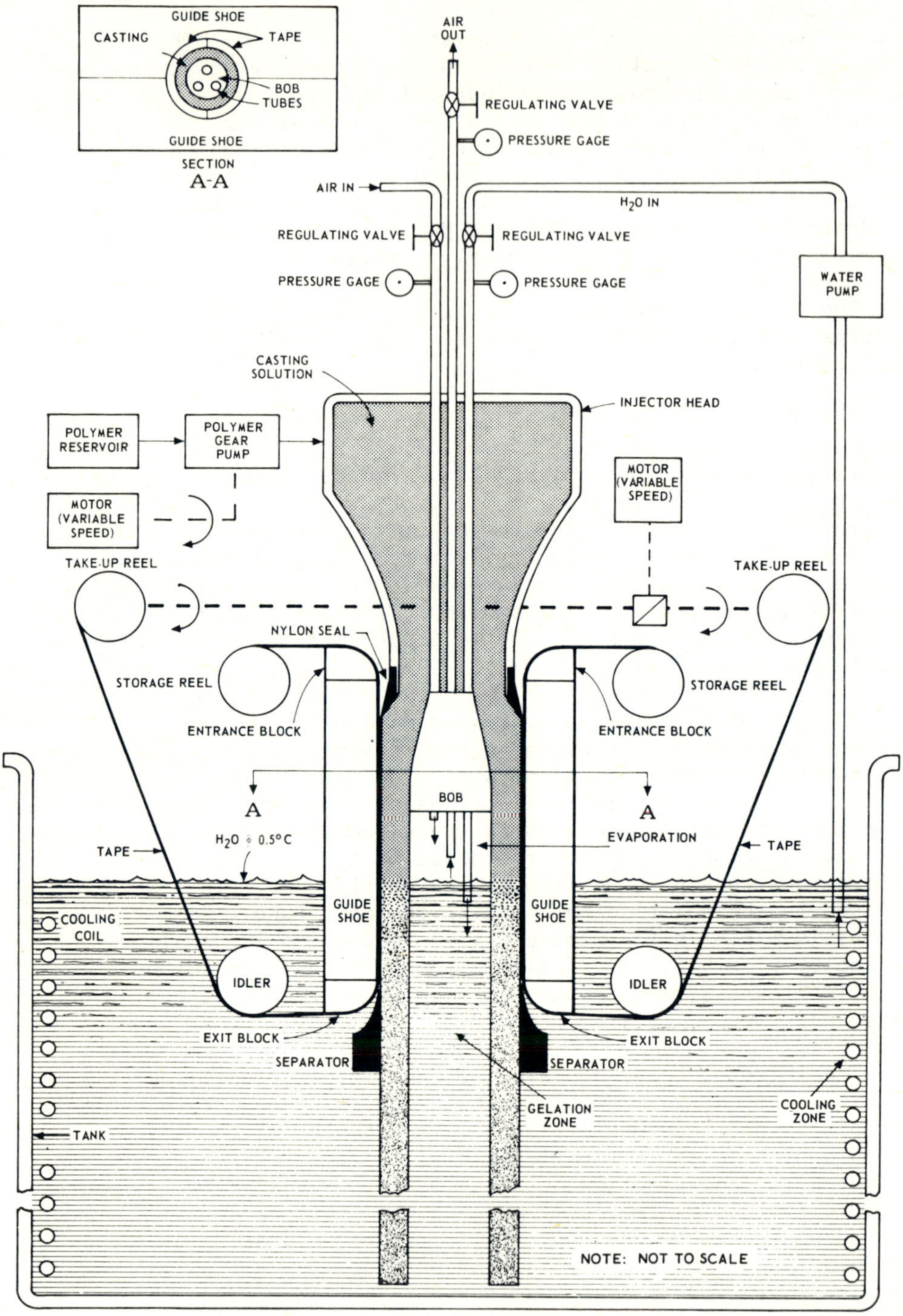

Fig. 10 - Functional Schematic of the Experimental Continuous Tubular Membrane Casting Device

TABLE 4

DESALINATION PERFORMANCE DATA FOR NOMINAL 1/4" TUBULAR MEMBRANES (CONTINUOUS CAST)

Evaporation Time, 1-3 sec. Gelation Time, 6 min.
Membrane Annealed at 60 psig and 85°C for 30 min.
Test Conditions: NaCl brine, 600 psig, 25°C, 0.4 gpm brine flow, N_{Re} = 8300.

C_{B_o} (ppm)	C_p (ppm)	J (gpd/ft^2)	S	R (%)	A $(\frac{10^{-5}\ cm}{sec\text{-}atm})$	B $(\frac{10^{-5}\ cm}{sec})$
5,455	1,831	18.0	2.98	66.42	2.25	39.6
5,455	1,940	22.1	2.81	64.42	2.76	52.1
5,709	914	10.2	6.24	83.98	1.30	8.74
5,600	743	8.4	7.53	86.73	1.08	5.86
5,600	983	13.6	5.69	82.44	1.74	12.9

the transition from a flat shape (when spooled on the storage reel) to a semi-circular shape (in the vertically oriented guide shoes - or forming channel). The bob and gelation medium injection tube are concentric with each other and with the tubular channel through the guide shoes. In Fig. 13, an overall view is shown of the continuous casting system mounted on the gelation tank. For ease in assembly and start-up, the entire system is mounted on a platform which, in turn, is attached to an electric hoist. Thus the system can be easily and quickly removed from the chilled (0.5 to 0.8°C) water bath.

This design of the injection zone portion of the system made possible the continuous extrusion of tubular reverse osmosis membrane with properties which were uniform in the longitudinal direction. The length of the produced membrane was limited only by the tape length and thus, since tape lengths of 50 feet were routinely used in our experiments, membranes up to 40 feet in length were extruded (approximately 10 feet of tape were required for start-up). The cross section of the extruded membrane was quite uniform. The overall membrane wall thickness variation at a given longitudinal position was contained within the range of 0.0065 to 0.0095 inches. At the higher lineal extrudate production rate (8 inch/minute), the membrane wall thickness variation in the circumferential direction was kept to within 0.007 to 0.008 inches. The observed opacity of the produced membrane was more constant with length at the higher extrusion rate. Also, the butted Teflon tapes produced minimal backside perturbation to the membrane wall in the vicinity of the butt-joint location when the higher extrusion rates were used. An example of such a cross section is shown in Fig. 14.

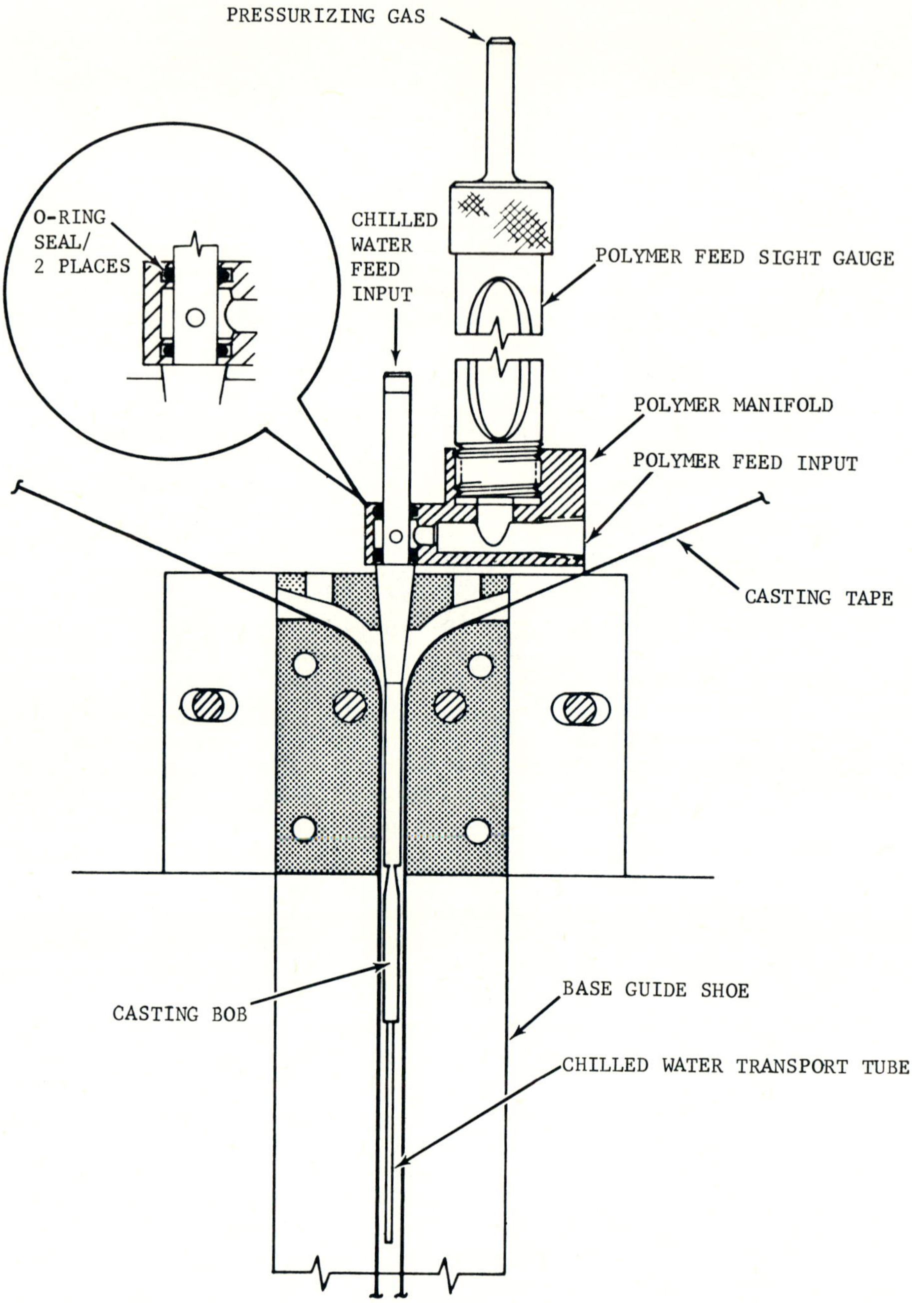

Fig. 11 - Front Section of Extrusion Zone in the Continuous Casting Device

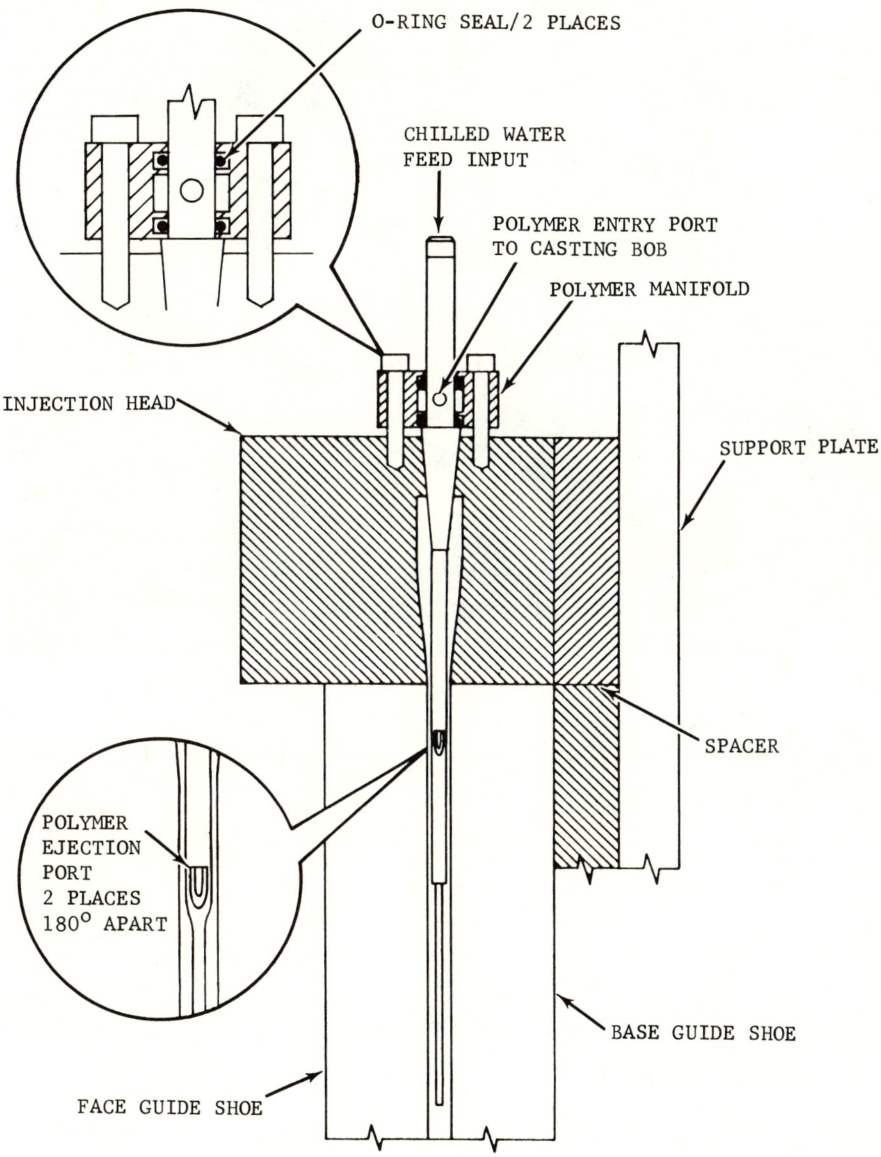

Fig. 12 - Side Section of Extrusion Zone in the Continuous Casting Device

Fig. 13 - Front View of the Top Portion of the Continuous Casting Apparatus with the Polymer Solution Injection System

The desalination performance of tubular membranes cast with the casting device using both butted and overlapped moving Teflon tapes is given in Table 5. The burst pressures of these membranes were comparable to those exhibited by batch cast membranes in the same type of pressure support. Fracture of the membrane tube wall under burst conditions (> 1200 psig) tended to occur in the region where the tape butt joints had occurred. An examination of the membranes' compaction characteristics (compaction slope = b) indicates that the continuously cast membranes compact in a manner similar to those which are batch cast for the same pressure support.

The permselective properties of the continuous cast membranes are similar, but not equal, to those of batch cast membranes tested under identical conditions. Reasons for this difference can be ascribed, at least in part, to the fact that, in contrast to batch

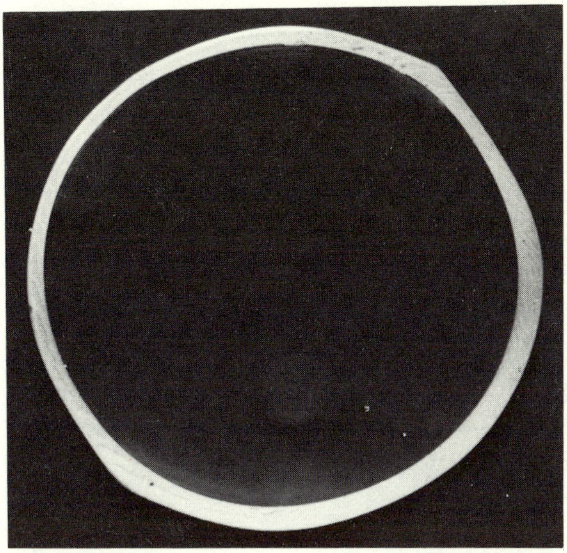

Fig. 14 - Enlarged Cross-Sectional View of Continuously Cast Nominal 1/4 Inch Tubular Membrane. Polymer Solution Cast on Moving, Longitudinally Butted, Teflon Tapes. Membrane No. 110 in Table 5.

casting, in continuous casting the polymer solution was chilled down to 4 to 10°C as it flowed into the extrusion annulus because chilled water is injected through the metal casting bob which is in turn in intimate thermal contact with the polymer solution in the extrusion annulus (Figs. 11 and 12). The data given for five different membranes in Table 6 indicate that decreasing the casting solution temperature of the standard cellulose acetate system yields more "open" (higher flux, lower salt rejection) membranes for the same heat treating temperature. Also, a slight loss in permselectivity (salt rejection) is incurred when the chilled polymer solution is used in the membrane casting process. Another reason for the difference between the permselective performance of batch and continuous cast tubular membranes is that the chilled water injected through the casting bob in the case of the continuous casting system is at a slightly higher temperature ($\sim$ 4°C) than usually used in the batch casting gelation bath (0.4 to 0.8°C). That this small difference in gelation medium temperature can tend to shift the membrane's performance "up" the standard curve (J versus S, or A versus B) towards the performance characteristic of more open and less highly salt rejecting membranes is shown by the data in Table 6.

TABLE 5

PERFORMANCE OF CONTINUOUS CAST TUBULAR MEMBRANES

Standard 1/4" Dia. Tubular Membranes, Heat Treated at 60 psig.
Test Conditions: NaCl brine, 600 psig, 25°C, 0.36 gpm brine flow, $N_{Re} \simeq 8000$.

Membrane Number	T_{HT} (°C)	C_{B_o} (ppm)	J (gpd/ft²)	S	R (%)	A $\left(\frac{10^{-5}\ cm}{sec-atm}\right)$	B $\left(\frac{10^{-5}\ cm}{sec}\right)$	t (hour)
109*	85.2	5560	29.6	3.35	70.12	3.74	51.5	2
110*	85.5	5560	41.1	3.11	67.81	5.21	75.4	1
110*	85.5	6100	31.9	3.61	72.29	4.09	48.7	32
		b =	- 0.0731	0.0430				
111*	85.3	4910	1.37	27.3	96.33	0.17	0.24	2
112*	75.4	5530	25.0	3.72	73.09	3.16	38.4	1
118**	85.6	5700	25.1	3.53	71.67	3.19	41.0	5
119**	85.3	5350	52.0	2.40	58.32	6.52	133	1
119**	85.3	5640	45.7	2.55	60.71	5.76	109	48
119**	85.3	5455	42.5	2.87	65.17	5.36	85.5	94
		b =	- 0.0406	0.0312				

*Butted Teflon tapes (Membranes 109-112), production rate = 8 in/min, q = 7.8 sec^{-1}, gelation temperature = 4 ± 0.2°C.

**Overlapped Teflon tapes (Membranes 118 and 119), production rate = 5 in/min, q = 4.9 sec^{-1}, gelation temperature = 2.3 ± 0.1°C.

In both modes, casting bob diam. = 0.144", δ = 0.017", polymer solution pressure = 20 psig, gelation water pressure = 4 psig, Teflon tapes 0.004" thick x 50' long, 30-38 ft lengths produced.

TABLE 6

EFFECT OF CASTING SOLUTION AND GELATION BATH TEMPERATURE ON THE DESALINATION PERFORMANCE OF BATCH CAST STANDARD 1/4" DIA. CELLULOSE ACETATE TUBULAR MEMBRANES

Membrane Heat Treated at 85.5°C and 60 psig.
Test Conditions: NaCl brine, 600 psig, 25°C, 0.4 gpm brine flow, $N_{Re} \simeq 8000$.

Casting Solution Temp. (°C)	C_{B_o} (ppm)	J (gpd/ft^2)	S	R (%)	A $(\frac{10^{-5}\ cm}{sec\text{-}atm})$	B $(\frac{10^{-5}\ cm}{sec})$	t (hr)	Gelation Bath Temp. (°C)	P_b^* (psig)
7	5180	42.6	8.70	88.51	5.53	20.8	3	∿1	1250
4	5180	45.3	8.33	88.00	5.88	22.9	3	∿1	1800
23	5190	36.1	14.1	92.92	4.70	10.7	1	∿1	1780
23	5690	39.6	12.4	91.93	5.22	13.3	1	3.8	-
23	5690	31.9	18.9	94.71	4.20	7.10	1	1.0	-

*"Burst" condition characterized by high leakage rate at end fittings.

As an aid in continuously collecting and storing the tubular membrane produced by the casting system, an automatic capstan take-up system was designed, built, and operated in conjunction with the caster. The make-up system was installed in a dual tank system. The speed of the motor driving the take-up spool could be continuously varied over a wide range (the rate of tubular extrudate production was $\sim$ 1/2 to 1 ft/min). The radii of curvature which the extruded tube had to experience in progressing from the exit of the caster/gelation bath system to the wind-up spool bath system were purposely kept larger than that which would promote tube wall collapse or kinking.

CONCLUSIONS

An investigation has been undertaken with the objectives of developing a sufficient understanding of the factors pertaining to the continuous casting of tubular membranes so that the inherent cost advantages derived from continuous casting may be fully exploited in future designs of reverse osmosis modules. Tubular membrane casting and heat treating techniques have now been improved to the point where reproducibility in product flux is within approximately one percent. A study of the effect of several fabrication variables on tubular membrane desalination performance has shown that performance is maximized at conditions not previously used. By using sufficiently short evaporation times during the casting process, the salt rejection properties of the membrane can be significantly increased. Performance data were obtained for single, batch-cast twenty-foot-long nominal 1/2" tubular membranes cast with a machine caster. The performance was comparable to the best results which have been obtained with two-foot-long tubular membranes. Viscosity data were obtained for standard cellulose acetate casting solutions over a wide range of shear rates, and at two different temperatures. The data were fit with a modified hyperbolic model and a power law model. A continuous tubular caster design was developed using the results of the fabrication variables study and the viscometric measurements. It is based on the principle of extrusion of the polymer solution through an annular orifice and onto moving split and butted tapes shaped into a cylindrical configuration. The continuous casting device was built and operated. Tubular membranes up to about 40 feet in length were produced with this experimental device. It produced tubular membrane whose compaction and burst pressure characteristics were similar to those of batch cast membranes and whose permselective properties were comparable to the results obtained from standard batch cast membranes. The apparatus proved to be capable of producing tubular membrane whose properties were essentially constant in the longitudinal direction while the length of produced tubing was limited only by the length of the tapes used to form a smooth, moving casting surface. The performance of the continuous casting apparatus was demonstrated at casting rates up to 2/3 ft/min.

ACKNOWLEDGMENT

The authors are grateful to the Office of Saline Water for support, under OSW Contract 14-01-0001-2206, of the work reported here. We are also appreciative of the assistance provided by Mr. Alexander O. Brodie in the design of the first version of the continuous casting apparatus. The viscosity data were obtained with the help of Drs. Robert A. Pett and Seymour Newman.

NOMENCLATURE

A product water permeation coefficient (two parameter model) (gal/day-ft^2-psi) or (cm/sec-atm)

b compaction slope defined by
$\log_{10} X(t) = \log_{10} X\ (1\ \text{hr}) + b \log_{10} t$
where t is expressed in hours and X is a generalized designation for J, S, A, or B

B salt permeation coefficient (two parameter model) (gal/day-ft^2) or (cm/sec)

CA cellulose acetate

C_{B_O} salt concentration of brine solution at module or test cell inlet (ppm)

C_P salt concentration in product solution (ppm)

D tube diameter (inch)

D_i inside diameter of extrusion annulus (inch)

D_o outside diameter of extrusion annulus (inch)

DS total degree of substitution

g_c gravitational constant = 32.17 lb_m-ft/lb_f-sec^2

J product permeation velocity or flux through the membrane (relative to a stationary observer) (gal/day-ft^2) or (cm/sec)

L	tube length or land length
$\dot{L}$	linear rate of production of extrudate (ft/min)
m	coefficient in power law model of viscosity-shear rate dependence (lb_f-sec^n/ft^2)
n	exponent in power law model of viscosity-shear rate dependence
N_{Re}	Reynolds number = $U_{Bo}\ D/\nu$
P_o	feed solution pressure at system inlet (psig)
ΔP	frictional static pressure drop in flow conduit (psi)
q	shear rate (sec^{-1})
Q_{Bo}	bulk brine volumetric flow rate at module system or test cell inlet (gal/min)
R	salt rejection = $(C_B - C_P)/C_B = 1 - 1/S$
R_i	inside radius of extrusion annulus (inch)
R_o	outside radius of extrusion annulus (inch)
s	$1/n$
S	salt reduction = $C_B/C_P = (1 - R)^{-1}$
t	time (hour)
T_o	feed solution temperature at inlet (°C)
T_{HT}	heat treating temperature (°C)
U_{B_o}	brine velocity at module or test cell inlet (ft/sec)
X_G	distance from bottom of casting bob to top of liquid level in gelation bath
α	R_i/R_o
δ	thickness of annular opening (inch)
μ	absolute viscosity (gm/cm-sec or poise)
ν	kinematic viscosity = μ/ρ (cm^2/sec)
ρ	total mass density of fluid (gm/cm^3)

τ shear stress (dyne/cm^2)

τ_E evaporation time (sec)

τ_G gelation time (sec)

τ_{HT} heat treating time (min)

τ_{WS} wall separation time - time for newly gelled polymer to separate from casting surface (sec)

REFERENCES

1. S. Manjikian, S. Loeb, and J. W. McCutchan, "Improvement in Fabrication Techniques for Reverse Osmosis Desalination Membranes." Proceedings of the First International Symposium on Water Desalination, Vol. 2, Washington, D. C.: U. S. Dept. of the Interior, Office of Saline Water (1967). pp. 159-173.

2. S. Loeb, Desalination, 1, 35 (1966).

3. B. Kunst and S. Sourirajan, J. Appl. Poly. Sci., 14, 723 (1970).

4. J. S. Johnson, J. W. McCutchan, and D. N. Bennion, "Preparation and Performance of Cellulose Acetate Semipermeable Membranes for Sea Water Service." Report UCLA-ENG-7139, University of California at Los Angeles (June 1971).

5. John L. Richardson, Gilbert Segovia, et al., "Reverse Osmosis Tubular Module Optimization." OSW Research and Development Progress Report No. 455 (December 1969).

6. John L. Richardson, G. Segovia, and Alexander O. Brodie, "Development of an Improved Tubular Reverse Osmosis Module for Water Treatment." OSW Research and Development Progress Report No. 576 (December 1970).

7. U. Merten (Ed.), Desalination by Reverse Osmosis, The M.I.T. Press, Cambridge, Mass. (1966).

8. R. B. Bird, W. E. Stewart, and E. N. Lightfoot, Transport Phenomena, John Wiley & Sons, Inc., New York, (1960), pp. 98-101, 119.

9. R. B. Bird, "Macromolecular Hydrodynamics." Chapter 1 in Lectures in Transport Phenomena, No. 4 in the AIChE Continuing Education Series. American Institute of Chemical Engineers, New York (1969).

10. Ulv Lohmander, Die Makromolekulare Chemie, 72, 159 (1964). Note also references 2 and 3 of this article.

11. Peter J. Wolfe, "Nonlinear Least Squares Curve Fit (GAUSHAUS)." UWCC ID Code C0017-00/S0017-00. University of Wisconsin Computing Center, Madison, Wisconsin (1965).

12. A. G. Frederickson and R. B. Bird, Ind. Eng. Chem. 50, 347 (1958).

13. Ernest C. Bernhardt (Ed.), Processing of Thermoplastic Materials. Reinhold Publishing Corp., New York (1959).

14. John L. Richardson, G. Segovia, J. W. Mason, W. J. Subcasky, et al. "Advanced Reverse Osmosis Membrane-Module Systems." OSW Research and Development Progress Report No. 661 (July 1971).

III. NON-CELLULOSIC MEMBRANES

THE PHENOMENOLOGICAL CHARACTERIZATION OF DP-1 MEMBRANES

Lynn E. Applegate and C. Richard Antonson
E. I. du Pont de Nemours & Company
Organic Chemicals Dept., Jackson Laboratory
Wilmington, Delaware 19899

INTRODUCTION

During the past few years, great interest and active research have been directed to desalination of brackish water and seawater via reverse osmosis. Until recently cellulose acetate membranes were the only practical desalinating membranes developed; however, with the discovery of permselective, aromatic, nitrogen-containing polymeric membranes by Richter and Hoehn[1], a whole new class of practical reverse osmotic membranes is available. These new membranes are superior to cellulose acetate membranes with respect to their permeation characteristics and polymer properties.

The phenomenological behavior of one of these new membranes (an aromatic polyamide-hydrazide; DP-1) has been examined and compared with the transport behavior of cellulose acetate membranes.

Typical permeation characteristics of DP-1 membranes are given in Table 1.

TABLE 1

PERMEATION CHARACTERISTICS OF DP-1 MEMBRANES

A. Seawater at 1000 psi
   Volume Flux = 10-15 gal/ft^2-day
   Rejection = 99.2-99.5%

B. Brackish Water (5000 ppm NaCl) at 1000 psi
   Volume Flux = 17-25 gal/ft^2-day
   Rejection = >99%

EXPERIMENTAL

The DP-1 polymer was prepared by reacting a mixture of 3-aminobenzhydrazide (80 mole %) and 4-aminobenzhydrazide (20 mole %) with a mixture of isophthaloyl chloride (70 mole %) and terephthaloyl chloride (30 mole %) in dimethylacetamide (DMAC). The polymer was isolated in a blender by addition of ice and water with

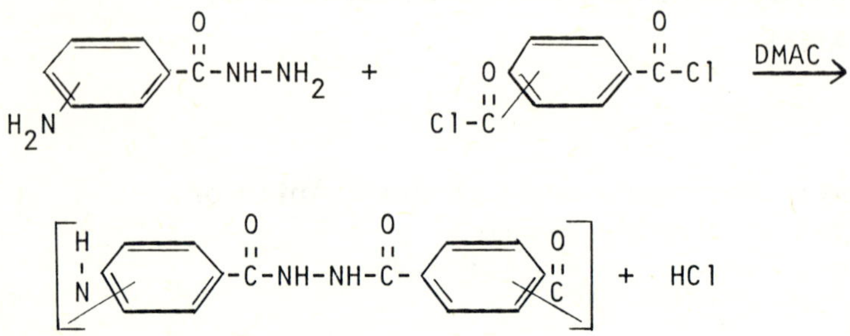

DP-1

vigorous stirring, washed with water, dried, and redissolved in DMAC for membrane casting. The polymer had an inherent viscosity of 0.85 (in DMAC at 25°) and a weight-average molecular weight ($\overline{M}_w$) of 90,000.

Membranes were cast from a 15% polymer solution in DMAC containing 30% lithium nitrate (based on the weight of the polymer). The membranes were cast with a doctor knife (15 mil setting), dried at 80° for 15-20 minutes, and then quenched in stirred distilled water. Membranes prepared by this procedure are anisotropic as evidenced by dyeing experiments, directionality of reverse osmotic experiments, and electron microscopy.

The reverse osmotic apparatus used to evaluate the membranes has been described in the literature[2].

The volume flux was determined by collecting a volume of liquid in a volumetric pipet over a known time period. The salt concentrations in the permeate and reject effluent streams were determined by conductivity measurements.

Rejections were computed from the feed and permeate salt concentrations assuming no concentration polarization on the feed side. In each set of runs, pressure was the independent variable while feed concentration was varied between sets. For each set, pressure-rejection data were analyzed in terms of several transport models. Rejection expressions derived from the models depend nonlinearly on the pressure, concentration, osmotic pressure, and ratios of the membrane mass transfer coefficients. In models involving only one ratio, the best estimate of that ratio (in a least squares sense) was obtained for each set of runs via a Fibonacci search; where more than one ratio was involved, a compromise between steepest descent and Taylor series expansion was employed[3]. Individual values of the mass transfer coefficients were computed from the best ratio estimates and the volume flux data.

With several sets of data, the rejection (R) and volume flux (J_V) as functions of pressure were analyzed simultaneously to obtain the best overall estimates of the mass transfer coefficients. Thus, for each model, the number of parameters was increased by one to include not only the coefficient ratios but also the water permeability. For the least-squares fit, it was necessary to define the sum square residuals (SSR) as the square of the experimental value minus the model-predicted value divided by the data standard deviation as defined below.

$$SSR = \sum_{i=1}^{N} \left(\frac{R^i_{exp} - R^i_{mod}}{\sigma_R} \right)^2 + \sum_{i=1}^{N} \left(\frac{J_{v_{exp}} - J_{v_{mod}}}{\sigma_{J_v}} \right)^2$$

Normalization via division by the standard deviation was necessary in order to account for the fact that rejection is moving in a very narrow range and is dimensionless while volume flux moves through a wide range of values and has units of liters/cm^2-sec. The resulting coefficient ratios did not differ significantly from the values obtained using rejection data alone while the water permeability was close to that obtained by analyzing the volume flux data separately. Thus, it was concluded that the simpler analysis outlined above was adequate.

THEORETICAL TRANSPORT MODELS

Sieve Transport Models

Sieve transport models depict the membrane as a network of capillaries in which flow is determined from Poiseuille's equation and solute rejection occurs by a sieving action. For partial rejection, the membrane must contain pores of various diameters, some pores being large enough to permit solute flow. Banks and Sharples[4] have derived phenomenological transport equations for various theoretical sieve models, and a brief review is presented below.

Sieve Transport Model I. If the effluent streams from rejecting and non-rejecting pores do not mix until they are at an appreciable distance from the membrane surface on the permeate side, the volume flux (J_v) and salt flux (J_s) are

$$J_v = k_1(\Delta P - \pi_o) + k_2\Delta P \tag{1}$$

$$J_s = k_2\Delta P C_o \tag{2}$$

where k_1 and k_2 are transport coefficients, ΔP is the pressure difference across the membrane, and C_o and π_o are the salt concentration and osmotic pressure on the brine side of the membrane. Since rejection (R) is defined by

$$R = 1 - \frac{C_e}{C_o} = 1 - \frac{J_s}{C_o J_v} \tag{3}$$

where C_e is the salt concentration in the permeate, the relation between rejection and pressure is

$$R = 1 - \frac{k_2\Delta P}{(k_1 + k_2)\Delta P - k_1\pi_o} \tag{4}$$

Sieve Transport Model II. If the stream from the rejecting pores mixes with the stream from the non-rejecting pores at the membrane surface on the permeate side, the driving force for the two types of pores will be $(\Delta P - \Delta\pi)$ where $\Delta\pi = \pi_o - \pi_e$ and π_e is the osmotic pressure of the permeate. For this model the transport equations are

$$J_v = k_1(\Delta P - \Delta\pi) + k_2(\Delta P - \Delta\pi) \tag{5}$$

$$J_s = k_2(\Delta P - \Delta\pi)C_o \tag{6}$$

and the rejection equation is

$$R = 1 - \frac{k_2}{k_1 + k_2} \tag{7}$$

Sieve Transport Model III. If the rejecting and non-rejecting streams mix at the membrane surface on the permeate side, and if the flow through the non-rejecting pores is controlled only by ΔP, the transport equations are

$$J_v = k_1(\Delta P - \Delta\pi) + k_2\Delta P \tag{8}$$

$$J_s = k_2 \Delta P C_o \tag{9}$$

Since $\Delta\pi = \pi_o - \pi_e$ and $R = 1 - \frac{\pi_e}{\pi_o}$, the rejection can be shown to be

$$R = \frac{k_2\Delta P + k_1\Delta P + k_1\pi_o - \{[(k_1+k_2)\Delta P + k_1\pi_o]^2 - 4k_1{}^2\Delta P\pi_o\}^{1/2}}{2k_1\pi_o} \tag{10}$$

Solubility-Diffusion Model

Lonsdale, Merten, and Riley[5] have derived transport equations based on a solubility-diffusion concept. The model assumes that water and salt dissolve in the membrane and diffuse through the membrane uncoupled. The transport equations for this model are

$$J_v = k_1(\Delta P - \Delta\pi) \tag{11}$$

$$J_s = k_2\Delta\pi \tag{12}$$

Using Equations (11) and (12) and the definition of rejection, the rejection can be shown to be[4]

$$R = 1 - \{k_1\pi_o - k_1\Delta P - k_2/B + [k_1{}^2\Delta P^2 + (k_2/B)^2 + 2k_1k_2\Delta P/B + k_1{}^2\pi_o{}^2 - 2k_1{}^2\pi_o\Delta P + 2k_1k_2\pi_o/B]^{1/2}\}/2k_1\pi_o \tag{13}$$

where B is C_o/π_o.

Solubility-Diffusion-Imperfection Model

Sherwood, Brian, and Fisher[6] have derived transport equations for a solubility-diffusion-imperfection mode. The model assumes that water and salt are transported through the membrane via a solubility-diffusion mechanism, but some imperfections are present which permit salt to pass undiluted. The transport equations for this model are

$$J_v = k_1(\Delta P - \Delta\pi) + k_3\Delta P \tag{14}$$

$$J_s = k_2\Delta\pi + k_3\Delta P C_o \tag{15}$$

It can be shown that rejection is

$$R = \{C_o\Delta P(1+\alpha) + \pi_o(C_o+\beta) - [C_o^2\Delta P^2(1+\alpha) + 2C_o\Delta P\pi_o (1+\alpha)(C_o+\beta) + \pi_o^2(C_o+\beta)^2 - 4\pi_o\Delta PC_o^2]^{1/2}\}/2\pi_oC_o \tag{16}$$

where $\alpha = k_3/k_1$ and $\beta = k_2/k_1$

Computer analysis of pressure-rejection data for the DP-1 membranes and cellulose acetate membranes (literature data[2]) was used to determine which equation [(4), (7), (10), (13), or (16)] best describes the transport behavior of the membranes.

RESULTS AND DISCUSSION

The computer analyzed pressure-rejection data for an isotropic cellulose acetate membrane[2] and an anisotropic DP-1 membrane (typical of several membranes) are shown in Figures 1-4 for two sodium chloride feed concentrations (o = experimental data). The pressure-rejection data were first analyzed in terms of the simple models which involve the estimation of only one parameter. Curves 1, 3, and 4 in these figures correspond to the sieve models I and III and the solubility model, respectively, and thus represent least squares fits to Equations (4), (10), and (13), respectively. Since sieve model II predicts rejection to be independent of pressure, which was not observed, this model was discarded. With low salt concentrations, the diffusive mechanism was clearly better than either of the pore flow mechanisms in a least square sense. However, with high salt concentrations, the sum square residuals

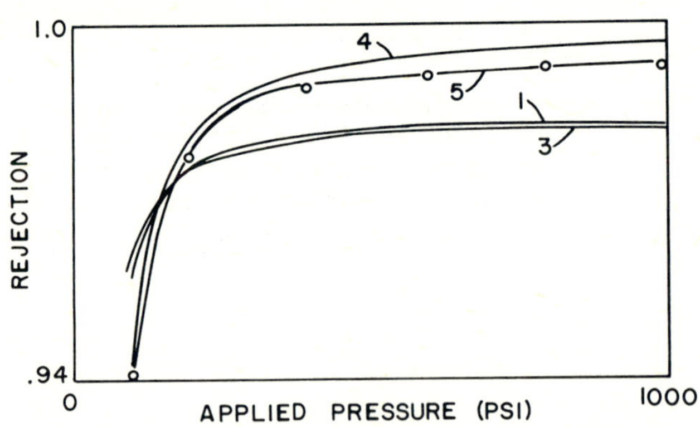

Fig. 1 - Salt Rejection vs. Applied Pressure for a Cellulose Acetate Membrane. 0.1 M NaCl Feed Solution.

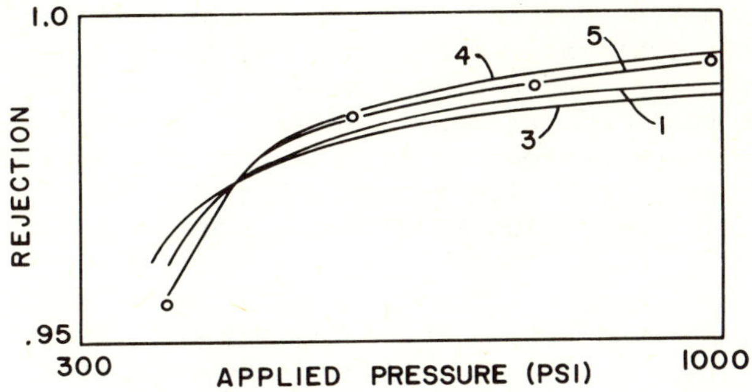

Fig. 2 - Salt Rejection vs. Applied Pressure for a Cellulose Acetate Membrane. 0.50 M NaCl Feed Solution.

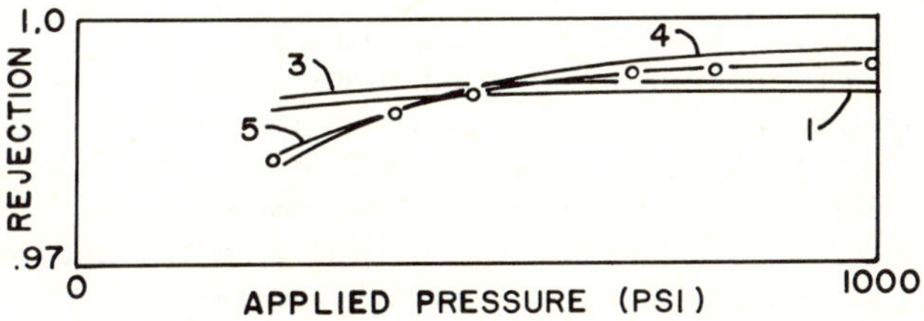

Fig. 3 - Salt Rejection vs. Applied Pressure for a DP-1 Membrane. 0.0564 M NaCl Feed Solution.

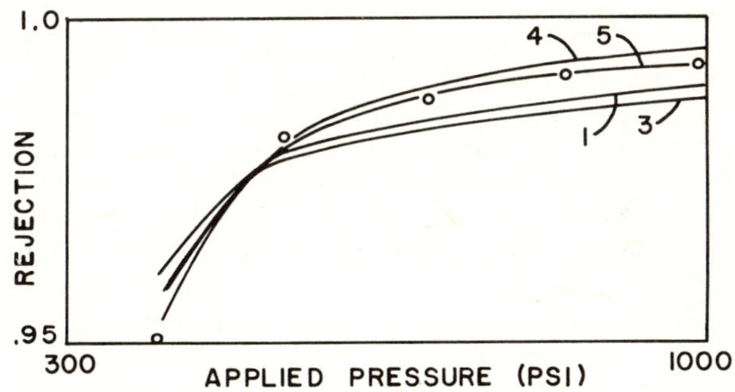

Fig. 4 - Salt Rejection vs. Applied Pressure for a DP-1 Membrane. 0.50 M NaCl Feed Solution.

were of the same order of magnitude for all simple models. Moreover, it was noted that in all sets of runs, all three simple models exhibited a systematic bias; that is, there was a definite trend in the deviations indicating a failure of the models to account for some phase of the mass transfer. The diffusive model was always below the experimental data in regions where the pore flow models were above and vice versa. This observation indicated a combined diffusion-pore flow model might provide a much better fit to the data.

The diffusion-imperfection model (line 5, corresponding to Equation (16)) requires the estimation of two parameters. In all cases, it was found that this model resulted in a reduction of the sum square residuals by one to two orders of magnitude when compared to the simpler models. This in itself is not surprising because two parameters are now being estimated instead of one. However, the deviations no longer exhibited a systematic bias; they were both small and random for the diffusion-imperfection model, thus establishing the model's efficacy for the prediction of rejection as a function of pressure over the range of pressure employed. However, to conclusively assert that the diffusion-imperfection model employed here is the true mass transfer mechanism would require comparison to alternative mechanisms of similar complexity. Nevertheless, the excellent fit achieved indicates that the diffusion-imperfection model is a good approximation of the true mass transfer mechanism in this operating region for both cellulose acetate and DP-1 membranes. The universality of the solubility-diffusion-imperfection model for various solutes has yet to be determined.

The computed values of the individual transport coefficients for both cellulose acetate and DP-1 membranes are given in Table 2.

As the salt concentration of the feed solution increased, k_1 always decreased. This observation has previously been noted [2,7] and appears to be a general phenomenon of reverse osmotic membranes. It has been postulated[7] that the membrane structure changes as the activity of water in the membrane changes. Thus, the activity of water increases as the salt concentration decreases, and the membrane structure undergoes orientation, i.e., it is "loosened". This phenomenon can be viewed as a hydration-dehydration process. Changes in k_2 and k_3 show that the structural changes which occur are complex, but collectively k_2 and k_3 decrease to produce a salt flux which is reduced by approximately the same extent as the volume flux. Thus, rejection at the same driving pressure ($\Delta P - \Delta\pi$) is approximately independent of salt concentration.

TABLE 2

TRANSPORT COEFFICIENTS FOR CELLULOSE ACETATE AND DP-1 MEMBRANES

Membrane	NaCl Conc.[a] (M)	$k_1{}^b \times 10^{10}$	$k_2{}^c \times 10^{12}$	$k_3{}^b \times 10^{12}$
Cellulose Acetate	0.0129	0.32	0.11	0.17
	0.10	0.29	0.11	0.09
	0.50	0.26	0.13	0.03
DP-1 (Membrane A)	0.056	6.80	2.96	1.07
	0.50	5.39	1.89	1.20
DP-1 (Membrane B)	0.056	4.14	2.11	0.97
	0.50	3.45	1.18	1.53

[a]Salt concentration of feed solution.

[b]In units of liters/cm^2-sec-psi.

[c]In units of moles NaCl/cm^2-sec-psi.

The effect of pressure on the DP-1 membrane structure was investigated by continuously increasing and decreasing the applied pressure and determining the transport coefficients for each pressure set. Pressure compaction of the membrane had the same effect on transport coefficients as increasing the salt concentration. Thus, k_1 decreased, and collectively k_2 and k_3 also decreased to yield a rejection which was not substantially different from that before compaction.

ACKNOWLEDGMENT

The authors wish to acknowledge Drs. L. A. Cescon and J. B. Ganci for many helpful discussions. Thanks is also due to Mr. M. M. Montgomery for assistance in the experimental work.

REFERENCES

1. J. W. Richter and H. H. Hoehn, U.S. Patent 3,567,632 (1971).

2. S. Rosenbaum and W. E. Skiens, J. Appl. Polym. Sci., 12, 2169 (1968).

3. D. W. Marquardt, S.I.A.M.J., 11 (2), 431 (1963).

4. W. Banks and A. Sharples, J. Appl. Chem. (London), 16, 153 (1966).

5. H. K. Lonsdale, U. Merten, and R. L. Riley, J. Appl. Polym. Sci., 9, 1341 (1965).

6. T. K. Sherwood, P.L.T. Brian, and R. E. Fisher, Ind. Eng. Chem., Fundam., 6 (1), 2 (1967).

7. W. Vieth, A. S. Douglas, and R. Bloch, "Study of Poly(Methacrylates) and Poly(Urethanes) as Reverse Osmosis Membranes. Effect of Water Clustering on Transport Properties", Office of Saline Water Research and Development Progress Report No. 352, August 1968.

PROPERTIES OF AROMATIC POLYAMIDE AND POLYAMIDE-HYDRAZIDE MEMBRANES

R. McKinney, Jr.

Chemstrand Research Center, Inc.
Durham, North Carolina 27702

INTRODUCTION

The search for new and improved membrane compositions has continued on a large scale since the discovery of cellulose acetate as a desalination membrane by Reid and Breton[1], and subsequent improvements in flux demonstrated by Loeb and Sourirajan [2,3]. Such efforts have generally been rewarded with only modest success, restricted primarily to modifications of cellulose esters.

As a result of this study, a novel class of membrane compositions have emerged based on a system of aromatic polyamide-hydrazide [4] and polyamide[5,6] polymers. Two polymers, characteristic of these groups, are described. They are: I. a polyterephthalamide of p-aminobenzhydrazide, and II. a polyterephthalamide of 1,3-bis (3-aminobenzamide)benzene. Their structures are given in Figure 1. The transport properties of asymmetric membranes prepared from each system are described. High selectivity to numerous dissolved salts at moderately high levels of water transport with excellent chemical and mechanical stability are the characteristic properties of these membranes.

EXPERIMENTAL

Polymer Synthesis

All polymers were prepared by low temperature solution polymerization in dimethylacetamide. Polymer I was prepared from terephthaloyl chloride and p-aminobenzhydrazide at 10°C. Polymer II resulted from the polycondensation of terephthaloyl chloride

Polymer I

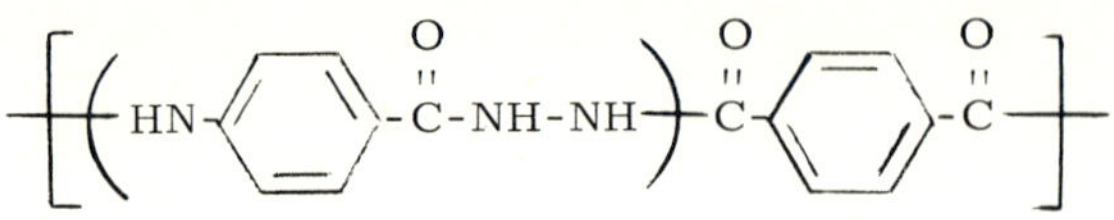

Polymer II

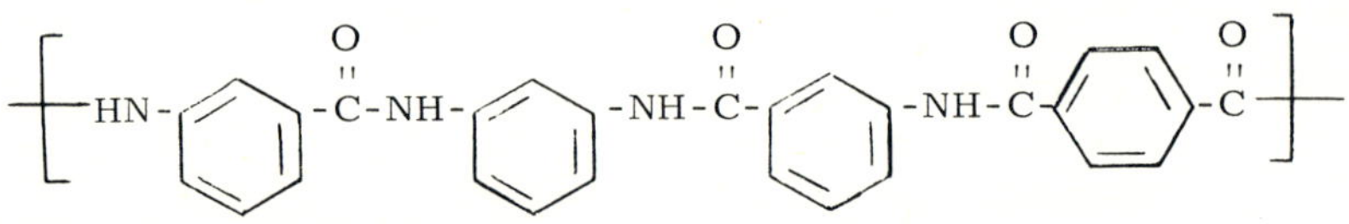

Fig. 1 - Structures of Aromatic Polyamide-Hydrazide and Polyamide Polymers

with the complex diamine 1,3-bis(3-aminobenzamide)-benzene at -20°C. Polymer in the molecular weight range 20,000 to 70,000 could be prepared in this manner. Once the polycondensation reactions were complete, calcium carbonate in quantities sufficient to neutralize 95 mole percent of the HCl liberated during the polymerization was mixed with the polymer solution. The mixture was heated to 75-80°C to achieve complete reaction of the $CaCO_3$ (producing $CaCl_2$) followed by vacuum degassing. The resulting solution was then pressure-filtered at about 1000 psi through a 5 micron membrane filter.

Membrane Preparation

Membranes of anisotropic structure were prepared from the polymer solution described above according to the following procedure. The freshly filtered solution (containing 6% Polymer I or 12% Polymer II) was poured on to clean dry glass plates and cast to a predetermined thickness by means of a Gardner[R] knife with an adjustable blade. The glass plate containing cast solution was immediately placed in a mechanical convection oven fitted with a filtered air intake and thermostated to ± 0.5°C. At the completion of the solvent evaporation period, the partially formed membrane was removed from the oven and coagulated in a deionized water bath. The membrane was then allowed to float free of the glass plate and remain in the water bath for a minimum of 12 hours to insure complete removal of solvent and $CaCl_2$. Membranes prepared from Polymer I required an annealing post treatment in water similar to that observed for cellulose acetate[2]. Polymer II membranes required no additional treatment to achieve high selectivity.

Membrane Testing

The reverse osmosis experiments were carried out in a membrane test system previously described[7]. The pressurized feed solution was pumped across the membrane surface at a linear velocity in excess of 200 cm/sec. Boundary layer phenomena are known to be negligible under these conditions. Test solutions employed in this study were 0.5, 3.5 and 6.0% NaCl. An operating pressure of 600 psi was employed for the 0.5% test solution while 1500 psi was used for the others.

RESULTS AND DISCUSSION

The formation of membrane anisotropy occurs during the oven drying phase of membrane preparation. During this period a membrane skin or thin layer of dense polymer is formed at the air surface as a result of the established solvent gradient. Coagulation preserves or fixes this structure. The matrix, or porous region on which this skin rests, is formed during coagulation by polymer precipitation resulting from the competing processes of coagulant diffusion and solvent extraction. In Figure 2, the structure of a membrane prepared from Polymer I is illustrated.

Fig. 2 - Electron Micrograph of Unannealed Polymer I (19000X)

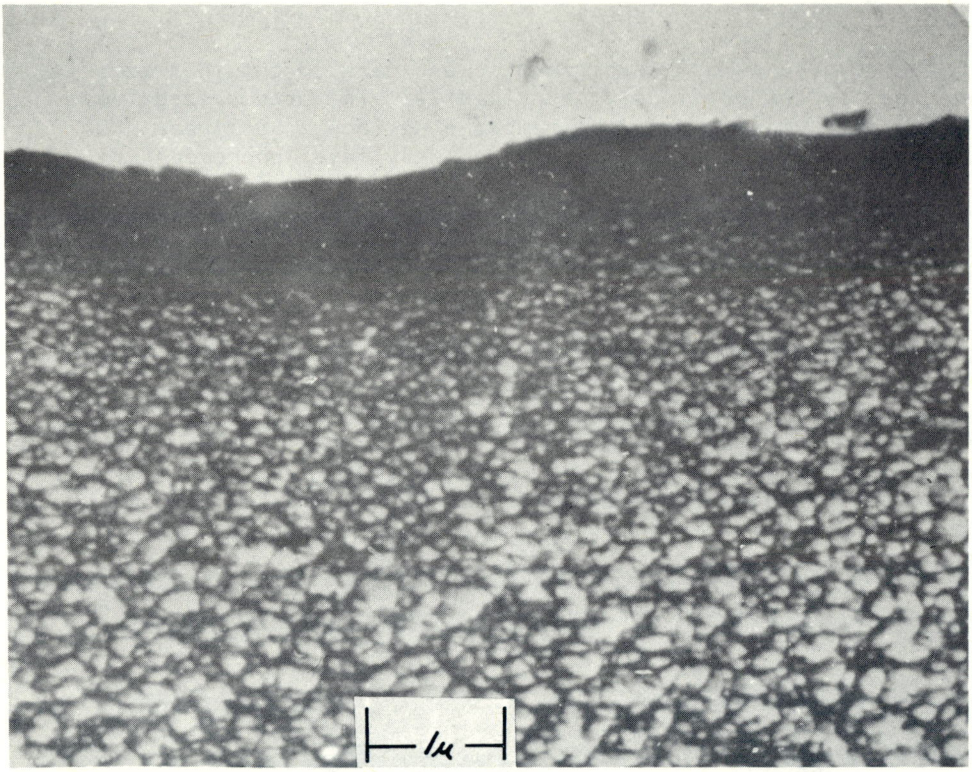

Fig. 3 - Electron Micrograph of Annealed Polymer I (19000X)

A rudimentary low density skin is evident when coagulated at 25°C. Annealing this membrane for 15 minutes in boiling water results in a densification of this layer (Figure 3). A corresponding change in membrane reverse osmosis properties is evident. The unannealed membrane provided a flux of 30 gal/ft^2-day (gfd)* at 60% rejection when tested at 600 psi with a 5000 ppm NaCl feed. The annealed membrane produced a flux of 2-4 gfd at 98% rejection when tested under the same conditions. A flux-rejection trade-off with annealing temperature exists for these membrane systems, i.e., higher annealing temperatures result in improved selectivity at a reduced flux. Electron microscopy indicated the skin thickness of the annealed sample to be 3000 to 5000 Å with a matrix pore-size range of 0.1 to 0.75 micron. These dimensions are comparable to that observed[8] for cellulose acetate although the structural sequence may be slightly altered. Membranes prepared from Polymer II do not require the annealing post treatment. A uniform continuous skin (Figure 4) is established during the oven evaporative treat-

*Divide gfd by 2.12 x 10^4 to convert to cm^3/cm^2-sec.

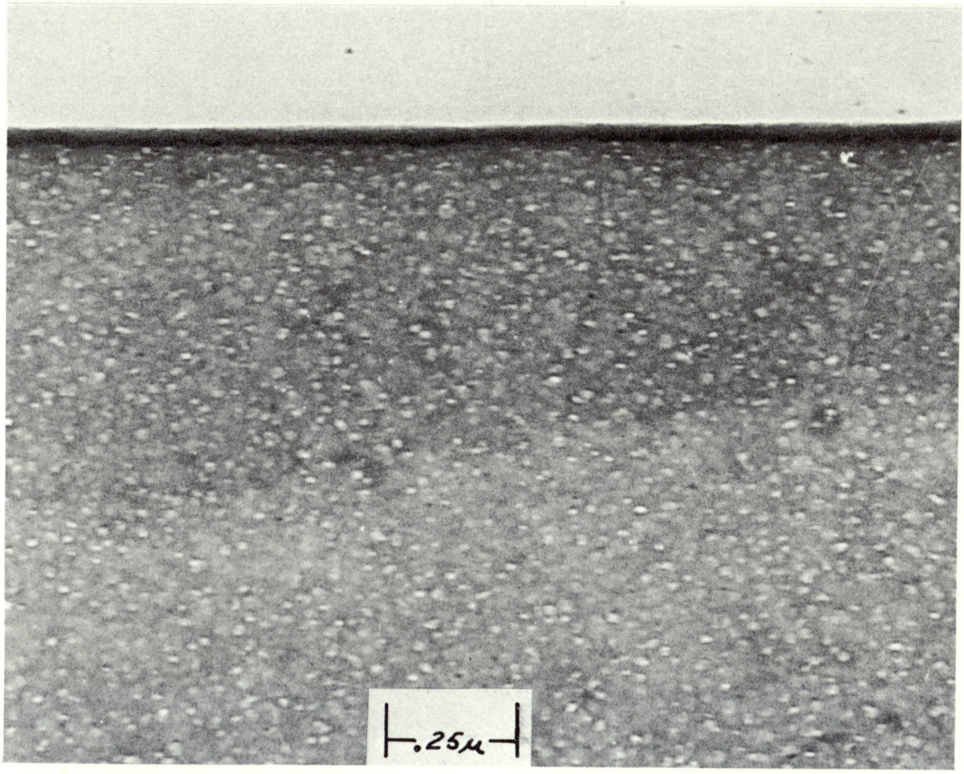

Fig. 4 - Electron Micrograph of Polymer II (71000X)

ment. Microscopy indicates this skin to be 300 Å thick although variations (ca. ± 100 Å), reflecting changes in membrane preparation technique, have been recorded. The matrix density (frequency and size of pores) is higher for Polymer II. Pore sizes range from around 14 to 150 Å for this system. Typical reverse osmosis performance for a Polymer II membrane with such structure would be 8 to 9 gfd at 99.6 to 99.8% rejection at 600 psi with a 5000 ppm NaCl feed.

The selectivity and product water flux of asymmetric membranes prepared from Polymer I and II are summarized in Table 1. Table 2 illustrates the selectivity of these membranes to several additional solutes. Highest rejection levels are evident for membranes prepared with type II polymer. Both systems provide high rejection levels for the aqueous salt solutions investigated.

Typical mechanical properties for asymmetric and homogeneous membranes prepared from these polymers are given in Table 3. The mechanical strength of each system is very high and can be appreciated when compared to that observed for cellulose acetate.

TABLE 1

REVERSE OSMOSIS DATA FOR AROMATIC POLYAMIDE-HYDRAZIDE AND POLYAMIDE MEMBRANES

Polymer	Operating Conditions[a] Feed (ppm as NaCl)	Operating Conditions[a] Pressure (psi)	Reverse Osmosis Data[b] Flux (gfd)	Reverse Osmosis Data[b] Rejection (%)
I	5,000	1,000	8.7	98.0
	35,000	1,500	8.3	92.8
	60,000	1,500	7.3	97.6
	Natural Seawater	1,500	7.0	95.7
II	5,000	1,000	15.4	99.4
	35,000	1,500	17.9	99.8
	60,000	1,500	11.4	99.8
	Natural Seawater	1,500	16.4	99.7

[a]Natural seawater sample from the Atlantic Ocean, Nags Head, North Carolina.

[b]All reverse osmosis data taken from a multipass test loop operated at zero percent recovery and 23 ± 1°C.

TABLE 2

SELECTIVITY OF POLYMER I AND II MEMBRANES TO VARIOUS SOLUTES[a]

Test Solution		Polymer I		Polymer II	
Solute	Concentration (ppm)	Product (ppm)	Rej. (%)	Product (ppm)	Rej. (%)
NaCl	5,000	100	98	10	99
$MgSO_4$	10,000	24	99	16	99
NH_4Cl	2,300	6	99	2	99
HCl	2,760	103	96	48	98
NaOH	2,000	925	54	430	83
NH_4OH	1,900	-	-	1,343	29
Boron[b] (pH=3.5)	67	-	-	11	83
(pH=10.2)	67	-	-	1	98
Urea	18,000	-	-	1,500	92
Fructose	2,000	21	99	18	99
Polyethylene glycol (400 mw)	10,000	-	-	30	97
Polyethylene glycol (1000 mw)	10,000	20	99	20	99

[a]Reverse osmosis data taken at 600 psi and 25 ± 0.05°C utilizing a Polymer I membrane with a flux of 2 gfd and a Polymer II membrane with a flux of 6-7 gfd. Intermediate checks with 5000 ppm NaCl verified a lack of membrane damage by any of the solutes tested.

[b]Solution concentrations in the product and feed are expressed as ppm of boron. Boric acid (with HCl) was used for the low pH tests, sodium borate (with NaOH) for the high.

TABLE 3

TYPICAL MECHANICAL PROPERTIES[a] OF POLYMER I AND II FLAT MEMBRANES AT 100% RELATIVE HUMIDITY

	Asymmetric Membrane			Homogeneous Membrane		
Polymer	Modulus (psi)	Yield Stress (psi)	Water Content (%)	Modulus (psi)	Yield Stress (psi)	Equilibrium Water Content (%)
I	250,000	7,100	28	348,000	9,700	10.4
II	82,000	2,000	54	322,000	9,100	12.3
Cellulose Acetate	33,000	780	-	162,000	3,700	15.6

[a]Mechanical property measurements carried out on an Instron tester at 24 ± 1°C with membranes positioned underwater. Water content measurements defined as weight of membrane water divided by weight of wet membrane. For asymmetric membranes it is an indirect measurement of porosity.

ACKNOWLEDGMENT

The author wishes to thank Dr. J. R. Holsten for his assistance in preparing the polymers used in this study. Thanks are also due Dr. V. F. Holland for the electron microscopy and to Messrs. J. H. Rhodes, L. C. Locust, P. B. Woodson, and Mrs. N. Wildman for their contributions in carrying out the measurements described.

This work was carried out under Contract 14-01-0001-1720 granted by the Membrane Division, Office of Saline Water, U.S. Department of the Interior.

REFERENCES

1. C. E. Reid and E. J. Breton, Jr., J. Appl. Polym. Sci., 1, 133 (1959).

2. S. Loeb and S. Sourirajan, "Sea Water Demineralization by Means of a Semipermeable Membrane," U.C.L.A. Report 60-60, July, 1960.

3. S. Loeb and F. Milstein, Dechema Monograph., 47, 805 (1962).

4. W. B. Black and J. Preston, Chemstrand Research Center, Inc., unpublished data.

5. J. Preston and F. Dobinson, Polymer Letters, 2, 1171 (1964).

6. B. M. Culbertson, Polymer Letters, 5, 805 (1967).

7. R. McKinney, Jr., Analytical Chemistry, 41, 1513 (1969).

8. G. J. Gittens, P. A. Hitchcock, D. C. Sammon, and G. E. Wakley, Desalination, 8, 369 (1970).

MORPHOLOGY AND REVERSE OSMOSIS PROPERTIES OF SULFONATED 2,6-DIMETHYL POLYPHENYLENE OXIDE MEMBRANES

Anthony B. LaConti, Paul J. Chludzinski, and Arnold P. Fickett
General Electric Co.
Lynn, Massachusetts 01910

INTRODUCTION

It is well known that ion exchange membranes are permselective and will pass water while rejecting dissolved salts under pressure. This is the case with the cation exchange membranes of sulfonated polyphenylene oxide (PPO), in which the anchored negative sulfonate ions repulse like charges (anions in solution) and, in order that electrical neutrality be preserved, cations must also remain behind. The advantage of using sulfonated PPO for reverse osmosis application compared to other commercially available ion exchange membranes, such as those discussed by Spiegler[1] and Merten et al.[2], is that it can be solvent cast as a thin (0.1 to 0.2 mil) unsupported membrane having excellent chemical and physical stability with good flux and rejection characteristics. Water contents are adjusted easily to give the desired permeability by controlling the ion exchange capacity of the sulfonated polymer. Techniques have also been developed for adjusting the water content of a sulfonated PPO membrane with a given ion exchange capacity. This work shows that for certain feeds sulfonated PPO may be comparable or superior to cellulose acetate.

EXPERIMENTAL

Polyphenylene oxide (PPO) is sulfonated according to the following equation. The general method for preparing sulfonated PPO is described by Plummer, et al.[3]. The sulfonation is conducted in chloroform at ambient conditions using chlorosulfonic acid ($ClSO_3H$). The PPO is dissolved in chloroform to form a

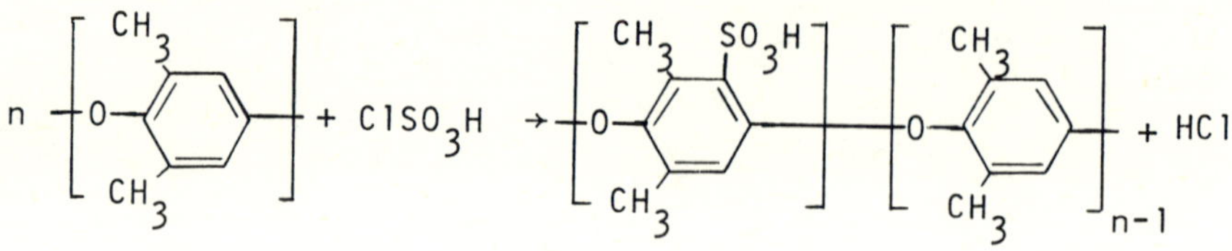

3 wt % solution. The proper amount of $ClSO_3H$ to sulfonate the PPO to the desired ion exchange capacity is added slowly with stirring over a 20-minute period. The polymer is separated and liquid discarded. Careful experimentation has shown that $ClSO_3H$ which is added to the PPO is the only active sulfonating agent. Fig. 1 shows the moles of $ClSO_3H$ needed to prepare polymers of different ion exchange capacity (IEC).

The polymer is dissolved in ACS certified methanol. It is poured into a Mylar tray, forming a film about 50 to 100 mils thick, and allowed to dry in air for 16-20 hours at room temperature. The dry polymer sheet is then shredded and washed with distilled water until the wash water shows no sulfate or chloride and has a pH above 4.0. The shredded polymer is then ground up in a Waring blender using ice to prevent heating. If the polymer overheats during grinding, it will degrade since the hydrogen ion form is unstable. The grinding is allowed to take place until

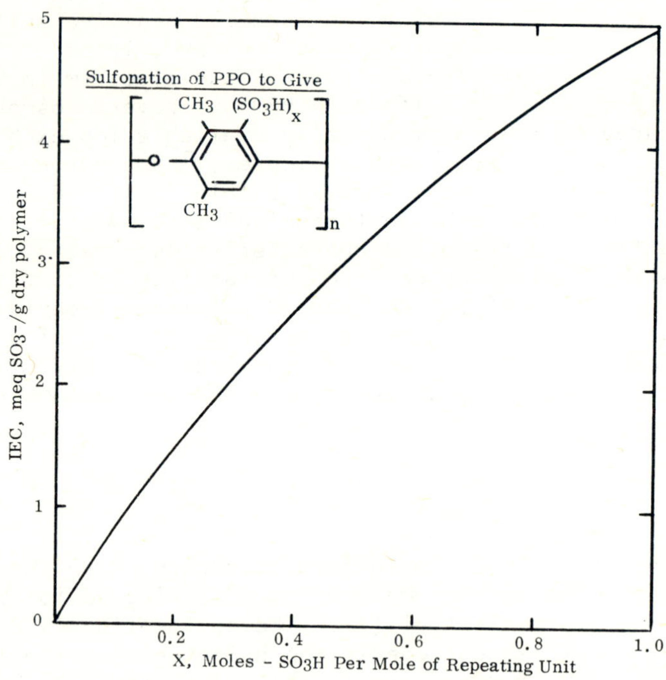

Fig. 1 - Sulfonation of PPO

the polymer particles measure about 50 mils. Larger particle size will cause the subsequent drying step to be inefficient and cause excessive water bound in the polymer. This in turn adversely affects membrane rejection. The polymer is filtered and washed with distilled water, after which it is spread out and air dried. The polymer is then stored in the hydrogen form after this step. If the polymer is desired in the sodium form, it is treated with 0.1 N NaOH, filtered and washed in distilled water. The polymer is stored after air drying.

The sulfonated PPO polymer is exhaustively dried for 72 hours under vacuum. The H^+ ion and Na^+ ion form are dried at 77 and 160°F, respectively. The dry polymer is dissolved in a 2:1 to 20:1 chloroform/methanol solvent, filtered through a 0.2μ Selas Flotronics porous silver disk using dry He gas, and a bead of casting solution (4 to 10% solids) is spread on a leveled glass plate. A draw-down bar at the desired setting is pulled across the bead to lay out the wet film. After casting the film, it is dried a minimum of 1 hour per 0.1 mil thickness of membrane at room temperature under controlled humidity. Half the drying time is under cover. The membranes are removed from the glass plates with 10% NaCl solution and stored wet in a plastic bag.

A polymer and membrane manufacturing procedure was devised based on the effects of various process variables on membrane performance[3,4]. Membranes were manufactured and found to be reproducible in flux to about ± 25% and in rejection to ± 5%. The reproducibility of flux within the same membrane sheet was ± 3%. Most membranes showed better than 25 gal/ft^2-day (gsfd) flux and 85% salt rejection when operated on natural Webster, South Dakota water at 1100 psi and 130°F.

A detailed description of the reverse osmosis test cell is given by Plummer, et al[3]. The membranes are placed on a suitable porous substrate such as Millipore RAWP, polysulfone, or polypropylene, and the cell is assembled. Feed water is circulated across a 1 x 3 inch piece of the membrane, with a feed channel depth of 1/8 inch. Two systems for developing pressure on the feed side of the assembly were used: one was a surge tank and nitrogen tank pressure (closed loop); the other was a circulating pump which developed the requisite pressure (open loop). A description of the two units is given in reference 4. Measurements of water flux are reported for pressures of 200 to 1200 psi. The reported water permeability values do not take into account osmotic pressure of the feed solution which acts in opposition to the applied pressure. The water permeability, Pr, is corrected for membrane thickness and pressure, and is expressed as (water flux) x (membrane thickness)/(applied pressure), in units of gsfd-mil/atm. Membrane thickness is measured in the dry form using an

Elcometer magnetic gauge with a 0 to 2 mil range.

RESULTS AND DISCUSSION

For the sulfonated PPO family of membranes, practical ranges of desired flux and rejection were first defined. Generally, a trade-off is made between the water permeability and salt rejection for each feed. A major effort was made to establish relationships between membrane properties and reverse osmosis performance. It was found for an invariant feed that a range of fluxes and rejections were obtained by varying the ion exchange capacity, thickness, and water content. Similarly, for a given membrane, a wide range of fluxes and rejections were obtained by varying pressure, temperature, and feed composition. Sulfonated PPO was found to be comparable to cellulose acetate for some feeds and inferior for others. Life tests (at least 100 hours) are described for certain feeds. In all instances, the slope of a plot of log flux vs. log time was quite small, indicating essentially no membrane degradation or compaction with time.

Effect of Membrane Morphology

Sulfonated PPO is readily cast into membranes of 0.1 to 1.0 mil in thickness. The membrane morphology can also be varied, thus yielding a wide range of performance. The data in Table 1 summarize some of the important physical, chemical, and reverse osmosis characteristics of typical membranes prepared under identical conditions. The reverse osmosis data were taken with a 1500 ppm NaCl feed at 77°F and 800 psi. The membrane properties which have the greatest impact on reverse osmosis characteristics are the ion exchange capacity (IEC) and water content (wt of water/total wt) of the membrane.

TABLE 1

MEMBRANE MORPHOLOGY AND RO PROPERTIES

IEC (meq/g)	Tensile Strength (psi)	Elong. (%)	Pr x 10^2 (gsfd-mil/atm) at 77°F	H_2O Content	Vol. Fraction H_2O	NaCl Rej. (%)	Pore Charge Density (meq SO_3-/cc H_2O)
1.82	4070	171	1.67	0.301	0.340	96	4.23
2.05	3612	171	4.24	0.383	0.427	92	3.31
2.28	3395	176	5.43	0.405	0.449	92	3.35
2.67	1765	126	11.0	0.512	0.560	89	2.54

The IEC (meq of SO_3^-/g dry polymer) affects water content and controls Pr. Higher values give membranes with higher water content with all other variables fixed. The physical stability of the membrane is also influenced by the IEC. Physical stability decreases with increasing IEC. The IEC function which appears to strongly influence salt rejection is the pore charge density (meq of SO_3^-/cc of pore fluid). Membranes with high values exhibit high rejections. Membrane water permeability and flux increase with increasing water content of the membrane. For membranes having an IEC of 1.8 to 3.0 meq/g the logarithm of the Pr observed during reverse osmosis exhibits a linear dependence with the volume fraction of water (Fig. 2). The membrane water content increases with increasing IEC, water content of the casting solution[4], $CH_3OH/CHCl_3$ ratio of casting solution[3], and temperature of water equilibration of the sulfonate membrane[3].

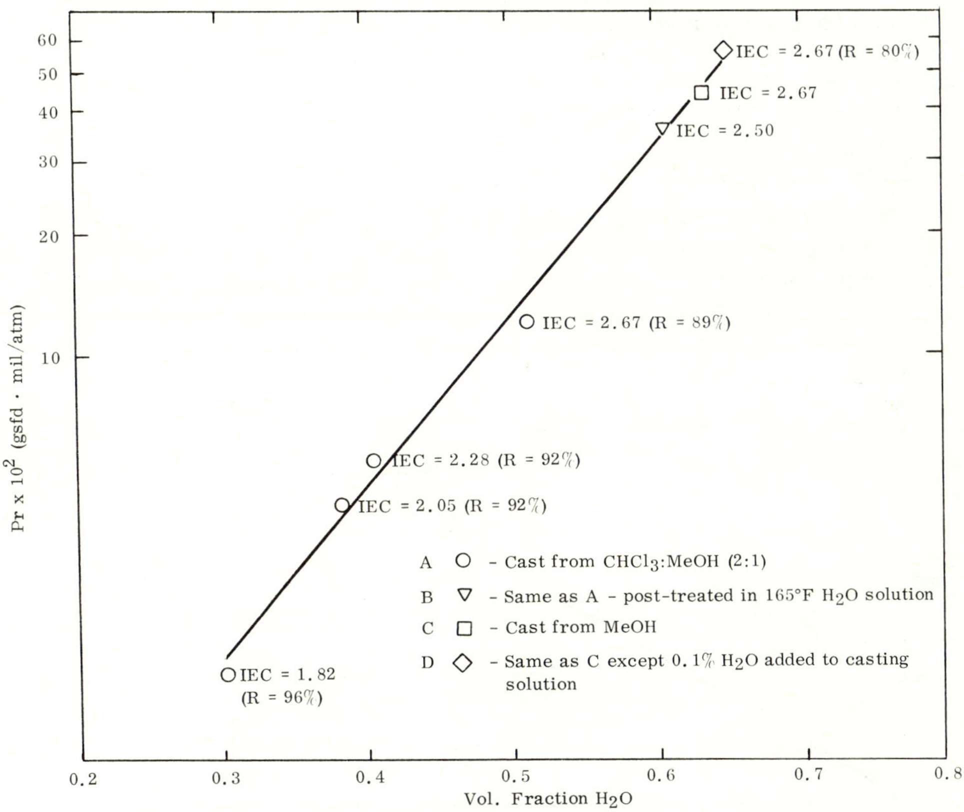

Fig. 2 - Effect of Membrane Water Content on Pr for 1000 ppm NaCl Solution at 77°F.

There are sufficient data to show that in the thickness range investigated (0.1 to 1.0 mil) the flux is inversely proportional to the membrane thickness[3,4]. Rejection of ionized species does not appear to be affected by membrane thickness.

Effect of Pressure and Temperature on Reverse Osmosis Properties.

For the sulfonated PPO family of membranes, an increase in pressure gives a direct increase in flux from 200 to 1200 psi at a given temperature (Fig. 3). The salt rejection, S, increases with increasing pressure according to the expression $S = X - C/P$, where X is the rejection at high pressure, C is a constant that depends on temperature and membrane properties, and P is the feed pressure.

Increasing the temperature of the feed solution greatly increases Pr, while rejection remains about the same or increases slightly. Thus, Pr will vary directly with feed temperature, all other conditions remaining constant. The extent of this dependence is best shown in Fig. 4, which is a plot of relative Pr values vs. feed temperature at 800 psi. As temperature increases from 50 to 170°F, Pr increases 2.5 fold.

Flux and Rejection Characteristics with Single Feed Systems.

The Pr and rejection characteristics for single component feeds at 77°F and 800 psi are shown in Table 2. Membranes MF-1 and MF-2 were identical films with an IEC of 2.04 meq/g. Membrane BF-811-1 had an IEC of 2.24 meq/g. All films were 0.2 mil thick.

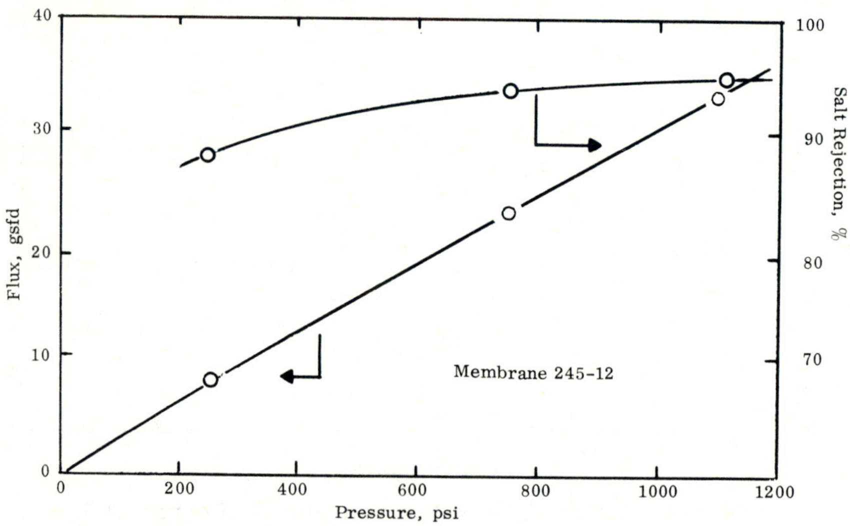

Fig. 3 - Effect of Pressure on Reverse Osmosis Membrane Performance, Webster Water, 130°F.

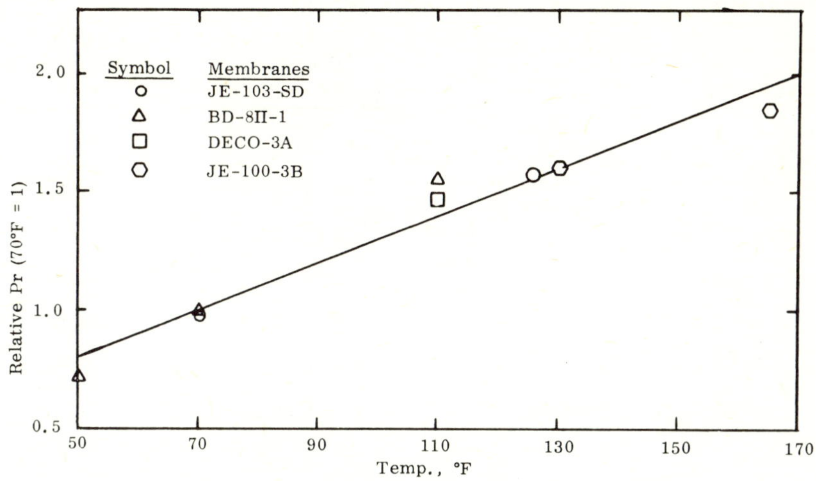

Fig. 4 - Effect of Temperature on Water Permeability at 800 psi.

TABLE 2

DEPENDENCE OF RO PERFORMANCE ON FEED COMPOSITION

Membrane	Feed	Conc. (ppm)	Pr x 10^2 (gsfd-mil/atm)	Salt Rejection (%)	Relative K
MF-1	$MgSO_4$	10300	1.7	98.1	5
MF-2	$MgCl_2$	8130	1.6	81.3	65
MF-2	NaCl	10000	2.8	92.1	30
MF-2	Na_2SO_4	12130	2.5	99.7	1
BF-811-1	$NaHCO_3$	1000	3.7	97.2	∿5

Rejections are consistent with the Donnan mechanism of rejection. The higher valence co-ions (anions) and the lower valence counter-ions (cations) are rejected best. Rejections decreased for different feed solutions in the following order:

$$Na_2SO_4 > MgSO_4 > NaCl > MgCl_2.$$

For salts which contain the same cation and common valence anions, the salt containing the larger anion exhibits the greatest rejection ($NaHCO_3$ > NaCl). For single salts, the rejection can be expressed by the equation S = 100 - KA where K is the relative rejection of each salt and A is a constant that depends on membrane, temperature and pressure. If Na_2SO_4 is assigned a K = 1,

TABLE 3

Pr OF SULFONATED PPO AS A FUNCTION OF THE COUNTER-ION

	Pr x 10^1 (gsfd-mil/atm)	Relative Pr
Na^+, Cs^+, H^+	1.5	1.0
Li^+	1.2	0.80
NR_4^+	<0.1	-
Hg^{++}	1.3	0.87
Cd^{++}	1.0	0.67
Ca^{++}, Ni^{++}, Pb^{++}, Mg^{++}, Ba^{++}, Fe^{++}, Cu^{++}	0.83	0.55
UO_2^{++}, Pb^{++} ($PbSO_4$ in membrane)	0.63	0.42
Al^{+++}	0.63	0.42
Cr^{+++}	0.58	0.39
Fe^{+++}(< 0.1 for NR_4^+)	0.55	0.37

All as Cl^- except Hg^{++}, Pb^{++}, and UO_2^{++}, which were as NO_3^-.

then A = 0.3 for the MF series of membranes when run at 800 psi and 77°F. The relative K values for single component feeds are shown in the last column of Table 2. The higher the K, the poorer the rejection.

The Pr values observed for Mg^{++} salts are significantly lower than those for Na^+ salts. More extensive data on the effect of cation type on Pr is given in Table 3 for a sulfonated PPO membrane with an IEC of 2.40 meq/g having a high H_2O content.

Water content measurements on the same membranes in different ionic forms indicate that Pr is not directly related to total water content of membrane. It appears to be related more to the "free" or unbound water content of the membrane. Membranes containing cations which have high waters of hydration or are of large size exhibit low Pr values.

Flux and Rejection Characteristics with Mixed Feeds.

For feeds with two or more salts, Pr is highly dependent on cation type and concentration. The effect on flux of adding 10, 20, 40, and 100 ppm Mg^{++} ion (as $MgSO_4$) to a 1000 ppm Na^+ ion (Na_2SO_4) feed with an initial flux of 30.4 gsfd is shown in Fig. 5. The corresponding decreases in flux are 19, 25, 28, and 32%. The addition of 100 ppm of Cu^{++} ($CuSO_4$), Cd^{++} ($Cd\ SO_4$), Ni^{++} ($NiSO_4$), Fe^{++} ($FeSO_4$), and Cr^{+3} ($CrCl_3$) to the Na_2SO_4 feed decreases the

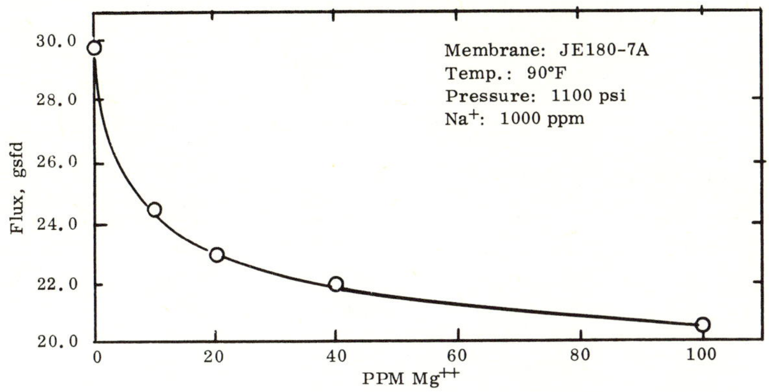

Fig. 5 - Effect of Mg^{++} on Flux with Na_2SO_4 Feed.

flux by 26, 26, 32, 32, and 49%, respectively. Recovery of original flux was achieved for all salts except $CrCl_3$ by subjecting the membrane to the pure Na_2SO_4 feed. The flux was decreased permanently by approximately 10% with $CrCl_3$. The relative Pr values obtained for the mixed feeds are in good agreement with values reported in Table 3. Cation rejection is 98 to 100% except with $CrCl_3$ whose addition lowers the Na^+ ion rejection to 93%.

Cation rejections for feeds containing mixtures of univalent and multivalent salts do not adhere strictly to the Donnan rejection mechanism. The multivalent cations preferentially exchange into the sulfonate membrane, resulting in a lowering of the Donnan potential through partial association of the anchored sulfonate groups with the multivalent cations. With the decreased Donnan potential, the rules pertaining to solution-diffusion membranes become important, being manifested by low univalent and high divalent ion rejection for mixed feeds. This is depicted in Table 4 where a typical ion rejection analysis is shown for Webster, South Dakota water for a 25 to 30 gsfd membrane. It is thus seen that for mixed feeds containing divalent cations and substantial quantities of univalent ions such as Na^+ or Cl^- ion a 25 to 30 gsfd membrane may exhibit relatively low rejections (80 to 85%). To achieve high rejections for mixed feeds containing substantial quantities of NaCl, lower water content membranes must be used. With synthetic seawater a 99% overall rejection was obtained with a sulfonated PPO membrane of IEC = 1.36 meq/g as shown in Table 5. Approximately 310 ppm of dissolved ions were found in the permeate. The flux for the 0.2 mil sulfonated PPO membrane remained invariant at 0.5 gsfd during a 170-hour life test. To achieve a flux of 5 to 10 gsfd, the thickness of this membrane must be decreased 10 to 20 fold for use at 77°F and 1100 psi.

TABLE 4

REJECTIONS OF NATURAL AND SYNTHETIC WEBSTER WATER AT 1100 PSI AND 130°F

Chemical Species	Individual Ion Rej. (%)	Chem. Comp. of Feed Water (ppm) Nat. Webster Water	Syn. Webster Water	Calc. Chem. Comp. of Permeate (ppm) Nat. Webster Water	Syn. Webster Water
Na^+	67	140	190	46	62
Cl^-	53	20	300	9	140
Ca^{++}	92	216	260	17	21
$SO_4^=$	94	700	830	42	50
Mg^{++}	90	80	80	8	8
Total ionic content (ppm)		1156	1660	122	281

Total salt rejection = 90% (natural Webster water)
83% (synthetic Webster water)

TABLE 5

REJECTIONS FOR SYNTHETIC SEAWATER AT 1100 PSI AND 77°F, WITH A LOW WATER CONTENT MEMBRANE

Chemical Species	Feed Conc. (ppm)	Permeate Conc. (ppm)	Rejection (%)
Na^+	7,900	80	99.0
K^+	390	5*	98.9*
Cl^-	16,000	203	98.7
Mg^{++}	1,340	13	99.0
Ca^{++}	440	2	99.5
$SO_4^=$	900*	6*	99.3*

*Best estimate

Life Characteristics

The most unique feature of sulfonated PPO is its resistance to compaction, hydrolysis, and degradation during life testing over a wide range of temperatures (77 to 165°F), pressures (600 to 1100 psi), and pH (5 to 13) with feeds containing inorganic and organic species. In most instances the compaction slope is $\leq$ 0.01 for the feeds investigated.

Brackish Water

Sulfonated PPO is comparable to conventional reverse osmosis membranes for treating water having a high hardness due to the presence of Mg^{++}, Ca^{++}, $SO_4^{=}$, HCO_3^{-}, $CO_3^{=}$, etc. For natural Webster water, membranes having an IEC of 2.15 to 2.25 meq/g appear to represent a good trade-off between Pr and rejection (Fig. 6). Life tests on natural Webster water at 130°F and 1100 psi, depicted in Fig. 7, demonstrate that fluxes of 30 gsfd and salt rejections of 90% can be readily achieved with 0.2 mil membranes. After approximately 50 hours of operation, the compaction slope decreases to 0.01 or less.

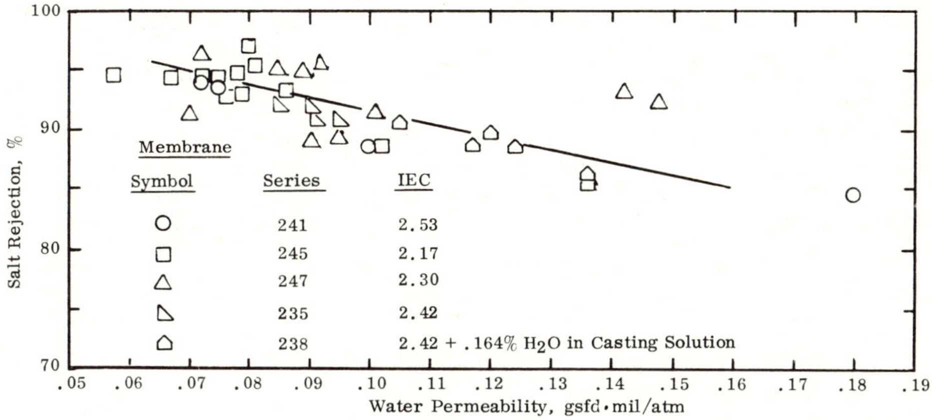

Fig. 6 - Performance of Sulfonated PPO on Webster Water at 1100 psi and 130°F.

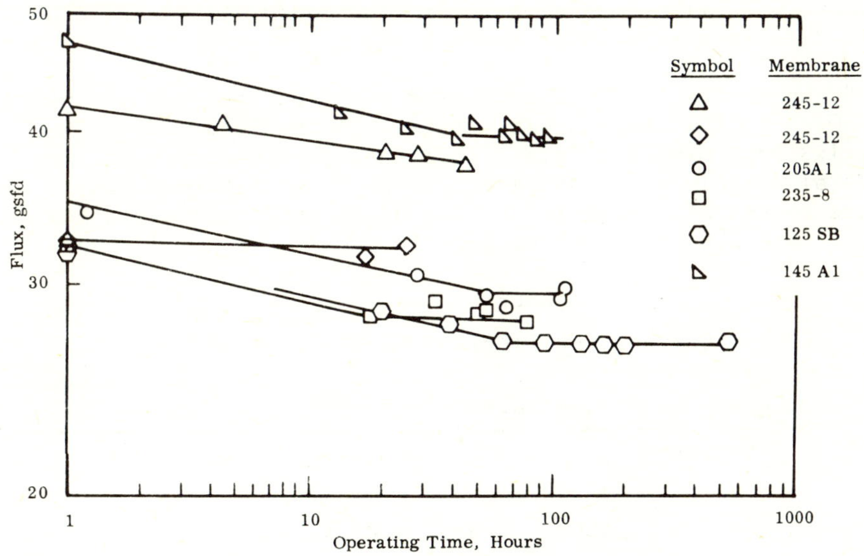

Fig. 7 - Life Tests of Sulfonated PPO Membranes on Natural Webster Water at 1100 psi and 130°F.

Caustic Feeds

Sulfonated PPO exhibits excellent stability in highly alkaline feeds where most conventional reverse osmosis membranes undergo rapid hydrolysis. A 100-hour life test on Na_2SO_4 (1000 ppm Na^+) at pH = 13 gave an invariant flux of 46 gsfd with a total salt rejection of 95% at 77°F. Some work was conducted also on treatment of high pH metal finishing waste water. A 100-hour life test on copper cyanide at pH = 11.4 gave a flux of 45 gsfd with copper and cyanide rejections of 98 and 93%, respectively[5]. Further evidence for the excellent stability in highly alkaline media is that the membranes exhibit essentially no change in chemical and physical properties when used as battery separators in 40% KOH at 105°F. Another important feature of being able to operate in alkaline media is that many species which undergo limited ionization in low and neutral pH feeds and are poorly rejected, e.g. weak acids such as acetic, lactic, etc., ionize to a greater extent in high pH media and are easily rejected by the sulfonated PPO membrane.

Secondary Sewage

Sulfonated PPO membranes exhibit good life characteristics in the presence of organic and bacterial impurities. Work was conducted to demonstrate that sulfonated PPO membranes were suitable for the purification of secondary sewage water by reverse osmosis. Samples of secondary waste water were obtained from the sewage treatment plant at Salem, New Hampshire. The rejection characteristics of a 0.2 mil membrane of IEC = 2.52 meq/g at 500 psi and 77°F were excellent as seen in Table 6.

A 350-hour life test was conducted with recirculated secondary waste flowing by the membrane at a rate of 1 gpm longitudinally across a 1x3 inch membrane face with a feed channel depth of 1/8 inch. Total dissolved solids rejection was approximately 86% during the course of the test. A flux of 22 gsfd was maintained during the last 280 hours of the life test.

TABLE 6

REVERSE OSMOSIS PURIFICATION OF WASTE WATER

	Na^+	K^+	Ca^{++}	Mg^{++}	$SO_4^=$	$HPO_4^=$	Total Carbon	Total Nitrogen
Secondary sewage feed (ppm)	73	14	20	3.5	31	32.2	42	22.8
Product water (ppm)	18	2.8	0.8	0.3	4.2	3.7	3.8	4.7
% Rejection	75	80	96	92	87	89	91	79

A life test using fresh secondary sewage was run on a higher rejecting sulfonated PPO membrane, coated on porous polypropylene, to determine the fouling and regeneration characteristics of the membrane. Initially, raw sewage was used, and the membrane fouled quickly in about 20 hours. The membrane was regenerated using two gallons of distilled water circulating at 160°F for about one hour, after which the initial flux of 11.5 gsfd was recovered. Thereafter the sewage water was filtered of suspended particles simulating a degree of filtration which may be used on practical units. Fig. 8 shows the fouling characteristics. After about a week the flux was between 7 and 8 gsfd. Overall rejection was approximately 90 to 95%. The membrane was again regenerated to its initial flux using the same hot water treatment. Thereafter, the flux vs. time characteristics followed the same behavior.

Synthetic Wash Water

Solvent-cast sulfonated PPO membranes of 0.2 to 1.0 mil thickness with ion exchange capacities of 1.36 to 2.28 meq/g were prepared, and reverse osmosis tests were carried out on synthetic wash water containing NaCl (1500 ppm), urea (1000 ppm), lactic acid (700 ppm), and soap or detergent (300 to 10,000 ppm). Several liquid soaps and detergents were evaluated. All testing was conducted at a sterilization temperature of 165°F and at 800 psi or less. The goal was to obtain a permeate which contained less than 10 microorganisms/cc and to achieve a flux of at least 5 gsfd with adequate rejections of NaCl (85.0%), urea (99.3%), lactic acid (99.0%), and detergent (99.9%). Reverse osmosis life tests of 1000 to 2250 hours (90 days) at 165°F were run on the most promising membranes.

Stability tests (test tube and reverse osmosis) at 165°F were conducted on sulfonated PPO membranes having an IEC of 1.36 to 2.28 meq/g with wash water containing 9 liquid soaps or detergents. Four levels of soap or detergent concentration were considered: 30, 300, 3000, and 10,000 ppm. The physical strength of membranes after test tube aging or reverse osmosis testing was found to decrease with increasing IEC of the membrane and increasing soap concentration. Loss in physical strength for most membranes was manifested by swelling, dimensional increase, poor salt rejection, and high flux. Membranes were found to be compatible with 30 ppm of Basic H, Miranol HS or Phisohex; 300 ppm of Miranol C2M-SF or Zephiran; 3000 ppm Approve; and 10,000 ppm Olive Leaf, Sano Septic, or modified Approve. (Sources in Appendix.)

Life tests were conducted on 0.2 mil membranes with IEC = 1.60, 1.82, 2.05 and 2.28 meq/g to determine the effect of membrane IEC on flux and rejection. For a given soap concentration (300 ppm), flux increased and rejection decreased with increasing IEC as can

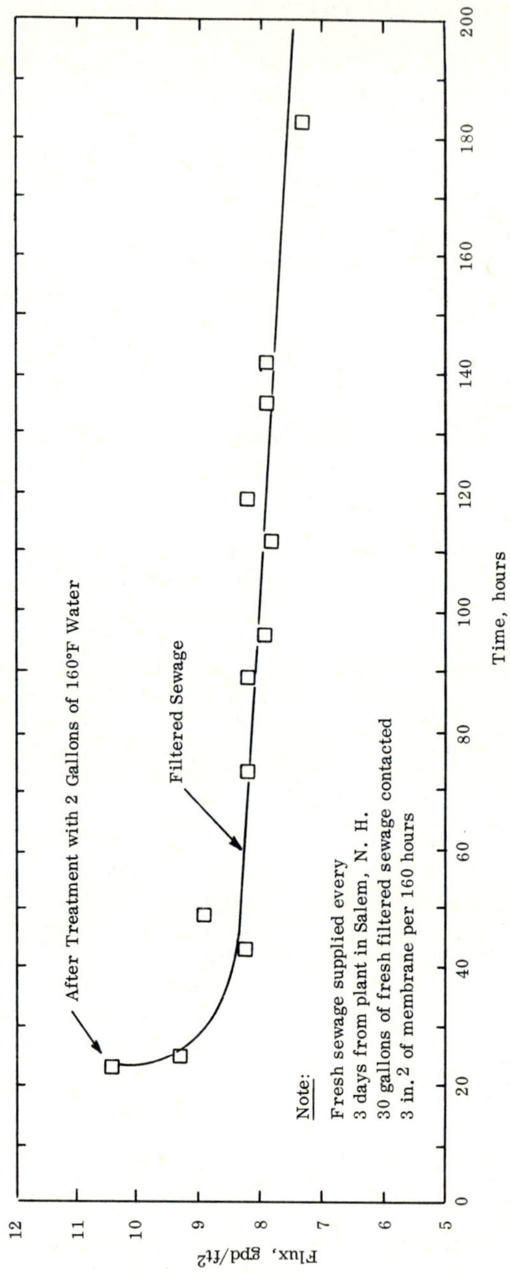

Fig. 8 - Fouling and Regeneration of Sulfonated PPO Membrane on Secondary Sewage be seen in Figs. 9 and 10. Steady state fluxes after 100 hours for membranes having an IEC = 1.60, 1.82, 2.05, and 2.28 meq/g were approximately 8.4, 19, 39, and 53 gsfd, respectively. Corresponding NaCl rejections were 95, 92, 78, and 69%, respectively. Organic rejections also decreased accordingly with increasing IEC: urea (35, 33, 28, 21%), lactic acid (99, 86, 85, 80%), and soap (99+, 98, 91, 81%).

The effect of soap concentration (300 to 10,000 ppm) was investigated with 0.2 mil membranes having IEC's of 1.60 and 1.82 meq/g. Flux increased (Fig. 11) while there was a slight decrease in rejection (5 to 10%) with increasing soap concentration. The membrane with the higher capacity completed 2160 hours (90 days) of life testing at 165°F, i.e., 1003, 523, and 634 hours, respectively, with 300, 3000, and 10,000 ppm soap. Corresponding steady

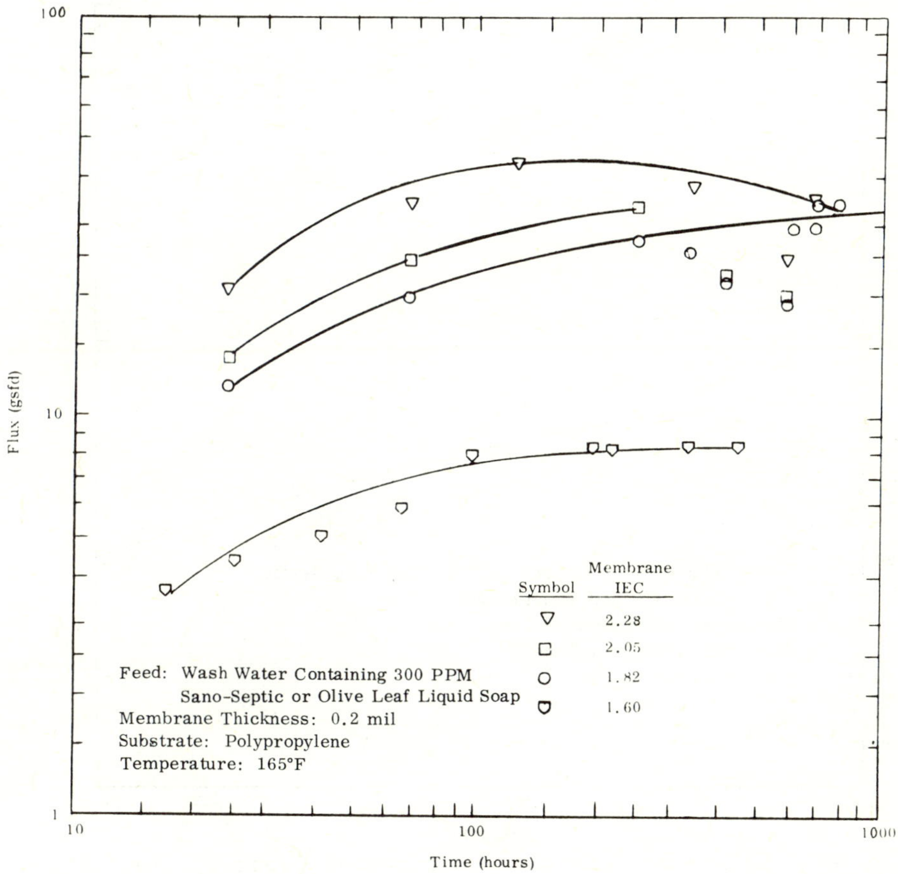

Fig. 9 - Effect of Membrane Ion Exchange Capacity (meq/g) on Flux

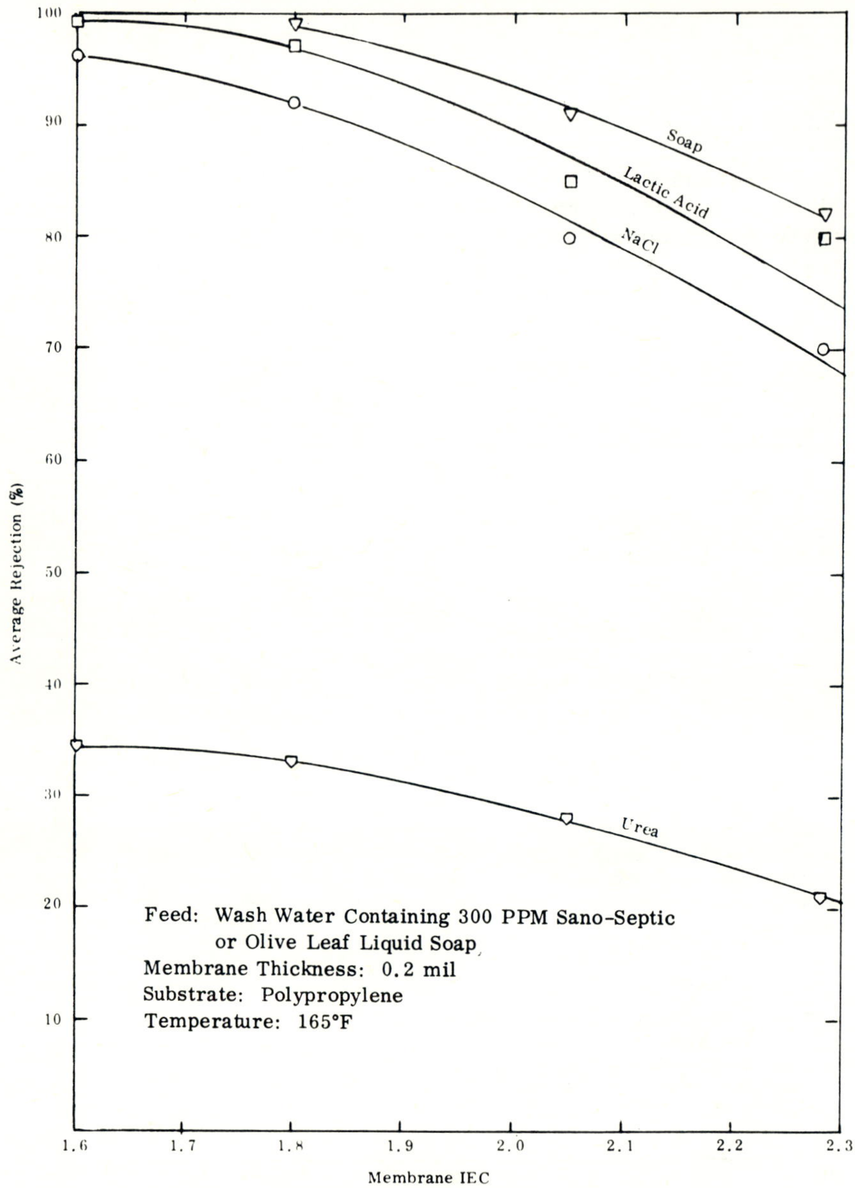

Fig. 10 - Effect of Membrane Ion Exchange Capacity (meq/g) on Rejection

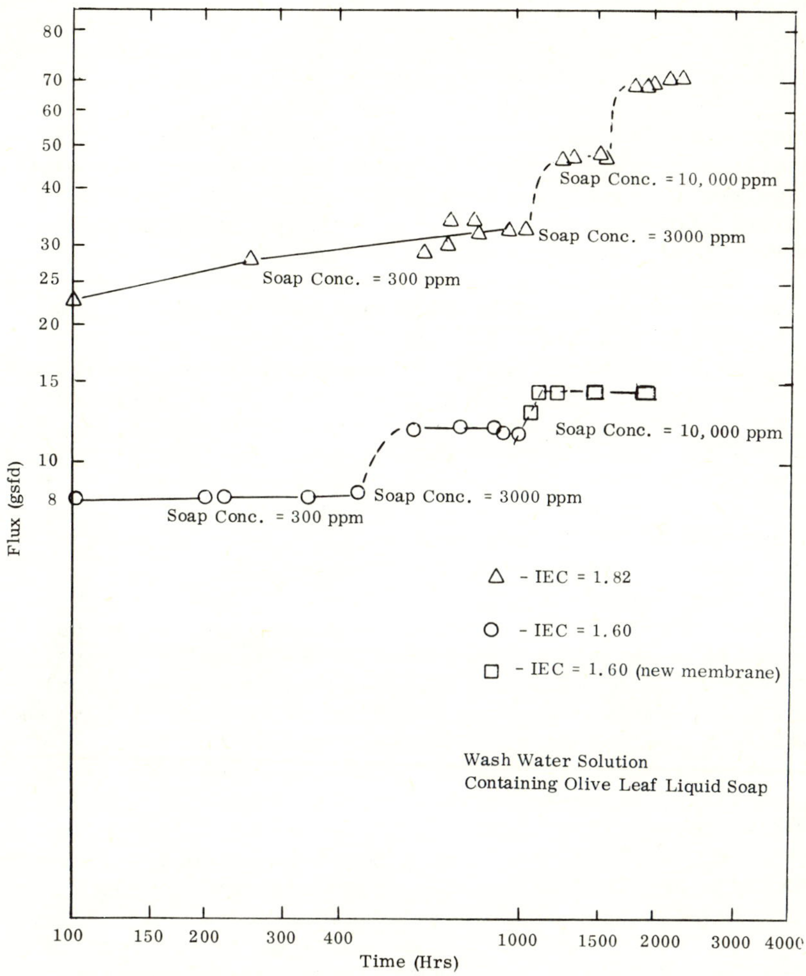

Fig. 11 - Effect of Soap Concentration on Flux at 165°F for 0.2 Mil Sulfonated PPO Membranes

state fluxes during those periods were 38, 50, and 70 gsfd. The other membrane completed approximately 2037 hours at 165°F, i.e., 503, 484, and 1050 hours, respectively, with 300, 3000, and 10,000 ppm soap. Corresponding steady state fluxes during those periods were 8.4, 12 and 15 gsfd. The life tests for these membranes are depicted in Figs. 9 and 10. The membrane with the lower capacity exhibited the best rejections. Its rejection decreased slightly in going from 300 to 10,000 ppm soap, i.e., NaCl (96 to 92%), urea (36 to 31%), lactic acid (99 to 96%), and soap (99+ to

94%). Similar decreases were observed for the membrane with capacity 1.82 meq/g.

Previous studies[3] with sulfonated PPO polymer showed that salt rejection increased with increasing $CHCl_3$:MeOH ratio of the polymer casting solution with all other variables fixed. Low ion exchange capacity membranes have a high solubility in $CHCl_3$. Thus the feasibility of increasing membrane rejection by using a casting solvent containing mostly $CHCl_3$ and minimal MeOH was investigated. Complete polymer solubility for sulfonated PPO with ion exchange capacity of 1.36 and 1.60 meq/g was achieved with a 20:1 $CHCl_3$:MeOH solvent. A comparison of rejections obtained with membranes cast from 20:1 and 2:1 $CHCl_3$:MeOH solvent systems is depicted in Table 7. The two membranes of 0.2 mil thickness (RO-68 and RO-69) were extremely stable and exhibited invariant fluxes of 3.1 and 6.6 gsfd, respectively, after 2250 and 1200 hours with wash water containing 10,000 ppm Olive Leaf soap. Rejections for the thin membrane of higher capacity (RO-69) were comparable to those observed for a 0.5 mil membrane of the same capacity cast from 2:1 $CHCl_3$:MeOH (RO-54). The lower capacity membrane (RO-68) exhibited generally higher rejections of each constituent in the synthetic wash water.

Bacterial and chemical analyses were conducted on actual shower water obtained using two subjects, each taking an entire body shower with 30 ml of Olive Leaf liquid soap (20% solids) and 1 gallon of water. Chemical analysis showed that the shower water contained approximately 1200 ppm soap, 57 ppm NaCl, $\leq$ 40 ppm urea, and 27 ppm lactic acid. The pH of the water was 7.7. Bacterial analysis (2 weeks incubation time) revealed there were 3.38×10^5 microorganisms/cc in the shower water from the two subjects. The microorganisms were identified as mostly Staphylococcus bacteria and Candida yeast. After exposure to 165°F for 1 hour, the reverse osmosis feed water contained 1.1×10^4 microorganisms/cc. The permeate which was obtained by passing the 165°F shower water through a 0.2 mil sulfonated PPO membrane at 800 psi (IEC = 1.6 meq/g) contained no microorganisms.

Tubular Hardware Design

Recent work has been aimed at incorporating the membrane into practical elements. It has been found that the membrane can be solvent cast onto a suitable microporous substrate which can then be inserted into standard tubular supports.

Briefly, the process involves:

a) Forming a tube from the porous substrate. This operation can be performed on conventional heat sealing equipment

TABLE 7

LIFE TESTS OF LOW IEC SULFONATED PPO MEMBRANES ON SYNTHETIC WASH WATER[a]

RO Life Test No.	Memb. IEC (meq/g)	Casting Solvent Ratio, $CHCl_3$:MeOH	Memb. Thickness (mils)	Time (hr)	Flux (gsfd)	Negative Compaction Slope[b]	Rejection (%)			
							NaCl	Urea	Lactic Acid	Soap
RO-54	1.60	2:1	0.5	2250	5.8	<0.001	95	36	98.1	97
RO-69	1.60	20:1	0.2	1200	6.6	<0.001	95	38	98.3	97
RO-68	1.36	20:1	0.2	1200	3.1	<0.001	97	47	99.4	98+

[a]Synthetic wash water containing 1500 ppm NaCl, 1000 ppm urea, 700 ppm lactic acid, and 10,000 ppm Olive Leaf liquid soap. Tests performed at 800 psi and 105°F.

[b]Slope of a plot of log flux vs. log time; does not include first 100 hours of life testing.

by forming a seam down the length of the tube. Longer tubes can be generated by conventional ultrasonic bonding techniques such as are used for other reverse osmosis substrates.

b) Coating the substrate with the membrane internally by a meniscus-coating technique whereby the polymer solution is coated through a meniscus-forming head as the tube is pulled over the head.

c) Inserting the coated tube into a standard porous Fiberglas support tube and sealing the ends of the membrane/substrate into this tube.

The coating process is simple and reliable, forming very thin films of membrane onto the support structure. The materials are standard production components and are inexpensive, the substrate costing about \$0.10-0.20/sq. ft. The resulting structures are mechanically strong and withstand rough handling.

Fig. 12 shows the principle of meniscus coating. A meniscus head contains a constant level of casting solution which is maintained about half way up the bore of a series of dispensing holes

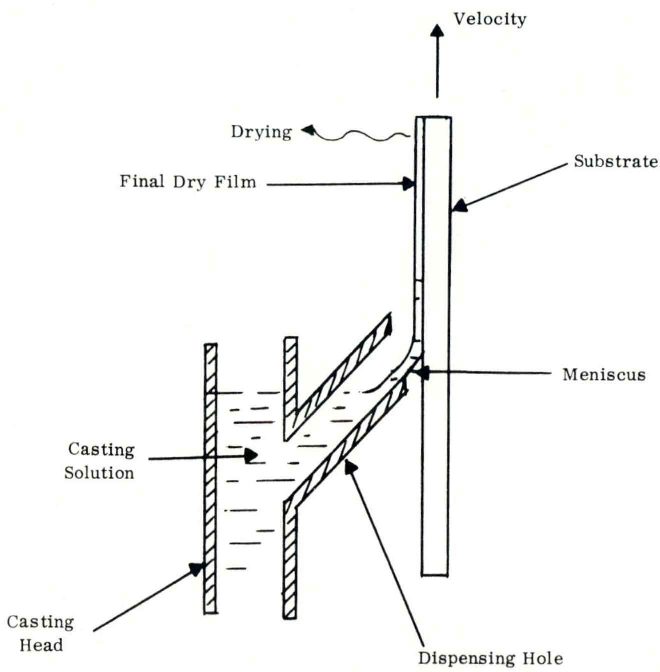

Fig. 12 - Meniscus Coating Configuration

(20 mils diameter). The outlets of the dispensing holes are held a fixed distance (10-15 mils) away from the surface of the substrate to be coated. In the static case, the polymer solution forms a meniscus which ends at the dispensing hole or just below it. A meniscus is struck between the substrate and the head by briefly touching the two together whereupon the polymer solution begins to flow from the meniscus head to the substrate. The fluid is then coated onto the substrate by moving the substrate upward at a constant velocity, allowing the solvent to evaporate quickly during the process. The thickness of the film is determined by the solids content, surface tension and viscosity of the casting solution, and the velocity of the substrate. An Instron tensile test machine is used to pull the substrate tube at a constant velocity of 50 inches/minute.

The best substrate for coating the membrane was found to be a porous polypropylene manufactured by Celanese Corp. and designated CDK 2400. This is a non-wettable material having a molecular weight cutoff in ultrafiltration of 20,000 to 40,000. This same material was successfully used as a substrate for sulfonated PPO membrane during 60 to 90 day life tests on 165°F wash water. A wettable type is also manufactured, but it does not provide good reverse osmosis performance for composite membranes, most likely because the wetting characteristics allow polymer to intrude into the pores causing plugging. Applying a masking layer prior to casting films of PPO may render the wettable material useful.

Sections of polypropylene tubes internally coated with sulfonated PPO were cut up and tested as flat sheets to observe the reproducibility of the process. Comparisons between sections of an individual tube as well as between different tubes showed good performance and reproducibility when tested on Webster water at 1100 psi and 130°F as shown in Table 8. The IEC of these mem-

TABLE 8

REVERSE OSMOSIS TESTS ON SECTIONS OF TUBULAR MEMBRANE/SUBSTRATE WITH WEBSTER WATER

Tube Batch Number	Tube Number	Flux (gsfd)	Salt Rejection (%)
1	1	40	92
1	2	48	94
2	1	48	92
2	2	45	89
2	3	48	90
2	4	57	87

branes was 2.1 meq/g. As flat sheets, the composite membranes prepared by meniscus coating exhibited approximately 6% higher rejection for a given flux than membranes prepared by manual casting on glass plates and tested in reverse osmosis with a porous polypropylene support.

ACKNOWLEDGMENTS

The authors wish to thank the Office of Saline Water, U.S. Department of the Interior, for support of much of the work presented here. The conscientious work of J. M. Amore, J. F. Enos and J. F. Austin in obtaining the bulk of the experimental results is appreciated.

REFERENCES

1. K. Spiegler, Chem. Eng. Progr. Symp. Ser., 55, 24 (1959).

2. U. Merten, H. K. Lonsdale, R. L. Riley, and K. D. Vos, "Reverse Osmosis Membrane Research," Office of Saline Water Research and Development Progress Report No. 265, General Dynamics, General Atomic Division, 1967.

3. C. W. Plummer, G. Kimura and A. B. LaConti, "Development of Sulfonated Polyphenylene Oxide Membranes for Reverse Osmosis," Office of Saline Water Research and Development Progress Report No. 551, General Electric Co., Lynn, Mass., 1970.

4. P. J. Chludzinski, J. F. Austin and J. F. Enos, "Development of Polyphenylene Oxide Membranes," Office of Saline Water Research and Development Progress Report No. 697, General Electric Co., Lynn, Mass., 1971.

5. L. T. Rozelle, North Star Research and Development Institute, personal communication.

APPENDIX

The sources of the soaps and detergents used in this study were: Basic H, Shaklee Products; Miranol products, Miranol Co.; Phisohex and Zephiran, Winthrop Labs.; Approve, Calgon Corp.; Sano-Septic and Olive Leaf, Rochester Germicide Co.

PBI REVERSE OSMOSIS MEMBRANES: AN INITIAL SURVEY

Frank S. Model and Leonard A. Lee
Celanese Research Co.
Summit, New Jersey 07901

INTRODUCTION

Research on reverse osmosis in these laboratories began in the mid-nineteen sixties, when work on cellulose acetate membranes was initiated. The promising results obtained at that time led to the continuation and extension of the studies under the sponsorship of the Office of Saline Water, U.S. Department of the Interior, beginning in 1968. These investigations determined the influence of polymer variables, notably the primary-to-secondary hydroxyl group ratio and acetyl value, and physical variables (e.g., casting speed, air exposure time) on the desalination performance of cellulose acetate flat membranes[1]. It was demonstrated that a 50% improvement in flux could be achieved in membranes exhibiting 90% salt rejection by proper manipulation of membrane preparation variables. At the same time, the difficulties inherent in this particular membrane system, i.e., poor compaction resistance and hydrolytic instability, were identified and minimized.

During the same period, techniques for the polymerization and spinning of polybenzimidazole were being developed. These programs have been sponsored by the Air Force Materials Laboratory, Air Force Systems Command, United States Air Force. This work has sought to investigate and optimize important variables in the manufacture of a thermally stable textile fiber based on poly-2,2'-(m-phenylene)-5,5' bibenzimidazole (PBI). The polymer is prepared from 3,3'-diaminobenzidine and diphenylisophthalate, according to the following scheme:

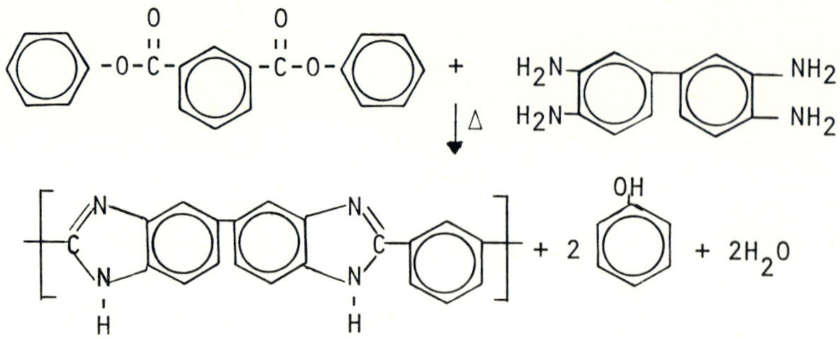

Studies performed in these laboratories have included modification of the original polymerization procedures of Vogel and Marvel[2], preparation of stable dope systems, definition of conditions for spinning, drawing, and crystallization of the textile fiber, and characterization of the final product and its precursors[3].

During these studies it was found that one of the unique properties of PBI was a very high moisture regain, about 13%, equal to or higher than that of the natural materials cotton and wool. This property has important ramifications in terms of textile end uses and also suggested the possible utility of PBI as a reverse osmosis membrane material.

The potential of PBI as a flat sheet reverse osmosis membrane was explored in these laboratories several years ago[4]. In this study, transport properties comparable to those of cellulose acetate were achieved at ambient operating temperature, whereas membrane performance characteristics significantly superior to cellulose acetate were observed at higher temperatures, up to 90°C. This work showed that the known stability of imidazole derivatives[5] could be translated into a PBI reverse osmosis membrane with outstanding potential for service over a wide range of conditions.

There are, however, general deficiencies of flat sheet membranes. In particular this configuration provides a very low ratio of active surface area to occupied volume - a parameter of critical importance in plant and system design. This shortcoming naturally led to the investigation of higher surface-to-volume ratio configurations, including the hollow fiber. The fiber geometry not only provides an extremely favorably surface area-to-volume ratio, but also greatly simplifies the hardware necessary for module fabrication. Whereas flat films are mounted on porous supports in specially designed modules, a bundle of hollow fibers is self-supporting and can be potted simply into lengths of standard size pipe.

The purpose of this paper is to present the initial body of results pertaining to the desalination performance of both flat sheet and, especially, hollow fiber PBI membranes, to describe the methods developed for fabricating laboratory test modules, and to introduce some preliminary information relevant to membrane structure and morphology.

EXPERIMENTAL AND RESULTS

An initial study was performed on flat sheet PBI reverse osmosis membranes prepared by techniques similar to those employed for cellulose acetate[1]. The membranes were tested at the standard conditions established previously for cellulose acetate (600 psi, 25°C, 0.5% NaCl, 65 feet/min flow rate). Typical performance data for flat sheet PBI are presented and compared to cellulose acetate in Table 1. These data show the flux to be comparable to that for cellulose acetate at temperatures near ambient. However, when the temperature was raised to 75°C and then 90°C, the advantage of PBI was dramatic and clear cut. Whereas the flux of the cellulose acetate membranes dropped rapidly to zero, that of PBI was seen to increase substantially. This behavior was no doubt a consequence of the fact that the operating temperatures employed were in the effective annealing range for cellulose acetate. This served to illustrate the excellent potential for PBI in elevated temperature applications and the corresponding lack of utility for cellulose acetate under like conditions.

TABLE 1

COMPARATIVE PERFORMANCE* OF PBI AND CA REVERSE OSMOSIS MEMBRANES

Operating Temperature (°F)	Testing Time (Hr.)	Flux (gal/ft^2-day)**			
		C.A. Membrane		PBI Membrane	
		Initial	Final	Initial	Final
70	5	15	12	16	14
120	5	22	20	21	17
167	20	10	5	24	19
194	5	0	0	29	25

*Test conditions: 600 psi, 0.5% NaCl feed stream, 65 fpm flow rate. Membranes were tested at temperature for the stated time period, then the temperature was raised to the next higher level and the membranes were re-tested.

**Salt rejection: ca. 95% throughout.

Hollow fiber reverse osmosis membranes have also been prepared from PBI. These membranes, spun by a novel, proprietary technique, had to be kept wet after spinning to maintain reverse osmosis activity. A satisfactory technique for storing membrane samples was devised wherein straight fiber bundles, three feet in length, were stored lengthwise in sealed plastic tubes filled with water. Fibers stored in this way were found to have undiminished desalination efficiency, even after months of storage.

The hollow fiber membrane configuration was found to be particularly amenable to uncomplicated module fabrication techniques. For testing purposes, bundles of 6-12 fibers, 18-24 inches in length, were potted into 3/4" stainless steel nipples using a suitable epoxy resin. One formulation found to be satisfactory consisted of 25 parts ERL 2795 epoxy resin (Smooth-On Mfg. Co., Gillette, N.J.), 6 parts Sonite 15 hardener (same source), and 5 parts Mod-Epox triphenyl phosphite modifier (Monsanto). This formulation, similar but not identical to one reported elsewhere [6], gave satisfactory compatibility with the water present in the fiber bundles. It also exhibited sufficient hardness and was found to cure quickly enough under ambient conditions so that samples could be tested soon after potting. The PBI-epoxy bond was excellent, as shown in Fig. 1. In a typical run, the fiber

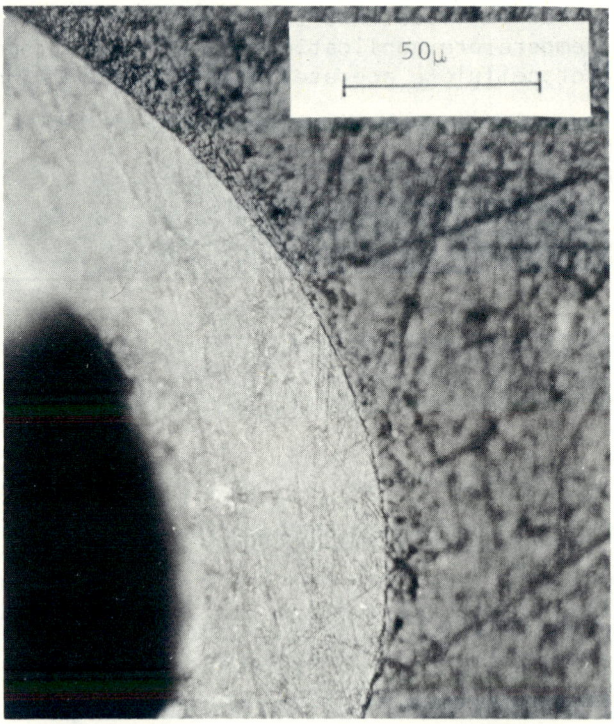

Fig. 1 - Fiber-Epoxy Bond.

bundle was looped to bring all ends together, these ends then being run through a cork borer which had been driven through a rubber stopper. When the cork borer was extracted, the rubber stopper expanded to hold the fiber ends sufficiently to prevent uncured epoxy from penetrating during potting. The stopper was inserted into a nipple and potting was effected. The fiber loops were immersed in water during this procedure to keep them from drying out. After potting, the assembly was simply lowered into an appropriately fitted stainless steel pipe, and the module was ready for testing.

Membranes were tested under static conditions using 5000 ppm NaCl, 25°C, and pressures of 200, 400, and 600 psi. A forerun was collected for at least an hour before each sampling run. The apparatus used, shown in Fig. 2, was pressurized with compressed nitrogen. It was found that a wide range of flux and rejection could be obtained with hollow fiber PBI membrane variants. This is illustrated in Table 2, which contains fiber dimensions and performance data for a number of variants representing the range of performance achieved to date. Apparently, fiber dimensions alone do not determine transport properties, since no universal correlation could be established. In general, thin-walled samples gave higher levels of flux than corresponding thick-walled variants.

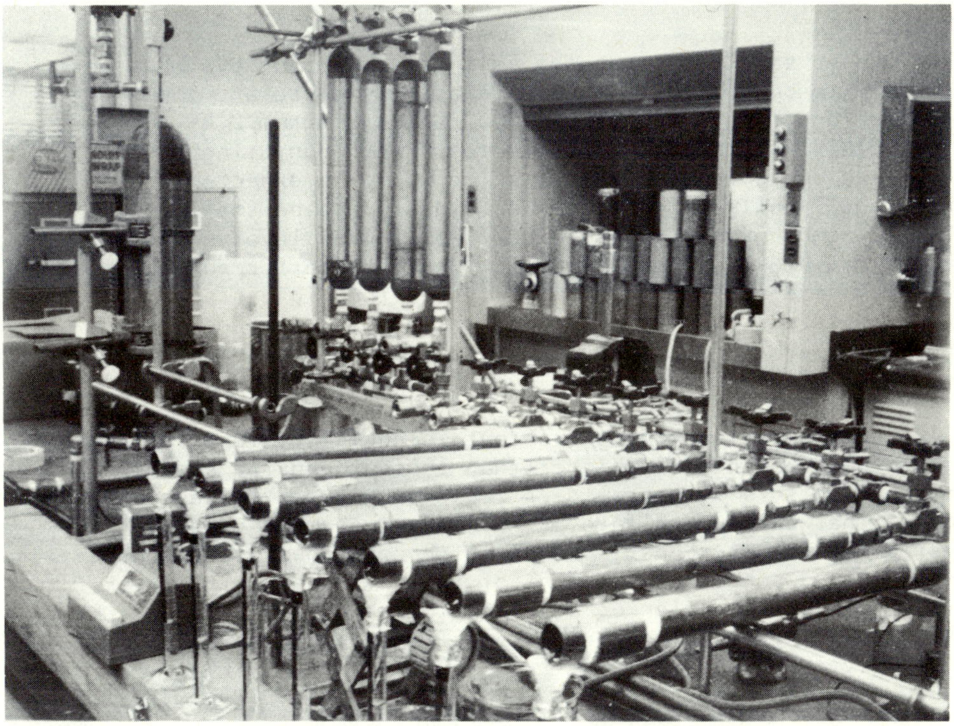

Fig. 2 - Static Test Apparatus.

TABLE 2

PERFORMANCE* OF TYPICAL PBI
HOLLOW FIBER REVERSE OSMOSIS MEMBRANES

Sample No.	Dimensions (microns) O.D.	I.D.	Wall Thickness	Flux(gal/ft^2-day)/Rejection** (%) 200 psi	400 psi	600 psi
1	250	95	78	1.43/55.4	-	4.38/68.3
2	292	123	85	1.56/70.9	3.26/84.4	4.61/88.1
3	220	91	65	1.48/72.8	3.52/83.1	5.17/85.3
4	199	94	53	1.13/83.6	2.51/91.3	3.89/96.7
5	296	153	72	1.02/87.0	2.44/91.8	3.40/91.5
6	330	123	104	0.55/96.8	1.40/98.5	2.13/98.6
7	236	131	53	6.31/43.1	12.40/52.7	14.00/40.2
8	281	140	71	0.83/94.9	2.04/97.2	2.91/96.7

*Test conditions: 0.5% NaCl, 25°C.

**Rejection determined conductometrically.

On the whole, samples possessing salt rejections of over 80% were also found to be uniform, to have relatively circular, concentric cross-sections, and to possess OD:ID ratios of from 2.25 to 3.4. It should also be noted that these tests were run using conditions under which concentration polarization would be expected to be significant. While this effect was expected to be smaller than with flat membranes because of the lower fluxes involved, it nevertheless exerts some adverse influence on performance. This is a likely explanation for the fact that the highest salt rejection observed during static testing was 98.6%. Salt rejections of up to 99.4% have been observed during experiments in which the feed stream was circulated over the film bundle.

Solute rejections for salts containing multivalent ions were found to be substantially higher than those found for sodium chloride. This is illustrated in Table 3. It can be seen, for example, that calcium sulfate is rejected much more readily than either of the other feeds, both of which contain at least one singly charged species. In light of these findings, it is believed that the rejection of divalent ions by hollow filament PBI membranes having sodium chloride rejections in excess of 95% is very nearly quantitative.

TABLE 3

PERMSELECTIVITY OF PBI HOLLOW FIBERS*
FOR VARIOUS DILUTE SALINE FEEDS

Feed	Flux (gal/ft^2-day)/Rejection** (%)		
	50 psi	100 psi	150 psi
500 ppm NaCl	0.71/6.2	1.79/17.1	3.10/21.8
500 ppm Na_2SO_4	0.81/37.9	1.98/55.1	3.18/58.6
500 ppm $CaSO_4$	0.73/-	1.83/68.9	2.87/75.6

*Membrane used had the following dimensions: ID = 150μ, OD = 265μ, wall thickness = 58μ.

**Rejection determined conductometrically.

As was the case with flat sheet PBI membranes, the high temperature performance of PBI hollow fibers was found to be outstanding. For example, in an experiment conducted at 75°C, three separate solutes were studied: sodium chloride, urea, and Santomerse SX (a detergent of the sodium alkyl aryl sulfonate type). The results of this study, summarized in Table 4, show that both flux and salt rejection with regard to sodium chloride and Santomerse SX are significantly higher at 75°C than at 23°C. The performance of the membranes with urea shows permselectivity to be lower than with the ionic solutes, though the level of rejection is nevertheless quite high for such a low molecular weight organic solute. It is important to note that a single module was used throughout this study, which was in continuous service at 70°C for several days. The level of performance during these experiments confirms the excellent high temperature performance of PBI hollow fiber membranes. The apparent drop in performance when the module was returned to room temperature is a consequence of concentration polarization, which results from the static testing mode used.

Long-term performance of selected PBI hollow fiber modules was determined under flow conditions, with the feed stream (0.5% NaCl, 25°C) being circulated over the fiber bundle at a linear flow rate of 16 feet/minute. The results of a three-month run at 600 psi are shown in Fig. 3. The rate of flux decline is seen to level off after the first few weeks of operation, suggesting that fiber membranes of this kind could be kept in service for at least a year with acceptable desalination performance.

The foregoing experiment involved a bundle of twelve fibers. In order to demonstrate the potential for scaling up to modules of practical size, a larger flow module containing more than 200 fiber

TABLE 4

ELEVATED TEMPERATURE SEPARATIONS WITH PBI HOLLOW FIBERS

Feed	Pressure (psi)	Temperature (°C)	Flux (gfd)	Rejection (%)
0.5% NaCl* (original static test)	200	23	0.57	90.5
	300	23	1.00	94.0
	400	23	1.43	95.2
	500	23	1.71	95.2
	600	23	2.07	96.0
0.5% NaCl*	300	75	1.06	97.9
	600	75	3.05	98.4
	600	23	0.63	97.9
0.5% Urea**	600	23	0.79	87.0
	300	75	1.67	52.2
	600	75	3.37	65.2
	600	23	0.59	80.4
0.5% Santomerse SX*	600	23	0.47	99.2
	300	75	0.92	95.6
	600	75	2.11	97.9
	600	23	0.31	98.5

*Rejections determined conductometrically.

**Rejections determined colorimetrically[7].

loops (i.e., 400 fiber ends) was constructed and tested under flow conditions at 400 psi. The initial salt rejection was measured to be 98.8% at a total module output of 0.43 gallons/day. Performance, illustrated in Fig. 4, was determined over a 2-1/2 month period. Flux and rejection stability were clearly superior to that of the smaller module, perhaps because of the lower operating pressure. The final output and salt rejection figures (0.43 gallons/day and 98.9%) are, in fact, well within experimental error of being the same as the initial figures. The potential for scale-up using these fibers is therefore deemed to be excellent, with indefinite membrane life being predicted at 400 psi operating pressure.

Morphological studies of individual PBI fiber variants have been conducted with the electron microscope, with a view toward establishing structure-property correlations. A scanning electron micrograph of a typical sample, sectioned in longitudinal perspective, is shown in Fig. 5. The striking structural characteristic of the PBI membrane shown in this micrograph is the presence of

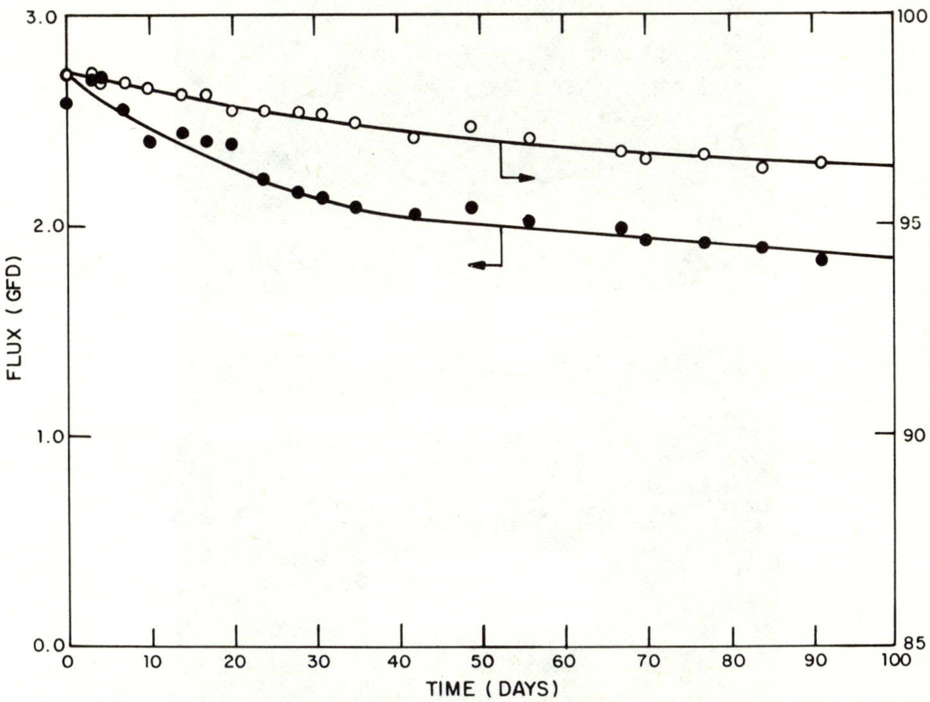

Fig. 3 - Long-Term Evaluation of Module Under Flow Conditions at 600 PSI.

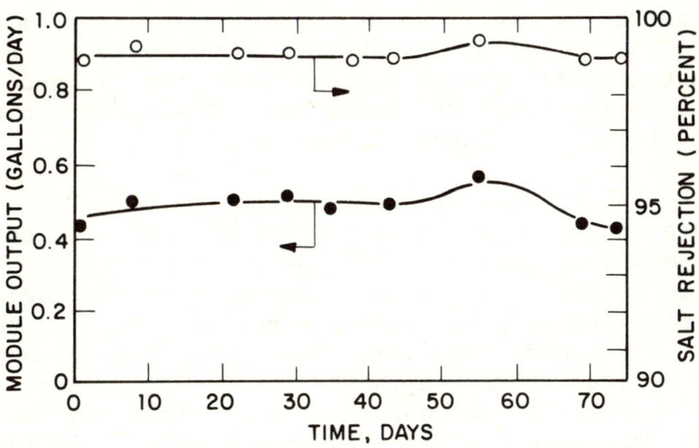

Fig. 4 - Long-Term Evaluation of Large Module Model (400 PSI).

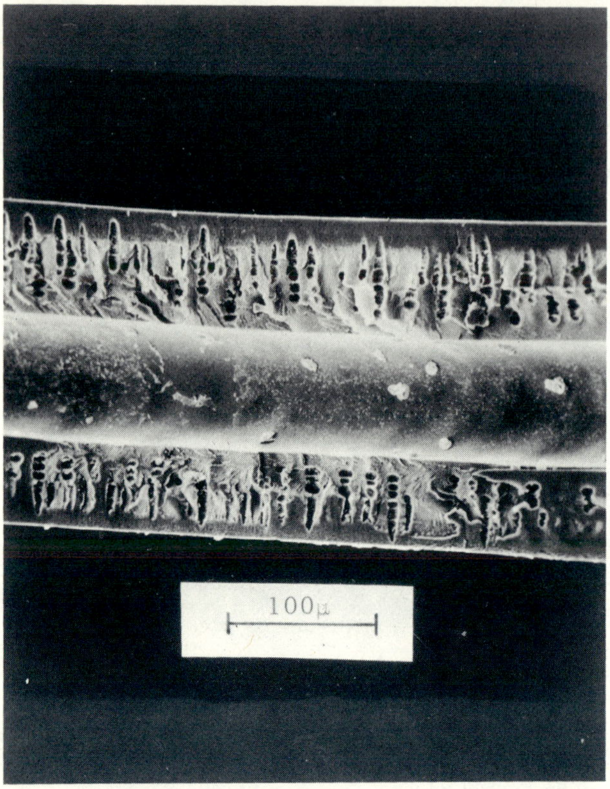

Fig. 5 - Scanning Electron Micrograph (Longitudinal Section) of Freeze-Dried Epoxy-Imbedded PBI Hollow Fiber Membrane.

a large concentration of radial voids. These voids are reminiscent of the finger-like intrusions observed during the formation of flat cellulose acetate reverse osmosis membranes[8], and thereby suggest that the hollow fiber PBI membranes possess a structure similar to that of the Loeb-Sourirajan membrane.

This hypothesis is strongly supported by the scanning electron micrographs shown in Fig. 6. While the sample used for examination was prepared in a somewhat different manner, the cross-sectional view nevertheless shows evidence of the radial voids. These voids are seen to extend to points very near the outer surface of the fiber. However, the longitudinal view shows a relatively smooth, void-free surface, implying the presence of a skin. Higher resolution micrographs will be necessary to further elucidate the nature of the membrane surface. It is believed that transport properties must also be related to void structure and void concentration. These areas of research are

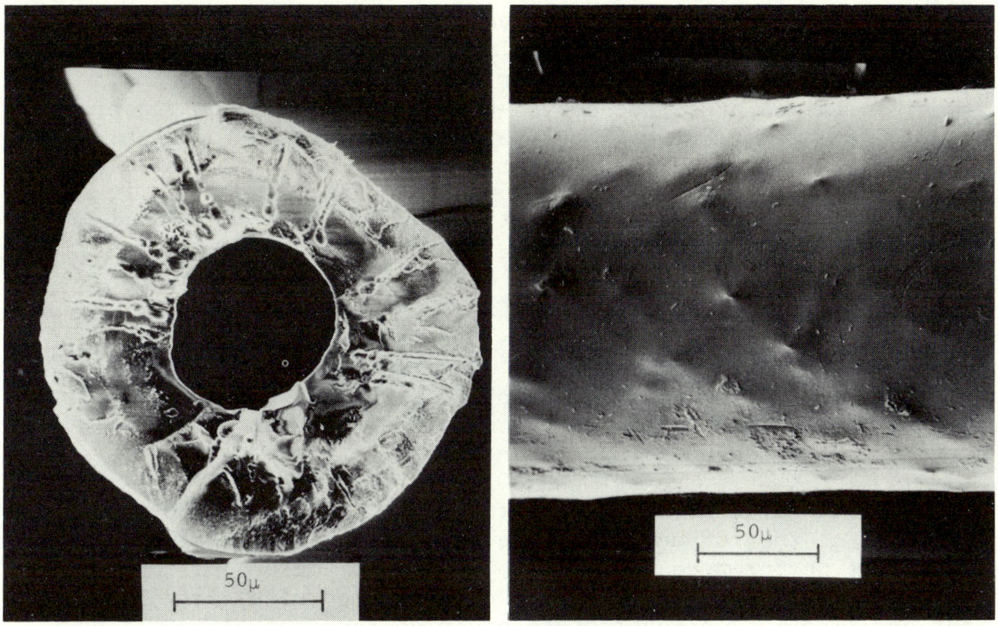

Fig. 6 - Scanning Electron Micrographs of Air Dried Membrane Sample. A. Cross-Section of Fracture Surface. B. Longitudinal View.

currently under active study using high resolution electron microscopy and standard thermal methods.

SUMMARY AND CONCLUSIONS

Initial work has been carried out to determine the reverse osmosis properties of the high performance polymer, polybenzimidazole. Although there has not yet been sufficient opportunity to explore its full performance range, enough work has been done to show PBI to be an extremely interesting, broadly applicable reverse osmosis material, useful in various configurations and systems.

Limited work with flat sheet PBI membranes indicates transport properties similar to those of cellulose acetate at room temperature, but which become increasingly superior as the operating temperature is raised to about 90°C.

A larger number of laboratory-scale results on hollow fiber PBI test modules and systems is likewise encouraging, a broad range of flux/rejection properties having been demonstrated. Sodium chloride rejections of up to 99% (at 2 gfd flux) and fluxes up to 14 gfd (at 40% rejection) have been observed during the course of this study. Equally as important, excellent flux stability has been obtained for periods of up to three months, even at sustained pressures above 400-500 psi. The high temperature performance of PBI hollow fibers has been found to be outstanding, as would be expected from the previously acquired flat sheet membrane data.

Preliminary information regarding morphological and structural features of PBI hollow fiber membranes has been obtained by scanning electron microscopy. Examination of the micrographs strongly suggests the presence of a structure characterized by radial internal water-filled voids or pores. Asymmetric layers are also suggested by the micrographs, as well as by the desalination performance data, so that the complete membrane is probably very similar to the Loeb-Sourirajan composite membrane structure.

The flat PBI membranes exhibit room temperature performance comparable to that of cellulose acetate and greatly superior high temperature performance. The hollow fiber PBI membranes show high versatility as evidenced by a demonstrated capability for water purification, high pressure and high temperature operation, process stream recoveries, special separations requiring good chemical resistance, and other applications. PBI membranes are being investigated more fully in current programs to develop their outstanding potential into practical, low cost systems to purify aqueous streams.

ACKNOWLEDGMENTS

The authors are indebted to Mr. Mitchell A. Sieminski who performed the microscopic studies. The authors also wish to thank Dr. Ian L. Hay who prepared the samples for electron microscopic examination and Mr. Norman S. Trouw who obtained most of the performance data. The hollow fiber membrane samples were provided by Dr. Michael J. Ram and flat films were prepared and tested by Mr. A. A. Boom. This work was performed in part under Contract No. 14-30-2810 with the Office of Saline Water, U.S. Department of the Interior, Washington, D.C.

REFERENCES

1. F. S. Model, H. J. Davis, A. A. Boom, C. Helfgott, and L. A. Lee, Office of Saline Water Research and Development Progress Report No. 657, June 1971.

2. H. A. Vogel, and C. S. Marvel, J. Polymer Sci., <u>50</u>, 511 (1961).

3. A. B. Conciatori, E. C. Chenevey, T. C. Bohrer, and A. E. Prince, Jr., J. Polymer Sci., Part C., 19, 49 (1967).

4. A. A. Boom, personal communication.

5. K. Hofmann, Imidazole and Its Derivatives, Inter-Science, New York, 1953.

6. J. W. Richter, and H. H. Hoehn, U.S. Patent 3,567,632, March 2, 1971.

7. H. H. Brown, Analytical Chem. 31, 1844 (1959).

8. M. A. Frommer, Hydronautics, Inc., personal communication.

REVERSE OSMOSIS PROPERTIES OF IONIC AND NONIONIC POLYMER MEMBRANES

H. Yasuda and A. Schindler
Research Triangle Institute
Research Triangle Park, North Carolina 27709

INTRODUCTION

The performance of a homogeneous polymer membrane in reverse osmosis may depend on various chemical and physical properties of the polymer and on the morphology of the membrane. Generalized criteria necessary for optimum reverse osmosis performance in terms of chemical structure of the macromolecule may not be evident. On the other hand, some physicochemical properties such as the equilibrium water content of the membrane may play a more important role than the chemical nature of the macromolecule. This seems to be particularly true for ionic polymeric membranes. Therefore, generalized criteria for an optimum reverse osmosis membrane may be sought from a study of the fundamental interaction between the polymeric membrane and water.

EXPERIMENTAL

The necessary crosslinks in membranes derived from ionic polymers can be introduced in different ways: (a) polymerization in situ in the presence of difunctional monomers, (b) crosslinking of membranes composed of polymers with functional groups by curing or vulcanization, (c) graft polymerization on hydrophobic polymer films, and (d) physical crosslinking by means of the hydrophobic domains of phase-separated block copolymer films. In order to obtain some generalizations regarding the reverse osmosis properties of ionic membranes, such membranes were prepared in three of the above four ways. Membranes polymerized in situ were not investigated, since the preparation of thin membranes, about 1 mil thick, by this method encounters some difficulties.

Vulcanized Copolymer Films[1]

These membranes consisted of random copolymers of butadiene with either styrene (BS-#), methacrylic acid (BMA-9), or 2- or 4-vinyl pyridine (BV-#). The copolymerizations were carried out in bulk with azobisisobutyronitrile (AIBN) initiation. The butadiene content of the copolymers was in the range of 20-40 mole-%. Films were cast from benzene or THF solution on glass plates by means of a doctor knife. The films were vulcanized by submersing the plates in a 5 vol.-% solution of freshly distilled S_2Cl_2 in dry CH_3NO_2 for 2 hours. The films containing vinyl pyridine were quaternized with gaseous CH_3Br at 70°C prior to their vulcanization. The styrene containing films were sulfonated with HSO_3Cl (2 vol.-% in $CHCl_3$) after vulcanization. The films containing sulfonated polystyrene or poly(methacrylic acid) were tested in their sodium salt forms.

Grafted Polyethylene Films[2,3]

The grafting of commercial low density polyethylene films of 0.5 and 1.0 mil thickness was carried out by post-polymerization of either γ-irradiated films[1] or UV irradiated films which were photosensitized with benzophenone[3]. In both cases the films were irradiated in air to promote the formation of active peroxide groups. The γ-irradiated films (PE-18, 20, ES-2) and the UV irradiated films (G-#) were all grafted at 70°C with diluted monomers.

Block Copolymer Films[4]

The block copolymers of styrene and vinyl pyridine were prepared under anionic polymerization conditions in THF with either butyl lithium or Na-naphthalene catalysts. The polymerizations were started with the formation of the polystyrene block and then changed to the polymerization of the vinyl pyridine. This sequence is necessary to avoid nucleophilic attack by the styryl carbanions on the poly(vinyl pyridine) with subsequent chain degradation and branching reactions. In order to obtain poly(vinyl pyridine) blocks sandwiched between polystyrene blocks, the living chain ends (pyridinyl carbanions) were coupled with 1,5-dibromopentane.

The block copolymer films (R-#) were prepared by casting from benzene or THF solution. The films exhibited good phase separation as revealed by transmission electron microscopy. In Fig. 1 is shown an electron micrograph of an ultrathin film cast from 0.6 wt.-% benzene solution. The polymer was an alternating block copolymer composed of several polystyrene and poly(vinyl pyridine) blocks and was prepared by Na-naphthalene initiation and coupling with 1,5-dibromopentane. The electron micrograph was obtained without film staining, the contrast very likely being a consequence of sample preparation. Poly(vinyl pyridine) swells slightly in

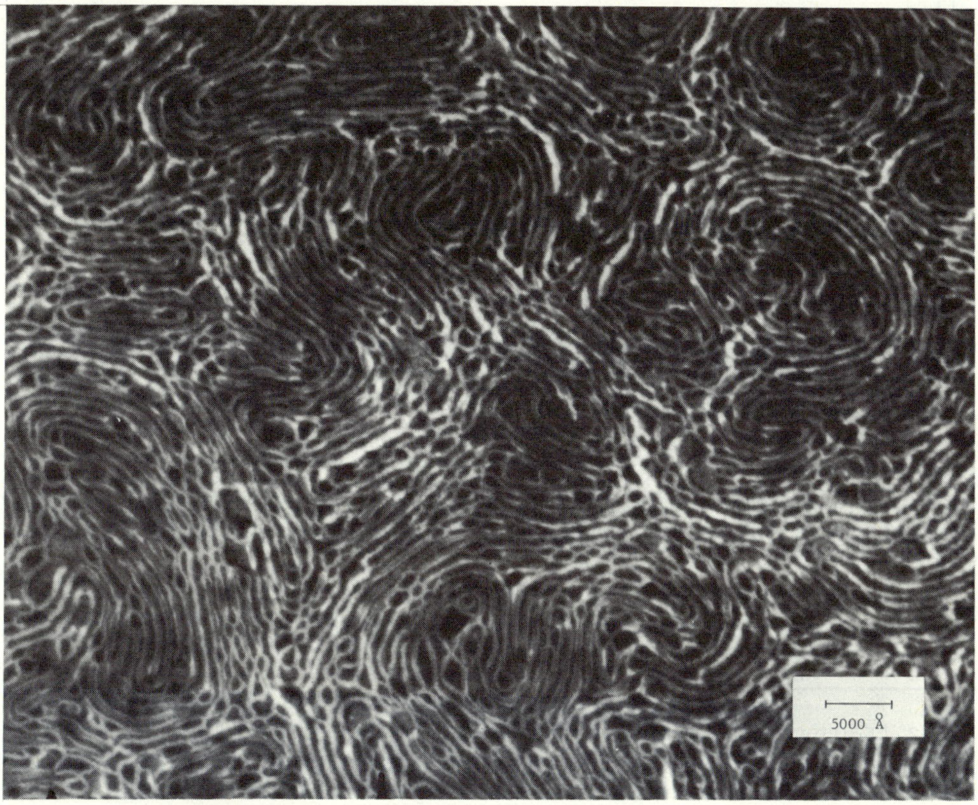

Fig. 1 - Transmission Electron Micrograph of an Alternating Styrene-2-Vinyl Pyridine Block Copolymer Film Cast From Benzene. Light areas: poly(2-vinyl pyridine); dark areas: polystyrene.

water while polystyrene does not. In the final state of sample preparation, the films are floated off on water and picked up by the supporting grid. Upon drying, the poly(vinyl pyridine) domains cannot shrink properly and consequently remain in a stressed state with a density lower than that of the polystyrene. The average domain width of the polystyrene phase is about 200-300 Å.

Sulfonated Polyethylene Films[2]

These films (PE-2,3) were prepared by sulfonation of commercial low density polyethylene films of 0.5 mil thickness with HSO_3Cl in CH_3Cl at 50°C.

Reverse Osmosis Tests

The reverse osmosis measurements were made in a test loop with stirred cells pressurized by an air-driven pump[5]. The applied pressure was adjusted so that the gross pressure minus the osmotic pressure of the feed solution, $\Delta p - \Delta\pi_o$, was maintained at 75 atm. The normalized water flux (in gallon-mil/ft^2-day) is calculated by multiplying the thickness of the membrane and the water flux. The thickness of the membrane was measured immediately after the reverse osmosis test, since some samples underwent noticeable compaction at high pressure. The reverse osmosis measurements were continued on each film until the water flux and salt rejection were stabilized. The concentration of feed and effluent were determined by conductivity measurements.

RESULTS AND DISCUSSION

A summary of the reverse osmosis performance of several types of ionic membranes under one set of experimental conditions is contained in Table 1. Additional results for styrene-vinyl pyridine block copolymer membranes, obtained under somewhat different conditions, are contained in Table 2. All of these results are summarized in Fig. 2 in terms of a plot of salt rejection against normalized water flux. Clearly the results for a wide variety of chemical systems fall on one fairly smooth curve.

Transport of Water in Polymer Membranes

The transport of water occurs in the intersegmental spaces within the membrane and, therefore, the transport is greatly dependent upon the amount of water existing in the membrane. A high water content at equilibrium means that the macromolecular segments are separated further by the presence of water. The permeation of water through homogeneous polymer membranes which have varying degrees of water content has been recently studied [6,7,8]. The results indicate that water in polymer membranes of low water content, e.g., less than 20%, is strongly associated with the macromolecules[8], and that the movement of water through these membranes can be visualized as occurring by activated diffusion regardless of whether or not hydraulic pressure is applied across the membrane[6]. On the other hand, in highly hydrated membranes, the flow of water occurs between more highly separated polymer segments[7], and the water flux is highly dependent upon whether or not hydraulic pressure is applied across the membrane.

The volumetric flux J_1 (cm^3/cm^2-sec) through a homogeneous polymer membrane of unit area can be given by the equation,

$$J_1 = K_1 \Delta p/\Delta x , \qquad (1)$$

where K_1 is the piezometric permeability coefficient and $\Delta p/\Delta x$ is the pressure gradient. The flux J_1 can be considered as the

TABLE 1

SALT REJECTION AND NORMALIZED WATER FLUX OF IONIC MEMBRANES
Reverse Osmosis Conditions: 0.2N NaCl at 1200 psi

Membrane Number	Water Flux (gfd-mil)	Salt Rej. (%)
Styrene-2-vinyl pyridine block copolymers (quaternized with CH_3Br):		
R-116	2.0	82.1
R-116	2.2	79.0
R-116	3.3	71.5
R-114	9.4	62.5
R-117	8.4	61.0
R-118	8.5	58.0
R-118	14.8	44.1
Random copolymers of styrene and butadiene (vulcanized with S_2Cl_2 and sulfonated):		
BS-6	2.9	78.0
BS-10	9.8	52.2
BS-9	19.8	41.3
BS-6	27.8	36.2
BS-9	39.2	25.9
Random copolymers of methacrylic acid and butadiene (vulcanized with S_2Cl_2):		
BMA-9	17.1	43.6
Sulfonated polyethylene films:		
PE-3	0.45	87.0
PE-2	1.0	80.0
Random copolymers of vinyl pyridine and butadiene (quaternized with CH_3Br and vulcanized with S_2Cl_2):		
BV-2	8.2	57.8
BV-5	8.4	54.7
BV-1	11.5	50.0
Polyethylene films grafted with 4-vinyl pyridine (quaternized with CH_3Br):		
PE-18	1.9	79.1
G-48	6.5	63.1
Polyethylene films grafted with methyl methacrylate (hydrolyzed with NaOH):		
PE-20	3.0	71.0
Polyethylene films grafted with styrene (sulfonated):		
ES-2	0.18	92.0
G-40	1.2	87.0
G-21	4.3	65.9
Polyethylene film grafted with vinyl toluene (sulfonated):		
G-20	1.5	80.2

sum of a diffusive flux, J_d, and a flux due to viscous flow, J_f, i.e.,

$$J_1 = J_d + J_f \quad (2)$$

Whether water moves by diffusion or by viscous flow at a given location and at a given time within a membrane under a pressure

TABLE 2

SALT REJECTION AND NORMALIZED WATER FLUX OF QUATERNIZED STYRENE-2-VINYL PYRIDINE BLOCK COPOLYMER MEMBRANES
Reverse Osmosis Conditions: 0.6N NaCl at 1500 psi

Membrane Number	Water Flux (gfd-mil)	Salt Rej. (%)	Membrane Number	Water Flux (gfd-mil)	Salt Rej. (%)
R-27	0.38	82.9	R-29	4.6	52.0
R-27	0.34	75.1	R-114	5.3	51.2
R-18*	0.85	71.6	R-9	6.3	52.1
R-18*	0.96	69.0	R-114	6.2	50.1
R-116	1.0	70.0	R-29	4.6	47.4
R-116	2.1	63.1	R-114	6.2	44.9
R-116	2.4	62.7	R-117	8.5	42.2
R-132	3.6	58.0	R-26	23.3	29.6

*Styrene-4-vinyl pyridine block copolymers

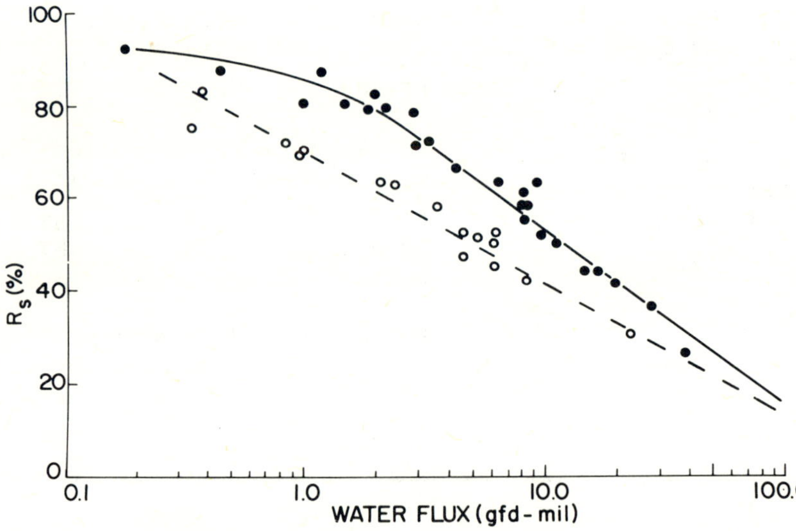

Fig. 2 - Relationship Between Salt Rejection, R_S, and the Normalized Water Flux (Gal-Mil/Ft^2-Day) for Ionic Membranes. The solid and dotted lines are for the 0.2N and 0.6N NaCl feed solutions, respectively. $\Delta p - \Delta\pi_o = 75$ atm.

gradient depends upon the conformation of the macromolecules at the location. The expression given by equation (2) does not necessarily assume independent paths in which either viscous flow or diffusion occurs exclusively, though this situation may occur in practical membranes. Regardless of the nature of the path, i.e., a fixed channel, visualized as an imperfection, or fluctuating intermolecular spaces within the homogeneous polymer, the total flux of water is the summation of two types of water transport.

The permeability coefficient K_1 can similarly be divided into the diffusive permeability coefficient, K_d, and the viscous flow conductivity coefficient, K_f, i.e.,

$$K_1 = K_d + K_f \ . \tag{3}$$

When the water flux is measured using radioactive water in the absence of a pressure gradient, the flux observed corresponds to J_d. Under such conditions, the mass flux, J_d' (g/cm^2-sec), is related to the diffusive permeability P_1 and to the concentration gradient of radioactive water, $\Delta c_1/\Delta x$, as follows:

$$J_d' = P_1 \ \Delta c_1/\Delta x \ . \tag{4}$$

From a consideration of the chemical potential of water, P_1 and K_d can be related as follows[6],

$$K_d = P_1 \ v_1/RT \ , \tag{5}$$

where v_1 is the molar volume of water. Consequently, K_1 is given by

$$K_1 = (P_1 v_1/RT) + K_f \ . \tag{6}$$

The viscous flow conductivity coefficient, K_f, derived from the frictional resistance of macromolecular segments, can be expressed by an equation of the form

$$K_f = \Phi \ (1/X - 1/X_c) \ ,$$

where Φ is a proportionality constant, X is a parameter which is related to the volume fraction, H, of water in the membrane [$X = (1 - H)/H$], and X_c is the critical value of X below which no viscous flow occurs in the membrane[7]. The calculated values of K_f [from eq. (6)] for various nonionic polymer membranes are shown in Fig. 3 in a plot of K_f versus 1/X.

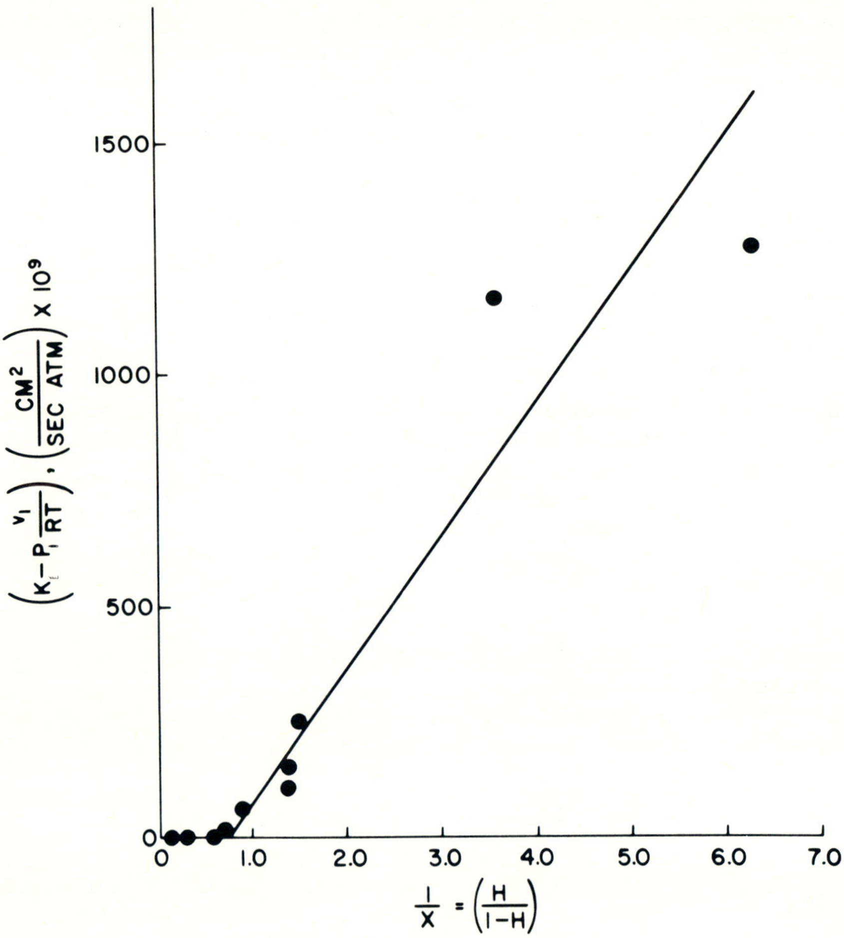

Fig. 3 - The Dependence of Flow Conductivity, $K_f = K_1 - P_1v_1/RT$, on the Parameter $1/X = H/(1-H)$, Where H is the Volume Fraction of Water in the Membrane. For details on the membranes used, see reference[6].

It is interesting to note that in the low hydration region, where $K_f = 0$, the water in the membrane does not freeze even below the normal freezing temperature of water (e.g., even at -100°C)[8]. At higher hydration, where K_f predominates in the overall value of K, freezing of water occurs and the observed heats of fusion indicate that some portion of the water in the membrane behaves as free water[8]. The amount of water which is highly associated with the macromolecules is nearly constant in highly hydrated membranes. The change of transport mechanism from activated diffusion in membranes of low hydration to viscous flow in highly hydrated membranes has a significant effect on the reverse osmosis properties of polymer membranes. In membranes of low hydration, the water moves by diffusion and, therefore, it is very likely that the solute (salt) also moves by diffusion. On the other hand, in highly hydrated membranes, water flows through the membrane and solute transport cannot be adequately represented by diffusion.

Salt Rejection by Polymer Membranes

Salt rejection is considered as the consequence of a depletion in salt flux relative to the water flux, rather than to exclusion of salt by the membrane. Therefore, regardless of the molecular level mechanism which causes the transport depletion, the salt rejection can be expressed by the following definition:

$$R_S \equiv 1 - C_2''/C_2' \ , \tag{7}$$

where C_2' is the concentration of the salt in the feed solution, and C_2'' is the concentration of salt in the effluent. Using the unit of water flux, J_1, as cm^3 water/cm^2-sec and for the salt flux, J_2, as grams salt/cm^2-sec, then we have that

$$C_2'' = J_2/J_1 \ . \tag{8}$$

Consequently, equation (7) is given by

$$R_S \equiv 1 - J_2/J_1 \, C_2' \ . \tag{9}$$

The mathematical expressions for J_1 and J_2 in relation to the driving forces are dependent upon the mode of water and salt flux.

The water flux, J_1, in reverse osmosis is given by

$$J_1 = K_1 \ (\Delta p - \Delta\pi)/\Delta x \ , \tag{10}$$

where $\Delta\pi$ is the osmotic pressure difference between the effluent and feed solutions.

As shown by equation (6), K_1 may or may not include a considerable contribution of the flow conductivity K_f, depending upon the hydration H of the homogeneous membrane. The diffusive permeability of water, P_1, on the other hand, can be expressed as a function of H for the entire range of H as follows,

$$\ln (P_1/D_{10}) = -\beta \, [X(1-\gamma)/(1+\gamma X)] - \ln (1+X) \;, \tag{11}$$

where D_{10} is the self diffusion constant of water, and β and γ are ratios of the parameters used in the free volume theory of diffusion, details of which can be found in reference[6].

If we express K_1 in terms of P_1 as follows,

$$K_1 = \omega \, P_1 v_1/RT \;, \tag{12}$$

the contribution of the viscous flow mechanism to the salt rejection can be conveniently expressed by the flow parameter, ω; i.e., $\omega - 1 = K_f/K_d$.

The salt flux through a membrane in reverse osmosis is the average flux of cation and anion of the salt. If either ion cannot move as fast as water the movement of the counter ion is accordingly restricted by the required electroneutrality and it leads to a net salt flux depletion. The mechanisms which cause the transport depletion of salt are different in ionic and nonionic membranes, as described in detail below.

Nonionic Membranes. In nonionic membranes, the transport depletion of salt can be expressed by the difference between the two diffusive permeabilities, P_1 and P_2, since the extent of salt depletion in viscous flow can be considered to be much smaller than in diffusion. In order to express this situation quantitatively, it may be assumed that whenever water moves by viscous flow no transport depletion of salt occurs. The supplementary data[8] obtained by the differential scanning calorimetry on the freezing phenomena and by wide line NMR on the mobility of the water-swollen polymer at low temperatures seems to justify this assumption. Then the salt flux, J_2, can be given in terms of the viscous flow and diffusion of water as

$$J_2 = K_f \frac{(\Delta p - \Delta\pi)}{\Delta x} C_2{}' + P_2 \frac{(C_2{}' - C_2{}'')}{\Delta x} \;, \tag{13}$$

where P_2 is the diffusive permeability of salt. Equations (6), (9), (10), (12), and (13) lead to

$$R_S = [\omega \, \{1 + \frac{P_2}{K_1 \, (\Delta p - \Delta\pi)}\}]^{-1} . \tag{14}$$

Equation (14) indicates that the mechanism of water transport plays a predominant role in the salt rejection of nonionic membranes since a high R_S requires a low value for ω ($\omega_{min} = 1$), regardless of the value of $P_2/K_1(\Delta p - \Delta\pi)$.

The diffusive permeability of salt, $P_2 = k_2D_2$, is not a function of H alone since the partition coefficient of solute in the membrane, k_2, depends on both H and the chemical nature of the polymer. The salt partition coefficient is generally defined as the ratio of the grams of salt per cc of polymer to the grams of salt per cc of solution. Thus, the partition coefficient decreases with decreasing H and does not describe the salt rejecting characteristics of the polymer[9]. The more specific parameter which is more closely related to the salt rejection can be derived by dividing k_2 by H, i.e., $\alpha = k_2/H$. With the introduction of the molal partition coefficient α, and consideration of the fact that salt ions cannot permeate without being hydrated by water, P_2 may be described as a function of H by using equation (11),

$$\ln P_2 = \ln D_{20} + \ln \alpha - \beta_2 \, [X \, (1 - \gamma)/(1 + \gamma X)] - \ln (1 + X) \,, \tag{15}$$

where D_{20} is the diffusion constant of salt in pure water.

Equation (14) can be written in terms of P_1 and P_2 as follows:

$$R_S = [\omega + \frac{P_2}{P_1} \cdot \frac{RT}{v_1 \, (\Delta p - \Delta\pi)}]^{-1} \,. \tag{16}$$

As ω becomes unity in the low H region, R_S depends entirely on P_2/P_1 at a given reverse osmosis condition.

The parameter α is highly dependent on the nature of the polymer and cannot be described by a simple function of H. However, the value of α generally decreases[9] with decreasing H, and the difference between P_2 and P_1 becomes more evident in the low H region as can be seen in Fig. 4. For an ideal case in which log P_1 is given by a linear function of X, all parameters involved in equation (14) can be given in terms of a common variable X, and the general trend of R_S as a function of K_1 can be estimated. The calculated dependence of R_S on K_1 is shown by a solid line in Fig. 5. Experimental points shown in Fig. 5 clearly indicate a rather abrupt decline of R_S with an increase of K_1 as expected from the increase of ω due to the change of water transport mechanism from diffusion to viscous flow as a result of excessive hydration.

The observed values of P_2/P_1 shown in Fig. 4 are much smaller than the value of α. This indicates that salt transport depletion is caused not only by the molal partition coefficient α but also by a kinetic factor described by the diffusivity. The decrease of

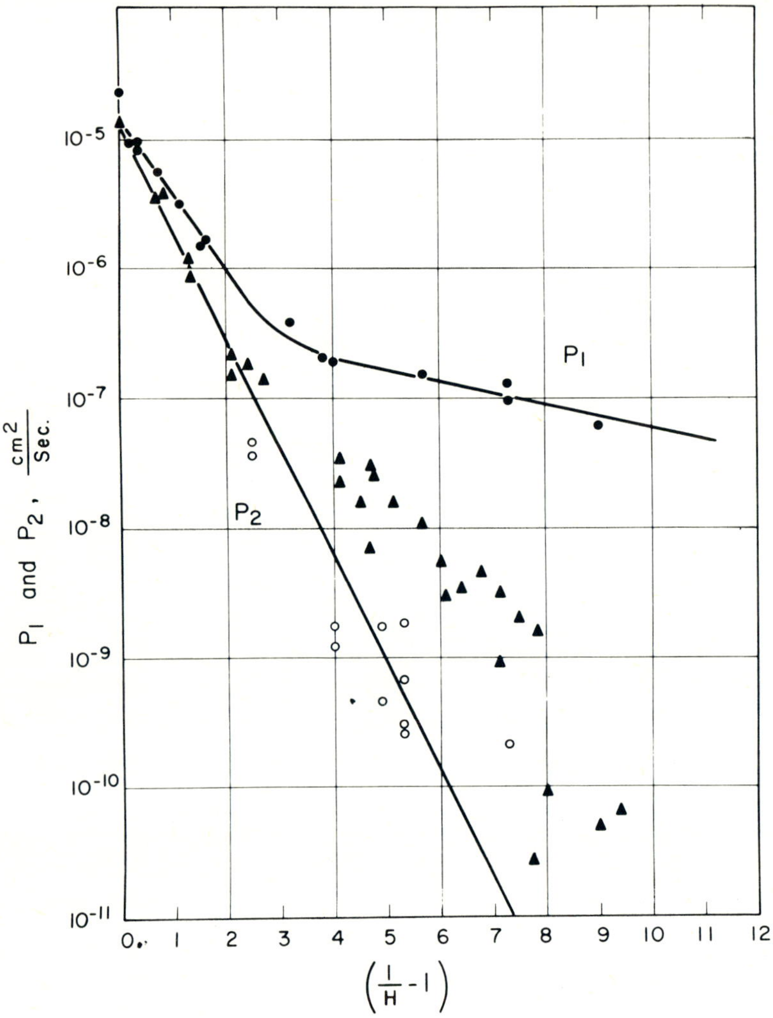

Fig. 4 - Comparison of the Diffusive Water Permeability P_1 and the Diffusive Salt Permeability P_2 of Homogeneous Polymer Membranes as Functions of the Parameter

$$X = \frac{1 - H}{H},$$

Where H is the Volume Fraction of Water in the Membranes.

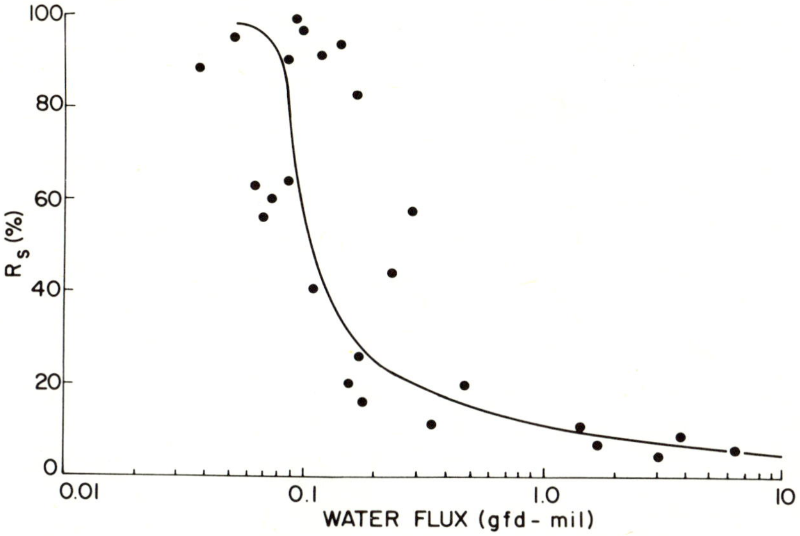

Fig. 5 - Relationship Between Salt Rejection, R_S, and the Normalized Water Flux for Nonionic Membranes. The solid line represents the dependence calculated from Eq. (14) with an empirical dependence of K_1, P_2, and ω on H. The details of the membranes and the calculation of the theoretical line are given in reference[9]. 0.6N NaCl, $(\Delta p - \Delta\pi_o)$ = 75 atm.

the diffusivity may be visualized by comparison of the parameters involved in equations (11) and (15). It has been shown[1,10] that the slope of a plot of log (P_2/D_{20}) against X is larger for large solutes than for small solutes. The hydrated ions are larger than water, and ionic solutes behave as relatively large sized solutes. The mode of distribution of water within the polymer phase is an important factor in the low hydration region[5].

Ionic Membranes. In most ionic membranes, the value of K_1 is much greater than P_1v_1/RT, indicating that water moves through these membrane by a viscous flow mechanism. The salt rejection of ionic membranes, on the other hand, is generally higher than that of nonionic membranes with similar K_1 values. It is clear that salt transport depletion due to Coulombic repulsion between the fixed ions of the polymer and the coions of the salt, often recognized as Donnan exclusion, is playing a role. This situation has been recently studied from the viewpoint of the volume element available for the transport of water and ions[11].

The water uptake in the ionic domains is governed by the number of fixed ions and by the film morphology, since swelling proceeds until the swelling pressure equals the stress generated in the hydrophobic matrix. Because of Coulombic repulsion, the fixed ions of the polyelectrolyte will arrange themselves into a diffuse ionic lattice without long-range order. Salt and water transport proceeds through this ionic lattice. For water transport, the total volume of hydrated water may be available, whereas for salt transport the available water volume is smaller, because the Coulombic repulsion between the fixed ions of the polymer and the coions of the salt will restrict their mutual approach to a minimum distance. Outside this salt-excluding water shell surrounding each fixed ion, the salt solution will invade the membrane at the feed concentration.

For the mathematical description of this model the following volumes have to be defined, all of them referring to one fixed ionic site: V_o, the total volume of water in the hydrated membrane per ionic site; V_1, the volume impermeable to water; and V_2, the volume impermeable to salt.

If we assume that the volume $(V_2 - V_1)$ will allow transport of pure water, that $(V_o - V_2)$ will allow transport of unmodified feed solution, and that the rates of volume flow in both elements are identical, then the total water flux can be represented by

$$J_1 = \Phi_1 + \Phi_2 , \tag{17}$$

where

$$\Phi_1 = J_1 (V_2 - V_1)/(V_o - V_1) \tag{18}$$

$$\Phi_2 = J_1 (V_o - V_2)/(V_o - V_1). \tag{19}$$

The salt flux, J_2, is then given by

$$J_2 = \Phi_2 C_2' , \tag{20}$$

where C_2' is the feed concentration of the salt.

From equations (8), (9) and (20), R_S is found to be related to V_o, V_1, and V_2 as follows:

$$R_S = (V_2 - V_1)/(V_o - V_1). \tag{21}$$

This equation can be expressed in terms of the volume fraction of water in the swollen membrane, H, and the hydraulic water permeability, K_1, for highly hydrated membranes can be approximated by an expression of the form[6,7],

$$K_1 = K_o \exp\,[-B\,(1 - H)/H]\;, \tag{22}$$

where K_o and B are constants.

The combination of equations (21) and (22) with a semiempirical assumption leads to the relation between R_S of an ionic membrane and K_1 as follows:

$$K_1 = A \exp\,(-B'\,R_S)\;, \tag{23}$$

where A and B' are constants[11]. This relationship observed with a variety of ionic membranes is shown in Fig. 2. It has important practical value for characterizing the reverse osmosis performance of ionic membranes in comparison to that of nonionic membranes, since the relation is expressed by K_1 and R_S which are the only parameters observable in ordinary reverse osmosis experiments. The salt rejection of ionic membranes increases with decreasing feed concentration, which is due to an increase in the salt-excluding-shell volume $(V_2 - V_1)$ at lower concentrations. This increase is more pronounced in the high R_S region as expected from equation (21). It is also important to note that the R_S of ionic membranes starts to deviate from the relation given by equation (23) as R_S increases, or K_1 decreases, and approaches the same point as for nonionic membranes. As the mode of water transport shifts from a predominantly viscous flow mechanism to diffusion, the mechanism of salt rejection accordingly changes to that for tight membranes in which water moves only by diffusion. The comparison of Fig. 2 with Fig. 5 clearly indicates an advantage of ionic membranes over nonionic membranes in terms of a higher water flux in the moderate R_S region.

In nonionic membranes, the difference in the characteristic size of the hydrated ions and the water molecule is responsible for the salt rejection in reverse osmosis. In ionic membranes, however, the impenetrable ionic field (for similarly charged ions) overrides the effect of size differences.

ACKNOWLEDGMENT

Studies reported in this paper were supported by the Office of Saline Water, U. S. Department of the Interior. The authors' special thanks are due to Dr. A. Peterlin for his stimulating and encouraging discussions. The authors are also indebted to Dr. K. Platt, Dr. J. L. Williams, and Mrs. C. E. Lamaze for their experimental contributions.

LIST OF SYMBOLS

ΔC_1 concentration difference of (radioactive) water across a membrane, g/cm^3

C_2' concentration of salt in feed solution, g/cm^3 water

C_2'' concentration of salt in effluent, g/cm^3 water

D_{10} self-diffusion constant of water, cm^2/sec

D_{20} diffusion constant of salt in pure water, cm^2/sec

H volume fraction of water in a membrane

J_1 water flux through a homogeneous polymer membrane, cm^3/cm^2-sec

J_1' water flux in g/cm^2-sec

J_d water flux due to diffusive flow, cm^3/cm^2-sec

J_f water flux due to viscous flow, cm^3/cm^2-sec

J_d' water flux due to diffusive flow, g/cm^2-sec

J_2 salt flux through a homogeneous polymer membrane, g/cm^2-sec

K_1 water permeability constant, cm^2/sec-atm

K_d diffusive water permeability, cm^2/sec-atm

K_f viscous flow water conductivity, cm^2/sec-atm

k_2 partition coefficient of salt, (g salt per cm^3 membrane)/(g salt per cm^3 solution)

P_1 diffusive water permeability, cm^2/sec

P_2	diffusive salt permeability, cm^2/sec
Δp	total pressure difference across a membrane, atm
R_s	salt rejection; $R_s \equiv 1 - C_2''/C_2'$
R	gas constant
T	absolute temperature
v_1	partial molar volume of water, cm^3
V_o	total volume of water per ionic site in a hydrated ionic membrane
V_1	the volume per ionic site which is impermeable to water
V_2	the volume per ionic site which is impermeable to salt
Δx	thickness of a homogeneous polymer membrane, cm
X	$(1 - H)/H$
α	molal partition coefficient of salt, $= k_2/H$
Φ_1	water flux in salt-rejecting volume of ionic membrane
Φ_2	water flux in non-salt-rejecting volume of ionic membrane
$\Delta\pi$	difference of osmotic pressure of feed solution and effluent
ω	K_1RT/P_1v_1

REFERENCES

1. H. Yasuda, L. D. Ikenberry, and C. E. Lamaze, Makromol. Chem., 125, 108 (1969).
2. K. L. Platt and A. Schindler, Angew. Makromol. Chem., in press.
3. A. Schindler and K. L. Platt, unpublished results.
4. A. Schindler and J. L. Williams, ACS Polymer Preprints, Vol. 10/2, 832 (1969).
5. H. Yasuda and C. E. Lamaze, J. Polymer Sci., A-2, 9, 1537 (1971).

6. H. Yasuda, C. E. Lamaze, and A. Peterlin, J. Polymer Sci., A-2, 9, 1117 (1971).

7. A. Peterlin, H. Yasuda, and H. G. Olf, J. Appl. Polymer Sci., in press.

8. H. Yasuda, H. G. Olf, B. Crist, C. E. Lamaze, and A. Peterlin, Proceedings of Symposium on Structure of Water at the Interface of Polymers and Water, ACS Los Angeles Meeting, March 1971, Plenum Press, in press.

9. H. Yasuda, C. E. Lamaze, and L. D. Ikenberry, Makromol. Chem., 118, 19 (1968).

10. H. Yasuda, A. Peterlin, C. K. Colton, K. A. Smith, and E. W. Merrill, Makromol. Chem., 126, 177 (1969).

11. H. Yasuda, C. E. Lamaze, and A. Schindler, J. Polymer Sci., A-2, 9, 1579 (1971).

IONIC BLOCK COPOLYMERS AS PIEZODIALYSIS MEMBRANES

G. Lopatin and H. A. Newey
Shell Development Company
Emeryville, California 94608

INTRODUCTION

Piezodialysis, or pressure dialysis, is a relatively new concept for desalting water by using membranes and pressure[1]. In contrast to reverse osmosis, where the product water is separated from a saline solution, piezodialysis has the feature that salt is preferentially removed from the feed. The process (though not the method) is quite analogous to electrodialysis except that the required ionic current is obtained by pressure and the consequent solution flow through the membrane, and not by an externally applied EMF as in electrodialysis. The most feasible approach to an actual technique incorporates a special membrane, the so-called charge mosaic membrane, which consists of small anionic and cationic exchange channels juxtaposed and each extending continuously from one membrane surface to the other as shown in the following diagram:

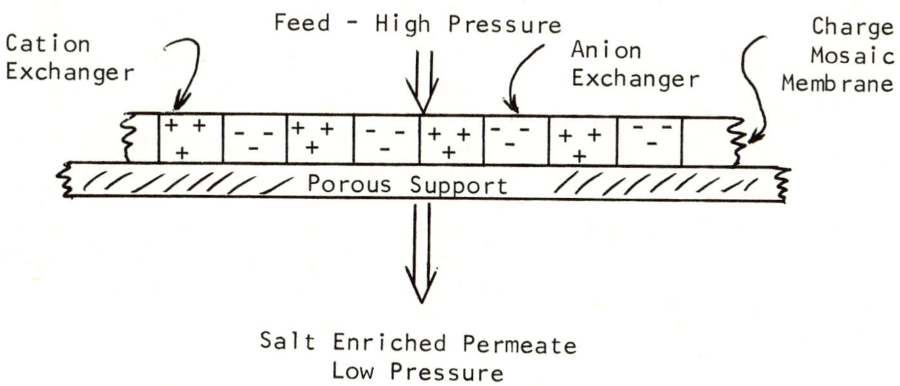

The flux produces opposite EMF's in the neighboring ionic channels, and the close proximity of the channels then enables local ionic currents to be set up which function to pass ions continually through the membrane. Several types of charge mosaic membranes have been made[2], and the piezodialysis effect has been demonstrated, although it is far from a practical reality. The main difficulty now is the construction of robust, efficient, charge mosaic membranes, so that data describing the process can be obtained.

The object of this investigation was to explore the potential for piezodialysis membrane construction of a new class of materials, i.e., ionic center-block substituted ABA block copolymers[3]. The parent hydrocarbon ABA block copolymers, KRATON thermoplastic elastomers*, e.g., polystyrene-polyisoprene-polystyrene, have been found to offer a versatile chemical structure which can be readily modified and controlled. In addition, the thermodynamic incompatibility between the end and center segments leads to microphase separation into discrete regions with a well defined morphological structure[4]. The particular structure depends primarily on the relative proportion of the different segments in the molecule. For example at about 25% weight fraction of polystyrene end blocks, and the remainder an elastomeric center segment, the polystyrene forms small (ca. 200 Å) spherical domains dispersed in the elastomeric matrix. The domains are connected to each other through the elastomeric segments. By locking the ends of the elastomeric segments in glassy regions, the domains, together with the topologically trapped entanglements, act as crosslinks in the rubber. Other morphologies, e.g. lamellar structures, are observed at higher proportions of polystyrene. With the proper choice of monomer, block lengths, and weight fractions of A and B segments, the ABA copolymers can have most of the useful properties of vulcanized, filled rubbers, yet they are soluble and flow as thermoplastics at elevated temperatures[5].

From these polymers, by selective placement of polar groups onto the center segments, a new family of ionic polymers was obtained. They swell in aqueous media and maintain high strength and extensibility in the swollen state. These features could be quite advantageous in membranes used for desalting processes. Further, the singular structure offers great versatility for membrane construction, a feature especially desirable for the development of charge mosaic membranes. Accordingly, the characteristics of these ionic polymers relevant to various facets of piezodialysis membrane development were investigated and the results of most interest are described below. In addition, novel charge mosaic membranes of two different types have been prepared from the polymers

---

*Shell Chemical Co.

and they are also described.

EXPERIMENTAL

The following three types of parent ABA block copolymers were studied:

Type I - polystyrene-polyisoprene-polystyrene (SIS)
Type II - polystyrene-polybutadiene-polystyrene (SBS)
Type III - polystyrene-EPR type copolymer-polystryene (SEPS)

Ionic groups, that is, sulfonic acid salts and quaternary ammonium salts, were substituted exclusively on the center block to give copolymers that could be represented as

SIS	SBS	SEPS
$(SO_3^-Na^+)_n$	$(N^+(CH_3)_3Br^-)_n$	$(SO_3^-Na^+)_n$

The total molecular weight of the parent copolymers was about 100,000 and except where noted the copolymers contained about 70% by weight elastomer.

The membranes were generally prepared by merely casting solutions of the polymer (10% by weight) onto glass plates with controlled evaporation. Tetrahydrofuran (THF) was the solvent for the Type I and Type III sulfonate polymers, toluene for the Type II quaternary ammonium salt. Draw knife-casting, spin-casting, and casting on aqueous substrates (water saturated with NaCl and THF) were also successfully carried out but these other methods had advantages for only specific uses.

The reverse osmosis testing was carried out using a flowthrough type cell, with a membrane area of 13.3 cm^2 active area. The membrane rested on a porous support, Tyvek* polyolefin paper, backed by a porous steel disk. The feed from a 20 liter reservoir was circulated by a multi-stage pump at the appropriate pressure (1500 psi maximum) with a typical flow rate of 250 cc per minute. The membranes were usually from 0.002" to 0.004" thick. Double membranes were used to avoid pinhole problems.

The swelling measurements were carried out by an optical swelling technique[6] and by weight increase measurements. The films did not swell isotropically so that the more convenient optical methods led to errors, but the major trends were similar in both methods.

---

*duPont de Nemours Inc.

The compressibility or deswelling under pressure of swollen sheets was measured by compressing the specimen, 0.050" thick, between two porous plates on an Instron tensile machine at a compression rate of 10% per minute. The decrease in thickness was measured directly with a feeler gage and the lateral expansion of the specimen was monitored with a traveling telescope.

RESULTS AND DISCUSSION

Solubility and Casting Characteristics

The various polymers could all be obtained in solution but usually only under specific conditions. The Type I sulfonates in the unionized (acidic) form are soluble in THF and other polar solvents and can be readily cast from these solutions into films. But as the sodium salt they are insoluble in all solvents commonly available. Films from the Type III sulfonates are prepared in two steps. The derivatized polymer is first obtained in the form of an acid chloride which is neutral, soluble in a wide variety of solvents, and is readily cast into a film. The neutral acid chloride film can then be readily hydrolyzed to the sulfonate salt in strongly basic aqueous media. The opportunity of handling these polymers in a neutral form proved to be an important advantage in charge mosaic membrane preparation, since the problem of premature adhesion between the cationic and anionic regions can be avoided by a neutral anionic precursor. The cationic polymers are prepared directly in the form of quaternary ammonium salts. These are soluble in toluene and pyridine-benzene mixtures and are directly cast into anion exchange films.

Solutions of all the ABA block copolymers studied spread well on glass. They readily form tough, coherent films under controlled evaporation conditions with or without a casting knife. The potential properties of the polymers for casting ultrathin films were also screened. Very thin films were obtained, less than 0.2 μ thick, when the Type I or Type III sulfonates were cast on water saturated with sodium chloride and THF.

Film Properties

Swelling. The films swell to a considerable extent in water, sometimes more than 1000%. However, the swelling is diminished in salt solution (Table I). This behavior is similar to the swelling of loosely crosslinked polyelectrolyte gels[7], and probably can be similarly explained, viz., the mutual electrostatic repulsions between the fixed ions, which tend to expand the network, are reduced by the screening effects of added salts and less swelling is observed. The swelling is very sensitive to ion exchange capacity, at least in the range from 0.8 to 2.7 meq/g and varies by more than a factor of 10. The sensitivity to relative block lengths is more difficult to ascertain from the data but there does not seem to be any strong effect. This is probably because the swelling is limited principally by the chain entanglements which are little affected by block lengths.

Tensile Behavior. Tensile tests on the swollen films reveal an important quality of these ionic block polymers, i.e., their high strength and flexibility. The breaking strength of swollen films (about 100% swelling) are typically in the range of 1500 psi or greater with about 500% extension at break. Values typical of other films, not block copolymers but similarly swollen, are 200 psi tensile strength and less than 50% elongation-at-break. The high strength of the ionic ABA block polymers should facilitate handling of thin films and the exceptional extensibility should be very important in helping membranes tolerate the large differential swelling stresses encountered in the preparation of high swelling mosaic membranes.

TABLE I

SWELLING CHARACTERISTICS OF SULFONATES DERIVED FROM TYPE I BLOCK COPOLYMERS

Elastomer Block Content (%)	Ion Exchange Capacity (meq/g)	Swelling Medium	Vol. Increase in 24 hr (%)
70	0.8	H_2O	35
		0.01 N NaOH	30
		0.1 N NaCl	15
		1.0 N NaCl	10
70	2.1	H_2O	760
		0.01 N NaOH	640
		0.1 N NaCl	245
		0.5 N NaCl	120
		1.0 N NaCl	80
88	2.7	H_2O	1350
		0.01 N NaOH	700
		0.01 N HCl	480
		0.005 N NaCl	500
		0.01 N NaCl	380
		0.10 N NaCl	160
		0.25 N NaCl	100
		1.0 N NaCl	55

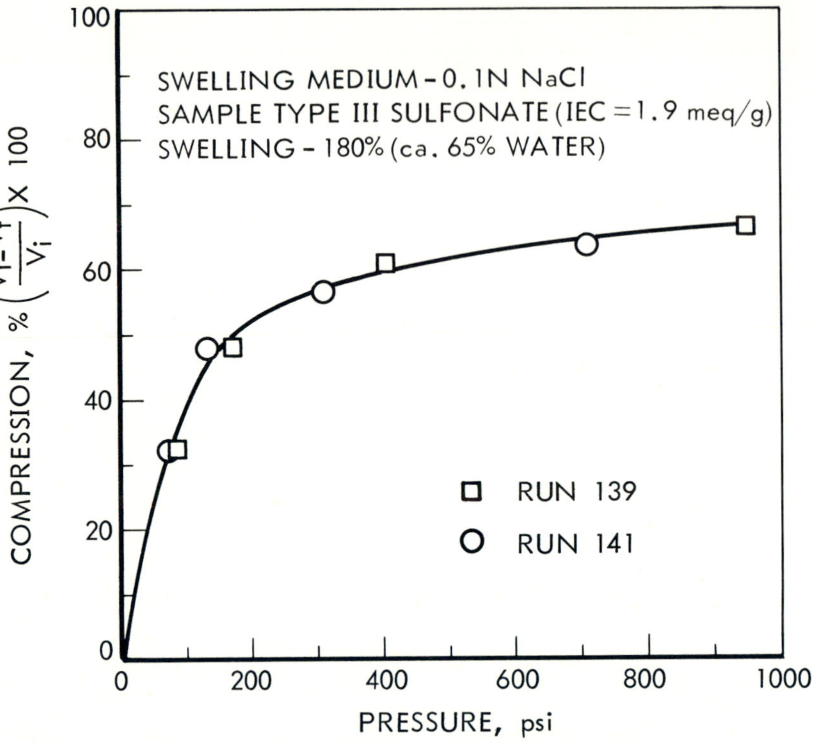

Fig. 1 - Compression of Swollen Sheet

Compressibility. Highly swollen crosslinked polyelectrolyte gels have been shown to be very compressible, i.e., relatively low pressures suffice to remove much of the water[7]. The ABA sulfonates are also highly swollen (although not as much as the polyelectrolyte gels) and therefore probably are also compressible. Since the membranes will be used under high pressures it is important to determine the compressive characteristics of these materials under the pressures expected in reverse osmosis or piezodialysis. Some experiments have been carried out on films swollen in 0.1N sodium chloride solution and the compression curve on a Type III sulfonate swollen to 180% is shown in Fig. 1.

It is seen that relatively low pressure brings about a considerable deswelling. This again resembles the behavior of the swollen slightly crosslinked polyelectrolyte gels mentioned above. For those materials the authors found a relation, an equation of state, between gel volume, pressure, and degree of ionization that at moderate swelling reduces to the following:

$$PV=\alpha nRT(1-B\frac{\alpha V}{n}), \quad (1)$$

where P is the pressure; V is the volume of swollen gel; α is the degree of ionization; n is the number of monomoles of crosslinked polymer; and B is a quite complex function of charge, reciprocal radius of ionic atmosphere, end-to-end distance of chains, and hydrodynamic length of monomer units. This function usually varies only slightly. At constant α, such as with the ABA sulfonates, the pressure and volume are nearly inversely proportional and therefore the degree of swelling will be very sensitive to the applied pressure. The near linearity of the curve in Fig. 1 up to 150 psi suggests that the above equation may hold for the swollen ABA sulfonates in the low pressure region. Thus, these systems, like swollen crosslinked polyelectrolytes and swollen rubbers[8], can be considered as compressible systems.

Measurements of the water exuded by the sample and absorbed in the porous plates were carried out and they confirmed the compression measurements (Table 2). The salt concentrations in the exuded water showed a 70% decrease in concentration with respect to the feed. This is only slightly lower in rejection than the films showed in reverse osmosis. A mechanical desalination technique based on mechanically deswelling a swollen mass of the ionic polymer is thus suggested by these observations. The high swelling and extensibility of these systems might make such a technique feasible where previous attempts have not proven successful[9].

Film Adhesion. A good adhesive bond between the anionic and cationic channels is a necessary condition for practical charge mosaic piezodialysis membranes. The membranes will be thin and used under high pressures so that poor adhesion would result in

TABLE 2

WATER LOSSES ON COMPRESSION OF SWOLLEN FILMS

Type III Sulfonate Polymers, 2.0 meq SO_3^-/g

	No. 141	No. 139
Initial water content (%)	65	65
Weight loss of specimen*	44	55
Water recovered (% of sample weight)	38	42
Salt rejection in recovered water (%)	70	--

*Weight changes were determined after release from compression.

very weak membranes and excessive leaking. In this regard the ionic ABA films display another unique characteristic. Strong irreversible adhesion between swollen films of opposite ionic type (ion exchange capacity = 2 meq/g) occurs on contact, or with slight pressure, under dilute saline solution. Adhesion does not occur when the salt concentration increases much beyond 0.6N NaCl, and lower ionic exchange capacity films require increased pressure.

An adhesion between films based on ionic bonding, apparently not seen in any other system, seems to be suggested by these data. Ordinarily, ionic bonding between oppositely charged surfaces will be prevented by the respective counterion clouds. Apparently, the special characteristics of poly-poly salt formation where the counterions are "squeezed" from between the chains, together with some aspect of the block polymer structure, perhaps its high flexibility, are critical for this adhesion.

Reverse Osmosis

An intensive reverse osmosis testing program has been carried out on the membranes. This was done for two reasons. First, the characteristics of each polymer, when a component in a mosaic piezodialysis membrane, can be roughly ascertained by its behavior as a reverse osmosis membrane. The membrane should be highly swollen and show high throughput and good rejection. If so, it would indicate that the high coupling between the water and the counterions necessary for piezodialysis had been obtained. The high salt rejection in highly swollen ionic systems would presumably be due to a rejection mechanism based on ionic interactions, e.g., Donnan equilibrium effects[10]. Second, since the membranes are very new materials with different morphological structures, it is of interest to know their behavior in types of desalination processes other than piezodialysis.

The reverse osmosis behavior of the Type III sulfonates and the Type II quaternary ammonium salts is shown in Table 3. The Type 1 sulfonates showed poorer reverse osmosis properties than the Type III sulfonates. The normalized flux and rejection of the latter are very high compared to other polymers. For example, the normalized flux is about 5 times greater than that for homogeneous cellulose acetate of similar rejection values. The favorable transport properties obtained, taken together with the other characteristics described, make these polymers excellent candidates for components in either reverse osmosis or piezodialysis membranes.

Charge Mosaic Membranes

Two approaches to making charge mosaic membranes from these materials were taken. Composite membranes, where particles of

TABLE 3

REVERSE OSMOSIS RESULTS ON IONIC A-B-A COPOLYMERS

Type	Ion Exchange Capacity (meq/g)	Pressure (psi)	Feed (N NaCl)	Rej. (%)	Normalized Flux (gal-mil/ft^2-day)
III-SO_3^-	2.0	1500	0.1	80	10
		800	0.1	66	8
		1500	0.6	70	5
		800	0.6	57	4
III-SO_3^-	0.9	1500	0.1	92	0.9
		800	0.1	78	0.7
		1500	0.6	81	0.5
		800	0.6	59	0.4
II-Pyr$^+$	0.7	1500	0.1	93	0.9
		800	0.1	84	0.4
II-$N(CH_3)_3^+$	2.0	1500	0.1	78	8.0
		800	0.1	74	5.4
		1500	0.6	61	4.5
		800	0.6	46	3.3

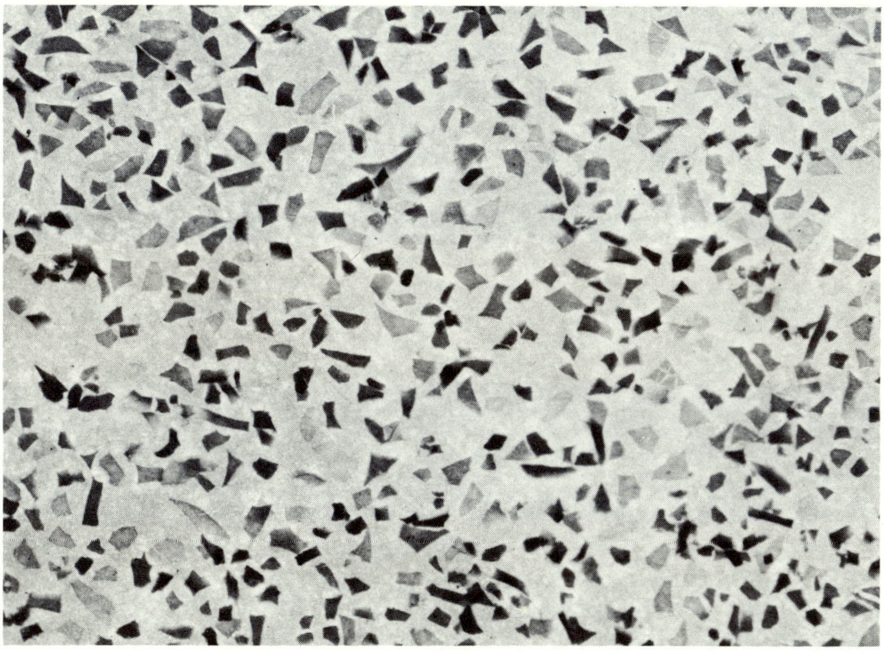

Fig. 2 - Composite Mosaic Membrane

anion exchange polymer (Type II, trimethylammonium salt) were suspended in a matrix of the cation exchange polymer (Type III sulfonate), were constructed to test the feasibility of fine grain, high swelling mosaics (Fig. 2). The membranes demonstrated that they can withstand the swelling and the pressure stresses and remain strong and coherent. They have been developed sufficiently to show strong indications of piezodialysis but actual salt enrichment was not quite attained. On the other hand, a different type of membrane did show positive piezodialysis. This membrane consists of two perforated films of opposite ionic type, placed one upon the other, with their holes displaced (Fig. 3). Thus, alternating regions of opposite charge are exposed and the regions between the holes merely act as reverse osmosis membranes. Up to 10% salt enrichment from very dilute solution (0.01N) was observed, and slight (2%) desalting of the feed was also obtained. It is expected that the perforated sheet membrane with its well defined geometry will be most useful for testing the effects of process variables but it will not be a path for a practical membrane. The composite mosaic approach, where much smaller interchannel distances are possible, would seem to hold the most potential for high piezodialysis efficiency and large scale production.

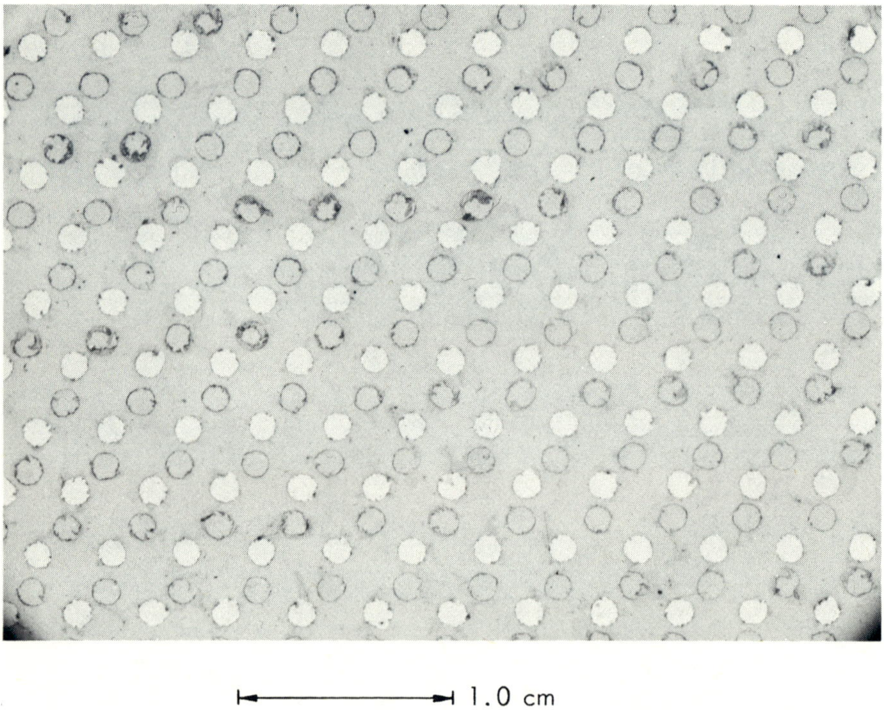

Fig. 3 - Perforated Sheet Membrane. Component Films: Type II Quaternary Ammonium Salt; Type III Sulfonate

ACKNOWLEDGMENT

The authors wish to acknowledge the support of this project by the Office of Saline Water, U. S. Department of Interior. They also wish to acknowledge the contributions made by W. P. O'Neill, D. E. Winkler, and O. D. Bergren.

REFERENCES

1. U. Merten, Desalination, 1, 297-310 (1966).

2. J. N. Weinstein and S. R. Caplan, Science, 161, 70-72 (1968).

3. G. Lopatin and H. A. Newey, Office of Saline Water Research and Development Progress Report No. 690, May 1971.

4. D. J. Meier, Polymer Preprints, 11, No. 2, 400 (1970).

5. G. Holden, E. T. Bishop and N. R. Legge, J. Polymer Sci., Part C, 26, 37 (1969).

6. E. T. Bishop and S. Davison, J. Polymer Sci., Part C, 26, 57 (1969).

7. A. Katchalski, S. Lifson, and H. Eisenberg, J. Polymer Sci., 7, 571 (1951). A Katchalski, et al., J. Polymer Sci., 8, 476 (1952).

8. L. R. G. Treloar, Physics of Rubber Elasticity, Clarendon Press, Oxford (1958), p. 196.

9. J. G. McKelvey, Jr., K. S. Spiegler, and M. R. J. Wyllie, J. Phys. Chem., 61, 174 (1957).

10. L. Dresner and K. A. Kraus, J. Phys. Chem., 67, 990 (1963).

IV. HOLLOW FIBERS AND SMALL TUBES

HOLLOW-FIBER REVERSE OSMOSIS MEMBRANES

Mark E. Cohen,* Michael A. Grable, and Billy M. Riggleman

Hercules Incorporated
Allegany Ballistics Laboratory
P.O. Box 210, Cumberland, Maryland 21502

INTRODUCTION

Hollow-fiber membranes for reverse osmosis desalination are typically of the dense wall type and exhibit water flux rates in the range of 0.1-1 gallons/square foot-day (gfd) [1,2]. These low flux rates are overcome by making the fibers extremely small (50 microns) in order to develop a relatively compact desalination system. Still, the need exists for hollow-fiber membranes with flux rates greater than those of dense wall fibers. Such membranes could be used to develop highly compact modules and further improve the economics of reverse osmosis desalination.

The hollow-fiber membranes developed in this study were of the asymmetric type with a structure comprised of a dense selective outer surface and a porous gelled substructure. This paper describes a procedure for making asymmetric hollow-fiber membranes of cellulose diacetate (CA) and reports the performance of these membranes.

EXPERIMENTAL

The CA asymmetric hollow-fiber membranes were prepared from a thermal gel precursor made by cooling, to its gelation point, a melt of 30 wt.-% CA (Eastman Grade E400-25) and 70 wt.-% plas-

*Present address - Gulf Environmental Systems Co., P. O. Box 608, San Diego, California 92112.

ticizer (mixtures of triethylene glycol and diethylene glycol monomethyl ether) [3-5]. The melt was spun at 120°C into hollow fibers of approximately 42 mils outside diameter (OD), which were drawn in the melted state to final dimensions ranging from 3-20 mils OD with the wall thickness being approximately one-third the OD. The melt was passed through a room temperature (25 ± 2°C) water bath to cause rapid gelation.

The CA hollow-fiber thermal gels were leached in water to remove plasticizers and then stored in water containing bactericidal agent. Before testing, the porous thermal gels were crystallized by heat treatment in 60-80°C water. Annealing, the final step in the preparation of asymmetric membranes, was carried out with a 10% aqueous dioxane solution at 60-70°C.

RESULTS AND DISCUSSION

The thermal gel process was employed to prepare CA hollow fibers. In this process a melt of CA and an appropriate combination of plasticizers is formed and spun into hollow fibers which are gelled on cooling by passing them through a water bath. The fibers are drawn to their final dimensions in the molten state. Gelation occurs almost immediately on contact with water, thereby making this process highly suitable for spinning hollow fibers.

In the thermal gel process, the solvent power of the plasticizers and hence the degree of solvation of polymer chains change with temperature. As the temperature of the heated plasticizer-polymer solution is lowered, the polymer chains presumably interact to form a quasi-crosslinked gel structure, and finally phase separation occurs, generating pores. A thermal gel is considered formed when the mixture sets, that is, develops a yield point. The thermal gel can be melted by heating and re-formed on cooling at or below the gelation point. The tendency to gel formation increases with increasing polymer concentration and molecular weight and, in the case of cellulose esters such as CA, with increasing degree of substitution.

If nonvolatile plasticizers are used to form a thermal gel, the tendency for surface "skin" formation is slight and the gels produced are relatively isotropic. In such structures, the plasticizers may be exchanged with water without substantially changing the dimensions, but these thermal gels will shrink irreversibly if allowed to dry. By the thermal gelation technique, CA structures can be prepared at almost any porosity level, but those of 70-75% porosity (equilibrium water content) have the best combination of flux rates and mechanical properties for application in hollow-fiber membranes [3,4].

Examination of the thermally gelled CA hollow fibers by electron microscopy revealed that the pore size of the outside surface was smaller than that in the substructure. Although the plasticizers used were relatively nonvolatile, apparently some removal of plasticizer occurred during spinning, effecting either a reduction in the pore size at the membrane surface or surface tension forces which promote "skin" formation.

The identification of a graded porosity in CA thermal gels suggested the possibility of forming an asymmetric membrane by annealing the outside surface. A general mechanistic model was devised as a basis for guiding the asymmetric membrane studies. The essential assumption in the model is that a graded porosity develops in a membrane as a result of the rapid evolution of plasticizer from the membrane surface in relation to the rate at which it can be replaced by diffusion from within the membrane or from a gelation bath. This rapid loss of plasticizer can occur either by evaporation or by diffusion. Under the proper conditions, therefore, a spun membrane is expected to form smaller pores on the surface than in the substructure.

According to this mechanism, such factors as spinning temperature, temperature difference between melt and gelation bath, volatility of the plasticizer, thermodynamic and kinetic solvent power of the plasticizer, rate of agitation of the gelation bath, and velocity of air past the membrane surface during spinning should affect "skin" formation. Within certain limitations, the higher the level of each of these factors, the greater should be the tendency to generate a graded porosity. Also, water gelation is more effective than air gelation in promoting a graded gel porosity.

The thermally gelled CA hollow fibers, like Loeb-Sourirajan type membranes, have essentially no salt-rejecting capacity in the unannealed state as shown in Table I. Attempts to form a selective surface on the hollow fibers by annealing using an aqueous heat treatment over the temperature range of 60-100°C did not yield a suitable asymmetric membrane. However, a short aqueous heat treatment such as 10 minutes at 60°C does increase the fiber's water permeability (see Table I), presumably via an increase in crystallinity.

Since crystallization of CA is a result in part of increasing the mobility of CA molecules, it was reasoned that a dilute aqueous solvent solution such as 10% aqueous p-dioxane [6] may be a proper medium for annealing CA thermal gels because this mixture would allow capillary forces to tighten the membrane surface, whereas water alone may not mobilize the polymer sufficiently.

TABLE 1

PROPERTIES OF THERMALLY GELLED CA HOLLOW-FIBER MEMBRANES
TEST PRESSURE, 600 PSI

Thermal Gel* Plasticizers (wt. ratio)	Fiber OD (mils)	Heat Treatment (water)	Solvent Annealing (10% aq. dioxane)	Feed Solution (ppm NaCl)	Water Flux (gfd)	NaCl Rejection (%)
TEG Only	15	None	None	35,000	3.6	0
TEG Only	15	10 min, 60°C	None	35,000	5.2	0
TEG Only	15	10 min, 60°C	1 min, 66°C	35,000	4.5	16.3
TEG/MeC(3/1)	15	None	None	35,000	8.1	0
TEG/MeC(3/1)	17.5	10 min, 60°C	None	5,000	14.2	0
TEG/MeC(3/1)	15	10 min, 60°C	1 min, 66°C	35,000	1.0	91.9
TEG/MeC(1/1)	15	None	None	35,000	6.9	0
TEG/MeC(1/1)	15	10 min, 60°C	None	35,000	10.6	0
TEG/MeC(1/1)	15	10 min, 60°C	1 min, 66°C	35,000	0.45	96.1
TEG/MeC(1/1)	15	10 min, 60°C	1 min, 66°C	5,000	5.6	96.0

*TEG = Triethylene glycol.

MeC = Methyl Carbitol (diethylene glycol monomethyl ether).

It was found that dilute solvent solutions were effective in inducing a selective surface layer formation when used at relatively low temperatures.

The data in Table 1 summarize the desalination properties of three groups of asymmetric CA hollow-fiber membranes. The composition of the plasticizer was varied from 100% triethylene glycol (TEG), a poor solvent for CA, to a 1/1 mixture of TEG and methyl Carbitol (MeC, diethylene glycol monomethyl ether), a much better solvent for CA than TEG alone. Each fiber group was tested three times: (1) as prepared, (2) after an aqueous heat treatment, and (3) after both an aqueous heat treatment and an annealing treatment in dilute p-dioxane solution.

The hollow-fiber membranes showed no significant salt selectivity, either as prepared or after the aqueous heat treatment. The aqueous heat treatment, however, did increase flux rates. At that stage the flux rates range from 5.2-14.2 gfd at 600 psi for fibers with 5-mil walls.

The annealing treatment in dilute aqueous p-dioxane dramatically improves the selectivity of the hollow-fiber membranes. For the three types of annealed membranes the salt rejections are 16.3, 91.9, and 96.1% when tested on a 35,000 ppm sodium chloride feed at 600 psi. As expected, the higher selectivities develop with the more effective plasticizer solvents. The membrane spun from a 1/1 TEG/MeC plasticizer shows the best performance of the three groups of membranes with a water flux rate of 0.45 gfd and 96.1% salt rejection of a 35,000 ppm sodium chloride feed at 600 psi; the performance is 5.6 gfd and 96.0% salt rejection at 600 psi on a 5,000 ppm sodium chloride feed (representative of a rather concentrated brackish water). Further, these asymmetric CA hollow-fiber membranes are capable of withstanding an external load of at least 1,000 psi, without collapsing.

The test results in Table 2 and Figures 1 and 2 show performance data on hollow-fiber membranes operating on brackish water feeds. Typical initial flux and salt rejection values are (1) 4 gfd and 94% rejection at 400 psi, and (2) 6 gfd and 96% rejection at 600 psi when operating on a 5,000 ppm sodium chloride feed. By varying annealing temperature, it is possible to trade off flux and salt rejection; for example, a flux rate of 8.2 gfd and 75% sodium chloride rejection resulted at 600 psi when the annealing temperature was reduced to 63°C from the normal 66°C level. Such a membrane would be suitable for producing potable water from a mildly concentrated sodium chloride feed or from a highly concentrated brackish water containing predominantly divalent ions since divalent ions are rejected by CA to a much greater extent than monovalent ions.

TABLE 2

PERFORMANCE OF ASYMMETRIC CA HOLLOW-FIBER MEMBRANES[a]

Annealing Temp.[b] (°C)	Fiber OD (mils)	Pressure (psi)	Flux[c] (gfd)	Rejection[c] (%)
		5000 ppm NaCl Feed		
66	21	200	1.8	88.5
65	20	200	3.0	80
64	20	200	3.3	76.4
63	20	200	3.7	60
66	21	400	4.0	93.8
65	20	400	6.0	88
64	20	400	6.6	85.4
63	20	400	7.2	71
66	21	600	5.6	96.0
65	20	600	6.5	92
64	20	600	7.5	89.2
63	20	600	8.2	75
		500 ppm $MgSO_4$ Feed		
66	20	250	4.4	98.0
63	20	250	5.5	89.9
66	20	400	6.6	97.7
63	20	400	8.7	92.3
66	20	600	8.1	98.3
63	20	600	9.5	95.2

[a] Membranes spun at 120°C from a 30% solution of CA (E400-25) in a 1/1 weight mixture of triethylene glycol and methyl Carbitol.

[b] Annealing carried out in a 10% solution of p-dioxane in water.

[c] Initial values.

The latter point is confirmed by the results in Figure 2 showing hollow-fiber membrane performance on a rather concentrated magnesium sulfate feed. Both the level of rejection and flux rate are higher on this feed than the corresponding values for a sodium chloride feed (Figure 1). At 600 psi, a flux rate of approximately 10 gfd and a salt rejection greater than 95% were achieved. At 250 psi, the performance was also good, illustrating the excellent suitability of these hollow-fiber membranes for treating brackish waters under a variety of operating conditions. The membrane flux

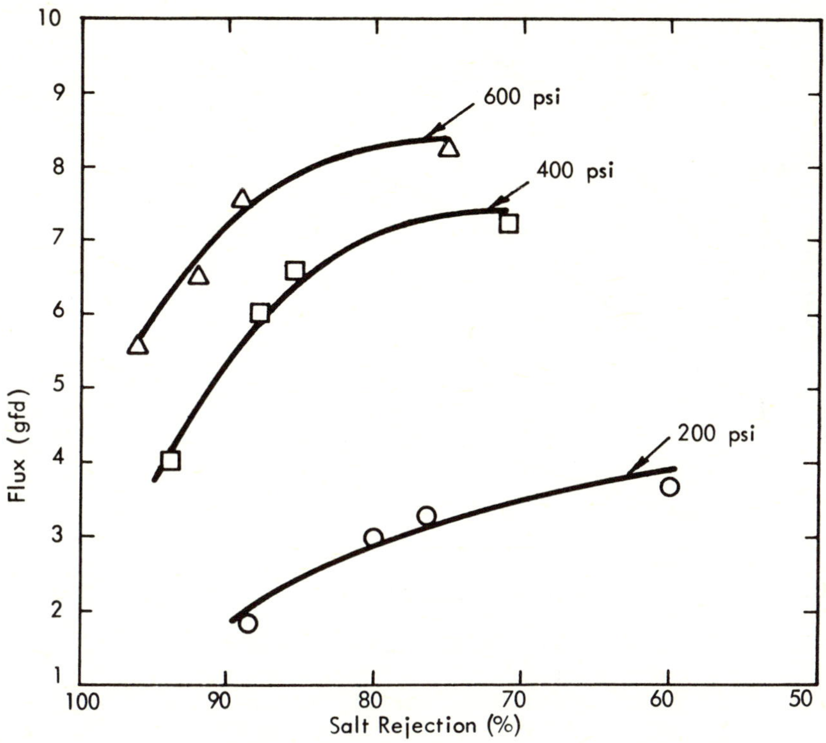

Fig. 1 - Flux Versus Salt Rejection Relationship of Cellulose Acetate Hollow-Fiber Membranes on 5000 ppm Sodium Chloride Solution

might be further improved by drawing the spun fibers into thinner wall membranes. Reducing the thickness of the fiber would lessen the resistance of the membrane to the passage of water.

The flux decline rates, i.e., the slope of a plot of log flux vs. log time, for the 66°C membranes in Table 2 were -0.11, -0.04 and 0.00 at 600, 400 and 200 psi, respectively. The flux-decline rates translate to useful module lifetimes of over 10 years at 600 psi if a lower flux limit of 1 gfd is established. The results favor an operating pressure as low as is consistent with feed concentrations and recovery levels to minimize compaction effects.

The high performance of the single asymmetric CA hollow-fiber membranes could be successfully translated to compact bench-scale modules of 100-300 gallons-per-day (gpd) capacity. As an example, a module containing a fiber bundle 2.7 inches in diameter (Figure 3) and 10.5 inches long with 10,734 fibers at a volumetric packing of 72% exhibited a fiber flux rate of 5.5 gfd (150 gpd output) at

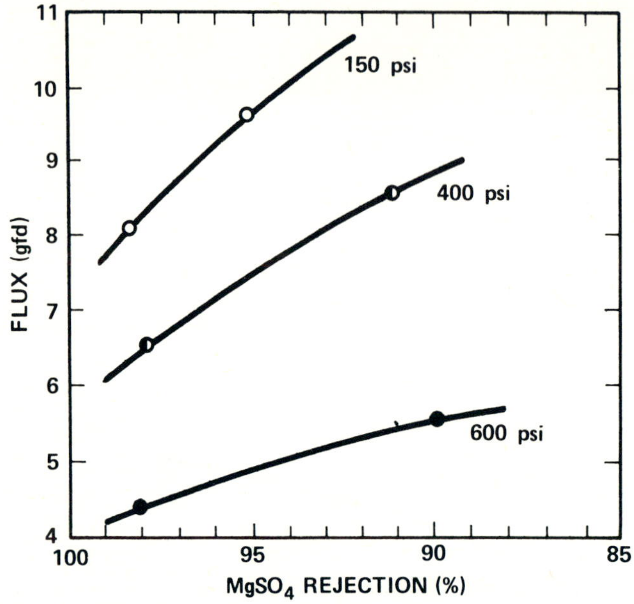

Fig. 2 - Flux Versus Salt Rejection Relationship For Cellulose Acetate Hollow Fiber Membranes on 5000 ppm Magnesium Sulfate Solution

90% salt rejection when operated at 400 psi on a 5,000 ppm sodium chloride feed at 30% water recovery. The high pressure capability of these hollow fibers is an important asset in modules operating at high water recovery levels or in modules treating a relatively concentrated feed brine. Further optimization of fiber preparation and treatment conditions should allow fabrication of even more compact, low-operating-pressure, hollow-fiber modules for treating hard or contaminated waters.

CONCLUSIONS

High-performance, hollow-fiber reverse osmosis membranes of unique structure were prepared by annealing a porous cellulose diacetate thermal gel containing a graded porosity to give a membrane comprised of a dense selective outer surface and a porous gelled substructure. These fibers have demonstrated low to moderate flux-decline rates and withstand external pressures of at least 1,000 psi without collapsing. By varying the annealing temperature, it is possible to trade off fiber flux and salt rejection.

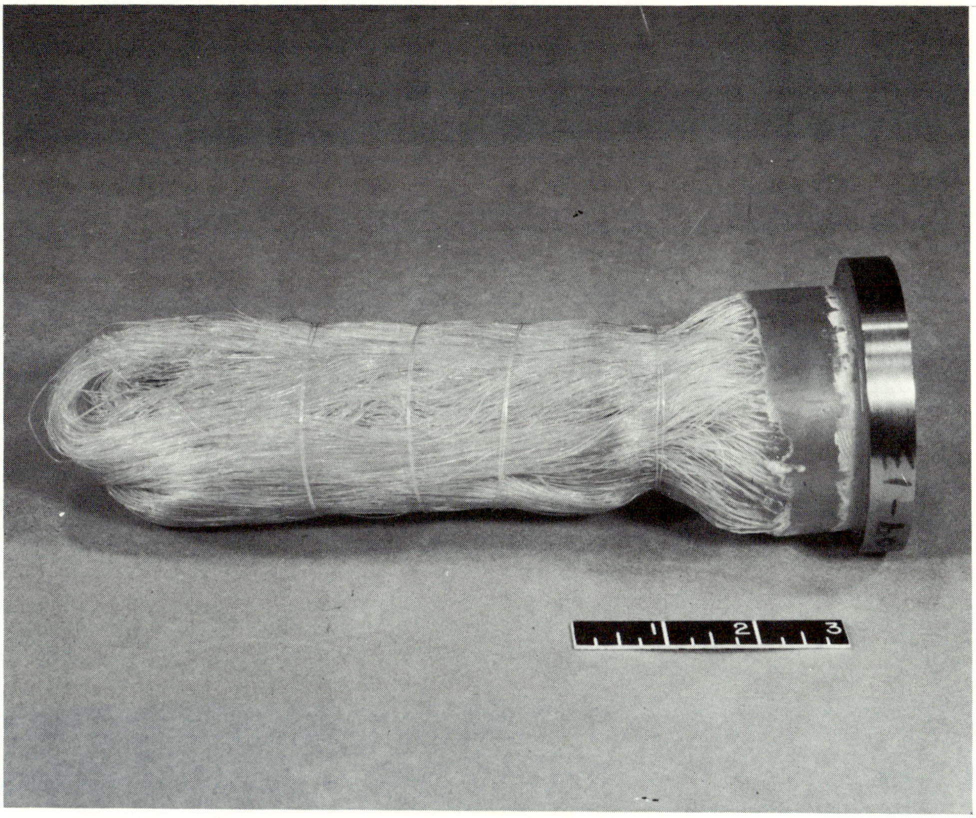

Fig. 3 - Hollow-Fiber Membrane Bundle Capable of Producing 100 gpd of Fresh Water

The performance of single hollow-fiber membranes was successfully translated to compact modules of 100-300 gpd capacity. The high pressure capability of these hollow fibers is an important asset in modules operating at high water recovery levels or in modules treating a relatively concentrated feed.

ACKNOWLEDGMENT

This research was supported by the Office of Saline Water, U. S. Department of the Interior, under Contract 14-30-2527.

REFERENCES

1. E. A. McLain, and H. I. Mahon, U. S. Patent 3,423,491 (January 21, 1969; assigned to Dow Chemical Company).

2. H. H. Hoehn and D. G. Pye, U. S. Patent 3,497,451 (February 24, 1970; assigned to E. I. du Pont de Nemours and Company).

3. M. E. Cohen and B. M. Riggleman, Office of Saline Water, U. S. Department of the Interior, Research and Development Progress Report No. 400 (February 1969).

4. M. E. Cohen, M. A. Grable, and B. M. Riggleman, Office of Saline Water, U. S. Department of the Interior, Research and Development Progress Report No. 518 (March 1970).

5. M. E. Cohen, M. A. Grable, and B. M. Riggleman, Office of Saline Water, U. S. Department of the Interior, Research and Development Progress Report No. 608 (September 1970).

6. F. Happey, J. Textile Inst., 41, T381-T403 (1950).

DEVELOPMENT OF IN SITU CASTING OF REVERSE OSMOSIS MEMBRANE TUBULES THROUGH THE HYDROCASTING METHOD

A. Gollan and M. P. Tulin
Hydronautics, Incorporated
Laurel, Maryland 20810

INTRODUCTION

Hydrocasting is a hydrodynamic process for the fabrication of long membrane tubules of small diameter (less than 3 mm) with skin on the interior wall and good desalination characteristics. It is based on a phenomenon described by Fairbrother and Stubbs [1] and later studied in detail by Sir G. I. Taylor[2]. It may be simply described as follows: when a gas is blown into one end of a tube containing a viscous fluid, forcing it out at the far end, a fraction of the fluid remains deposited on the wall of the tube in the form of a uniform annular layer. Tulin[3] first proposed to utilize this phenomenon for the casting of small diameter tubular reverse osmosis membranes.

The advantage (in terms of compactness and economy) of small tubular membranes for liquid separations has become steadily clearer during the last few years. The main problems inherent in small tubules or hollow fibers are clogging and limitations in operational strength. Clogging may be much relieved and cleaning facilitated through the circulating of feed inside the membrane tube. Hydrocasting was originally proposed as a possibility for the economic fabrication of small diameter tubules in situ within suitable porous supports and for the replacement (cleaning out and refabrication) of the membrane tubules when necessary. A completely automated hydraulic process was envisioned as a final goal.

A particular advantage of hydrocasting in comparison to most spinning and wet-extrusion techniques for fiber manufacture is that it allows the use of standard or nearly-standard cellulose acetate solutions of the Loeb-Sourirajan type. Another important advantage is that the resulting tubules are provided with skin on the interior walls, resulting in good desalination characteristics. The technique is not, however, limited to cellulose acetate (CA) membranes, being generally applicable to membranes produced by gelation of viscous casting solutions through the process of wet deposition, air curing (evaporation of solvent), liquid leaching, and annealing. It has, for instance, been used to produce salt-rejecting tubules with good flux characteristics, cast from a fully aromatic polyamide (Hydronautics, Inc., unpublished work).

It seems at the present time that the importance of the hydrocasting process is not only for in situ fabrication within porous tubes, but also for the potentially low cost fabrication of membrane tubules for use, unsupported, at low pressures (around 100 psi) for brackish water desalination and similar purposes. In this case, the tubules are hydrocast within impermeable support tubes, like glass, and then withdrawn.

Systematic studies of the hydrocasting process have been carried out and some of the results will be described here. Some of these studies have been concerned with phenomena occurring during the process and which effect its applicability. Others relate to membrane morphology, strength, and desalination characteristics of hydrocast membranes. The status of the development at this time can be summarized very briefly as follows, emphasizing the practical results: (1) good quality membrane tubules (diameters between 0.5 mm and 1.5 mm, length 4') have been hydrocast in glass tubes from standard Loeb-Sourirajan solutions; (2) hydrocast tubules, wrapped in nylon backing cloth and inserted into perforated metal support tubes, have demonstrated good desalting characteristics at pressures as high as 900 psi; (3) unsupported hydrocast tubules have been tested without failure for 2000 hrs. at 100 psi; (4) the performance of the unsupported tubules is good enough to suggest the feasibility of their use for brackish water desalination at low pressures; (5) a multitude of good quality membrane tubules have been hydrocast in glass tubes from standard Loeb-Sourirajan solutions in multiple quantities (up to 10 per casting); (6) membrane tubules have recently been hydrocast within a suitable porous nylon tube and successfully tested at pressures up to 300 psi; and (7) the feasibility of hydrocasting membrane tubules of fully aromatic polyamides has been very recently demonstrated.

The discussion here is limited to items (1)-(4) above and is organized so that a description of the physical properties of hydrocast membranes and their performance is preceded by a brief description of the hydrocasting technique and of phenomena accompanying the process, including the results of theoretical analysis and experiments.

HYDROCASTING

The Process

There are four essential steps involved in hydrocasting membrane tubules: (i) the introduction of the casting solution; (ii) the deposition on the tube wall of an annular layer by a gas bubble forced into the tube; (iii) the simultaneous evaporation into the gas bubble of solvent from the deposited casting solution; (iv) liquid leaching and final gelation by a liquid forced into the tube following the gas; and (v) annealing with a temperature controlled liquid bath.

These steps may obviously be accomplished with a variety of mechanical arrangements, requiring more or less manual operation. The schematic of a simple but typical hydrocasting system is shown in Fig. 1. The casting tube could be varied in size and orientation and could readily be replaced by a number of parallel tubes (a tube bundle) all served at once by the same casting system. With reference to Fig. 1, the following procedure is employed for hydrocasting with this system:

> Casting solution from container (1) is forced by compressed gas (6) into support tube (2) to a desired length and valve V-3 is closed. The gas bubble tube (3), is flushed by the gas to a predetermined length of tube and this later becomes

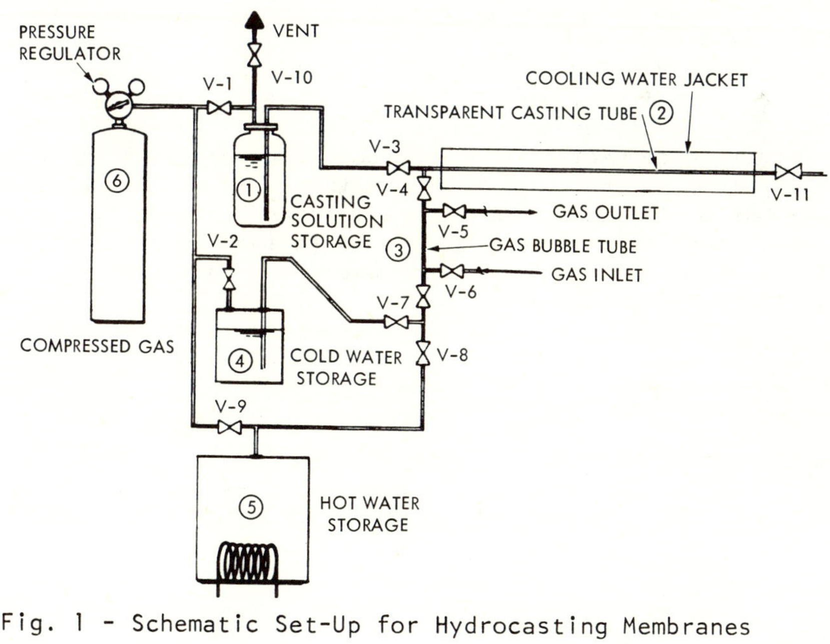

Fig. 1 - Schematic Set-Up for Hydrocasting Membranes

the forming gas bubble. When the gas flushing is completed valves V-5 and V-6 are closed trapping the gas bubble. The cold water container (4) is then pressurized to the predetermined blowing pressure and valve V-7 is opened compressing the gas bubble. At this stage, valve V-4 is opened and the gas bubble, followed by the gelling water column, is introduced into the support tube (2), starting the membrane forming process.

When gelling is completed, annealing hot water from container (5) replaces the cold water stream.

Phenomena Occurring During Hydrocasting

There are three phenomena which occur during bubble travel which are important and will be further discussed. These are: deposition of the annular layer; deformation (slumping) of the deposited layer under the action of gravity; and contraction of the gas bubble. In addition, there are phenomena occurring during evaporation and leaching which effect membrane morphology and strength, and which play a vital role in determining the best diameters for hydrocasting membrane tubules. All of these will subsequently be discussed.

Filling. The quantity of casting solution required to fill a given casting tube in order to obtain a final membrane length L depends on the thickness of the annular layer formed during the air blowing process and therefore on the fraction, m, deposited. The latter, as discussed below, reaches an asymptotic value at sufficiently high blowing velocities. For typical hydrocasting conditions a value m = 0.56 may be assumed, so that the casting tube must be filled to a length 0.56L. The time required to fill is of interest in connection with the possible intrusion of casting solution into the pores of a porous tube. It may be calculated on the assumption that the flow is quasi-steady and highly viscous (i.e., neglecting inertia). For a non-Newtonian fluid satisfying the power relation between stress and shear rate as given in the Appendix, the filling time is*:

$$t = \left(\frac{3n+1}{n+1}\right) \left(\frac{2mL}{d}\right)^{\frac{n+1}{n}} \left(\frac{2\mu_o}{\Delta p}\right)^{\frac{1}{n}} . \qquad (1)$$

For a standard casting solution (25% Eastman Chemical Products, Inc. E398-10 cellulose acetate, 30% formamide, 45% acetone by wt.), n = 0.86. Note that the corresponding relation for a Newtonian fluid (n = 1) is:

*See List of Symbols at the end of the paper for definition of terms.

$$t = 16 \frac{m^2 \mu L^2}{\Delta p d^2} . \qquad (1a)$$

Equation (1) is based on the following relation for the average (over the tube cross-section) velocity of the casting solution of plug length x:

$$\overline{u} = \frac{n}{3n+1} \left(\frac{\Delta p}{2u_o x}\right)^{1/n} \left(\frac{d}{2}\right)^{\frac{n+1}{n}} . \qquad (2)$$

Quite clearly the plug velocity increases without limit as the plug becomes smaller, unless the driving pressure is at the same time suitably reduced. In practice, typical plug velocities are 0.5 cm/sec and driving pressures are 25 atm/meter of filled length.

Equation (1) for the total filling time has been compared with measurements and the agreement is good.

Deposition. It is a curious fact that a tube cannot be emptied of a viscous liquid by flowing gas through it at moderate velocities, the amount of liquid left deposited on the wall actually increasing with the gas speed in the tube. This was first noted by Fairbrother and Stubbs[1]. The actual hydrodynamic flow is rather complicated, and it was not until the experimental and theoretical work of Taylor[2] that the problem was properly elucidated. Taylor showed that for Newtonian fluids the fraction, m, of fluid remaining in the tube was a function of a single dimensionless parameter combined of the fluid viscosity, μ, the gas blowing velocity, $\overline{u}$, and the fluid-gas surface tension, σ. That parameter is $\mu\overline{u}/\sigma$. Taylor further showed that for sufficiently large values of this parameter, the fraction of deposited fluid approached 56%, see Fig. 2. In most cases the dimensionless parameter is well above a value of 1.5 during hydrocasting. This results in the advantage that the as-cast thickness ($t/d \approx 0.15$) is not sensitive to moderate changes in blowing velocity, as it would be if the parameter had a value less than about 0.2. The shape of the Taylor curve is substantiated by experiments using standard CA casting solutions, but the experimental and theoretical values differ slightly as shown in Fig. 2. This is presumably because of the non-Newtonian nature of the casting solution (see the Appendix).

Casting Speed. The speed of bubble travel is simply related to the plug speed and fraction deposited:

$$m = \frac{u_b - \overline{u}}{u_b} . \qquad (3)$$

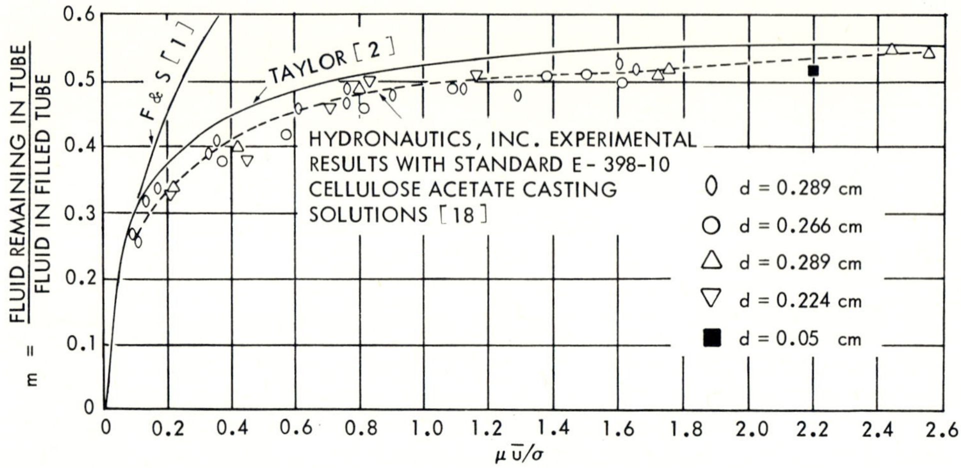

Fig. 2 - Dependence of the Parameter m on the Dimensionless Parameter $\mu\bar{u}/\sigma$

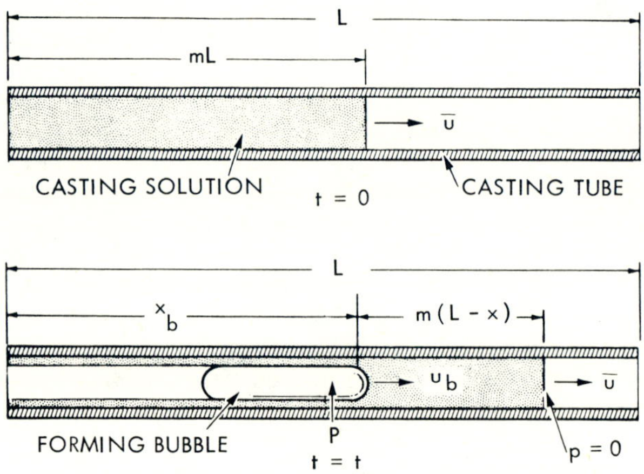

Fig. 3 - Longitudinal Relations During Hydrocasting

Referring to Fig. 3 and substituting for $\bar{u}$ from eq. (3) into eq. (2), the bubble speed is found to be:

$$u_b = \frac{dx_b}{dt} = \frac{n}{3n+1}\left[\frac{\Delta p}{2m\mu_o(L-x_b)}\right]^{1/n}\frac{(d/2)^{\frac{n+1}{n}}}{1-m}, \tag{4}$$

and integration yields the following relation for the time required for the bubble to reach a distance x_b from the entrance of the tube at constant driving pressure:

$$t = \left(\frac{3n+1}{1+n}\right) \frac{1-m}{(d/2)^{\frac{n+1}{n}}} \left(\frac{2m\mu_o}{\Delta p}\right)^{1/n} \left[L^{\frac{n+1}{n}} - (L-x_b)^{\frac{n+1}{n}}\right] . \tag{5}$$

The total time to hydrocast a tubule of length L is given by substituting x_b for L in eq. (5).

For a Newtonian fluid the appropriate equations reduce to:

$$u_b = \frac{\Delta p \ d^2}{32m\mu_o(1-m)(L-x_b)} \tag{4a}$$

$$t_b = \frac{16m(1-m)\mu_o(2x_bL-x_b^{\ 2})}{\Delta p \ d^2} . \tag{5a}$$

As the time to hydrocast is a very important parameter in connection with deformation of the deposited layer and air curing effects, these relations are useful. The predictions of hydrocasting time are in good agreement with measurements.

Bubble Contraction. During preliminary hydrocasting experiments in glass and nylon tubes it was observed that the inert gas bubble used for membrane formation contracted while traveling along the casting tube. This phenomenon is of particular importance when small bubbles are utilized, or when relatively long membranes are desired, since complete disappearance of the bubble results in imperfect membranes. It was finally concluded that the gas was dissolving into the deposited casting solution. In order to estimate and minimize bubble contraction under different casting conditions, an analysis was made. As a result it was possible indeed to achieve improvements by changing certain parameters as suggested by the theory. The analysis follows, together with comparison with measurements.

During deposition, the leading edge of the gas bubble cuts into the casting solution, pushes part of it ahead, and leaves behind an almost stationary annulus. The gas in the bubble may almost certainly be assumed to be well mixed because of the relatively high diffusion coefficient in the gas and because of flow mixing effects. The mathematical solution described below will be based upon the above assumption, which leads to an unsteady problem for radial diffusion of gas into the casting solution.

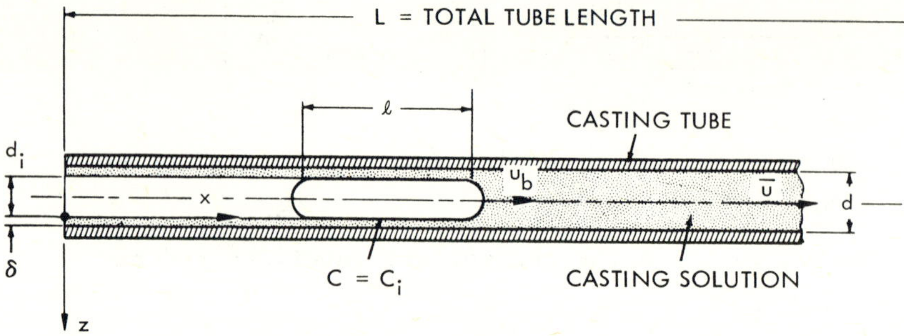

Fig. 4 - Schematic Presentation of Coordinates for Hydrocasting

With reference to Fig. 4 we will assume the following: a) laminar flow prevails throughout the cast solution plug; b) the casting solution annulus thickness is small compared to the radius of the tube, but large enough with reference to the penetration depth (i.e., Higbie's[4] penetration concept applies); c) diffusion at the bubble ends is negligible compared to total mass transferred; d) the forming gas bubble is saturated with acetone vapor; e) the initial concentration of the gas comprising the gas bubble in the casting solution is zero; f) axial diffusion of the absorbed gas in the casting solution is negligible; and g) the length of the forming bubble, ℓ, is at any time small compared to the overall casting tube length, so that variation in velocity over a bubble length is negligible.

Concentrating on an infinitesimal annular section of casting solution just formed around the gas bubble at any location, x, the unsteady radial diffusion into the casting solution is governed by the following equation to an approximation implicit in the above assumptions:

$$\frac{\partial c}{\partial t} = D \frac{\partial^2 c}{\partial z^2} , \tag{6}$$

subject to the boundary conditions:

$$\begin{array}{lll} c = 0 & t = 0 & 0 \le z \le \delta , \\ c = c_i & t > 0 & z = 0 , \\ c = 0 & t > 0 & z = \delta = \infty , \end{array}$$

and where $t = \int_{x_b - \ell}^{x_b} \frac{dx}{u_b}$.

The solution of eq. (6) is well known; it is:

$$\frac{c(z,t)}{c_i} = \text{erfc}\left(\frac{z}{\sqrt{4Dt}}\right) . \tag{7}$$

On the basis of this solution, the instantaneous flux of gas into the casting solution is readily calculated and, integrating over the length of the bubble, one can calculate the rate of loss of gas. This loss results in the contraction of the bubble. Relating the bubble contraction to the rate of gas loss results in a prediction for the bubble length as a function of the distance traveled by the bubble, x_b, and the other significant parameters. This relation is:

$$\ell^{1/2} = \ell_o^{1/2} - \frac{12}{d} K^* \sqrt{D} \frac{n}{(2n+1)\sqrt{\Psi\pi}} \left[L^{\frac{2n+1}{2n}} - (L-x_b)^{\frac{2n+1}{2n}} \right], \tag{8}$$

where

$$\Psi = \frac{n}{(3n+1)(1-m)} \left(\frac{\Delta p}{2m\mu_o}\right)^{1/n} (d/2)^{\frac{n+1}{n}} . \tag{9}$$

The distance traveled at which the bubble disappears is the maximum possible hydrocasting length; it is:

$$\frac{L_{max}}{d} = \left[\left(\frac{\Delta p}{\mu_o}\right)^{\frac{1}{2n}} \sqrt{\frac{\ell_o d}{D}} \frac{1}{\phi K^*} \right]^{\frac{2n}{2n+1}} , \tag{10}$$

where

$$\phi = \frac{12}{\sqrt{\pi}} \frac{(3n^2+n)^{1/2}}{(2n+1)} 2^{\frac{n+2}{2n}} m^{\frac{1}{2n}} (1-m) . \tag{11}$$

Of course, the appropriate equations for a Newtonian fluid can be obtained by setting n = 1 in the above equations. Further details and a more exact analysis of bubble contraction in the case of a Newtonian fluid have been presented elsewhere[5].

It may be seen, according to eq. (8) above, that the maximum hydrocasting length increases rapidly with casting tube diameter; this is because the volume of gas available increases with tube size more rapidly than the wall area through which gas diffusion occurs. The influence of $\Delta p/\mu_o$ is in controlling the bubble velocity; higher velocities allow greater hydrocasting distances before bubble disappearance. The influence of the gas appears through the product $K^* \sqrt{D}$. Based on appropriate data, it appeared that helium possesses a value of this product only 0.31 times the value for nitrogen and experiments were subsequently performed with both gases.

Bubble contraction experiments were carried out with the standard E398-10 casting solution in non-porous 3/16'' and 1/8'' ID glass and nylon tubes. Assuming m = 0.56, and n = 0.86 in equation (8):

$$\ell^{1/2} = \ell_o^{1/2} - 6.4 \frac{K^*}{d^{2.08}} \sqrt{D} \left(\frac{\mu_p}{\Delta p}\right)^{0.582} \left[L^{1.58} - (L-x_b)^{1.58}\right]. \quad (12)$$

In order to correlate the experimental results with this theory, data on the solubility constant and the diffusion coefficient of the gas in the bubble into each specific casting solution should be known. Unfortunately, such data are not available for the pertinent ternary system. Nevertheless, a semi-empirical comparison can be drawn by calculating $K^* \sqrt{D}$ from one experimental run and then applying the result to other experimental conditions.

Fig. 5 shows the results obtained in tube 1 with N_2 as the bubble gas medium. In that case the value $K^* \sqrt{D} = 4.3 \times 10^{-4}$ (in the cgs unit system) was calculated from the data on the bubble size at x = 515 cm while the whole curve was calculated from equa-

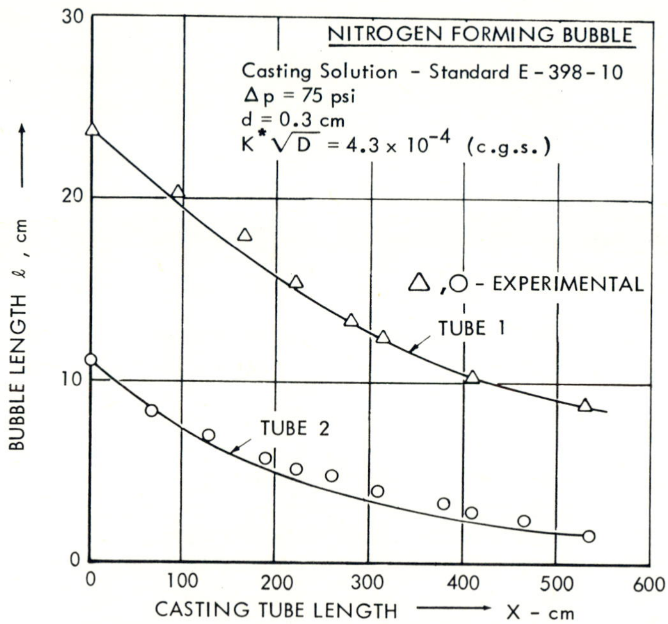

Fig. 5 - Bubble Contraction - Experimental and Calculated (Equation 12) Results tion (12). Very good agreement in the shape of the curve between theory and experiment is observed. The calculated $K^* \sqrt{D}$ value from tube 1 was further used to predict the bubble contraction under other operating conditions. In tube 2, the initial bubble size was changed and in tube 3 (Fig. 6) the blowing pressures and the support tube diameter were changed. Very good agreement between theory and experimental data was again obtained. The same was true in the case of helium and, as expected, longer blowing lengths were obtained than in the case of nitrogen. Since the ratio of $K^* \sqrt{D}$ for helium to that for nitrogen is 0.31 we conclude from equation (10) that under similar casting conditions the use of helium will allow more than doubling the cast length. The results on bubble contraction are combined below with results pertaining to deformation of the casting solution in order to arrive at conclusions regarding maximum hydrocasting lengths in the horizontal orientation of the casting tube.

Deformation of Cast Solution. Unless substantial gelation of the cast solution has taken place, the deposited solution will flow around the gas bubble resulting in non-uniformity of the produced membrane. This deformation seems to be most noticeable when the casting tube is in a horizontal orientation, as might sometimes be required, in particular at the casting tube entrance where the bubble velocity is the lowest and more time for creep

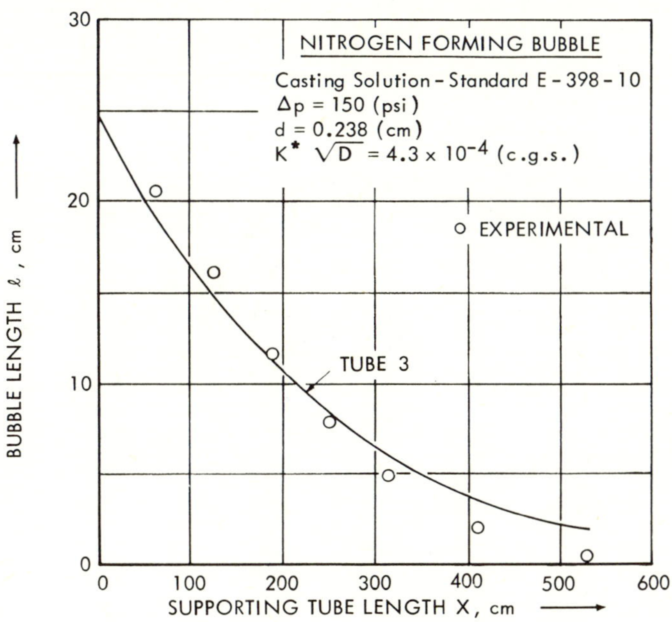

Fig. 6 - Bubble Contraction - Experimental and Calculated (Equation 12) Results

is available. For this reason considerable attention has been given to this phenomenon, including a theoretical analysis.

Under the influence of gravity the casting solution begins to creep in a transverse plane, thinning the annulus at its top and thickening it below (see Fig. 7). The creep commences with the arrival of the forming bubble and is substantially halted by gelation of the casting solution; this is assumed here to take place with the arrival of the gelling water. Clearly the resulting eccentricity of the tube increases with increase in casting time, and this in turn is a function of many parameters including tube diameter and driving pressure.

In order to understand the phenomenon better and to aid in correlating results, a theoretical analysis was carried out[6]. The basic assumptions involved are: a) viscous and gravity effects dominate (inertia and surface tension effects are small); b) the velocity distribution in the θ direction is half-parabolic (zero on the tube wall and a maximum at the edge of the annulus); and c) the initial layer thickness, δ_o, is small compared with the annulus diameter. Then the continuity equation can be transformed into a quasi-linear first order equation defining the variation of the layer thickness in time and around the circumference. The equation has been solved only with certain approximations and the results for a Newtonian liquid are essentially summarized in Fig. 8. Following the theoretical analysis given in reference [6] we readily obtain for a non-Newtonian liquid the following expression for the deformation at the top (θ = 0):

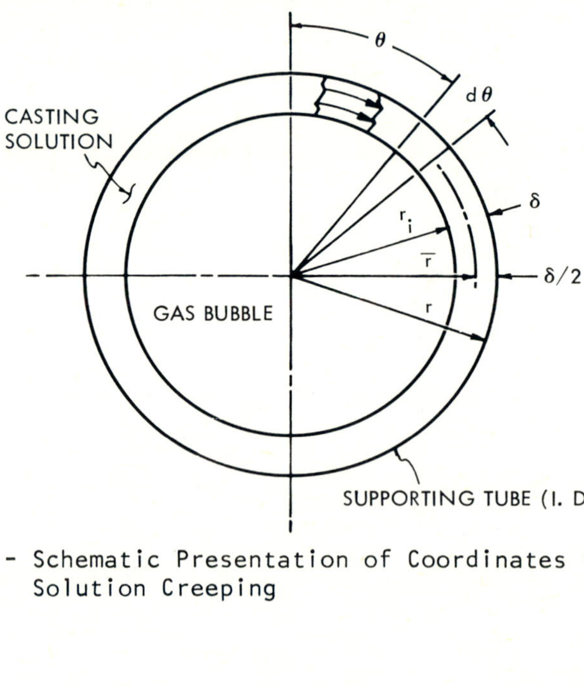

Fig. 7 - Schematic Presentation of Coordinates for Casting Solution Creeping

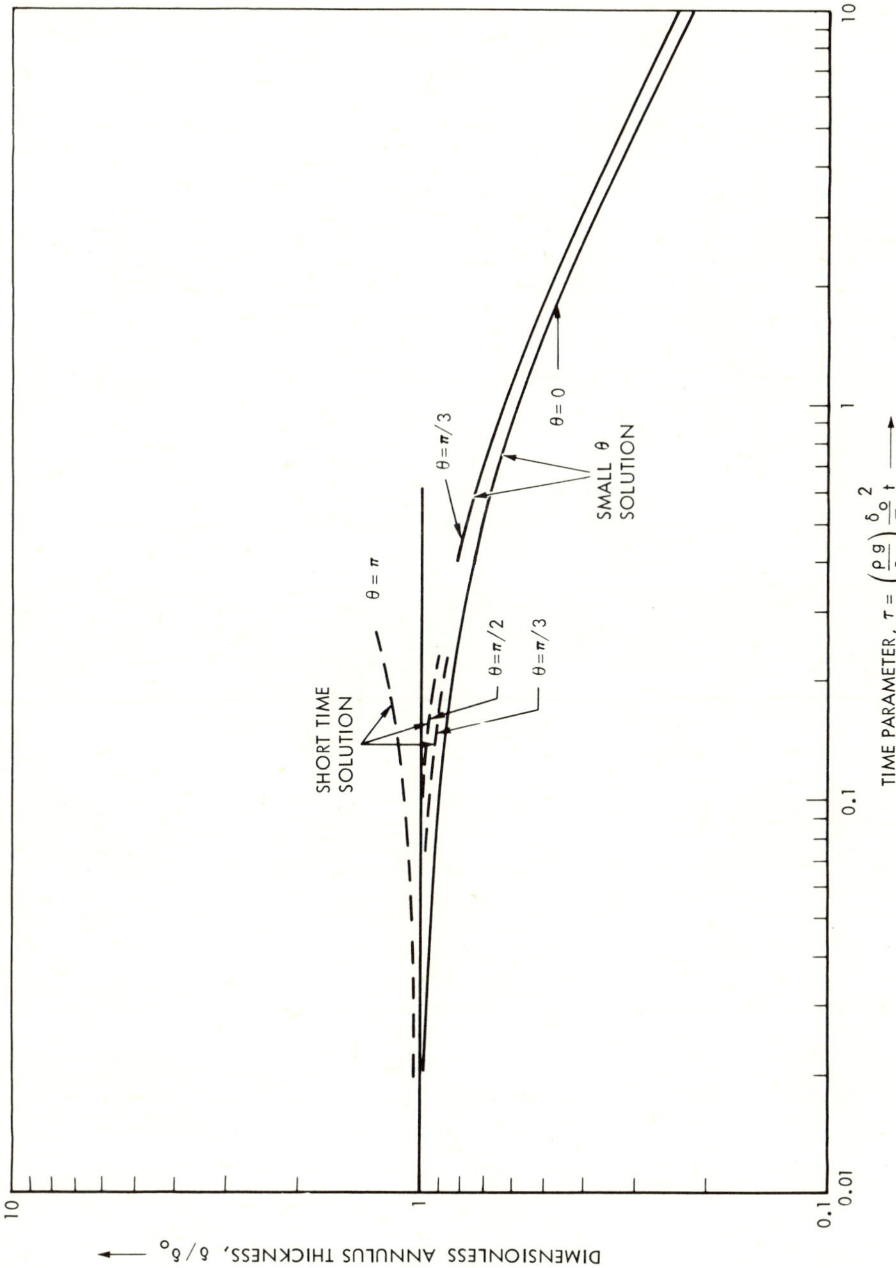

Fig. 8 - Thickness Variation of the Casting Solution Annulus With Time and θ

$$\frac{\delta}{\delta_o} = \left[\frac{1}{(\frac{n+1}{n})\tau + 1}\right]^{\frac{n}{n+1}} , \tag{13}$$

where

$$\tau = \frac{n}{2n+1} \left(\frac{\rho g}{\mu_o}\right)^{1/n} \frac{\delta_o^{\frac{n+1}{n}}}{\overline{r}} t . \tag{14}$$

The important facts to be derived from the results in Fig. 8 are: a) the time of deformation scales with the parameter $(\rho g/3\mu)\delta^2{}_o/\overline{r}$, relatively small values of this parameter corresponding to relatively large permissible times of hydrocasting; and b) if eccentricity of the tubule is to be kept less than about 10%, then the time parameter τ must be limited to a value of around 0.1.

It should be noted that small values of τ do not necessarily mean small values of real time. For the standard casting solution and $\overline{r} = 0.05$ cm, $\tau \approx 3.5 \times 10^{-3}t$. Therefore, a value of $\tau = 0.1$ corresponds to $t \approx 30$ sec. This demonstrates the "creeping" nature of the deformation.

Longer bubbles are required if long tubules are to be cast. However, if only a certain eccentricity factor $\chi = \delta/\delta_o$ is tolerable the maximal bubble length, and hence the minimal tubule length that can be cast horizontally, is limited. Since maximal creep occurs at the cast tube entrance the initial bubble length is readily calculated from

$$\ell_o = u_b(0)t , \tag{15}$$

where $u_b(0)$ is obtained from equation (4) at $x_b = 0$ and t is derived from equation (14) for a given χ value.

Substitution of equation (15) into equation (10) with $\delta_o = 0.168$ d and $\overline{r} = 0.417$ d (for m = 0.56) yields the maximal hydrocast tubule that can be cast within specified eccentricity limits.

For the standard casting solution, $\rho \approx 1$ g/cm^3 and $\chi = 0.9$; the following expression for the maximal hydrocast length is obtained:

$$L_{max}[\text{cm}] = \zeta\, d^{1.195}\, \Delta p^{0.538} , \tag{16}$$

where $\zeta = 0.79$ for a nitrogen bubble and $\zeta = 1.36$ for a helium bubble.

For hydrocasting in tubules with diameter of about 1.0 mm and Δp = 300 psi the maximum length is between 10 to 20 ft, using nitrogen, and 20 to 40 ft using helium. These lengths are considerably larger than generally optimum for use in tube bundles. Most of the authors' recent experience has been with tubules of length about 4', cast in the vertical orientation and no problems with deformation have been encountered. It is noteworthy that equation (16) is valid only when hydrocasting is performed with the casting solution at room temperature. If precooling of the casting solution is employed the equation may serve only for a first approximation.

MEMBRANE TUBULE MORPHOLOGY AND STRENGTH

Systematic study of the various process parameters affecting tubule performance has indicated that good quality membrane tubules can be produced in the range of 0.5 to 1.5 mm in diameter. When unsupported, these tubes have exhibited surprising high burst and ultimate strengths, and good creep and desalination properties. The improvement in membrane performance has been obtained primarily through the control of macropore formation within the membrane structure.

In the following sections some of the parameters affecting macropore formation in the membrane as well as tubule strength will be summarized.

Control of Macropore Formation

The existence of finger-like pores in flat cast or tubular membranes has been reported in the literature[7-9]. One should realize, however, that the presence of macropores in flat cast membranes is much less critical than in small tubules where compaction is also associated with an increase in the inside diameter of the membrane. This situation is aggravated when pressurization of the tubular membrane is also associated with membrane creeping. Whatever the mechanism of macropore formation may be it is apparent that either a reduction in the physical size of the pores or their elimination is highly desirable. Discussion of the predominant parameters affecting macropore formation in hydrocast membranes follows.

Membrane Diameter. The influence of casting tube inner diameter on macropore formation may be inferred from Fig. 9. Plotted are gelled membrane dimensionless thickness, porosity, and shrinkage versus casting tube inner diameter. Here we define membrane shrinkage as the ratio of the gelled thickness to the casting solution annulus thickness as formed by the gas

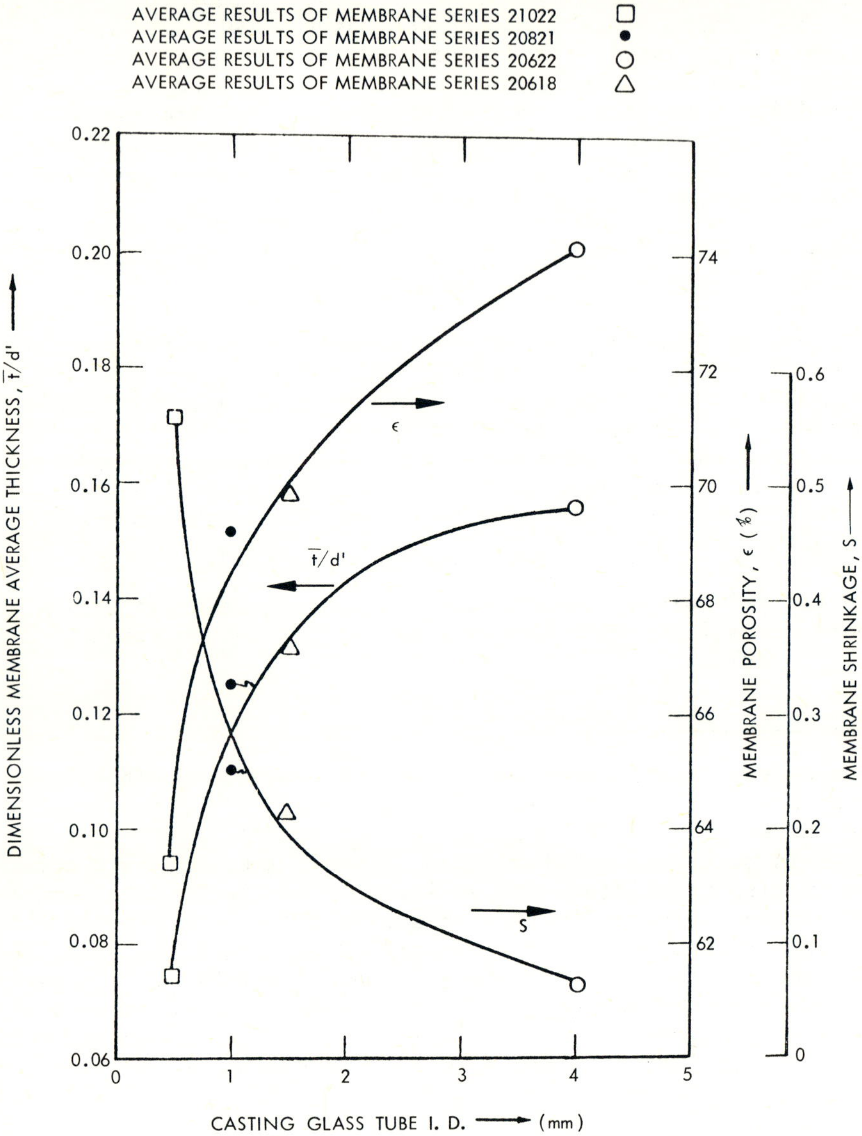

Fig. 9 - Effect of Casting Tube Inner Diameter on the Dimensionless Membrane Thickness, Shrinkage, and Porosity of Water Gelled Membranes Cast From E400-25 Cellulose Acetate Casting Solutions (CA:formamide:acetone = 22.7:27.3:50)

bubble. In these tests the casting solution composition and other casting parameters were held constant. The effect of reducing diameter on membrane shrinkage occurs during evaporation and gelation, not during the casting process itself. The reduction in porosity, which leads to the smaller dimensionless thickness of the membranes cast at diameters below 2 mm, is apparently a direct consequence of the reduced cast thicknesses. This reduced thickness allows much faster thickening (sol-gel transformation) of the entire cast solution thus inhibiting finger formation. In the case of the smallest diameter studied (0.5 mm) the gelled membrane thickness is about 50% of the thickness as cast. It is interesting to note that despite the large dimensional changes in the membrane skin circumference which occur during evaporation and gelation, a healthy salt-rejecting skin is formed.

Gelling Water Activity. Gelling with low activity water solutions has been shown[10] to cause marked changes in the morphology of flat cast membranes. It has been suggested that the presence of solutes in the gelling solution increases the ratio of solvent-to-water flow during membrane precipitation (consequently decreasing the gelation rate), thus avoiding macropore formation.

The role of salt gelling in inhibiting the formation of finger cavities or voids in the porous sub-structure of gelled membranes was easily visualized with the aid of the so-called "drop" technique first utilized in connection with the development of the hydrocasting process[11]. In this method a drop of casting solution is placed between microscope slides squeezed together, placed in an optical microscope, and then gelled around the outer edge of the drop by inserting a drop of water between the slides. The dramatic effect of salt gelling as demonstrated by this technique is shown in Fig. 11 in the paper by Frommer and Lancet in this volume.

The effect of salt gelling on pore size can also be inferred from membrane thickness, shrinkage, and porosity measurements. Fig. 10 summarizes these results for 1 mm O.D. hydrocast membranes. In spite of the scatter in the results the trends are quite clear. As can be observed, the increase in salt content in the gelling water has an effect on membrane morphology equivalent to reducing the casting tube diameter.

Although aqueous NaCl solutions up to 5 wt.% were used exclusively in the tests as a matter of convenience, other aqueous salt solutions are expected to yield similar results[12].

Air Bubble Length. The term "air evaporation time" or "air exposure time" is frequently referred to as one of the prerequisite steps in the preparation of reverse osmosis membranes according to

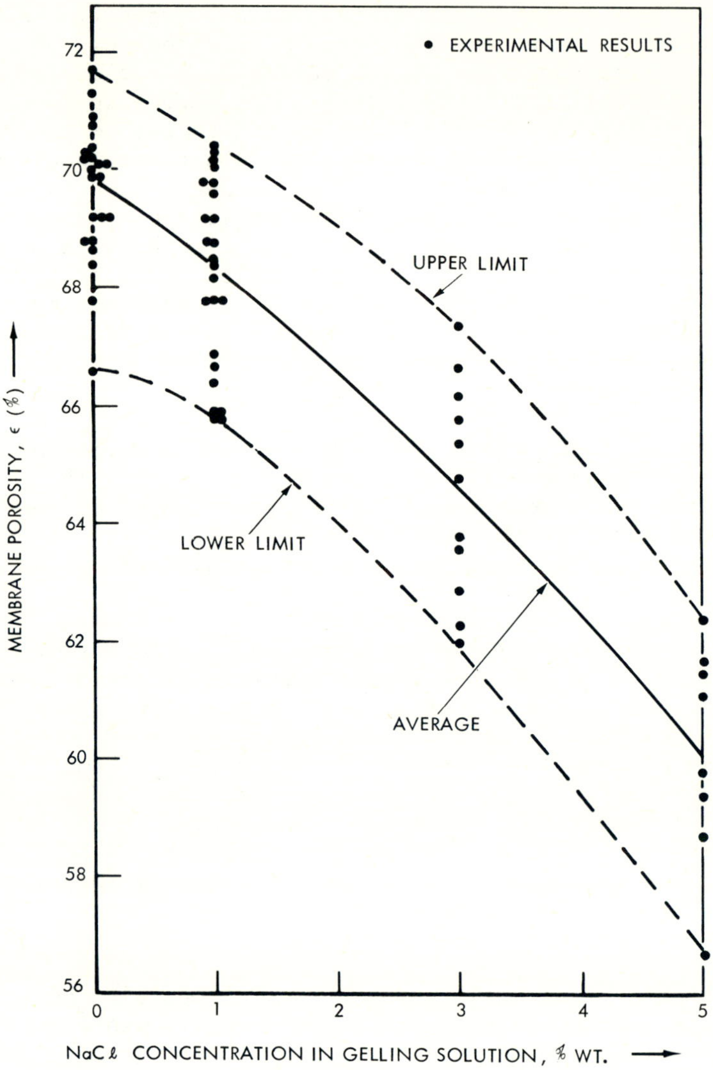

Fig. 10 - Effect of Salt Gelling on the Porosity of 1 mm O.D. Membranes, Series 20821

the Loeb-Sourirajan technique. In the hydrocasting process, exposure time is provided by the forming gas bubble and controlled by its length. As in flat cast membranes longer air bubbles were effective in reducing pore formation. Fig. 11 shows a typical view through the skin side of 1.5 mm O.D. membranes. Although in the membrane formed by the short bubble a large number of macropores can be seen clearly, considerably fewer pores are observed in the one formed with the longer bubble. In relatively large diameter membranes, however, larger forming bubbles were found less effec-

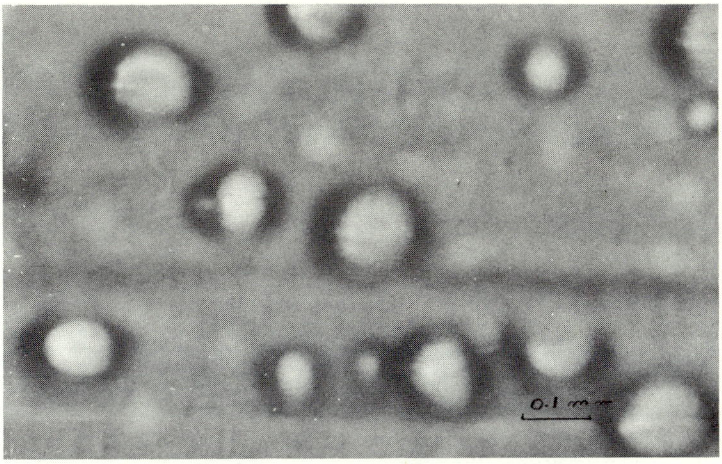

a. FORMING BUBBLE LENGTH 2.5 cm

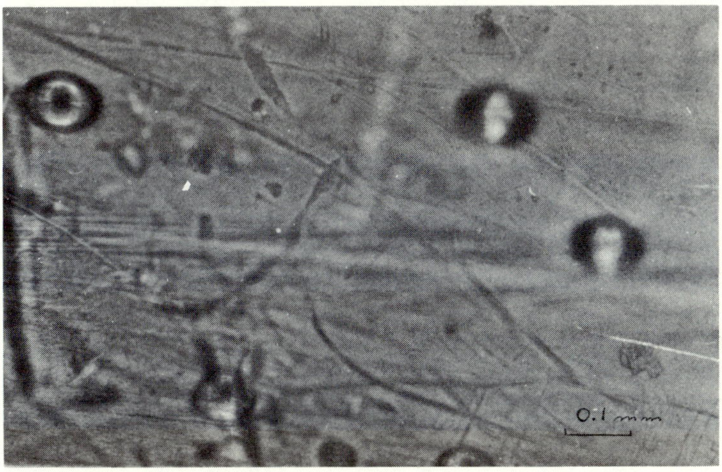

b. FORMING BUBBLE LENGTH 130 cm

Fig. 11 - Effect of Forming Bubble Length on Pore Formation in the Membrane Structure (Looking Through the Skin)

tive in preventing macropore formation as a result of thicker cast layers. (See above discussion on effect of membrane diameter on morphology.)

<u>Burst Pressures, Ultimate Tensile Strength, and Creep of Unsupported Tubules</u>

The maximum operating pressure of hydrocast membranes is primarily controlled by the Taylor relation[2] if we assume that the final membrane strength is proportional to the amount of cellulose acetate in the casting solution.

From the force balance inside a tube we have

$$pd' = 2(\beta\sigma_T)t \ , \tag{17}$$

where p is the operating pressure, d', t and σ_T are the initial I.D., thickness, and tensile strength of the membrane, respectively, and β is a correction factor to account for the cellulose acetate content, porosity, and pore structure of the gelled membrane. Since t and d' are proportional to d we obtain from equation (17)

$$p \sim \beta\sigma_T \ . \tag{18}$$

Equation (18) indicates that the operating pressure is independent of the tube diameter and is set by the tensile strength of the cellulose acetate and the membrane pore structure as expressed by β. Improvement in membrane strength can therefore be obtained through the utilization of inherently stronger polymers in the casting solution, elimination or suppression of macropore formation, and by the utilization of higher polymer content in the casting solutions. In the case of cellulose acetate, fluidization limitations require that only small increases be made in the polymer content. A certain increase in strength may also be anticipated when higher molecular weight cellulose acetate is utilized[13]. One should remember however that in the low pressure range, say 150 to 200 psi, even a slight increase of 50 to 100 psi in the maximum operating pressure is substantial, both from the water and salt transport points of view.

Burst pressure experiments were carried out on representative test sections cut from each hydrocast membrane. The membranes were exposed to a gradual increase of pressure at a rate of 5 psi/sec, until rupture took place. The ultimate tensile strength, σ_u, was arbitrarily defined as:

$$\sigma_u = \frac{p_b d'}{2\bar{t}} \tag{19}$$

where p_b and $\bar{t}$ are the burst pressure and average membrane thickness given by $\bar{t} = 1/2(t_{min} + t_{max})$, respectively.

It is noteworthy that σ_u, as calculated from equation (19), is lower than the actual one, since at pressures above the yield point d' starts to increase quite rapidly, while the membrane thickness is decreasing. Thus, at the burst pressure, the ratio of the inside diameter to the membrane thickness is considerably higher than the initial value of that ratio taken into account in equation (19).

Representative results indicating the effect of membrane diameter and NaCl concentration in the gelling water solution on the ultimate tensile strength and burst pressure, for water-gelled membranes, are given in Figs. 12 and 13, respectively. As can be seen, a substantial increase in the ultimate tensile strength and burst pressure was observed with a decrease in the membrane diameter and an increase of salt content in the gelling water. While the ultimate stress is rising rapidly as tubule diameter decreases below 1 mm the net effect on burst strength is not pronounced because of the decrease in cross-sectional area due to membrane shrinkage. The data presented in Figs. 12 and 13 compare very favorably with the data given in references [14] and [15].

Heat treatment has been known to strengthen the physical properties of membranes. Hydrocast membranes, when compared with non-heat treated membranes, have shown an average of 10 to 15% higher burst pressure.

Creep behavior of unsupported membrane tubules under load is of great importance for practical applications. In a series of long term tests, unsupported hydrocast membranes were subjected to stepwise pressure increases while both diameter changes with time at given locations and membrane performance were recorded. Typical creep data are presented in Figs. 14 and 15. According to these data the CA tubules can endure 100 psi for periods as long as 400 hours without undergoing creep, while pressures in excess of 160 psi can cause continuous creep. Other long term creep tests in excess of 2000 hours at 100 psi have also indicated a minimal strain of less than 5%. A remarkable conclusion to be drawn from these tests is that strains as large as 10% could be endured without rupture or catastrophic loss in performance. Several membranes exhibited dimensional stability even at a system pressure of over 200 psi after 500 hours of continuous testing. In spite of an appreciable loss in rejection of NaCl by these membranes, reasonable rejection (over 90%) was still recorded for $MgSO_4$ at close to 18% diameter change. The increase in the membrane ultimate strength, as affected by salt gelling, is also reflected in their creep behavior as can be seen by comparing Figs. 14 and 15.

PERFORMANCE OF HYDROCAST MEMBRANES

The improvements in membrane tubule morphology resulted in a notable change in their performance. Because of favorable effects on burst strength, compaction, and economics, work has concentrated

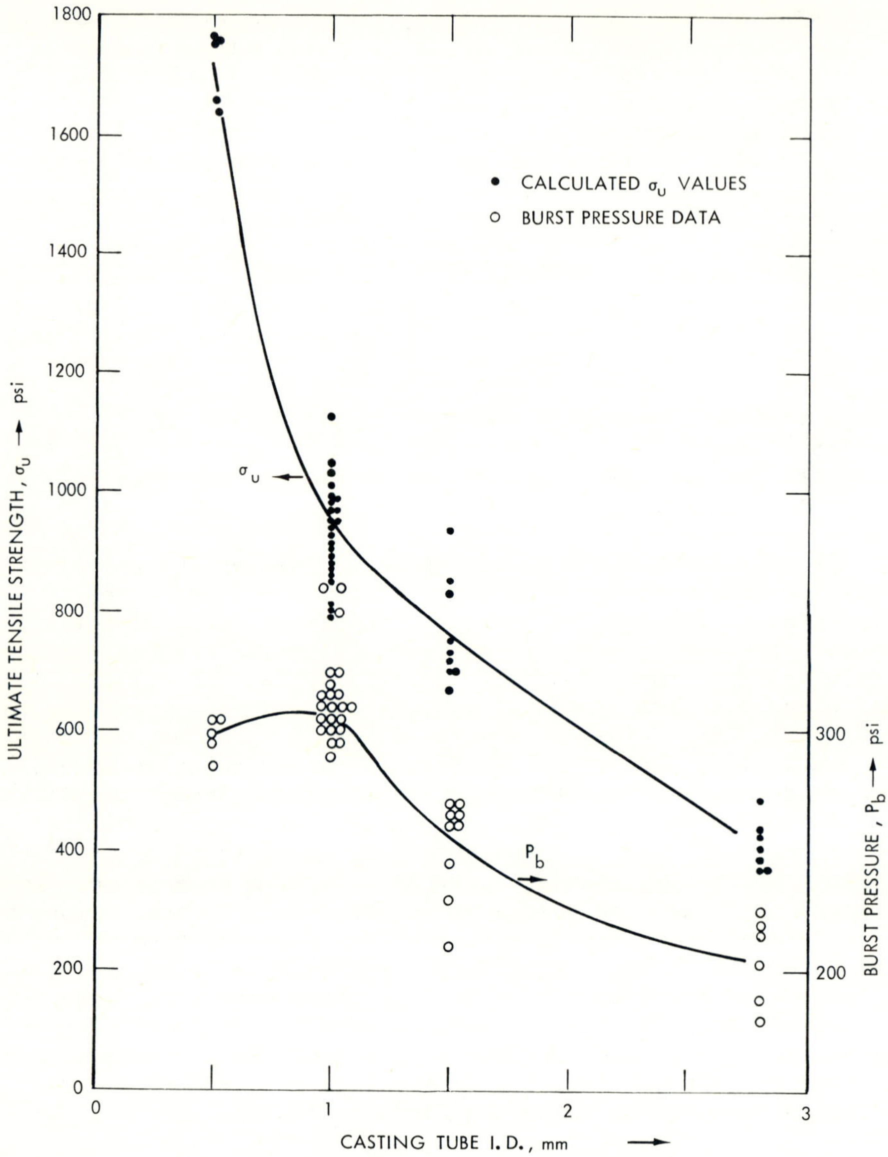

Fig. 12 - Effect of Glass Casting Tube Diameter on the Burst Pressure and Ultimate Tensile Strength of Unannealed Water Gelled Membranes Cast From E400-25 Cellulose Acetate Solutions (CA:formamide:acetone = 22.7:27.3:50)

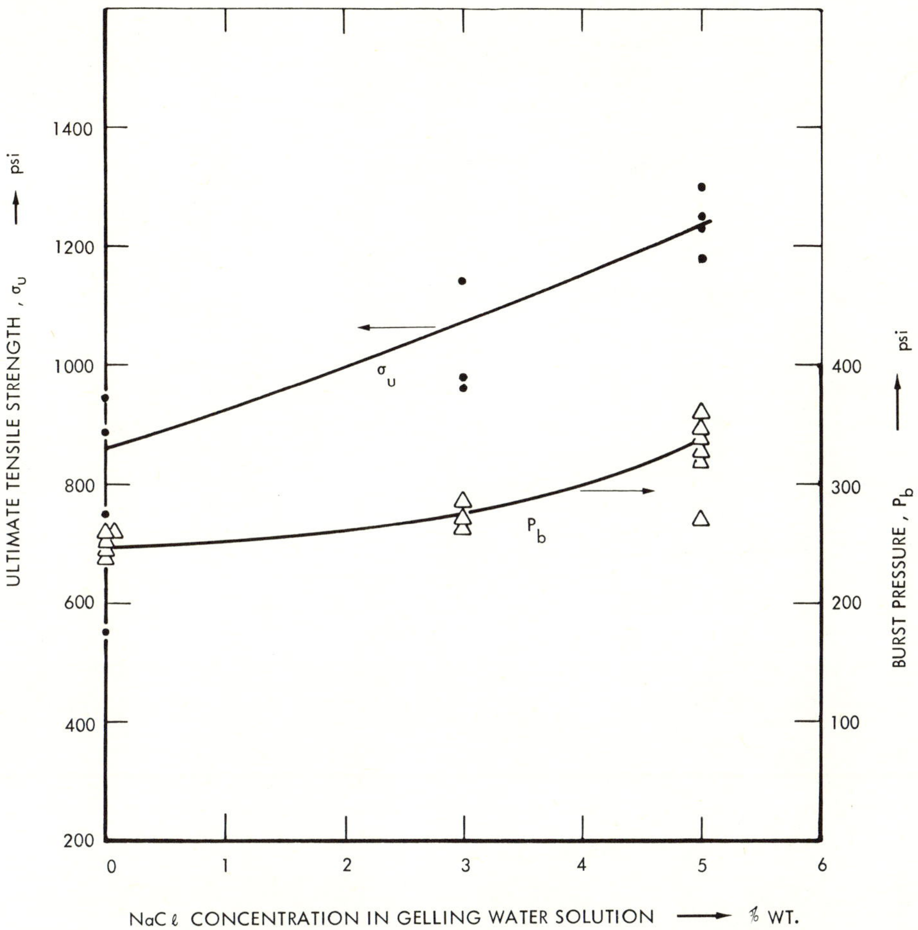

Fig. 13 - Effect of Salt Gelling on Burst Pressure and Ultimate Tensile Strength of 1 mm O.D. Membranes Cast From Standard E398-10 Cellulose Acetate Casting Solution (CA:formamide:acetone = 25:30:45)

on hydrocasting tubules of diameters below 1.5 mm. The fact that unsupported tubules can endure internal pressures as high as 300 psi for short times suggested the study of their performance at intermediate pressures with a view toward determining their applicability for brackish water desalination at relatively low pressure.

Characterization of the unsupported membranes was carried out in a pressurized loop, with between 50 and 220 psi system pressure, under laminar flow conditions in tubules up to 2-1/2 ft long. The water flux and salt rejection for a variety of salts (KCl, $NaNO_3$,

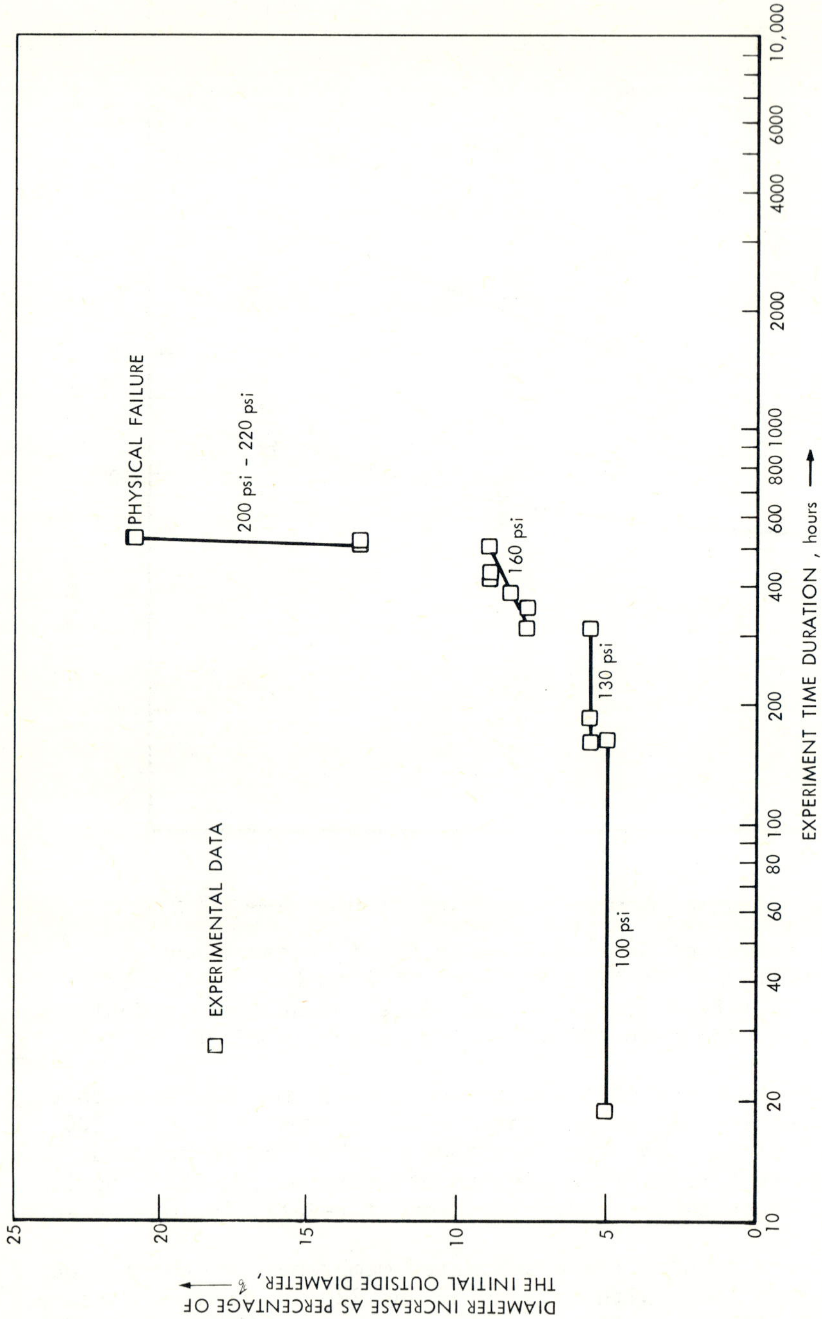

Fig. 14 - Creep Data of a 1 mm O.D. Water Gelled Hydrocast Membrane as a Function of Time and Inside Pressure

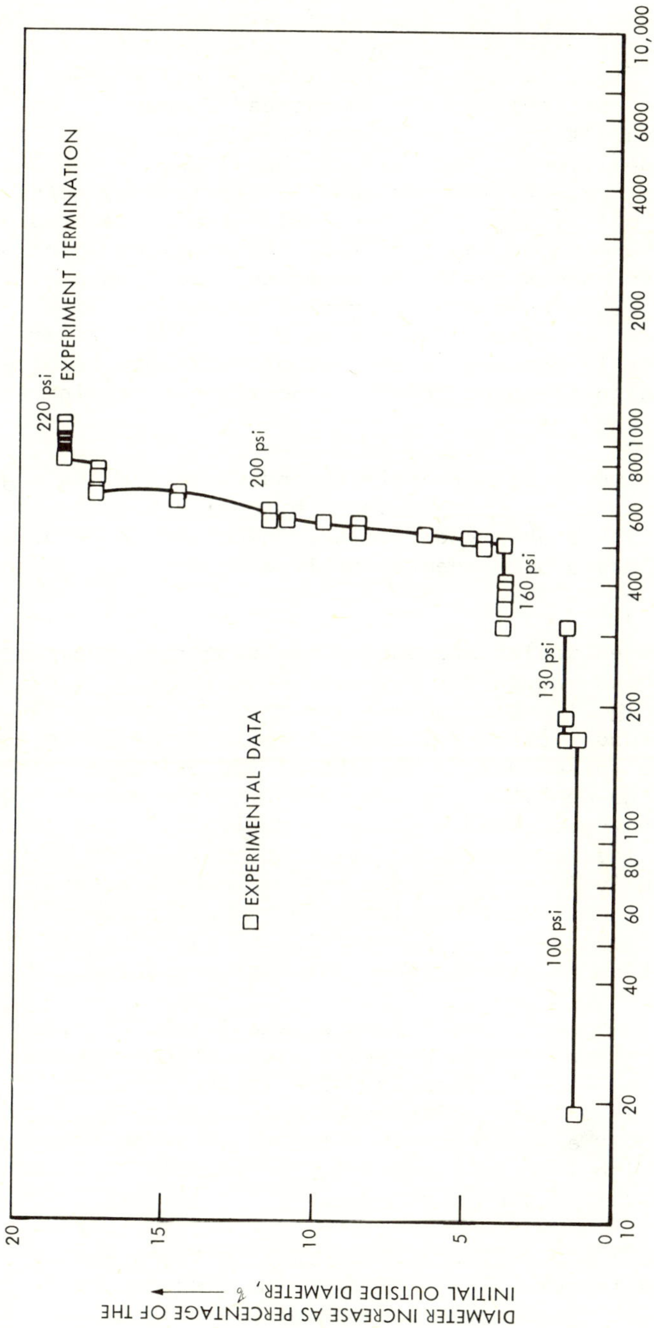

Fig. 15 - Creep Data of a 1 mm O.D. Hydrocast Membrane Gelled in 5% NaCl as a Function of Time and Inside Pressure

$CaCl_2$, $MgSO_4$, and NaCl) have been studied. The results are summarized in Table 1. The dependency of the over-all water transport at 100 psi net pressure calculated from data for NaCl and $MgSO_4$ solutions (obtained with feed concentrations of 500 to 2000 ppm) is presented in Figs. 16 and 17. The data points in each figure represent a different tubule, and the spread is mostly real and represents the variation in performance resulting from different manufacturing conditions. The best of these tubules would provide a flux at 100 psi net pressure of about 5 gal/ft^2-day (gfd) at 90% rejection and 2 gfd at 95% rejection, while the corresponding mean values of flux seem to be about half of these values. The performance for $MgSO_4$ is much superior as expected. Most notably, fluxes at 100 psi net pressure of about 2 gfd were measured at 99% rejection of $MgSO_4$. It is clear from the data that the performance of hydrocast tubules is comparable to that of membranes reported in the literature also prepared from Loeb-Sourirajan casting solutions [7,13].

In terms of membrane permeability and rejection, salt gelling has an effect similar to that produced by an increase in annealing temperature as can be observed from examining the data in Table 1. It therefore offers another degree of freedom for controlling membrane performance.

When membrane tubules are adequately supported they can sustain pressures high enough for seawater desalination. This is illustrated in Fig. 18 which shows test results of over 900 hours of a supported tubule at pressures increasing from 230 to about 810 psi. During these tests pressures of 900 psi were reached for short periods. No membrane rupture took place, although clogging is believed to have occurred especially during the last few hundred hours on test. The tested tubule was wrapped in nylon backing cloth and carefully inserted into a perforated brass tube. These results, in terms of strength, flux, and rejection, demonstrate the potential of operating integrated tubule systems, obtained via in situ casting in porous tubes, at high pressures.

High flux membranes in the form of small tubules offer an excellent combination for low pressure applications such as home drinking water and water softening. Due to their relatively small inside diameter, hydrocast membranes can operate at low velocities and high yield over a short length without suffering from excessive polarization. With 1 mm O.D. tubules, for example, a packing density of 750 ft^2/ft^3 can be achieved.

Several small modules were prepared by potting a tube bundle at both ends and tested. Tests of these modules produced essentially the same results as were obtained with single tubes. A picture of one of the modules is presented in Fig. 19.

TABLE 1

TYPICAL PERFORMANCES OF E394-60 AND E400-25 CELLULOSE ACETATE HYDROCAST MEMBRANES WITH VARIOUS SALT SOLUTIONS

Tubule O.D. (mm)	NaCl in Gelling Sol'n. (%)	Anneal. Temp. (°C)	Feed Solution		Time on Test (hr)	Press. p (psi)	Net[(1)] Press. $p-\Delta\pi$ (psi)	Lp $\left(\frac{10^{-5}\ cm}{sec\text{-}atm}\right)$	Water Flux[(2)] (gfd)	Salt Rej.[(1)] R (%)	Desal. Ratio, $\alpha=\frac{1}{1-R}$
			Solute	Conc. (ppm)							
0.86	0	70	NaCl	1830	47	100	78	10.5	11.9	28.6	1.4
"	"	80	"	"	93	"	77	5.7	6.4	52.4	2.1
"	"	85	"	"	232	"	78	3.2	3.9	61.5	2.6
"	3	70	"	"	71	"	77	2.8	3.1	79.6	4.9
1.00	3	N.H.T.[(3)]	$MgSO_4$	1970	8	"	77	5.1	5.7	89.9	9.9
"	5	"	"	"	47	"	82	5.9	7.0	96.6	29.7
"	5	"	"	"	671	135	110	4.5	7.3	97.1	34.9
"	5	"	"	"	47	100	87	5.0	6.3	98.4	64.1
"	5	"	"	"	527	"	87	4.9	6.3	97.9	46.2
"	5	"	"	"	671	135	117	4.4	7.4	98.5	68.6
0.86	3	"	"	1980	93	100	91	6.6	8.8	98.5	66.0
"	3	"	"	"	93	"	91	2.8	3.7	99.4	17.9
"	0	"	NaCl	1720	2	"	80	20.5	24.0	8.1	1.1
"	0	"	$NaNO_3$	1910	4	"	85	19.6	24.4	0.0	1
"	0	"	$MgSO_4$	1800	25	"	92	12.6	17.1	75.9	4.1
"	0	"	$CaCl_2$	900	29	"	92	16.8	22.8	41.2	1.7
"	0	"	KCl	980	73	"	91	15.6	21.0	10.4	1.1
"	0	"	NaCl	1720	77	"	80	17.8	21.0	10.1	1.1
"	0	80	"	1720	82	"	79	2.2	2.5	75.9	4.1
"	0	"	$NaNO_3$	1910	101	"	84	2.3	2.8	56.9	2.3
"	0	"	$MgSO_4$	1800	104	"	93	2.3	3.1	96.1	25.9

(1) Average values over membrane length. gfd = gal/ft^2-day. Salt rejection = 1-(feed concentration)/(product concentration).

(2) Based on the tubule I.D.

(3) Non-heat treated

TABLE 1 (Cont'd)

Tubule O.D. (mm)	NaCl in Gelling Sol'n. (%)	Anneal. Temp. (°C)	Feed Solution		Time on Test (hr)	Press. p (psi)	Net Press.[1] $p-\Delta\pi$ (psi)	Lp $\left(\frac{10^{-5}\ cm}{sec\text{-}atm}\right)$	Water Flux[2] (gfd)	Salt Rej.[1] R (%)	Desal. Ratio, $\alpha=\frac{1}{1-R}$
			Solute	Conc. (ppm)							
0.86	0	80	$CaCl_2$	900	106	100	84	2.5	3.1	89.7	9.7
"	0	"	KCl	980	125	"	91	2.6	2.8	76.6	4.2
"	0	"	NaCl	1720	128	"	79	2.4	2.8	73.1	3.7
"	0	86	NaCl	1720	197	"	80	1.4	1.7	85.0	6.6
"	0	"	$NaNO_3$	1910	200	"	85	1.4	1.7	69.0	3.2
"	0	"	$MgSO_4$	1800	221	"	93	1.3	1.8	96.5	28.2
"	0	"	KCl	980	124	"	91	1.4	1.8	83.9	6.2
"	0	"	$CaCl_2$	900	145	"	93	1.5	2.0	95.2	20.7
"	0	"	NaCl	1720	149	"	80	1.4	1.7	84.0	6.4
"	0	N.H.T.	$MgSO_4$	2000	25	"	83	17.0	20.7	87.8	8.2
"	0	80	$MgSO_4$	2000	96	"	91	4.2	5.6	93.9	16.3
"	0	"	NaCl	1720	145	"	73	3.4	3.6	60.0	2.5
"	0	85	NaCl	1720	29	"	70	2.9	3.0	71.4	3.5
"	3	N.H.T.	$MgSO_4$	2000	25	"	47	15.2	10.5	98.0	49.4
"	3	75	$MgSO_4$	2000	25	"	86	2.3	2.9	99.3	152.0
"	3	75	NaCl	1720	169	"	76	2.8	3.0	80.4	5.1

(1) Average values over membrane length. gfd = gal/ft^2-day. Salt rejection = 1-(feed concentration)/(product concentration).

(2) Based on the tubule I.D.

(3) Non-heat treated

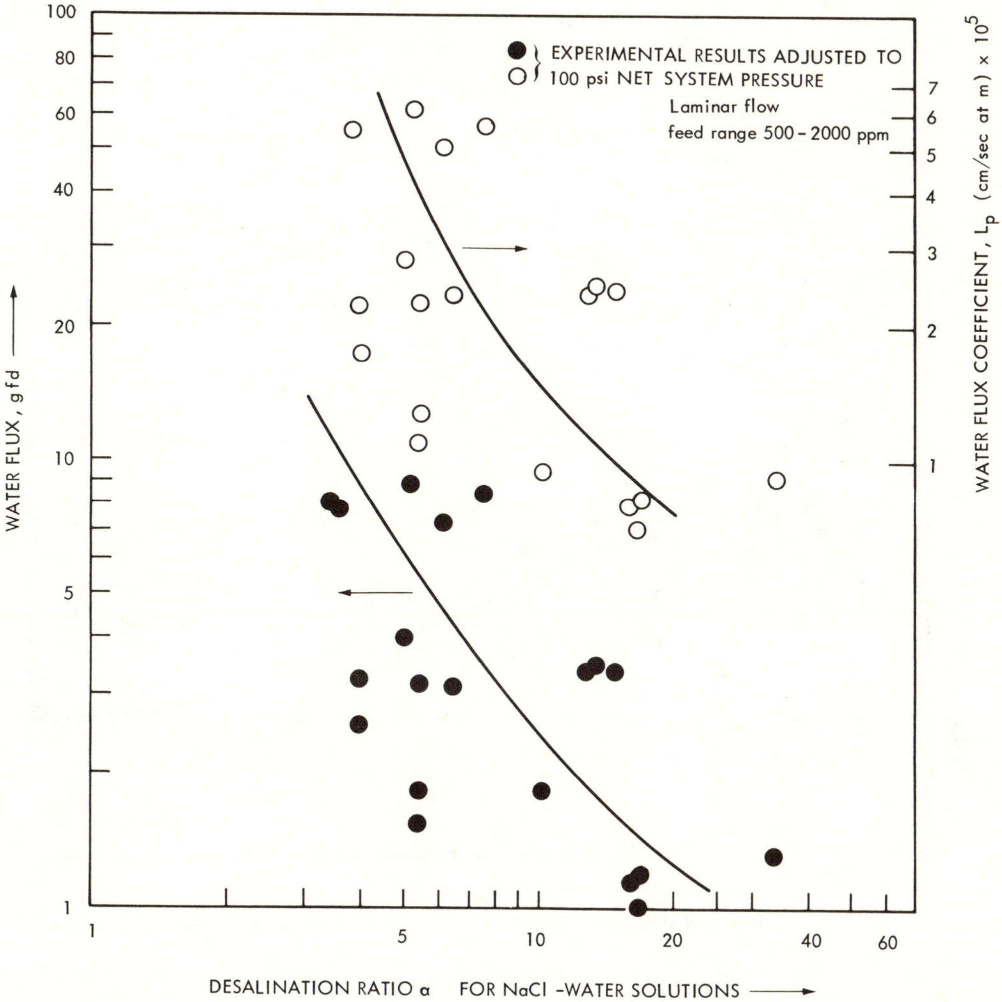

Fig. 16 - Water Flux and Water Flux Coefficient of E400-25 Cellulose Acetate 1 mm O.D. Hydrocast Membranes as a Function of the Desalination Ratio for NaCl-Water Solutions

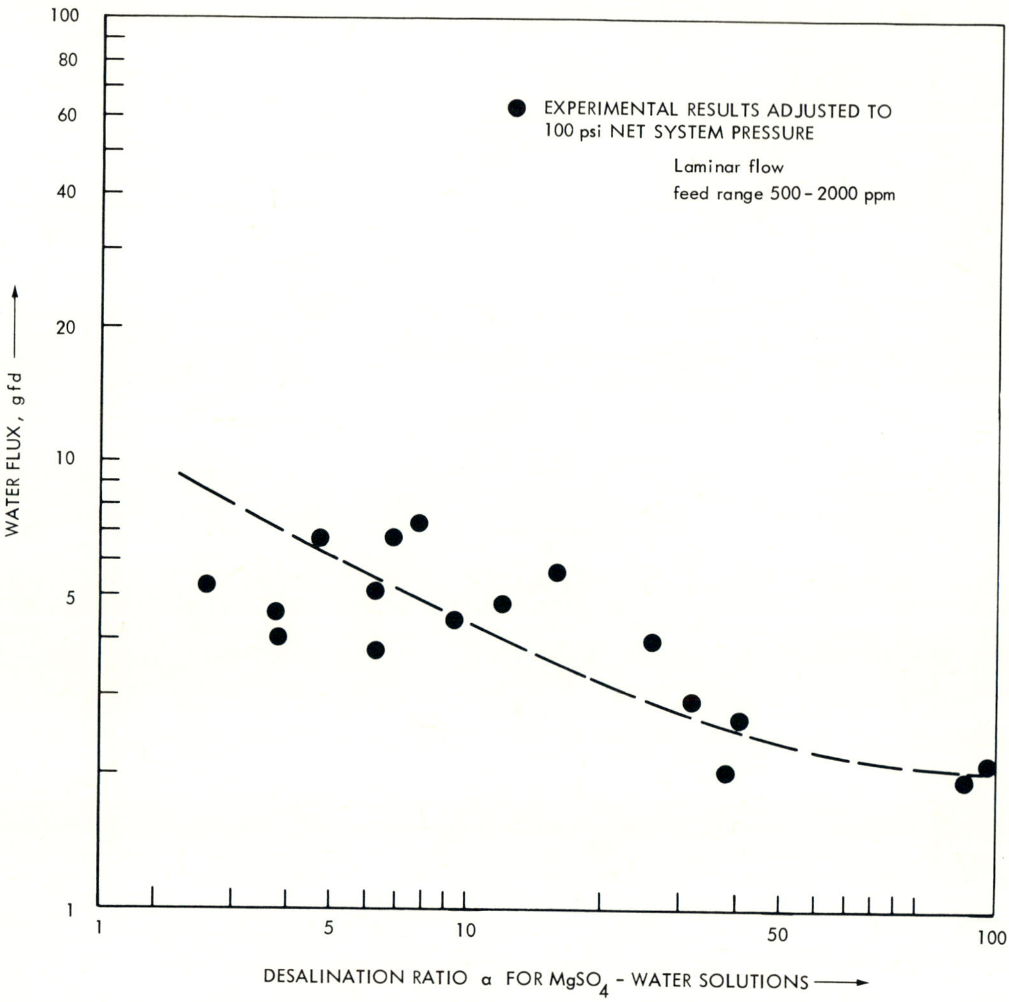

Fig. 17 - Water Flux of E400-25 Cellulose Acetate 1 mm O.D. Hydrocast Membranes as a Function of the Desalination Ratio for $MgSO_4$-Water Solutions

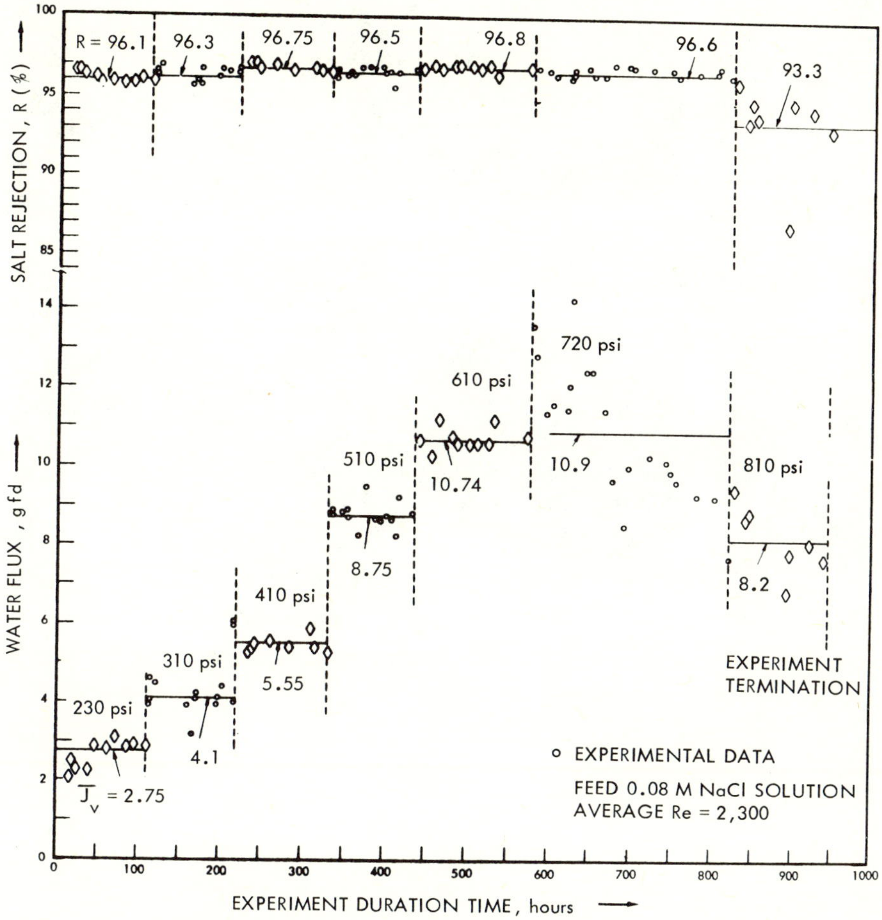

Fig. 18 - High Pressure Water Flux and Salt Rejection Data of a Hydrocast Membrane in a Support Tube

SUMMARY AND CONCLUSIONS

Hydrocasting is a viable process through which membrane tubules (cellulosic or non-cellulosic) may be formed. It has been demonstrated that cellulose acetate membrane tubules of diameters between 0.5 and 1.5 mm can be produced with skin on the interior surface and with good desalting properties. Properly supported hydrocast tubules have been tested with satisfactory results at pressures up to 900 psi. Unsupported tubules have been creep- and performance-tested for upwards of 2000 hours at 200 psi. Fluxes between 1 and 10 gfd have been measured at practical desalination

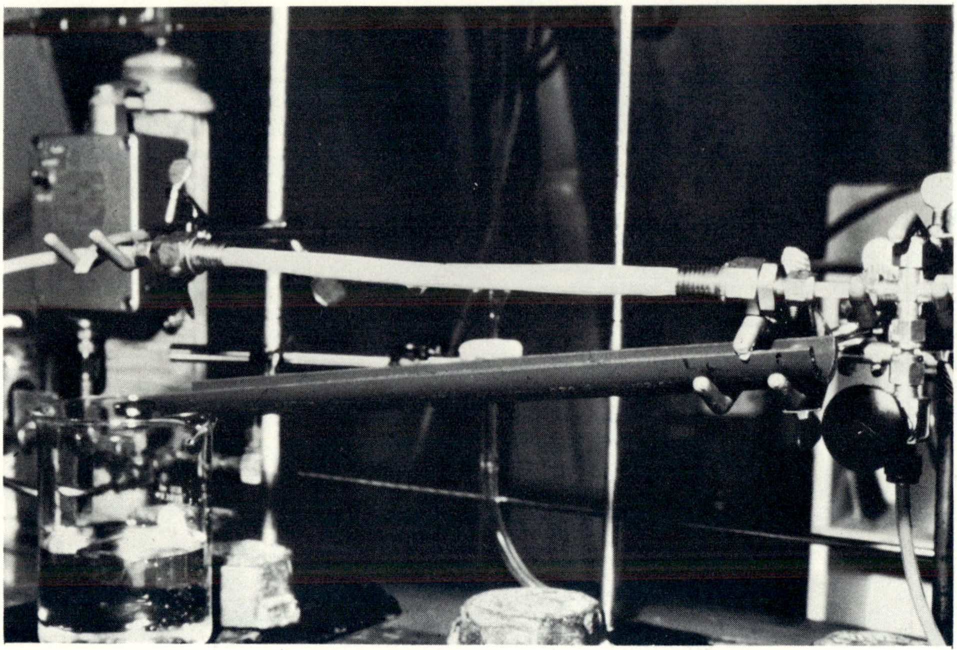

Fig. 19 - Unsupported 1 mm O.D., 10" Long, 62-Tube Hydrocast Membrane Module

ratios for both NaCl and $MgSO_4$ solutions, suggesting the use of unsupported multi-tubule modules for low pressure brackish water desalination and for water softening.

ACKNOWLEDGMENT

The authors wish to thank the Office of Saline Water, U. S. Department of the Interior for the support of this work and Dr. M. A. Frommer for his helpful discussions.

APPENDIX

NON-NEWTONIAN CHARACTER OF THE STANDARD CASTING SOLUTION

Scant information concerning the rheological behavior of Loeb-Sourirajan cellulose acetate casting solutions is available in the literature. Richardson et al.[16] have measured the viscosity of a standard E398-10 casting solution (25% cellulose acetate, 30% formamide, 45% acetone by wt.) with a Ferranti-Shirley viscometer at room temperature. They represented their data with a hyperbolic model. On the other hand it might be expected that a power law[17] representation of the shear-dependent viscosity data should be adequate for most practical purposes. On plotting the shear-stress versus shear-rate data as given in reference[16] for a standard E398-10 casting solution, the power law model was found to represent the data reasonably well over the most interesting range, as follows:

$$\tau = \mu_o \left(\frac{du}{dx}\right)^n = 350 \left(\frac{du}{dx}\right)^{0.86} . \tag{A1}$$

In addition, the parameters given in this equation, when substituted in the filling time equation [eq. (1)], gave excellent agreement with measurements.

LIST OF SYMBOLS

c Concentration

c_i Concentration at the bubble-casting solution interface

d Casting tube I.D.

d_i Casting solution annulus I.D.

d' Membrane tubule I.D.

D Diffusion coefficient

g Gravitational acceleration

K^* Solubility constant at S.T.P.

ℓ Bubble length at location x

L Length of support tube

Lp Water flux coefficient

m Fraction of layer left in the tube, defined by eq. (3)

n Parameter defined in eq. (A1)

p Pressure

Δp Driving pressure in hydrocasting

p_b Burst pressure

$\bar{r}$ Average radius, Fig. 7

R Salt rejection

s Membrane shrinkage defined as $\bar{t}/\delta_o$

t Time or membrane thickness

$\bar{t}$ Average membrane thickness

t_b Time corresponding to x_b

$\bar{u}$ Average velocity

u_b Velocity of leading edge of bubble

x Longitudinal coordinate measured from the inlet of the support tube

x_b x location of the bubble's leading edge

z Transverse coordinate measured from the casting solution interface

α Desalination ratio (feed concentration divided by product concentration)

β Constant, eq. (17)

δ Thickness of the casting solution annulus

ε Void fraction or porosity

θ Angle in cylindrical coordinates, Fig. 7

μ Viscosity

μ_o Parameter defined in eq. (A1)

ζ Constant, eq. (16)

$\Delta\pi$ Osmotic pressure difference across membrane

ρ Density

σ Surface tension

σ_T Tensile strength

σ_u Ultimate tensile strength defined in eq. (19)

τ Time parameter in eq. (14). Shear stress in Appendix.

ϕ Dimensionless parameter, eq. (11)

ψ Constant, eq. (9)

χ Eccentricity factor, δ/δ_o

Subscripts: b, Bubble o, Initial

REFERENCES

1. F. Fairbrother and A. E. Stubbs, J. Chem. Soc., 1, 527 (1935).
2. G. I. Taylor, J. Fluid Mechanics, 10, 161 (1961).
3. M. P. Tulin, Hydronautics, Inc., U. S. Patent 3,592,763 (to Hydronautics, Inc.) July 1971.
4. R. Higbie, Trans. Am. Inst. Chem. Eng., 31 (1935).
5. A. Gollan and M. P. Tulin, Proceedings of the 3rd International Symposium on Fresh Water from the Sea, Dubrovnik, Yugoslavia, Vol. 2, p. 423-442, 1970.
6. A. Gollan, Hydronautics, Inc., Office of Saline Water Research and Development Progress Report No. 542, February 1970, Appendix 3.
7. S. Sourirajan, Reverse Osmosis, Academic Press, New York, 1970.
8. C. W. Saltonstall, D. L. Hoernschemeyer, D. L. Schaeffler, et al., Office of Saline Water Research and Development Progress Report No. 434, U. S. Government Printing Office, June 1969.
9. University of California at Los Angeles, Saline Water Research Progress Summary No. 29, January 1969.
10. M. Frommer, R. Matz, and U. Rosenthal, Ind. Eng. Chem. Prod. Res. Develop., 10, 193 (1971).
11. Hydronautics, Inc. Progress Report 7007-2 to the Office of Saline Water, 1970, to be published.

12. E. Zisner and S. Loeb, Proceedings of the 3rd International Symposium on Fresh Water from the Sea, Dubrovnik, Yugoslavia, Vol. 2, 615, (1970).

13. U. Merten, (Ed.) Desalination by Reverse Osmosis, The MIT Press, Cambridge, Mass., 1966, p. 119.

14. See reference 13, page 149.

15. See reference 7, page 102.

16. J. L. Richardson, et al., Office of Saline Water Research and Development Progress Report No. 576, December 1970.

17. R. B. Bird, W. E. Stewart, and E. N. Lightfoot, Transport Phenomena, John Wiley and Sons, Inc., New York, 1966, p. 11.

18. M. Tulin, M. Matz, and A. Gollan, Office of Saline Water Research and Development Progress Report No. 542, February 1970.

V. COMPOSITE MEMBRANES

POLYELECTROLYTES IN AQUEOUS SOLUTIONS—FILTRATION, HYPERFILTRATION, AND DYNAMIC MEMBRANES*

James S. Johnson, Jr.
Chemistry Division
Oak Ridge National Laboratory
Oak Ridge, Tennessee 37830

INTRODUCTION

Filtration is one of the oldest separation techniques, apparently known to the Chinese and Hebrews in ancient times[1]. It remains one of the most important, and new methods of carrying it out, as well as new applications, continue to be devised. In the last century, it expanded beyond the separation of solid particulates from liquids or gases to separations of colloids from fluids, by ultrafiltration. In recent decades, filtration of low-molecular solutes has been found possible, by "reverse osmosis." Thinking of removal of dissolved solids as a filtration process is helpful on occasion, and the alternate name "hyperfiltration" [2] is sometimes used.

Polyelectrolytes enter this discussion in two distinct, though related, ways. First, the hyperfiltration membranes of concern here are of the ion-exchange class, and thus are composed of polyelectrolytes. Second, polyelectrolytes often are added to influence filtration by interaction with the material to be removed, and they are so used in some of this work. We include inorganic polymers and hydrous oxides, which have ion-exchange properties[3], in the polyelectrolyte class.

---

*Most of this work was supported by the Office of Saline Water, U.S. Department of the Interior, under Union Carbide Corporation's contract with the U.S. Atomic Energy Commission. Certain aspects were sponsored by the Water Quality Office-EPA and the National Science Foundation.

HYPERFILTRATION BY DYNAMIC MEMBRANES

Salt filtration by membranes having ion-exchange capacity was observed in the course of ultrafiltration[4] well before neutral membranes such as cellulose acetate were proposed for desalination[5]. The contribution of equilibrium distributions of salt and water between membrane and solution phases was understood qualitatively, and for idealized models, aspects of transport had also been analyzed[6]. Since the activity of a salt depends on the product of ionic activities, the high concentration of mobile ions in an ion exchanger having a charge sign counter to the fixed charge will tend to exclude mobile co-ions from the exchanger phase. In addition to this equilibrium aspect, the concentration of salt in the product of hyperfiltration will depend on the flux of water relative to diffusion of salt in the membrane, and on the coupling of salt and water fluxes[7].

Ion-exchange membranes have received far less attention for hyperfiltration than neutral membranes. Partly, neglect has stemmed from emphasis on seawater desalination; rejections by ion-exchange membranes are expected to decrease with increasing concentration and are not likely to be adequate for producing potable water in a single stage from the ocean. For low-salinity feeds in desalting and for pollution control, however, rejections might be adequate. Further, the fact that rejection depends on a coulombic mechanism might allow use of a membrane having a looser structure and, for a given thickness, higher permeability[8] than one dependent on the short-range forces presumably active with neutral membranes. The problem with ion-exchange membranes has been to prepare them of thickness comparable to the submicron active layers of cellulose acetate, in order to realize potentially high production rates.

In 1965, an observation made in the course of research under the ORNL Water Research Program pointed a way to do this[9]. A sodium chloride solution, containing a Th(IV) salt at about the millimolar level, was circulated under pressure past a porous support, in this case a 0.2 μ Selas Flotronics Co. silver filter. After operation of the order of an hour, analysis of the effluent indicated that an appreciable fraction of the salt was being filtered, at fluxes much higher than those reported at the time for membranes having any significant salt filtration capabilities.

We concluded that a thin layer of hydrous Th(IV) oxide had in some way formed at the interface between pressurized solution and the porous support. On hydrolysis, many metal ions form polymers in solution, and the hydrous oxide precipitates in many respects resemble a solid ion-exchanger formed by cross-linking an organic polyelectrolyte[3]. In equilibrium with acidic media, they tend to have fewer hydroxides, and the other anions present to maintain charge neutrality frequently are readily exchangeable; in basic solution, the opposite situation holds, and hydrous oxides tend to be cation exchangers.

Attribution of salt filtration to a hydrous oxide film was confirmed in a number of ways. Salts of many other hydrolyzable ions were found to form rejecting membranes dynamically, and membrane formation was easier with colloidal dispersions of hydrous oxides than with solutions of hydrolyzable salts. Some other oxides found to form membranes were Zr(IV), Fe(III), Sn(IV), and U(VI). Hyperfiltration properties as a function of pH are qualitatively as predicted from known ion-exchange properties. With Zr(IV), which has received most attention, results over a wide range of conditions--pH of feed, salt concentration, and valence of ions of feed salt--were successfully correlated[10].

There would appear to be no reason why formation of salt-filtering layers, which we call "dynamic membranes," should be limited to inorganic ion exchangers, and we found that they were indeed formed of such polyelectrolytes as: polyvinyl benzyl trimethyl ammonium chloride, polystyrene sulfonate[11], and dispersions of low crosslinked ion-exchange beads of these polymers; polyvinyl pyridine[12]; polyacrylic acid[13]; polyvinyl sulfonic acid, polymaleic acid, and polyvinyl amine; and humic acid[12], and polyelectrolytes already present in pulp mill wastes[14], sewage[15], and textile dyeing wastes[16]. Membrane formation from particulates has also been observed elsewhere[17].

The promise of high fluxes is frequently realized; values of hundreds of gallons per square foot per day (gpd/ft^2 or gfd) are sometimes observed. An exceptional example is given in Fig. 1, a hydrous Zr(IV) oxide membrane formed at high circulation velocity past the membrane surface. Use of turbulence promoters has increased flux through membranes[18,19], and high turbulence levels associated with high circulation velocities seem to have a similar effect during membrane formation. The support in the case of interest was porous polyvinyl chloride-acrylonitrile on a nylon fabric, nominal pore size 0.45 μ. This commercial filter material (Acropor AN, from Gelman Co.) was wrapped around a 5/8" O.D. porous stainless steel tube (5 μ), and 0.05 M NaCl containing 10^{-4} M hydrous Zr(IV) oxide was circulated in the annulus between the outside of the tube and a pressure jacket[20]. When observed rejection, $\underset{\sim}{R}_{obs}$ (percentage of feed salt removed), of over 50% was attained, flux was still over 1400 gfd. At this flux, concentration polarization of filtered salt is high; if the solute concentration at the membrane-feed interface is taken into account [7,21,22], intrinsic rejection, $\underset{\sim}{R}$, is estimated to be about 80%.

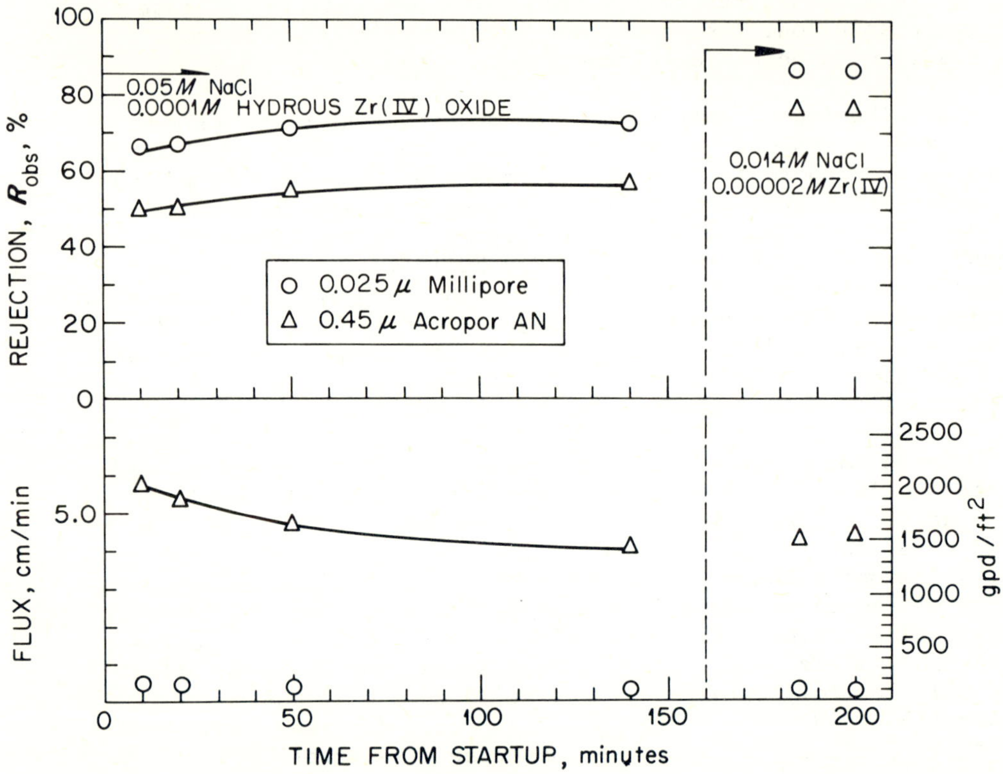

Fig. 1 - Hyperfiltration with Hydrous Zr (IV) Oxide Membrane Formed at High Circulation Velocity (50 ft/sec circulation velocity, pH 4 to 4.5, 1000 psig)

Table 1 lists the zirconium contents of a set of membranes formed in a single run. For a given circulation velocity, there appears to be a rough trend toward higher zirconium content with increasing pore size. For a given pore size, zirconium content is higher at lower circulation velocity, but the correlation of flux with zirconium content and circulation velocity is the opposite of what might be expected in the case of 0.025 μ Millipore. Fluxes through hydrous oxide membranes formed on supports of half-micron pores are usually higher at higher circulation velocities, but, for reasons not understood, are observed to be erratic at the small pore size end of our experimental range.

In another experiment, formation of a polyacrylate layer on a hydrous Zr(IV) oxide membrane, in a manner to be discussed below, did not affect the zirconium content significantly.

TABLE 1

ZIRCONIUM CONTENTS* OF DYNAMIC MEMBRANES FORMED AT 950 PSIG AND pH 3.8 (25°C)

Support	Velocity (ft/sec)	[Zr] ($\mu g/cm^2$)	R_{obs} (%)	Flux (gfd)
0.025 μ Millipore**	35	15	55	171
0.05 μ Millipore	35	31	62	220
0.1 μ Millipore	35	7	64	131
0.45 μ Acropor AN	35	160	64	480
1.2 μ Acropor AN	35	126	62	510
0.025 μ Millipore	14.5	99	57	178
0.45 μ Acropor AN	14.5	232	62	300

*Analyses were by a spectrophotometric method using Arsenazo III as the chromogenic agent. We thank W. R. Laing of the ORNL Analytical Chemistry Division for these results.

**Millipore Corp.

Hydrous oxide membranes exemplify some of the problems, as well as the promise, of dynamic membranes. Hydrous Zr(IV) oxide membranes are easily formed in clean apparatus. In anion-exchange form, their rejection of salt from a NaCl-water solution is good, and from $MgCl_2$-water, even better. Unfortunately, the optimum solution acidity is below pH 4, which does not match well most natural saline waters. Further, their rejection is seriously decreased by the presence in feeds of common polyvalent counterions, such as sulfate.

Much of our effort has been directed toward finding a dynamic membrane for which these disadvantages would be minimized. Of organic polyelectrolytes tested, polyacrylic acid formed membranes of relatively good rejection in the neutral pH range[13], but reproducibility was poor, apparently because membrane formation with this additive was sensitive to pore size distribution of the support. In addition, performance was poor with natural water composition, which typically includes calcium and magnesium.

Hydrous Oxide-PAA Membranes

About two years ago, we found that, if a hydrous oxide membrane is exposed to polyacrylic acid at acid pH, these difficulties are ameliorated[23]. Fig. 2 summarizes changes in hyperfiltration properties subsequent to such exposure as NaOH was added to a 0.05 M NaCl feed containing 25 ppm of polyacrylic acid (PAA, Rohm and Haas Co. Acrysol A-3). Rejections of over 90% were attained in the neutral pH range. In experiments in which PAA was introduced at

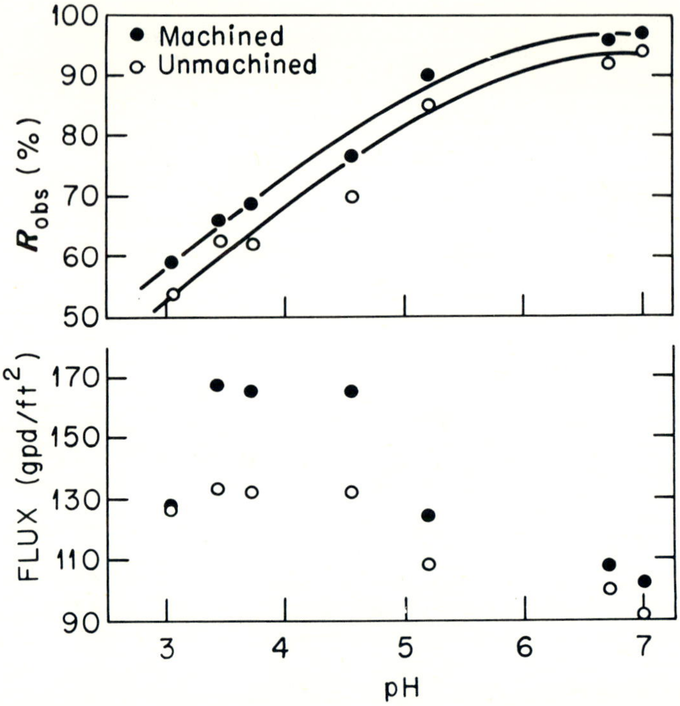

Fig. 2 - Formation Curves for Zr(IV) - PAA Hyperfiltration Membranes on Machined and Unmachined 6C Carbon Tubes. Acrysol A-3 PAA. (0.05 M NaCl, 900 psig, ∿ 60°C, 8 ft/sec circulation velocity)

neutral pH, it had little effect until the feed was cycled through acid pH. Apparently the polyacrylic acid becomes attached to the anion-exchange form of the oxide, and remains so, unless the pH is raised to high values, over 10. The rise in rejection reflects the increase in ion-exchange capacity as the carboxyl groups are neutralized.

The supports in the experiment of Fig. 2 were carbon tubes designated 6C by the manufacturer[12]. The extrusion skin had been machined from one of them, a process which improves performance[24]. These tubes seem more likely in the near future to be practical as components of modules for application[25] than wrapped porous metals, even though production rates with the carbon tubes are not usually as high. Many other tubes appear promising for practical application, particularly ceramic tubes[26]. In Fig. 2, fluxes were about 100 gfd at 60°C. Production rates for hydrous Zr(IV) oxide-polyacrylate (Zr(IV)-PAA) membranes depend roughly on the inverse of solution viscosity and would therefore be about half as high at room temperature as at 60°C.

These dual-layer Zr(IV)-PAA membranes have in many respects more favorable properties than single-layer polyacrylate membranes [13]. Advantages include more reproducible properties, less sensitivity to type of porous support (a characteristic presumably contributed by the hydrous oxide sublayer), and better performance with simulated natural waters. Fluxes are also usually higher than with single-layer polyacrylate membranes.

Most effort with hydrous oxide-polycarboxylate membranes has been centered on Zr(IV)-PAA[23,27,28]. Other oxides, such as those of Sn(IV), Fe(III), and Th(IV), can be used as sublayers, and it may well be that one of them would be as good or better than Zr(IV) for this purpose. Examples of results illustrating various properties of Zr(IV)-PAA membranes follow. Not all were obtained with membranes prepared by the most favorable procedures delineated to date. Consequently, there are some differences in rejections and fluxes, even when porous supports and experimental conditions at the time of the tests were similar.

Concentration Dependence of Rejection. Rejections by ion-exchange membranes are expected to decrease with increasing concentration of the feed solution. Fig. 3 illustrates the dependence for NaCl solutions. In these experiments, values of R_{obs} were measured as a function of circulation velocity, and values of R, the fraction of salt present at the pressurized solution-membrane interface not in the effluent, were estimated by an extrapolation to infinite circulation velocity[22]. This procedure is designed to determine the rejection, undistorted by the effect of concentration polarization.

Ideal "Donnan exclusion" theory predicts that a plot of log $(1 - R)$ vs log c_{NaCl} should approach a slope of 1 at low concentration, behavior which is observed with hydrous Zr(IV) oxide anion-exchange membranes[10]. In common with other dynamic membranes of organic polyelectrolytes, rejection does not decline with increasing concentration as rapidly as simple theory predicts. A likely explanation is a decrease in membrane water content as the concentration of salt in the contacting solutions increases.

Pressure Dependence. Fig. 4 illustrates the effect of pressure on hyperfiltration properties. At the highest pressure of this series, 65 atm., the flux is not high enough to realize the asymptotic rejection[7], though other measurements indicate that a further increase in pressure will cause only a slight increase in rejection. Fluxes are roughly proportional to pressure.

Effect of Molecular Weight of Polyacrylic Acid Used in Membrane Formation. Two illustrations of the influence of molecular weight of the organic constituent are given. Fig. 5 illustrates the change in properties on raising pH after exposure of the hydrous

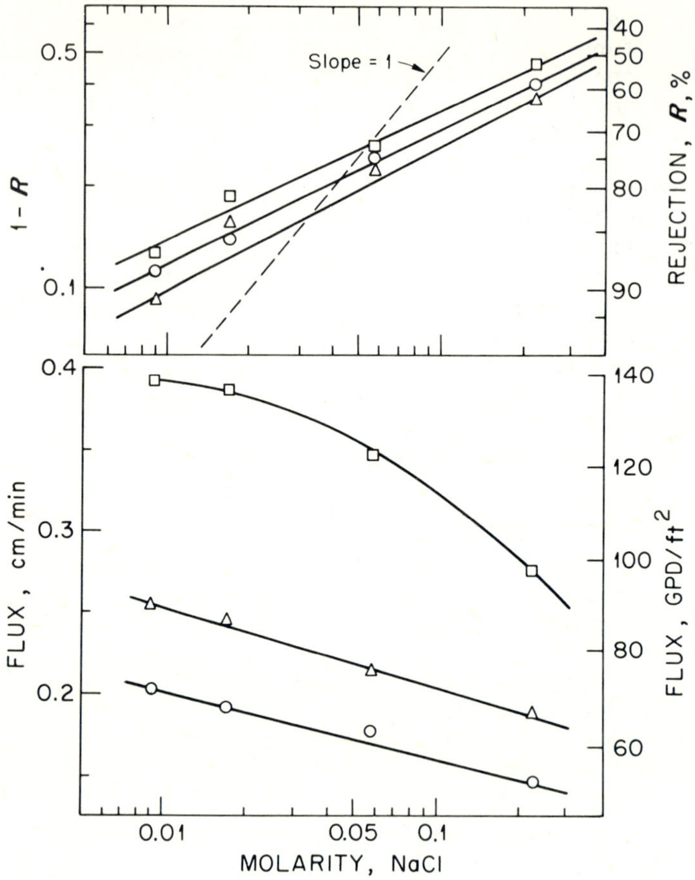

Fig. 3 - Effect of Concentration on Hyperfiltration by Hydrous Zr(IV) Oxide-PAA Membrane
Pressure 1000 psig
□ 1.2μ Acropor
△ 0.05μ Millipore
○ 0.025μ Millipore

oxide layer to polyacrylic acid (in the case of Nalco Chemical Co. A-1, a high molecular weight polycarboxylate of high charge density) in acid solutions. Results of two series of measurements are given. The reproducibility is fair, except for the highest molecular weight, which gave erratic results. Fig. 6 summarizes results of tests with a simulated natural water. These and other results [23] indicate that best rejections are realized with PAA in the molecular weight range of 50,000 to 150,000.

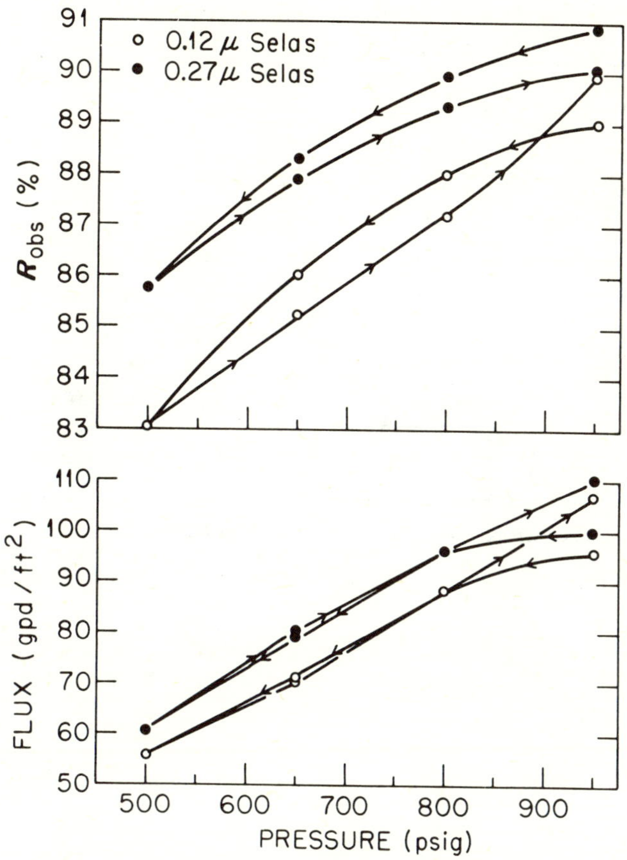

Fig. 4 - Effect of Pressure on Hyperfiltration Properties of Dual-Layer Membranes on Selas Ceramic Tubes. Arrows indicate chronological order of pressure excursion. (0.05 M NaCl, 35 ft/sec circulation velocity, 25°C, pH ∿ 7)

<u>Performance with Simulated Natural Waters</u>. For desalination, the usefulness of Zr(IV)-PAA membranes hinges on how well they perform with natural waters. Laboratory experience with compositions typical of waters in the western U. S. have been encouraging. Fig. 7 summarizes tests with several synthetic solutions, approximating natural (1X) compositions[29]; tests were also carried out with twice (2X) or three times (3X) the nominal concentrations, insofar as solubilities of various components permitted, in order to determine the effect of concentration increases incidental to recovery of product water. The best overall measure of rejection performance is probably "total cation," determined by passing

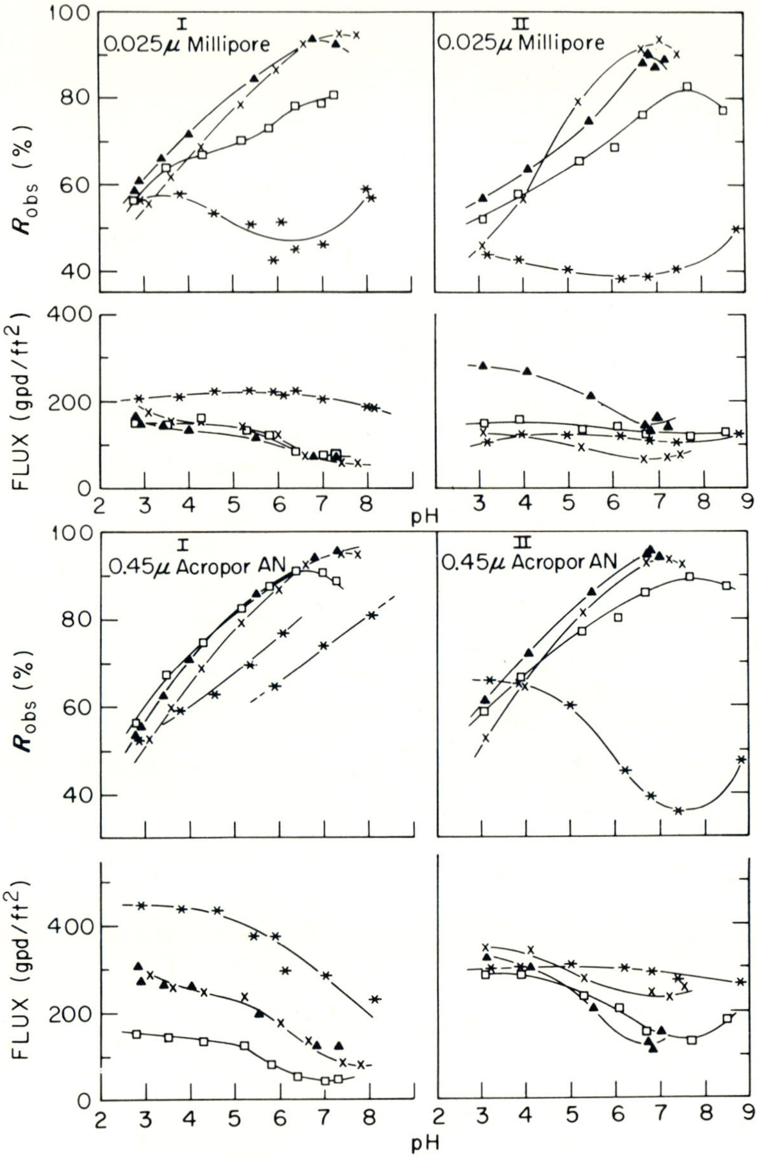

Fig. 5 - Effect of Polycarboxylate Molecular Weight Upon pH Response During Dual-Layer Membrane Formation

I and II are two independent series of experiments.

X Acrysol A-1 ($<50 \times 10^3$)
□ Acrysol A-5 ($<300 \times 10^3$)
Δ Acrysol A-3 ($\sim 150 \times 10^3$)
* Nalco A-1 ($6\text{-}10 \times 10^6$)

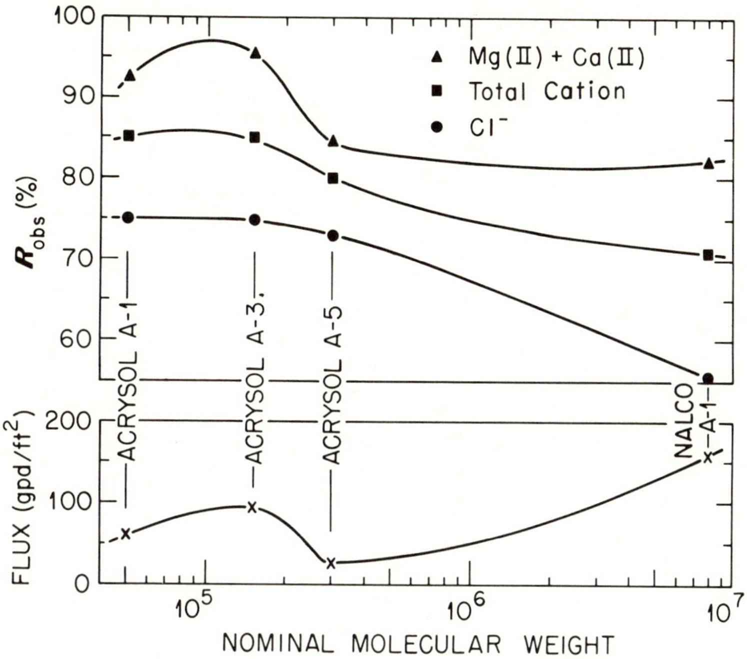

Fig. 6 - Effect of Polycarboxlate Molecular Weight on Hyperfiltration of Coalinga, Calif. Water by Dual Layer Membranes (950 psig; 0.45μ Acropor AN; 35 ft/sec circulation velocity)

samples of feed and product through columns of cation-exchangers in the hydrogen form and titration of displaced acid.

An aspect of the results in Fig. 7 which is not readily explained is the relative insensitivity of rejection to total salinity, in comparison with the dependence for NaCl alone (Fig. 3). Since the rejection of $MgCl_2$ from aqueous $MgCl_2$ solutions is lower than that for NaCl from aqueous NaCl solutions at the same normality, the fact that Mg(II) and Ca(II) are more highly rejected from mixtures than Na^+ is also inexplicable. Similar behavior has been observed with cast PAA membranes[30]. Both rejections and flux were higher for the Zr(IV)-PAA membranes than for dynamic single-layer PAA membranes (half-shaded points of Fig. 7, taken from reference 13), though part of the flux difference may arise from differences in circulation velocity and porous support.

With both the dual-layer and single-layer membranes, presence in the feed of the divalent counterions--Ca(II) and Mg(II)--generally lowers permeability. However, membrane formation procedures frequently result in higher fluxes than those in Fig. 7.

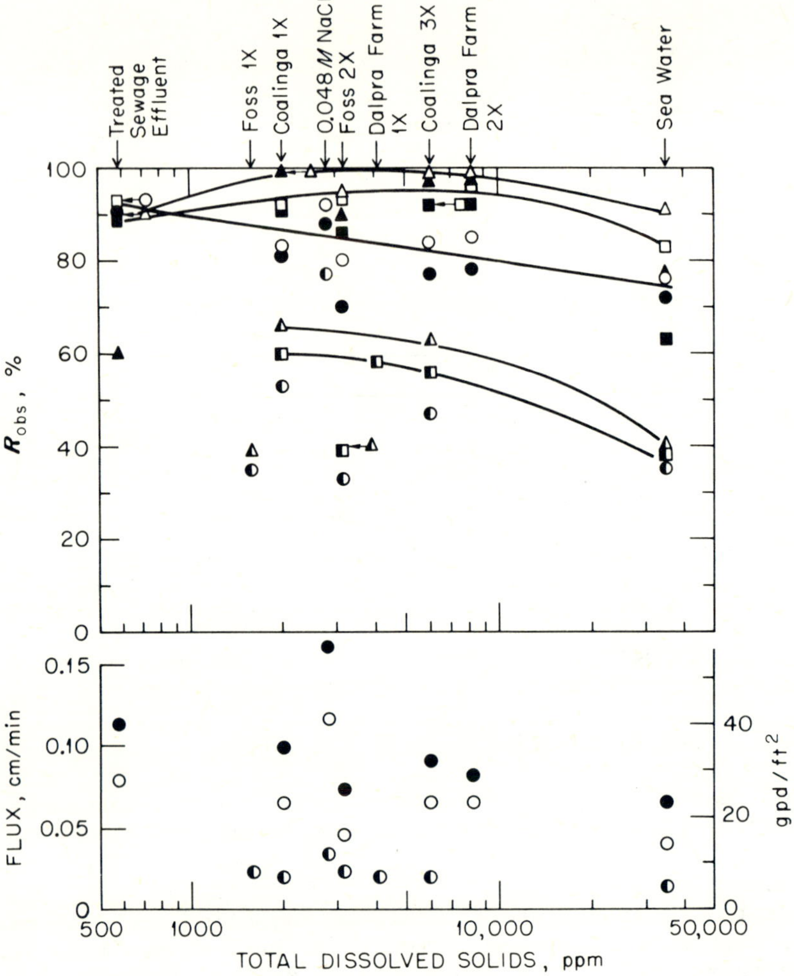

Fig. 7 - Hyperfiltration of Natural Water Compositions With Dynamically Formed Polyacrylate Membranes (1000 psig)

	Chloride	Total Cation	Mg^{2+} + Ca^{2+}
Dual Layer Hyd. Zr(IV) Ox.-PAA (35 ft/sec) 0.05μ Millipore	○	□	△
0.45μ Acropor AN	●	■	▲
Single-Layer PAA (5 ft/sec) 0.3μ Ceramic Tube	◐	◧	half-filled △

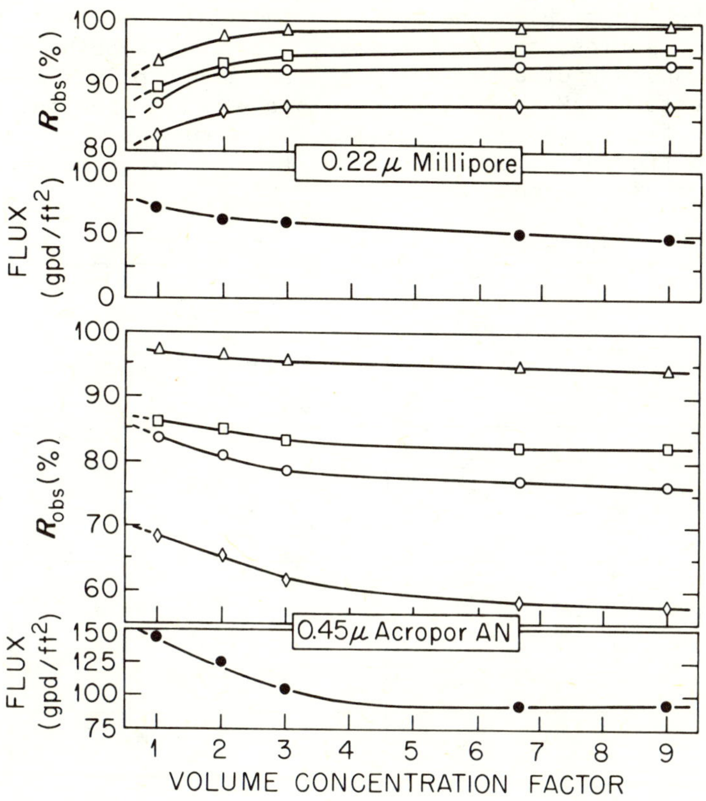

Fig. 8 - Hyperfiltration of Coalinga Water by Dual-Layer Hydrous-Zr(IV) Oxide-Acrysol A-3 Membranes as a Function of Volume Concentration Factor. (950 psig, circ. vel. 35 ft/sec, 25°C)

○ Conductivity △ Mg(II) + Ca(II)
◇ Cl^- □ Total Cation

Fig. 8 summarizes an experiment with a solution initially similar in composition to that of water at Coalinga, California. Product was discarded, and the feed solution allowed to concentrate to about 1/9 of its original volume. Because of incomplete rejection by these membranes and lower rejections by others present in the same test system (not reported in Fig. 8), concentration of various components was not as great as indicated by the reduction in volume--about a factor of five for chloride, 6 for total cation, and 7 for Ca(II) plus Mg(II). Rejections were comparable to those in Fig. 7 for the membrane on the 0.22 μ support, were somewhat lower for the 0.45 μ, and both were also relatively insensitive to concentration. Fluxes in both cases were higher than those in Fig. 7.

A limited trade-off between rejection and flux is possible by adjusting pH (Fig. 9). The symbol α indicates the first measurement of the series, and ω, the last, after tests both in the presence and absence of added PAA (left and right halves of Fig. 9). The effects of pH seem to be reversible. The open and closed points in the figure represent results for membranes which had been operated at different circulation velocities in experiments carried out prior to those in the figure.

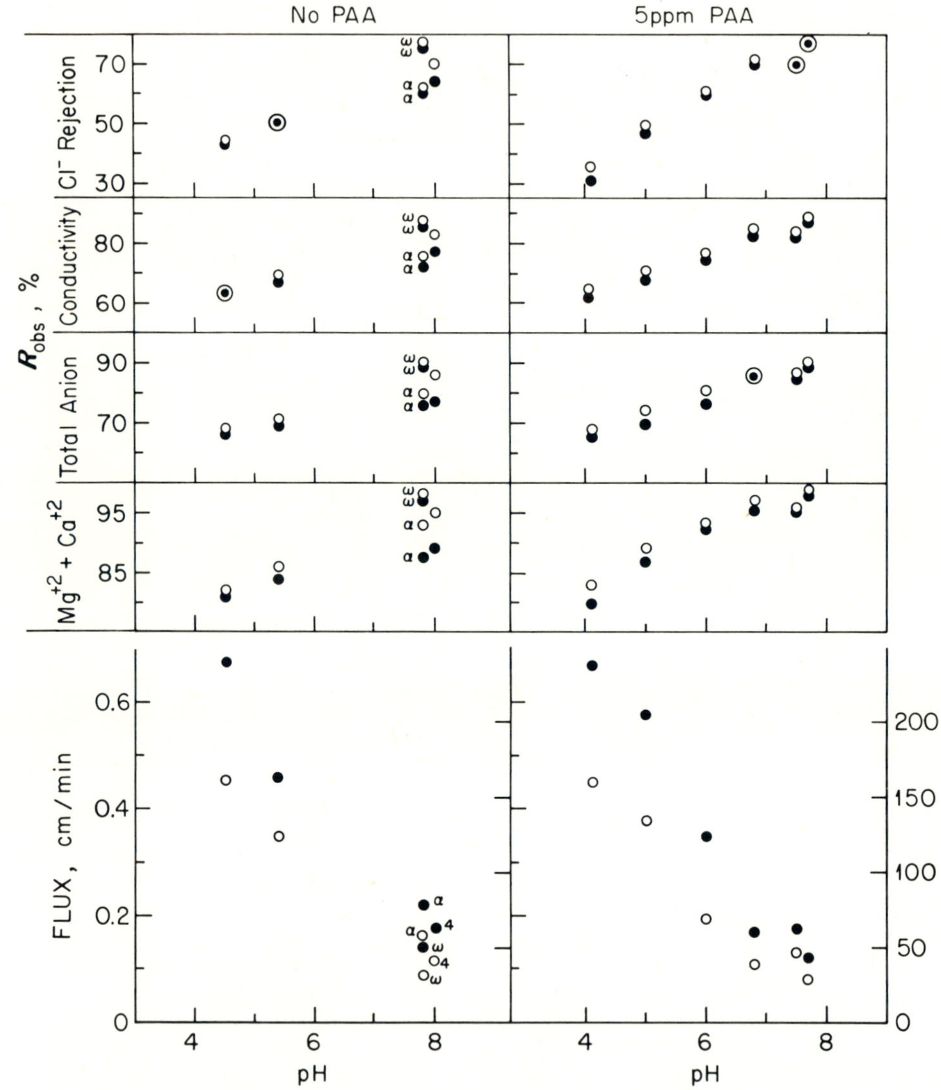

Fig. 9 - Effect of pH on Hyperfiltration of Coalinga Water With Hydrous Zr(IV) Oxide-PAA Membrane. (0.45μ Acropor AN; circ. vel. 35 ft/sec)

Flux Decline. Decrease of production rate with time is a problem with all hyperfiltration membranes. With the dynamic class, there is the possibility of regeneration in situ, without disassembly of equipment. The best way to accomplish this may be by backwashing, which is most readily accomplished if membranes are formed on supports of large pore size, pretreated with filteraids to give a thin layer of fine pore size[31]. While suitable filteraid procedures for Zr(IV)-PAA membranes have not been developed as yet, reasonably good recovery of properties can be attained by chemical washes.

Fig. 10 summarizes fluxes obtained with several membranes in a 13-day run with synthetic Coalinga water, followed by regeneration and further operation for about four days. During most of the run, product was returned to the feed reservoir, and the experimental conditions were those given in the figure legend. From 200 to 317 hours, however, variations in circulation velocities and in concentration were investigated.

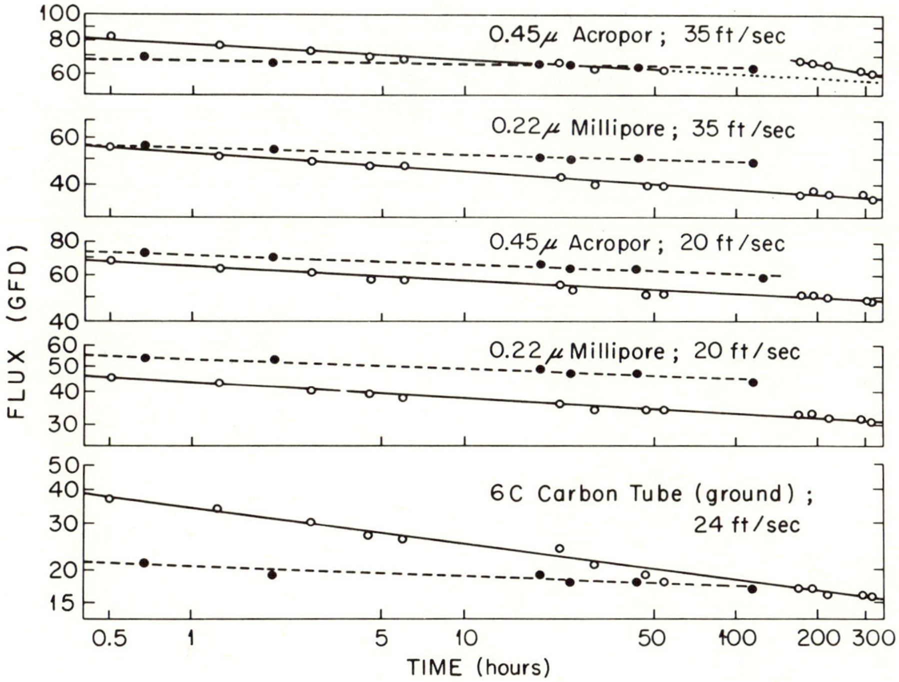

Fig. 10 - Flux Decline Plots for the Hyperfiltration of Coalinga Water by Hydrous Zr(IV) Oxide-PAA Dual-Layer Membranes. (950 psig, 25°C, circulation velocities as shown)
○ Initial Run ● After Regeneration

Regeneration was by a five-minute 1 M NaOH wash, followed by circulation of 0.05 M NaCl containing 50 ppm of PAA initially at pH 2.2, with subsequent base additions to neutrality. This amounts to re-forming the polyacrylate layer by the usual procedure. It was not necessary to form the hydrous Zr(IV) oxide layer a second time.

Rejections were similar to others reported here for Coalinga; examples are listed in Table 2. In Fig. 10 the fluxes are plotted according to the common practice, logarithm of flux vs log time. It is an empirical observation that such plots are frequently linear[32], and the negative of the slope, b, is taken as a measure of the rate of flux decline. Slopes are listed in Table 3.

The values of b are in the range usually found for conventional cellulose acetate membranes. In all cases values of b are less after regeneration than before.

TABLE 2

COMPARISON OF HYPERFILTRATION OF COALINGA WATER BY MEMBRANES FORMED AND OPERATED AT DIFFERENT CIRCULATION VELOCITIES. DATA TAKEN AT END OF INITIAL 317 HOUR RUN AND AT END OF POST-REGENERATION SEGMENT OF RUN (115 HOURS) (950 PSIG, 25°C)

			$\underset{\sim}{R}_{obs}$ (%)		
Support*	Velocity (ft/sec)	Flux (gfd)	Cl^-	Mg(II) + Ca(II)	Total Cation
	Before regeneration, t = 317 hours				
0.45 μ AP	35	60	71	97	86
	20	48	77	98	90
0.22 μ MP	35	35	92	>99	97
	20	31	92	>99	97
6C Carbon	24	16	70	95	85
	After regeneration, t = 115 hours				
0.45 μ AP	35	63	76	98	89
	20	59	67	96	85
0.22 μ MP	35	49	87	>99	95
	20	44	82	99	93
6C Carbon	24	17	74	92	84

*AP = Acropor; MP = Millipore.

TABLE 3

FLUX DECLINE PARAMETERS BEFORE AND AFTER MEMBRANE REGENERATION FOR THE HYPERFILTRATION OF COALINGA WATER BY DYNAMIC DUAL-LAYER MEMBRANES (950 PSIG, 25°C). J_1 IS FIRST FLUX IN FIG. 10

Support	Circulation Velocity (ft/sec)	Initial Run		After Regeneration	
		J_1 (gfd)	b	J_1 (gfd)	b
0.45 μ Acropor AN	35	83	0.059	69	0.013
0.22 μ Millipore	35	55	0.065	56	0.024
0.45 μ Acropor AN	20	68	0.064	73	0.037
0.22 μ Millipore	20	46	0.056	54	0.040
6C Carbon (ground)	24	37	0.145	21	0.037

Treatment of Waste Streams

Hyperfiltration, both with conventional and dynamic membranes, is also promising for treatment of waste solutions. In many of these problems, a high degree of salt removal is not necessary. The advantages of high flux and the availability of membranes able to operate in a variety of chemical environments may be favorable to the dynamic class relative to conventional membranes.

With some waste streams, the ability of many dynamic membranes to operate at temperatures considerably above 25° may be advantageous, particularly if the product water can be immediately recycled. Present commercial cellulose acetate and polyamide membranes are limited in this regard. Neutral organic solutes frequently are contaminants, and these are filtered to a considerable extent by dynamic membranes[33,34], the rejection being largely dependent on molecular size. Results of considerable promise have been obtained in preliminary tests with sulfite[14] and kraft[35] pulp mill wastes; municipal sewage[15,20]; and textile dyeing wastes [16].

Two examples will be cited here, both taken from a series of tests of a prototype carbon tube module, supplied to us by Union Carbide Corp. The first was with a 50:50 cotton-polyester dye waste, used in other experiments reported elsewhere[16]. The solution contained* about 4000 ppm total dissolved solids, 600 ppm total carbon, 1600 Pt-Co color units, and 2400 ppm chloride, with pH about 12. It was neutralized with HCl before introduction into the loop. Results are summarized in Table 4. The experiment was

*Measured at time of run. Somewhat different from values in Ref. 16.

TABLE 4

HYPERFILTRATION OF 50:50 COTTON-POLYESTER DYE WASTE BY Zr(IV)-PAA MEMBRANES ON UNION CARBIDE PROTOMODULE (950 PSIG, 63°C, 12.4 FT/SEC, pH 6.8)

Time (hr)	Flux (gfd)	R_{obs} (% carbon)	R_{obs} (% color)	R_{obs} (% Cl^-)
0.67	76	-	-	-
1.0	72	98	99	84
2.0	59	99	99	87

carried out at a temperature near that of the dyeing process, the objective being to test the possibility of recycle of the hot product water in order to conserve energy. No dyeing tests were carried out with the product of this experiment. In another, however, producing a comparable water, the effluent was reused successfully, and it was further found possible to dye cloth with the concentrated brine [36].

Fig. 11 summarizes another test, on the caustic extract generated in washing kraft pulp between bleaching steps. The solution was neutralized with acidic kraft bleaching waste before use. Temperature of operation was approximately that of the bleaching process, and the feed was concentrated by discarding product. The effects of two regenerations are indicated. In the first, a 1 M NaOH solution, followed by 1 M HNO_3, was circulated. Both hydrous Zr(IV) oxide and polyacrylate layers were then again formed on the support. The second was by a $NaHCO_3$-Na_2CO_3 wash, and it was not necessary to re-form the membrane afterwards. As with the experiment of Fig. 10, flux decline was less rapid after regeneration.

CROSS-FLOW FILTRATION

Placing "reverse osmosis" in the category of filtration processes suggests that the use of a flow past the membrane surface might be useful. Circulation is necessary in salt filtration, where rejection is never complete, to combat concentration polarization, and it also inhibits fouling[37,38]. One might expect by analogy that a cross flow in filtration of particulates and colloids would tend to keep the filter cake thin and help maintain flux[39]. The idea has been incorporated in commercial equipment for some time[40], and has been embodied abroad in a tubular configuration[41], similar to our arrangement. Recently, cross flow has been used in "membrane ultrafiltration,"[42] which utilizes an asymmetric preformed detachable membrane.

	REJECTION R_{obs} (%)	CONCENTRATE	PRODUCT
COLOR (Pt-Co Units)	99.9	9000 - 45,000	5 - 35
TOTAL CARBON (ppm)	94 - 97	775 - 3200	25 - 100
CHLORIDE (ppm)	BELOW	1000 - 5000	150 - 1000

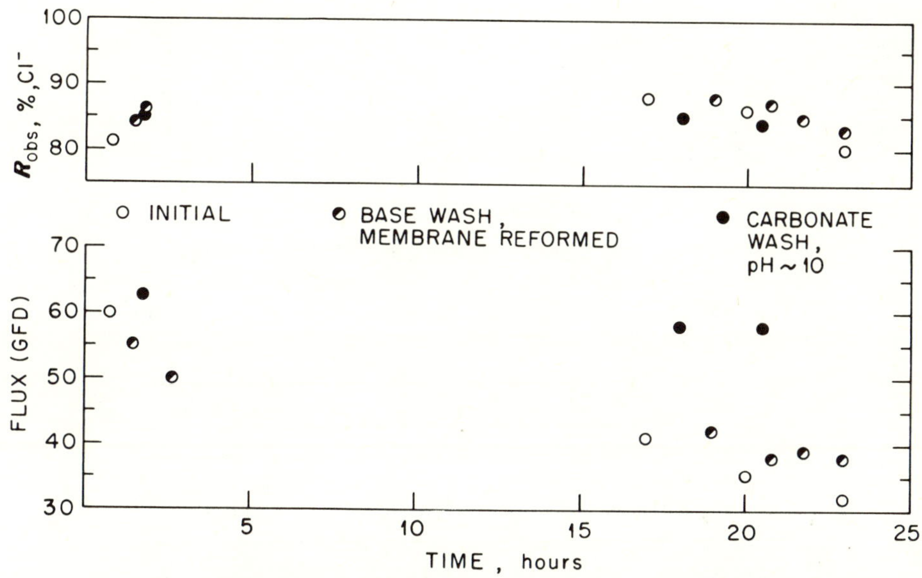

Fig. 11 - Hyperfiltration of Kraft Pulping Wastes with Zr(IV)-PAA Membrane Dynamically Formed in Prototype Module. Feed, caustic extract, acid bleach added to pH 7-7.5. Support, 6C carbon tubes, Union Carbide (950 psig, 12.4 ft/sec circ. vel., 58-67°C)

Most of the effort in this labroatory toward developing cross-flow filtration has been directed toward treatment of municipal sewage, specifically of the effluent from primary treatment [43]. Fig. 12 indicates that use of cross flow indeed slows flux decline. The characteristics of the product indicated in the legend are typical, except that turbidity in this experiment was unusually high; normally it is one JTU or less, about equivalent to local tap water.

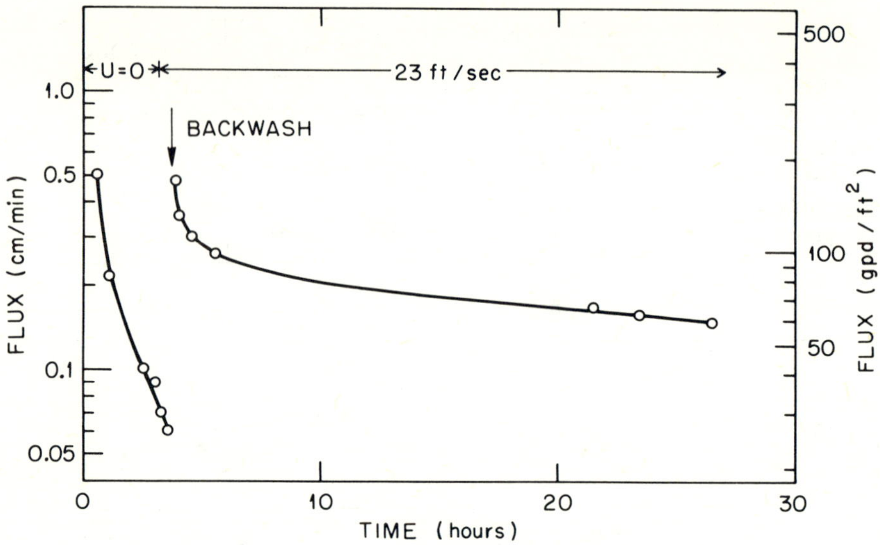

Fig. 12 - Filtration of Sewage With and Without Cross-Flow. U is the linear velocity over the membrane surface. (15 psi, 325 mesh S.S. screen, PSE with Fe(III) and carbon at 95% water recovery, pH 5.9-6.4)

	Average Removals with Cross-Flow		
	Org C	Phosphate	Turbidity
Product Removal	6 ppm 85%	0.3 ppm 99.3%	3.5 JTU 97.5%

Cross flow alone is, however, not adequate to maintain good performance. Apparently material in sewage which would tend to form dynamic salt-filtering membranes[15] must be hindered from doing so. Since in the first few minutes product is usually cloudy, some sort of dynamic filter cake "membrane" is necessary for the process, but its characteristics must be controlled. This is accomplished by addition of a hydrolyzable salt, usually of Fe(III) or Al(III), and sometimes powdered carbon. Fig. 13 shows that increasing Fe(III) concentration increases both efficiency of contaminant removal and flux.

The process, as applied to primary sewage effluent, is thus a physical-chemical treatment, with cross-flow filtration used to separate purified water. The product of the treatment is in many respects the equal of that from conventional activated sludge secondary treatment combined with some procedures classed as

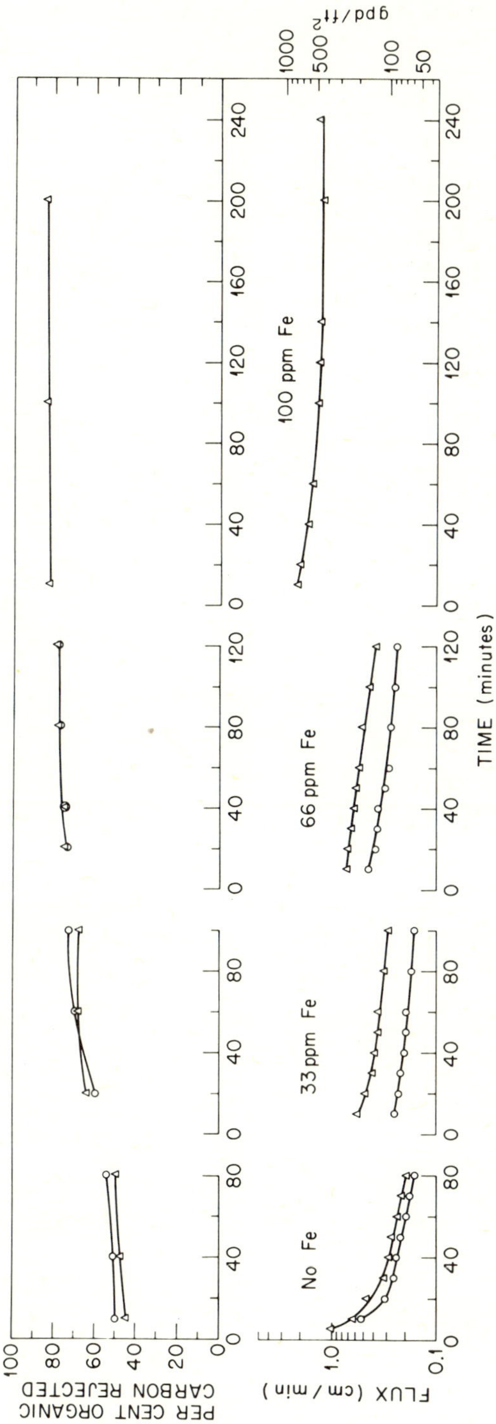

Fig. 13 - Effect of Fe(III) Concentration on Cross-Flow Filtration of Effluent From Primary Treatment of Municipal Sewage. Membrane supports were fire hose jackets: feed inside, o; outside, Δ. (50 psi; 15 ft/sec)

tertiary. A preliminary economic analysis projects that costs will be reasonable, in view of the quality of the product[44].

ACKNOWLEDGMENTS

Contributions of several teams from the Massachusetts Institute of Technology School of Chemical Engineering Practice, Oak Ridge Station, have been especially important. These teams were under R. H. Mayer, Director, and M. Bautista, Assistant Director, in 1969-70, and John T. Day, Director, and Parvez Wadia, Assistant Director, in 1970-71. Individuals of various teams are noted in references 27, 28, 34, and 35.

R. E. Minturn has directed recent work in hyperfiltration. K. A. Kraus and H. A. Mahlman were in charge of the cross-flow filtration of sewage program; K. A. Kraus and J. A. Dahlheimer were in charge of sewage hyperfiltration[20]. C. G. Westmoreland, W. G. Sisson, L. L. Fairchild, J. Csurny, J. W. Myers, and J. R. Love contributed technical assistance. I am also grateful for the advice and support of H. E. Podall, Office of Saline Water.

REFERENCES

1. George D. Dickey, Filtration, Reinhold Publishing Co., New York (1961).

2. K. S. Spiegler, Salt Water Purification, Wiley, New York (1962).

3. K. A. Kraus, H. O. Phillips, T. A. Carlson, and J. S. Johnson, Proc. Second Intern. Conf. Peaceful Uses of Atomic Energy, Geneva, 28, 3 (1958).

4. J. W. McBain and R. F. Stuewer, J. Phys. Chem., 40, 1157 (1936); J. G. McKelvey, K. S. Spiegler, and M. R. J. Wyllie, Chem. Eng. Prog. Symp. Ser., 55, No. 24, 199 (1959).

5. C. E. Reid and E. J. Breton, J. Appl. Polymer Sci., 1, 133 (1959).

6. T. Teorell, Proc. Soc. Expt. Biol. Med., 33, 282 (1935).

7. J. S. Johnson, L. Dresner, and K. A. Kraus, Chapter 8 in Principles of Desalination, K. S. Spiegler (Ed.), Academic Press, New York (1966).

8. L. Dresner and K. A. Kraus, J. Phys. Chem., 67, 990 (1963); L. Dresner, J. Phys. Chem., 69, 2230 (1965).

9. A. E. Marcinkowsky, K. A. Kraus, H. O. Phillips, J. S. Johnson, and A. J. Shor, J. Am. Chem. Soc., 88, 5744 (1966).

10. A. J. Shor, K. A. Kraus, W. T. Smith, Jr., and J. S. Johnson, Jr., J. Phys. Chem., 72, 2200 (1968).

11. K. A. Kraus, H. O. Phillips, A. E. Marcinkowsky, J. S. Johnson, and A. J. Shor, Desalination, 1, 225 (1966).

12. K. A. Kraus, A. J. Shor, and J. S. Johnson, Desalination, 2 243 (1967).

13. S. B. Sachs, W. H. Baldwin, and J. S. Johnson, Jr., Desalination, 6, 215 (1969).

14. J. J. Perona, F. H. Butt, S. M. Fleming, S. T. Mayr, R. A. Spitz, M. K. Brown, H. D. Cochran, K. A. Kraus, and J. S. Johnson, Environ. Sci. Technol., 1, 991 (1967).

15. H. C. Savage, N. E. Bolton, H. O. Phillips, K. A. Kraus, and J. S. Johnson, Water and Sewage Works, 116, 102 (1969).

16. C. Aurich, C. A. Brandon, J. S. Johnson, Jr., R. E. Minturn, K. Turner, and P. A. Wadia, J. Water Pollution Control Federation (in press).

17. E. R. Brownscombe and L. R. Kern, U. S. Patent 3,331,772, July 18, 1967.

18. D. G. Thomas and J. S. Watson, Ind. Eng. Chem. Proc. Des. Develop., 7, 397 (1968).

19. D. G. Thomas, P. H. Hayes, W. R. Mixon, J. D. Sheppard, W. L. Griffith, and R. M. Keller, Environ. Sci. Technol., 4 1129 (1970).

20. "Application of Hyperfiltration to Treatment of Municipal Sewage Effluents," Water Pollution Control Research Series ORD 17030 EOH 01/70, Federal Water Quality Administration, U. S. Department of the Interior.

21. T. K. Sherwood, P. L. T. Brian, R. E. Fisher, and L. Dresner, Ind. Eng. Chem. Fundamentals, 4, 113 (1965).

22. A. J. Shor, K. A. Kraus, J. S. Johnson, Jr., and W. T. Smith, Jr., Ind. Eng. Chem. Fundamentals, 7, 44 (1968).

23. J. S. Johnson, Jr., R. E. Minturn, and P. H. Wadia, submitted to J. Electroanal. and Interfacial Chemistry.

24. W. R. Mixon, M. S. Thesis, University of Tennessee, ORNL-TM-2150 (1969), p. 22; D. G. Thomas, ORNL, private communication.

25. J. B. Cabellon, A. K. Padia, and L. E. Whitesides, Jr., Mass. Inst. of Technol. School of Chemical Engineering Practice, Oak Ridge Station, Report ORNL-MIT-124 (1971).

26. W. R. Mixon and D. G. Thomas, ORNL, unpublished results.

27. J. M. Ganzer and A. K. Mehta, Mass. Inst. of Technol. School of Chemical Engineering Practice, Oak Ridge Station, Report ORNL-MIT-100 (1970).

28. I. Onal, C. S. Goff, and J. M. Aslam, Mass. Inst. of Technol. School of Chemical Engineering Practice, Oak Ridge Station, Report ORNL-MIT-104 (1970).

29. Mason-Rust, Office of Saline Water Research and Development Progress Report No. 134, U. S. Department of the Interior (1966). Foss Reservoir water, H. E. Podall, private communication.

30. S. B. Sachs and H. K. Lonsdale, J. Appl. Polymer Sci., 15, 797 (1971).

31. J. S. Johnson, Jr., K. A. Kraus, S. M. Fleming, H. D. Cochran, Jr., and J. J. Perona, Desalination, 5, 359 (1968).

32. U. Merten, H. K. Lonsdale, R. L. Riley, and K. D. Vos, Office of Saline Water Research and Development Progress Report No. 208, U. S. Department of the Interior (1966).

33. J. R. Kuppers, A. E. Marcinkowsky, K. A. Kraus, and J. S. Johnson, Jr., Separ. Sci., 2, 617 (1967).

34. A. J. Heman and J. C. Bronfenbrenner, M.I.T. School of Chemical Engineering Practice, Oak Ridge Station, Report ORNL-MIT-112 (1970).

35. J. Acosta, J. P. Jacks, and S. H. Rubin, M.I.T. School of Chemical Engineering Practice, Oak Ridge Station, Report ORNL-MIT-127 (1971).

36. C. A. Brandon, J. S. Johnson, R. E. Minturn, and J. J. Porter, to appear in Proceedings of New Orleans Meeting of the American Society of Mechanical Engineers, March 27-29, 1972.

37. J. D. Sheppard and D. G. Thomas, Applied Polymer Symposia, No. 13, "Membranes from Cellulose and Cellulose Derivatives," A. F. Turbak (Ed.), (1970), p. 121.

38. J. D. Sheppard and D. G. Thomas, Desalination, 8, 1 (1970).

39. J. A. Dahlheimer, D. G. Thomas, and K. A. Kraus, Ind. Eng. Chem., Proc. Des. Develop., 9, 565 (1970).

40. "Shriver Continuous Thickener," T. Shriver & Co., Inc., Harrison, N. J.; C. A. Jahreis, U. S. Patent 2,364,366 (1944).

41. A. I. Zhevnovatyi, Int. Chem. Eng., 4, 124 (1964).

42. C. W. Desaulniers and R. W. Hausslein, "Ultrafiltrative Dewatering of Spent Powdered Carbon," Federal Water Quality Administration WPCRS 17020DA03/70 (PB-197865) (1970).

43. H. A. Mahlman and K. A. Kraus, ORNL, unpublished results.

44. P. H. Wadia, K. A. Kraus, A. J. Shor, and L. Dresner, Oak Ridge National Laboratory Report, ORNL-4729 (1971).

POLYACRYLIC ACID COMPOSITE MEMBRANES FOR BRACKISH WATER DESALINATION

C. E. Milstead and M. Tagami
Gulf General Atomic Company
San Diego, California 92112

INTRODUCTION

In recent years, considerable interest has developed concerning the use of membranes containing fixed charged groups for water desalination applications[1-6]. The transport properties of these charged membranes have been indicated to be qualitatively consistent with the Donnan ion exclusion mechanism[7]; i.e., salt rejection is a decreasing function of concentration, divalent coions are rejected better than monovalent coions, and monovalent counterions are rejected better than divalent counterions. Earlier studies in this laboratory[8] involved the preparation and properties of polyacrylic acid (PAA) films cast directly onto a finely porous support membrane composed of mixed cellulose esters, cellulose nitrate-cellulose acetate (CN-CA). These studies showed that in dilute salt solutions the PAA-CN-CA composite membranes exhibited a desalination performance comparable to asymmetric CA membranes of the Loeb-Sourirajan type[9]. In general, (1) water flux and salt rejection were increasing functions of pressure, (2) rejection was a function of feed concentration and charge on the counterion, and (3) the membrane properties were relatively independent of pH above about pH 6, with increasing water flux at lower pH values accompanied by declining NaCl rejection. Typical long-term flux and rejection of 3000 ppm NaCl feeds at 1500 psi ranged from 50 gal/ft^2-day (gfd) and 80% rejection to 4-5 gfd and 99.5% rejection. The increased salt rejection, at the expense of reduced water flux, was obtained by heat-treating the membranes in air at 110°C for 1 to 3 hr.

As part of a recent study[10], several commercially available polyelectrolytes and several porous supports were investigated for possible use as membrane materials for sewage treatment. Among this group, PAA showed the most promise as a potential membrane for wastewater treatment, and membranes cast on polysulfone (PS) from a very high molecular weight PAA (MW > 10^6) prepared in this laboratory showed transport properties that were superior to those of membranes prepared from lower molecular weight commercial PAA (Rohm and Haas Acrysol A-5, MW < 300,000) cast on either PS, CN-CA, or CN.

In the present work, the preparation and properties of cast PAA membranes have been investigated in an effort to produce membranes with acceptable transport properties for brackish water desalination. Much of this effort was directed toward the preparation of a finely porous PS support material. In addition to being a good support for PAA, this material offers good strength and excellent resistance to hydrolysis not available with cellulosics.

EXPERIMENTAL

Membrane Preparation

High molecular weight PAA was prepared by stirring one mole of vacuum-distilled acrylic acid in 200 cc of distilled water with 0.125 g potassium persulfate as a catalyst for about 1 hr at 23°C. The highly viscous product was diluted by incremental additions of water and rolled for several days until a final concentration of about 3.5 wt-% solids was attained. The solution was used at this concentration for membrane casting. Viscosity measurements on dry PAA films dissolved in p-dioxane indicated a molecular weight > 10^6. Viscosity measurements on several batches of PAA prepared in this manner indicated a range of molecular weight of 3 to 10 x 10^6.

The PAA film was cast onto the surface of a sheet of porous support membrane (8 in. by 11 in.), taped to a glass plate, using a casting knife with the blade gap set at 10 mils. After overnight drying, the membranes were heat-treated in air before use. The thickness of the PAA layer was typically about 5 μ and the layer could not be removed from the support membrane. During reverse osmosis experiments, most of the PAA is washed from the support membrane surface in a short period of time, leaving a thin layer on the surface and in the pores. The resultant membrane is somewhat similar to the dynamically formed polyelectrolyte membranes being studied at Oak Ridge National Laboratory (see the paper by James S. Johnson, Jr. in this volume).

Reverse Osmosis Testing

The reverse osmosis experiments were performed in a flow system that consisted essentially of a solution reservoir (or sump) maintained at atmospheric pressure, a pair of 5-μ cartridge filters, a high-pressure pump with a rated capacity of 3.5 gal/min, an accumulator, two parallel sets of three reverse osmosis cells, and a back-pressure regulator, as well as gauges and temperature controls. Almost all metal parts were 316 stainless steel. The volume of solution in the system was typically 80 liters. The pump permitted feed solution to be circulated at a sufficiently high velocity (> 150 cm/sec) over the surface of the membrane to minimize concentration polarization phenomena at the water fluxes observed in this work. Pressure fluctuations were damped to less than ± 5 psi by an accumulator, and pressure drift during the measurements was essentially nil.

Each cell contained a rectangular membrane approximately 1 in. by 3 in. (20-cm^2 area). The membranes were tested in reverse osmosis experiments at 25° ± 1°C at an applied pressure of 800 psi. Solute concentrations in feed and product water were measured either by conductivity or by use of specific ion electrodes (Orion Research, Inc.) in conjunction with an Orion Model 801 digital pH/millivolt meter.

Polysulfone Porous Support

Preparation. In view of the desirable properties of PS, i.e., strength and hydrolytic stability, and the superior performance of PS substrates as compared with CN and CN-CA[10], considerable effort was directed toward the preparation of support membranes of this material with a surface porosity and water permeability that would provide an optimal substrate for the PAA. Initially, PS membranes were prepared according to the procedure developed at North Star Research and Development Institute[11] by dissolving PS (Union Carbide P-3500) in dimethylformamide (DMF) and adding a nonsolvent, methyl cellosolve (MC), in the weight ratio of 5 DMF:1 MC. A film was cast at 10 mils clearance of the doctor blade on a glass plate and immersed in water at room temperature to gel the membrane.

Subsequent modifications in the casting procedure markedly improved both surface appearance and membrane reproducibility from one casting to the next. The room-temperature water bath was replaced with a 15 wt-% NaCl quench bath to alter the rate of solvent removal, and the membranes were cast at a speed of 18 ft/min on 4-ft-long glass plates. Exposure of the wet film to room atmosphere was limited to about 1 sec before immersion of the plate into the salt water. The PS membranes were then soaked in distilled water to leach out residual NaCl prior to drying.

The water permeability of PS membranes was found to be strongly dependent on both the solids and MC content of the casting solutions. Water flux through the membranes ranged from about 55 to 10,000 gfd (with distilled water at 40 psi) over a solids content range of 6 to 19 wt-% PS. The water permeability decreased by about a factor of 3 as the MC content was reduced from 16 to 0 wt-%.

Over-all, the best support material was produced from a casting solution containing 12.5 wt-% PS and 12.5 wt-% MC in DMF. The dry membrane thickness was about 70 μ and the strength and water permeability of the membrane (about 4500 gfd at 40 psi) appeared to be desirable for optimum PAA membrane performance. In this study, this membrane is considered the "standard" support material for preparation of PAA membranes.

Porosity. Fig. 1 shows electron photomicrographs of the finely porous surfaces of the standard PS support and the support cast from a solution with the MC omitted. Also shown, for comparison, is a CN-CA support that has approximately the same water permeability (flux per unit pressure of approximately 8 x 10^2 g/cm^2-sec-atm) as the standard PS. The difference between the surface porosity of these three membranes is striking. (The bulk porosity of all three membranes is approximately 70%.) A quantitative analysis of the pore-size distribution yields a surface porosity of 13.9% with an average of 66 pores/square micron for the standard polysulfone, 4.2% and 21 pores/square micron for polysulfone without methyl cellosolve, and 22.5% and 33 pores/square micron for CN-CA.

A bar chart showing the pore-size distributing for the standard polysulfone and CN-CA supports is shown in Fig. 2. It is readily seen that the CN-CA has much larger pores, with an area-median pore diameter of about 1500 Å, as compared with 600 Å for the polysulfone. The polysulfone support shows a very narrow range of pore diameters with all pores aetween 200 and 1200 Å. The range of pore diameters for the polysulfone support prepared without methyl cellosolve was also 200 to 1200 Å with a median of 600 Å.

RESULTS OF PAA MEMBRANE EVALUATION

Effect of Postcasting Heat Treatment

The effect of postcasting heat treatment on membrane properties was studied using PAA-PS membranes annealed in air over a

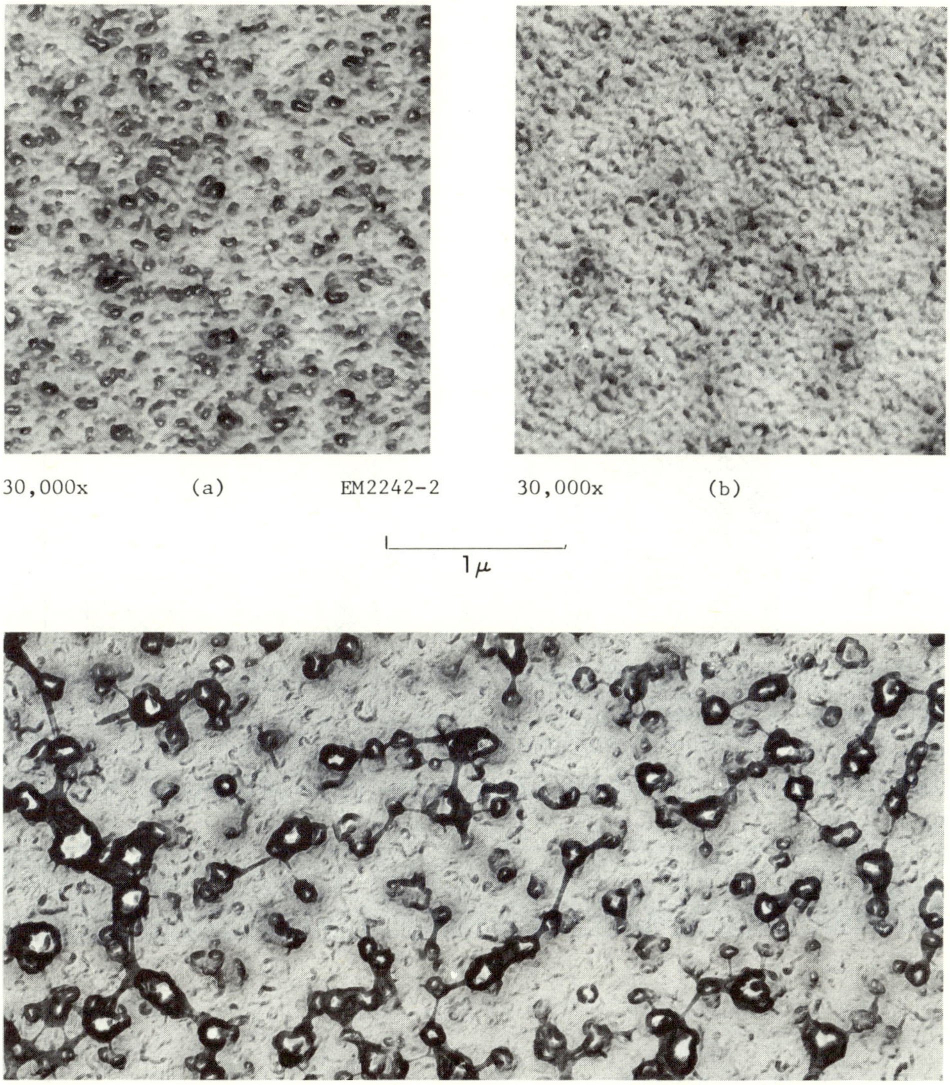

Fig. 1 - Electron Photomicrographs of Palladium-Shadowed Carbon Replicas of the Finely Porous Surface of (a) Standard Polysulfone Support Membrane, (b) Polysulfone Support Membrane Cast From a Solution With Methyl Cellosolve Omitted, and (c) CN-CA Porous Support

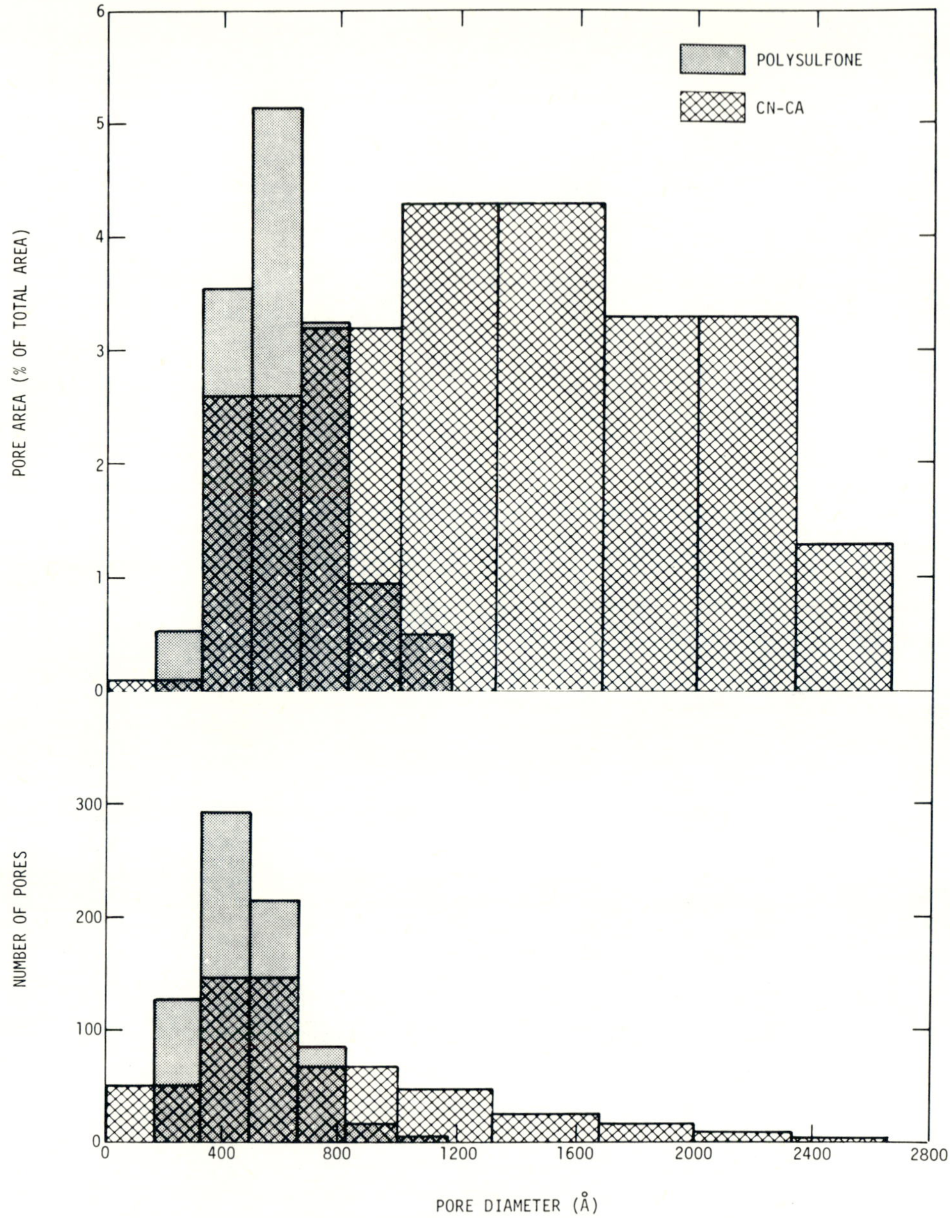

Fig. 2 - Pore Size Distribution (Number per Square Micron) and Pore Area of the Finely Porous Surface of CN-CA and Standard PS Support Membranes temperature range of 110° to 150°C. Flux and rejection characteristics of the membranes were evaluated in reverse osmosis tests at 800 psi using a 3000 ppm NaCl feed. These tests showed that little improvement in membrane performance was obtained as annealing conditions were increased beyond 30 min heat treatment at temperatures up to 120°C, with increasing rejection accompanied by decreasing water flux between 120° and 135°C, and finally severe flux decline with little improvement in rejection above 135°C. Optimum desalination performance for the "standard brackish water membrane," designated PAA-PS 125-60, resulted from a 1-hr heat treatment at 125° ± 1°C. Flux and rejection data taken after about 5 days operation are shown in Fig. 3. A curve showing the transport properties of asymmetric CA membranes of the Loeb-Sourirajan type under the same operating conditions is presented for comparison. The solid circles represent PAA-PS membranes where the PS support was prepared with no methyl cellosolve in the PS casting solution. It is readily noted that these membranes give substantially higher rejections but lower water fluxes than membranes prepared from the standard PS casting solution. The superiority of the PAA-PS membranes over the CA membranes is clearly evident in the high-flux region. CA membranes with water fluxes in the range of 50-60 gal/ft^2-day give salt rejections of only 45-60% as compared to 75-80% for the PAA-PS membranes.

The reproducibility within a set of membranes is generally good; however, variations are frequently noted from one test to another. Generally, however, the water flux vs. rejection data fall along a broad curve, i.e., higher flux is accompanied by lower rejection. The variability in the membrane performance is attributed more to variations in the test procedure than to the properties of the membranes. As discussed below, fouling of PAA membranes by test loop contaminants is a major problem. Even in the case of nominally identical tests, the level of contaminants may vary slightly from one test to another, primarily influencing the observed water flux. With increasing flux, concentration polarization becomes more severe, resulting in lower salt rejections. Thus, it is a combination of membrane fouling and concentration polarization that is believed to be the cause of the variability in membrane performance rather than lack of reproducibility in membrane preparation.

Change of Membrane Properties with Time of Operation

Initial evaluations of membrane performance were made using data obtained after overnight operation, nominally about 20 hr. Several longer-term tests were conducted to observe changes in flux and rejection properties of the standard PAA-PS 125-60 membranes at 800 psi and 25°C with 1% and 0.3% NaCl feeds. Results of these tests are shown in Fig. 4, with each curve representing

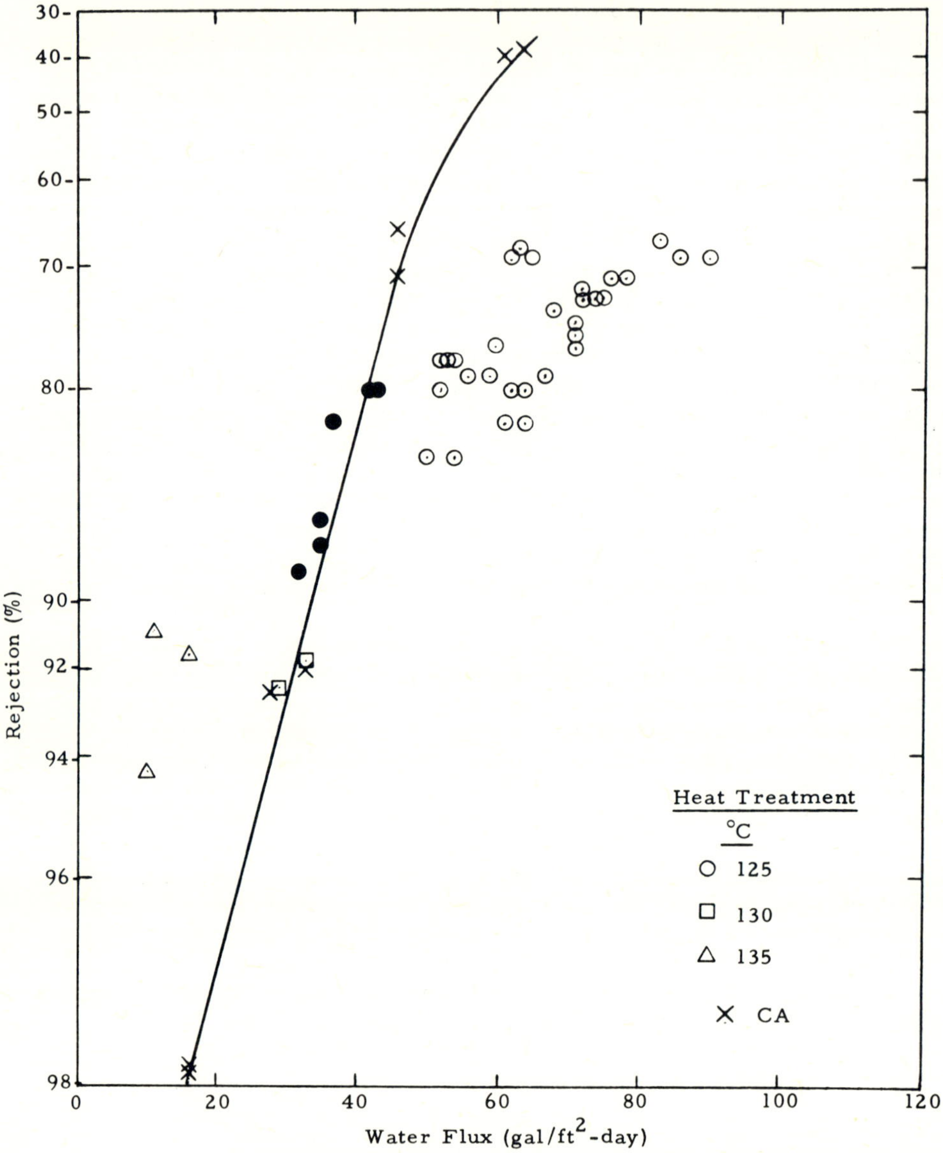

Fig. 3 - Water Flux vs NaCl Rejection for PAA-PS and Asymmetric CA Membranes at 800 psi and 25°C With 3000 ppm NaCl (5-Day Data)

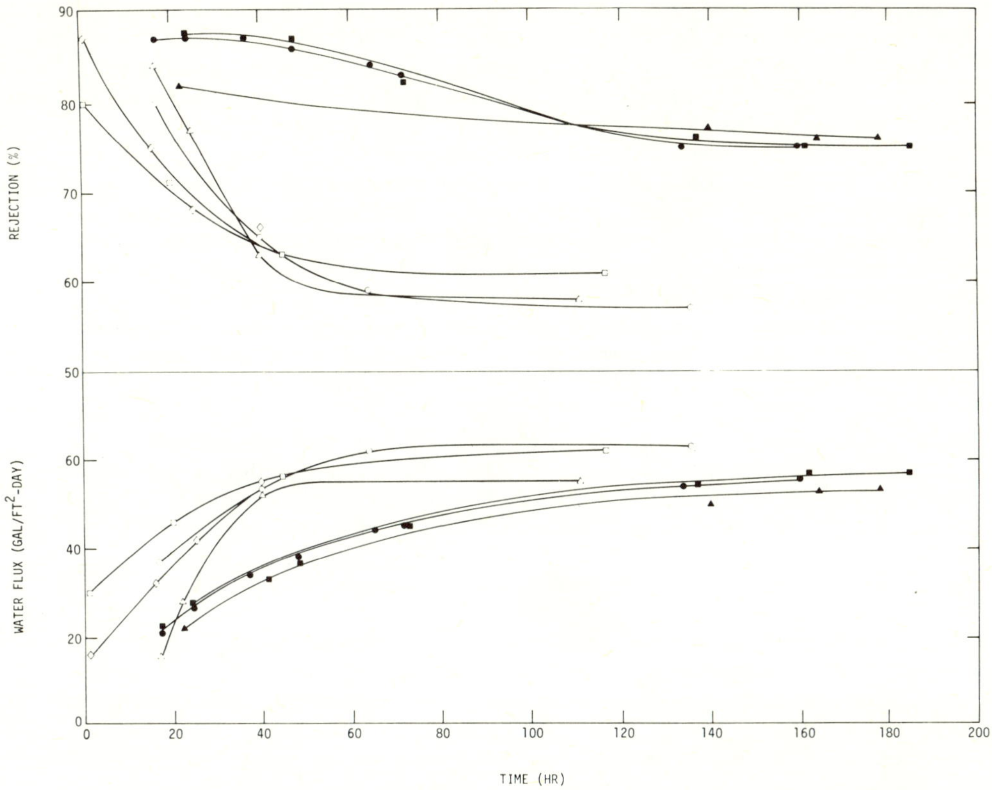

Fig. 4 - Change in Flux and Rejection of Standard PAA-PS Membranes with Time of Operation at 800 psi and 25°C with 1% (Open Symbols) and 0.3% (Closed Symbols) NaCl

the average value measured for three membrane samples. Reproducibility in performance among the sets of membranes tested at each feed concentration was generally quite good. Although the rejection curves look quite dissimilar, a plot of (100-% Rejection) would show that the NaCl concentration in the product water increased by about a factor of 2 in each case before a constant value was established. These results clearly show that about 4 to 5 days of operation are required before "typical" membrane performance characteristics can be determined. This is in contrast to the behavior of CA membranes, for which the typical performance level is attained within a few hours.

The behavior of the PAA-PS membranes during the first few days of operation is attributed to the removal of the PAA layer from the membrane surface. In some tests in the present study, the PAA membranes were initially flushed with distilled water (not returning the water to the sump after passing over the membranes) for up to 1 hr before placing the system in the recirculation mode of operation and adding NaCl to the sump. In other tests, the membranes were immediately exposed to the recirculating NaCl feed. No significant difference in membrane performance was noted between the two methods of membrane treatment, suggesting that the presence of PAA in the feedwater has no appreciable effect on membrane properties. A simple calculation based on the estimated thickness of the PAA layer, the membrane sample surface area, and the volume of water in the test loop indicates a maximum concentration of about 0.6 ppm PAA in the sump during each test, assuming complete removal of all PAA from the support membrane.

Additional longer tests were conducted for periods up to 710 hr to observe the performance of PAA membranes under a variety of conditions and with solutes other than NaCl. In general, flushing the system with distilled water prior to testing with a different solute results in a transient effect such that several days are required before "typical" membrane performance is restored.

Four long-term tests were conducted to determine the flux decline of PAA-PS membranes with time. Since these membranes are easily susceptible to fouling from loop contaminants, the results of these measurements are presented merely to indicate the approximate degree of flux decline. Three of these tests were conducted with a 3000 ppm NaCl feed on membranes operating at 800 psi; the fourth test was carried out using synthetic Coalinga, California water. Flux-decline slopes [determined from a plot of log (water flux) versus log (time)] of -0.074, -0.15, and -0.28 were observed in these tests using the NaCl feed. A flux-decline slope of -0.15 was observed using the Coalinga water. These values are somewhat higher than the -0.086 flux decline measured under the same conditions with asymmetric CA membranes of comparable desalination characteristics, i.e., 72% rejection and 46 gfd (heat-treated in water for 30 min at 60°C). The wide range of the flux-decline slopes for the PAA membranes suggests that the observed performance may have been influenced by fouling contaminants in the test loop. The charged surface of the PAA membrane would be expected to make these membranes more susceptible to fouling than CA, which is generally considered to be essentially uncharged.

Solute Rejection

The performance of PAA-PS 125-60 membranes was evaluated at 800 psi using several selected sodium salts at 0.05 N concentrations (Table 1). At the conclusion of each solute test, the system was flushed with distilled water for 1-1/2 to 2 hr to ensure com-

TABLE 1

REJECTION OF SODIUM SALTS BY PAA-PS MEMBRANES
(800 PSI, 25°C, 0.05 N FEED)

Solute	Length of Run (hr)	Water Flux (gfd)	Rejection (%)
NaCl	113	70	74
Na_2SO_4	70	61	97.8
$NaNO_3$	71	70	82
$NaHCO_3$	21	107	86
$NaC_2H_2O_3$	70	79	80
$Na_2Cr_2O_7$	46	36	68

plete removal of the previous solute from the system before the new solute was added. With the exception of $NaHCO_3$ (pH 9.3) and $Na_2Cr_2O_7$ (pH 4.9), all solutes were tested at pH 6.5 to 7. The unusually high water flux with $NaHCO_3$ is the result of the shortness of the run. Subsequent tests with a $NaHCO_3$ feed showed that a reduction in water flux occurs with time of operation, stabilizing after about 72 hours.

The low $Na_2Cr_2O_7$ rejection, which would not be expected for the divalent anion $Cr_2O_7^{=}$, suggests that a significant fraction of the solute was probably present as $HCrO_4^-$ or even H_2CrO_4. With increasing pH the average charge on the chromate species increases. The rejection of Cr^{VI} ions by CA membranes has been shown to be dependent on pH, with the lowest rejection occurring in the acidic region[8].

Qualitative agreement with the ion-exclusion mechanism and dependence of rejection on ionic charge is illustrated by data presented in Table 2. These results were obtained on a single set of PAA-PS 125-60 membranes at 800 psi with a 0.05 N feed concentration. The divalent coion ($SO_4^{=}$) is rejected much better than the monovalent coion (Cl^-) and, in turn, the rejection of the monovalent counterion (Na^+) is higher than that of the divalent counterion (Ca^{++}).

The performance of PAA-PS 125-60 membranes with a synthetic mixture representative of a naturally occurring brackish water (Coalinga) was evaluated at 800 psi. These membranes were initially tested for 166 hr with 3000 ppm NaCl, giving 70% rejection

TABLE 2

SOLUTE REJECTION BY PAA-PS MEMBRANES
(800 PSI, 25°C, 0.05 N FEED)

Solute	Length of Run (hr)	Water Flux (gfd)	Rejection (%)
NaCl	264	55	76
Na_2SO_4	96	44	98.6
$CaCl_2$	72	49	40

---

with a water flux of 69 gfd. After 209 hr with a Coalinga water feed (pH 8.3), the membranes showed 70% conductivity, 50% chloride, and 94% divalent cation rejections with a water flux of 70 gfd.

Stability of PAA-PS Membranes

Some of the advantages expected of PAA-PS membranes over asymmetric CA membranes for desalination were the resistance of these membranes to hydrolysis with operation at extremes in pH, and to other operational fluctuations such as uncontrolled shutdowns (caused, for example, by power failure). An example of such stability was shown during a 710-hr test at 800 psi with a 3000 ppm NaCl feed during which both PAA and CA membranes with comparable transport properties were subjected to (1) 473 hr at approximately pH 7, (2) membrane fouling by test loop contaminants which reduced the water flux of the PAA membranes to less than 12 gfd, (3) exposure to 1.5% oxalic acid (pH 3.8) for 30 min, (4) cycling of the system from pH 2.8 to 11.5, (5) depressurization by turning the pump off, followed by rapid repressurization to 800 psi, and (6) further cycling of pH from approximately 3 to 11 over a 5 day period. At the conclusion of this test, with the pH restored to approximately 7, the PAA membranes showed a rejection of 69% with a water flux of 70 gfd, in contrast to a 15% rejection and 26 gfd for the CA membranes.

Although the PAA membranes were clearly more resistant to adverse operational conditions than the CA membranes, the CA membranes were considerably less susceptible to fouling from contaminants in the test loop than the PAA membranes, the latter showing severe flux decline at various times during the cycling of feed pH.

Membrane Fouling

Throughout the program to evaluate the transport properties of PAA membranes, fouling was a problem due to contaminants in the reverse osmosis test loop. Membrane fouling was revealed by examination of the membrane samples at the conclusion of the test period. Frequently the membranes were coated with a light-brown precipitate resembling iron hydroxide, although at other times the fouling layer was almost colorless. It was also noted that the degree of fouling was inconsistent from one test period to another; however, the level of fouling appeared to be uniform within a set of membranes. Spectrographic analysis of deposits on the membrane surface after reverse osmosis testing showed the major contaminant to be iron, with appreciable quantities of copper, chromium, silicon, nickel, and molybdenum, all corrosion products of stainless steel.

Fouling by organic contaminants was also evidenced. Tests showed that very rapid fouling of PAA membranes results from leachable components from the 5-μ filter cartridge installed in the test loop. CA membranes in the same tests were unaffected by these contaminants in the feed. It was determined that about 6 hr of flushing of the filters (at a rate of 210 gal/hr) with distilled water was required to reduce the level of leachable organics to the point where rapid fouling was not observed.

On the other hand, on several occasions a surprising resistance of PAA membranes to fouling and flux decline has been observed. For example, experiments with spent sulfite pulp liquor at 1.2% solids concentration, which would be expected to produce severe fouling, gave a flux decline of only 43% (from 37 to 21 gfd) in 340 hr at 800 psi, whereas the rejection of solids, conductivity, and total organic carbon remained in the 90% to 95% range.

CONCLUSIONS

PAA-PS membranes have exhibited flux and rejection characteristics superior to the high flux asymmetric CA membranes under brackish water desalination operating pressures. Long-term reverse osmosis measurements on PAA-PS membranes have shown water fluxes greater than 50 gfd with rejections of about 75% from a 3000 ppm NaCl feed at 800 psi. PAA membranes have shown a decided tendency to foul, due to both organic and inorganic contaminants in the feed. Further study of the factors affecting fouling of these membranes is in progress, as well as additional characterization of membrane performance in mixed salt systems and on naturally occurring brackish waters.

ACKNOWLEDGMENT

The authors wish to thank the Office of Saline Water, U. S. Department of the Interior, for the support of this work under Contracts 14-01-0001-1778 and 14-30-2609.

REFERENCES

1. W. H. Baldwin, D. L. Holcomb, and J. S. Johnson, J. Polymer Sci., 3, Part A, 833 (1965).

2. S. B. Sachs, E. Hoffer, and O. Kedem, Office of Saline Water Research and Development Progress Report No. 324 (Part II), May 1968.

3. A. E. Marcinkowsky, K. A. Kraus, H. O. Phillips, J. S. Johnson, Jr., and A. J. Shor, J. Am. Chem. Soc., 88, 5744 (1966).

4. K. A. Kraus, A. J. Shor and J. S. Johnson, Jr., Desalination, 1, 225 (1966); Desalination, 2, 243 (1967).

5. A. S. Hoffman, M. Modell, and P. Pan, J. Appl. Polymer Sci., 13, 2223 (1969); J. Appl. Polymer Sci., 14, 285 (1970).

6. S. B. Sachs, W. H. Baldwin, and J. S. Johnson, Desalination, 6, 215 (1969).

7. F. G. Donnan, Z. Electrochem., 17, 572 (1911).

8. H. K. Lonsdale, R. L. Riley, C. E. Milstead, L. D. LaGrange, A. S. Douglas, and S. B. Sachs, Office of Saline Water Research and Development Progress Report No. 577, March 1970.

9. S. Loeb and S. Sourirajan, Advan. Chem. Series, 38, 117 (1963).

10. A. S. Douglas, M. Tagami, and C. E. Milstead, Federal Water Quality Administration, Water Pollution Control Research Series ORD-17040EF006/70, (1970).

11. L. T. Rozelle, J. E. Cadotte, R. D. Corneliussen, and E. E. Erickson, Office of Saline Water Research and Development Progress Report No. 359, October 1968.

TUBULAR ULTRATHIN CELLULOSE ACETATE MEMBRANES FOR WATER DESALINATION

L. T. Rozelle, J. E. Cadotte, A. J. Senechal, W. L. King, and B. R. Nelson

North Star Research and Development Institute
Minneapolis, Minnesota 55406

INTRODUCTION

The polymer used most often as a membrane for water desalination by reverse osmosis has been cellulose acetate. Reid and Breton[1] originally showed that this material had excellent potential as a reverse osmosis membrane. Loeb and Sourirajan[2] later developed the asymmetric cellulose acetate membrane that is used most extensively at the present time in either a flat sheet or tubular form.

Composite ultrathin membranes provide a promising new approach for effective water desalination through high flux, low flux decline, and adequate salt rejection. The composite ultrathin membranes developed in this laboratory were originally prepared by pouring a solution of cellulose acetate in a suitable solvent such as cyclohexanone onto water and allowing it to spread over the water surface. Evaporation of the solvent left behind an ultrathin film of polymer. Flat sheets of ultrathin cellulose acetate, with thicknesses under 2000 angstroms, were placed upon a suitable microporous support film, e.g., polysulfone or Millipore Corp. VFWP filters, and were shown to exhibit reverse osmosis performance equivalent to asymmetric cellulose acetate membranes[3]. Other polysaccharide derivatives have also been investigated, and some have shown considerable promise as reverse osmosis desalination barriers[4,5].

A simple process for casting ultrathin membranes directly into commercially available fiberglass reverse osmosis tubes containing a microporous support has been developed. The composite desalination membrane discussed in this paper consists of an ultrathin cellulose acetate membrane, which may be as thin as 200 angstroms, supported on a microporous polysulfone film. The polysulfone support film has high chemical resistance and strength and is particularly applicable for tubular in situ replacement of the ultrathin membrane.

This paper describes the preparation, properties, and reverse osmosis performance of tubular ultrathin cellulose acetate membrane-polysulfone support composites. The reverse osmosis performance includes long-term laboratory and on-site testing of brackish waters, wet-dry performance, and in situ regeneration studies.

MATERIALS AND EXPERIMENTAL TECHNIQUES

Rigid Tube Modification

Fiberglass reverse osmosis tubes with an inside diameter of 1/2-inch were obtained in lengths of 25-1/2 inches (Abcor, Inc.) and appropriate fittings attached. The tube was filled with a 12-1/2 percent solution of Dow PVC 111-4 in dimethylformamide (DMF). The PVC solution was drained, a 0.520-inch diameter aluminum bob was passed through the tube, and the tube was immersed in deionized water to gel the PVC coating. The tube was air-dried prior to the installation of the membrane-support composite. This procedure yielded tubes with a relatively narrow pore size distribution.

Polysulfone Support

Casting. The polysulfone support lining was prepared by filling a 0.555-inch ID stainless steel tube with a 15-percent solution of Union Carbide P3500 polysulfone in DMF. The tube was drained and a 0.547-inch diameter aluminum bob was passed through the tube to provide a uniform film of casting solution on the inside wall. The tube was then lowered mechanically in a smooth, continuous motion into a one-percent DMF solution which gelled the polysulfone.

Properties. Microporous polysulfone films are used as the supports for the ultrathin desalination membrane because of their unique combination of properties. These include: (1) ease of preparation and handling; (2) resistance to pressure compaction; (3) resistance to biological and chemical degradation; and (4) ability to be air-dired without significant loss of water flux.

Electron micrographs of the polysulfone supports showed that the air-dried or glossy surface had a much finer porosity than did

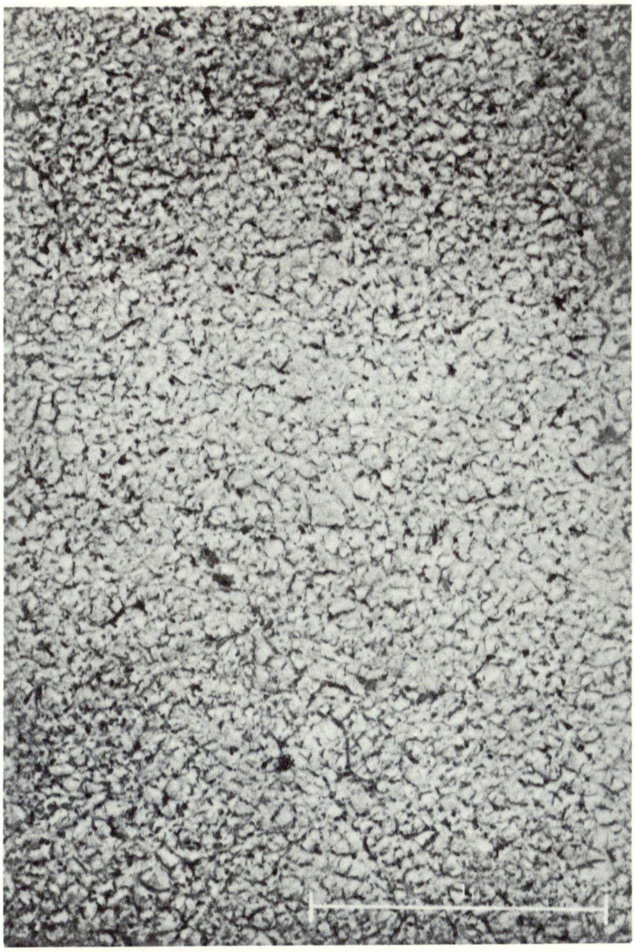

Fig. 1 - Electron Micrograph of the Glossy Surface of a Flat-Cast Polysulfone Support Film. (Magnification - 50,000 x, from a Carbon-Palladium Replica)

the contact or dull surface. Fig. 1 is an electron micrograph of the glossy side of the film, flat-cast on a glass plate. The surface contained both pores and connecting channels with pore diameters and channel widths generally under 200 Å. Approximately 16 percent of the surface area of this sample was open. Larger pore sizes (up to 300 Å) have been observed for the tube-cast polysulfone.

Fig. 2 is an electron micrograph showing the dull surface of the flat-cast film. On this surface, the pores are much larger, ranging from 3000 to 20,000 Å in diameter. (The long, light areas in this micrograph were caused by the fracturing of the replica

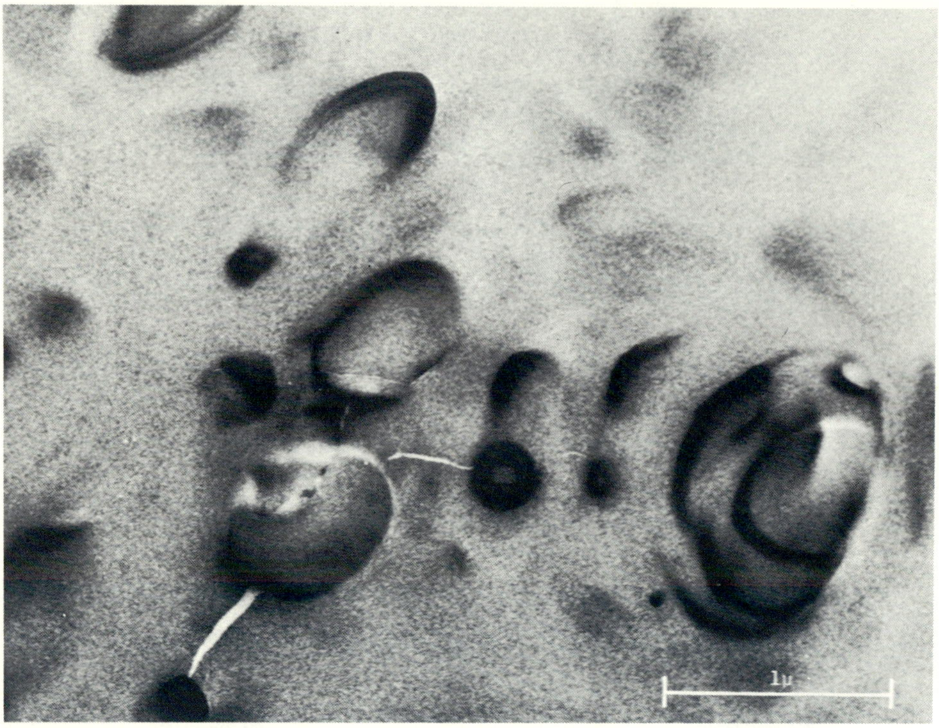

Fig. 2 - Electron Micrograph of the Dull Surface of a Flat-Cast Polysulfone Support Film. (Magnification - 40,000 x, from a Carbon-Platinum Replica)

during the removal of the replicating plastic.) Small pores which appear to lead into the film can be noted at the base of many of the larger pores. The fine porous structure on the glossy side of the support film is believed to extend only a short distance into the film. There, the porous structure probably expands to form enlarged channels which lead to the large pores on the dull surface of the film.

Polysulfone supports have been characterized with respect to tensile properties, water permeability, and creep. A comparison of these properties with those of Millipore VFWP filters are given in Table 1. The tensile strength of the 1.8 mil-thick microporous polysulfone films was approximately one-half that of the Millipore film. Also, a significantly higher elongation and lower modulus of elasticity was observed for the polysulfone film compared to the Millipore film. These Millipore filters are known to consist of mixed cellulose esters[6] which exhibit stress-strain characteristics similar to cellulose acetate[7]. Dense (non-porous) polysulfone films have a greater tensile strength than corresponding dense cellulose acetate films[7,8], and the data in Table 1 suggest

TABLE 1

COMPARISON OF PROPERTIES OF MICROPOROUS POLYSULFONE FILMS* WITH MILLIPORE VFWP FILTER

Property	Polysulfone	Millipore VFWP
Tensile Strength	950 psi	2050 psi
Elongation	25%	8%
Modulus of Elasticity	28×10^3 psi	42×10^3 psi
Water Transport at 13.5 psi	0.84 ml/cm^2-min	0.40 ml/cm^2-min
Creep at 1000 Hours	2.1% (2.3%)**	3.2%

*Flat cast from 15 percent DMF solution

**Tube cast

---

that these microporous polysulfone films have a higher bulk porosity than that of the Millipore film. This higher degree of porosity is supported by the higher water transport rate of the polysulfone film.

The creep properties of tube- and flat-cast polysulfone and the Millipore filters are presented graphically in Fig. 3. The initial deformation of both microporous polysulfone films was greater than that of the Millipore filter. Because the initial

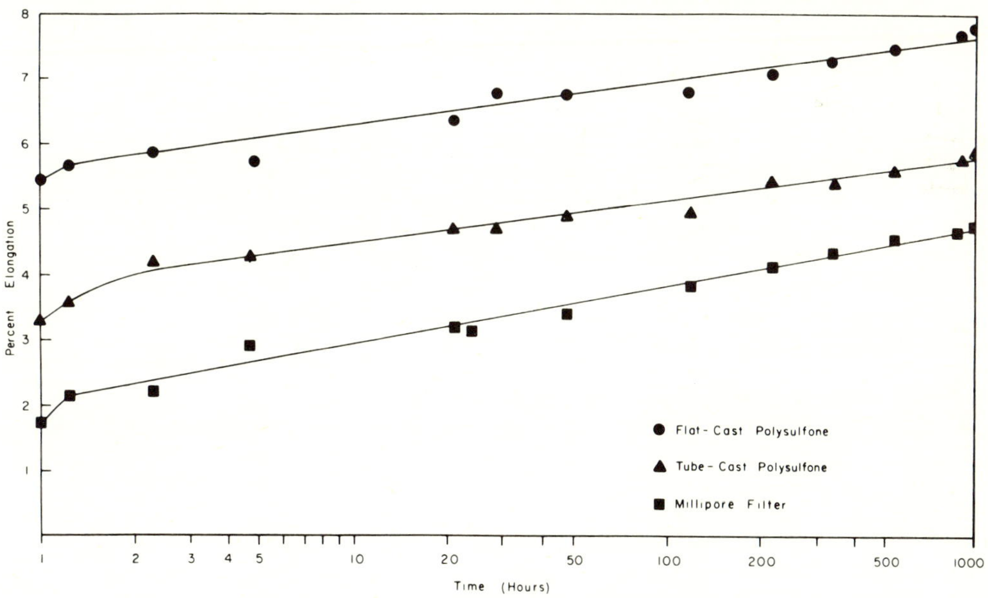

Fig. 3 - Creep of Microporous Support Films Under Stress of 200 psi

deformation of dense cellulose acetate films is significantly higher than that of dense polysulfone films[7,8], this observed reversal further suggests that the microporous polysulfone films contain considerably more open volume than the Millipore filter. In addition, for these particular cases the flat-cast polysulfone was less dense and had more open volume than the tube-cast polysulfone.

The creep rates after 1000 hours of testing (from Fig. 3, listed in Table 1), show that the polysulfone exhibited about two-thirds the creep rate of the Millipore filter. The lower creep for the polysulfone support films than for the Millipore filter indicates that the long-term flux decline during reverse osmosis should be less for a system using polysulfone for the membrane support than it would for a system using a Millipore filter or cellulose acetate support.

Ultrathin Membranes

Casting. The ultrathin cellulose acetate (Eastman Chemical Co. E398-10) membrane was cast on a wet polysulfone support liner while the liner was still in the steel casting tube. The tube was filled with a dilute aqueous polyacrylamide (Reten 210) solution, on the surface of which was placed a small amount of casting solution consisting of a 2.5-percent solution of cellulose acetate (E398-10) in a 7:1 ethyl acetate : acetic anhydride solvent. The tube was then drained at a given rate (draw rate) controlled by a microvalve and flowmeter. A draw rate of 0.45 inch per second yielded a membrane with a thickness of about 200 Å.

The wet polysulfone support-membrane composite was then pulled from the stainless steel casting tube and air-dried. The dried composite was attached to a dowel and drawn into the fiberglass tubes. The ends of the composite were sealed to the tube wall with Shell Chemical Co. Thermoplastic 125 (or rubber grommets if modules were used). Fig. 4 is a schematic representation of the membrane casting procedure. As Fig. 4 indicates, the ultrathin membrane can also be cast directly into the fiberglass tube containing the polysulfone support liner sealed inside.

The membrane thickness was measured using a Reichert metallograph equipped with a Nomarski polarization interferometer.

Casting Variables. The reverse osmosis performance of ultrathin cellulose acetate membranes is dependent, to a great extent, on the thickness of the membrane and the degree of annealing. It has been shown that the water flux is inversely proportional to thickness for membranes 400 Å to 1000 Å thick[9]. Salt rejection is not significantly affected by membrane thickness, as would be

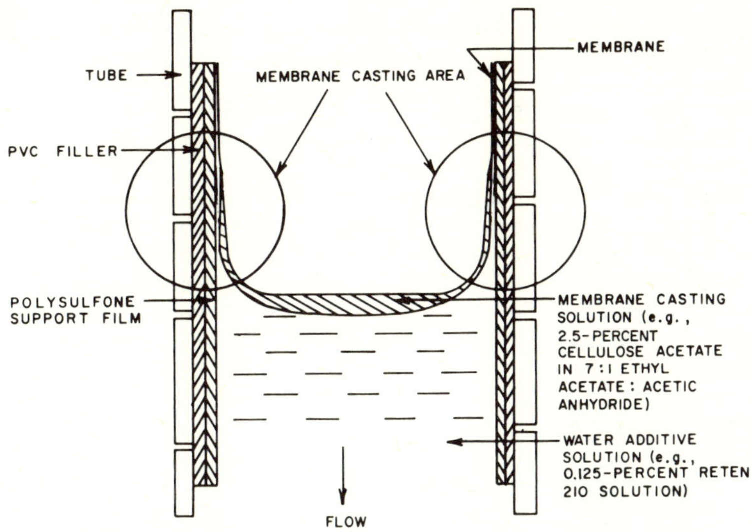

Fig. 4 - Schematic of Ultrathin Membrane Tubular Casting Procedure

expected from the solution-diffusion model for mass transport by reverse osmosis[10]. The concentration of the cellulose acetate in the casting solution and the rate of membrane formation (draw rate) have been shown to be major variables in controlling the thickness and hence the flux of tube-cast ultrathin membranes.

Membrane thickness is directly proportional, and water flux is inversely proportional, to the concentration of the casting solution. For example, ultrathin membranes prepared under a given set of conditions from casting solutions containing from 2.5- to 7.5-percent cellulose acetate, varied in thickness from 400 Å to 1300 Å respectively, and exhibited corresponding water fluxes which varied from 28 gallons per square foot of membrane per day (gfd) down to 10 gfd.

Membrane thickness is inversely related, and water flux is directly related, to the membrane draw rate. For example, varying the draw rate from 0.10 to 2.0 inches per second resulted in membrane thicknesses varying from 1000 Å to 200 Å, respectively. The corresponding water flux varied from 9 to 40 gfd.

Another variable which controls thickness is the viscosity of the polyacrylamide (Reten) solution used to cast the membranes [9]. Increasing the viscosity of the solution from 2 to 10 centipoise caused an increase in the water flux from 12 to 22 gfd. Thickness measurements showed that this increase in viscosity resulted in a decrease in the thickness of the membranes. This effect is attributed to the ability of the higher viscosity solution to provide a smoother flow of polymer solution down the tube wall.

Annealing the membrane by drying at room temperature or at higher temperatures, effectively decreases the water flux and increases the salt rejection. The higher the annealing temperature, the lower the flux and greater the salt rejection. It was found previously that air-drying a membrane was equivalent to annealing at about 60°C[3]. Air-drying of the membrane was effective in increasing the salt rejection to over 95 percent without substantial loss in water flux. By controlling the thickness of the membrane, it was possible to prepare high rejecting membranes exhibiting water fluxes of about 50 gfd.

Reverse Osmosis Measurements

The reverse osmosis measurements were carried out in a closed-loop high pressure system with a constant flow of brine through the system. Single-tube reverse osmosis measurements were made under brackish water test conditions, i.e., 1.0-percent sodium chloride brine at 800 psi and 25°C. The brine flow rate was 2800 ml per min. Product water flux was determined gravimetrically and the amount of salt rejected by the membrane was determined with a calibrated conductivity cell. With a one-percent saline solution, 95% salt rejection was necessary to achieve product water with 500 ppm salt or less, the standard maximum concentration allowable in drinking water[11].

RESULTS AND DISCUSSION

Short term reverse osmosis testing has shown that the ultrathin membrane tubular composites have considerable promise for reverse osmosis applications[5]. In addition to high water fluxes and rejections, coupled with low flux decline, the feasibility of in situ membrane regeneration was indicated. The possibility of using the composite for applications that involve drying has been shown to be feasible because the composite is normally dried for annealing purposes. The results of the ultrathin membrane tubular composite studies given below include: long term reverse osmosis performance, both in the laboratory and on-site; initial in situ membrane regeneration studies; and wet-dry cycling properties.

Long-Term Testing

Laboratory Results. Typical long term reverse osmosis performance of single tubes two feet in length are given in Table 2. For membrane No. 1 after an initial flux of 23.6 gfd, the flux

TABLE 2

LONG TERM REVERSE OSMOSIS TESTS OF ULTRATHIN CELLULOSE ACETATE MEMBRANES IN TUBES
Test Conditions: 25°C, 800 psi, 1% NaCl, 2.8 liters/min flow rate

Membrane Test Number	Cellulose Acetate Membrane Casting Conditions*		Reverse Osmosis Test Data						Cleaning Procedure
	Casting Solution Concentration (%)	Draw Rate (in/sec)	Test Duration (hr)	Water Flux (gfd)**		Salt Rejection (%)		
				Initial	Final	Initial	Final	
1	5.0	0.5	3000	23.6	20.0	98.2	97.2	5% $NaHSO_3$
								5% Citric acid
2	5.0	0.7	3000	24.5	22.6	97.6	96.4	2% Citric acid
3	2.5	0.25	4400	24.4	21.0	98.1	97.0	5% Citric acid
								2% Citric acid
4***	1.25	0.25	1500	43.0	31.2	94.5	94.4	2% Citric acid

*Membranes 1-3 were approximately 400 ± 50 Å thick and membrane 4 was approximately 200 ± 50 Å thick.
**Final water fluxes were measured 500 hours after the last citric acid wash.
***Test 4 used American Standard plastic ball turbulence promotors.

decreased to 18.6 gfd after 1000 hours of testing. At that time, the tube was washed with a five-percent sodium bisulfite solution in an attempt to remove iron oxide deposits on the membrane. This resulted in an increase in flux to 20.5 gfd. After 1320 and again after 1800 hours of testing, the tube was washed with a five-percent citric acid solution, which increased the flux to 21.5 gfd. These three cleaning operations caused a slight progressive decline in the salt rejecting properties of the membrane. The salt rejection prior to cleaning was 98.2 percent, and after three cleanings, was 97.0 percent.

Similar results were obtained during the testing of membranes 2 and 3. Membrane 2, with an initial flux of 24.5 gfd, exhibited a flux of 22.0 gfd at 1000 hours. After two washes with two-percent citric acid solutions at 1700 and 2500 hours, the flux increased to 24.1 gfd and 23.8 gfd, respectively. The rejection prior to the two cleanings was 97.6 percent, and after each cleaning was about 96.5 percent.

The reverse osmosis performance of membrane 3 of Table 2 is plotted in Fig. 5. The flux decreased rapidly for about four days and then more slowly until, after 1000 hours of testing, it had declined to 21 gfd. At that time, the tube was washed with a citric acid solution and the flux increased to 25 gfd. In this particular test subsequent washes did not result in substantial flux increases; however, the flux decline was very low after 2600 hours.

Thinner, higher flux membranes were cast to have an average thickness of 200 Å. Membrane 4 in Table 2 exhibited an initial water flux of 43 gfd, which declined to 30 gfd after 1500 hours of testing. The highest salt rejection obtained using the higher flux membrane was 94 percent. This was probably due to the increased concentration polarization caused by the higher water flow rate through the membrane. At 600 hours the membrane was cleaned with a two-percent solution of citric acid with a negligible effect on the water flux.

In these laboratory tests the flux decline during the first 200 hours of testing apparently was due largely to deformation and compaction of the composite. After 200 hours, fouling appeared to be the major factor in flux decline. In general, the over-all flux declines were low. For the six-month test (Fig. 5) the flux declined about 14 percent from the initial value.

The over-all flux decline of the higher flux membrane (membrane 4, Table 2) was considerably higher than the other membranes. The citric acid wash was not as effective in removing iron oxide from this high flux membrane and thus did not increase the water flux significantly. The higher flow may have caused relatively

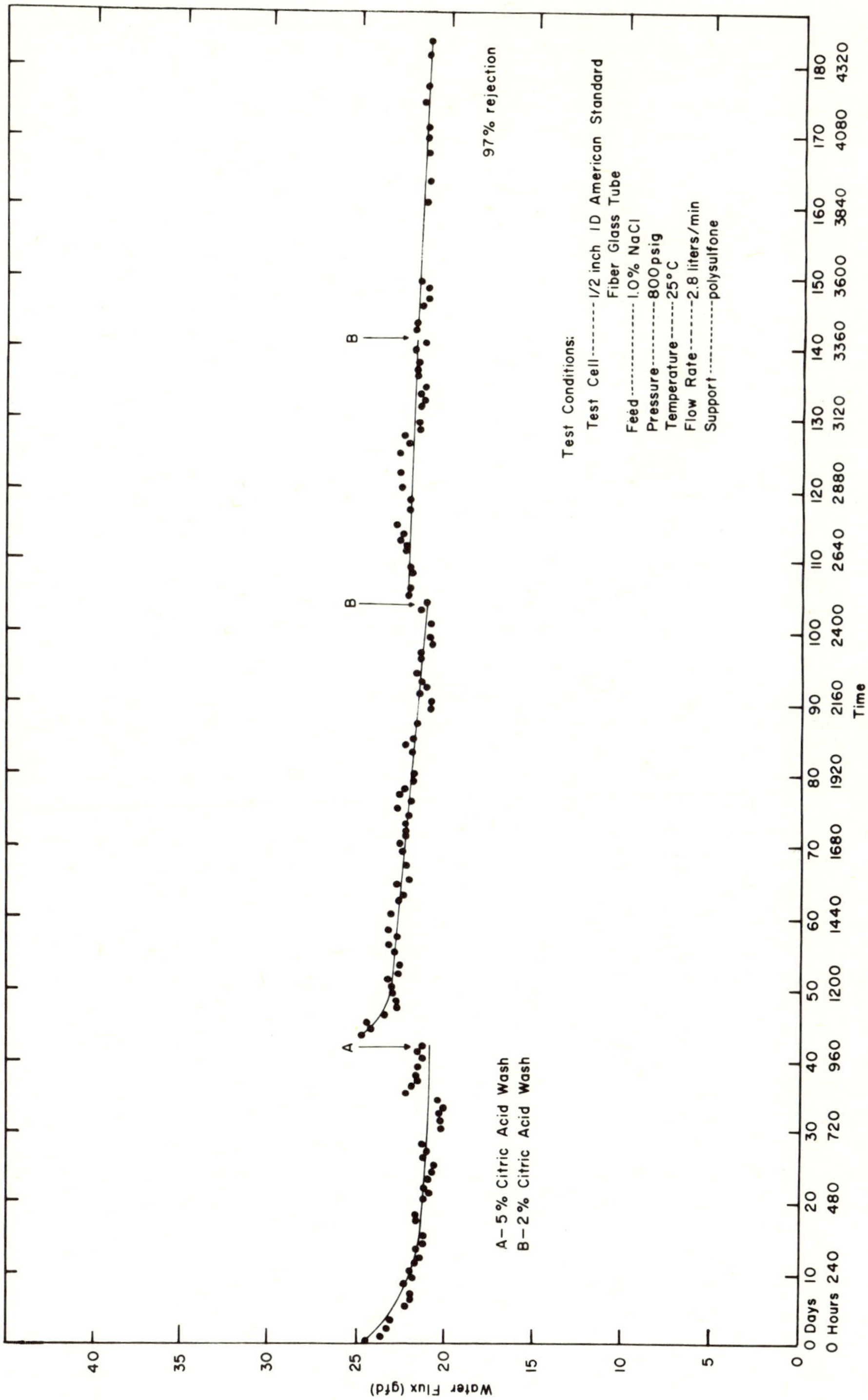

Fig. 5 - Reverse Osmosis Performance of an Ultrathin Cellulose Acetate Membrane in a Tube deep penetration of the iron oxide in the membrane. Although the flux decline of the high flux membrane was considerably greater in the first 200 hours than those of the lower flux membranes, the actual flux at the end of 1500 hours of testing was still signficantly higher for the higher flux membrane.

The results of long-term, single-tube tests indicate that:

1. Ultrathin membranes as thin as 200 Å can exhibit stable and predictable reverse osmosis performance over an extended period of time.

2. Higher flux membranes (40 gfd) exhibit larger overall flux declines than do moderate flux membranes (20 gfd).

3. For the moderate flux membrane, the initial steep decline in flux is believed to be due mainly to composite deformation and the subsequent more gradual decline to fouling, e.g., with iron oxide.

4. The loss in flux due to such fouling can be restored by washing the membrane with a two- or five-percent citric acid solution.

On-Site Testing. Ultrathin cellulose acetate membranes cast in American Standard TM 2-4 modules have been tested at the Office of Saline Water Webster Test Facility, Webster, South Dakota. The feed consisted of sand-filtered well water containing approximately 2000 ppm dissolved solids at temperatures between 10 and 11°C. Turbulence promoters were used in the reverse osmosis tubes. The reverse osmosis performance of one of the modules is plotted in Fig. 6.

The initial water flux of this module was 47 gfd. After 5500 hours of testing, the flux was maintained at slightly under 20 gfd. Fouling of the membrane was due to iron oxide which was removed by citric acid washes during the first 3000 hours of testing. At 4000 hours, sodium hexametaphosphate was added and the flux was then relatively level at 18 to 19 gfd up to 5500 hours. System shutdowns also produced a flux increase which was probably due to a cleaning action at start-up and elastic recovery of the membrane support composite due to the pressure release. The dissolved solids rejection was generally over 97 percent throughout the test.

These test results show that the ultrathin membranes are capable of long-term use with relatively high flux and adequate rejection without damage to the membrane. The high flux decline due to iron oxide fouling can be reversed significantly by membrane cleaning techniques.

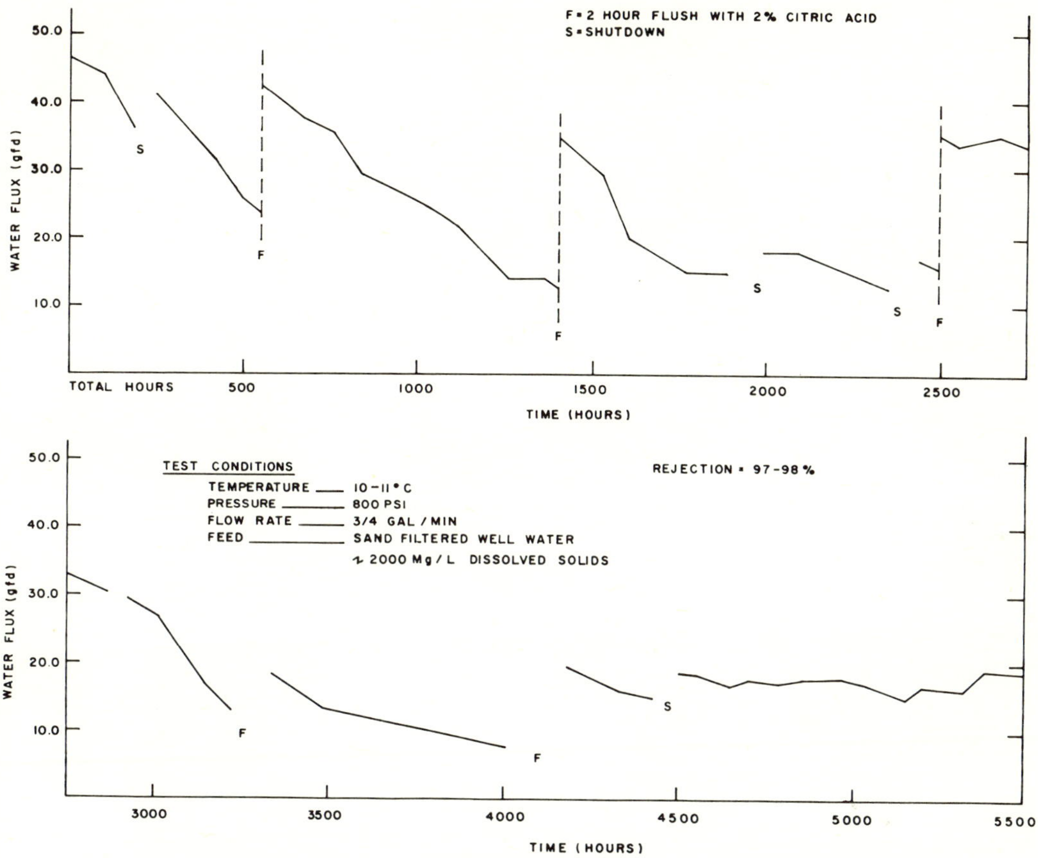

Fig. 6 - Long Term, On-Site Reverse Osmosis Performance of American Standard TM 2-4 Module Containing Ultrathin Membrane-Polysulfone Composites

In Situ Regeneration

The use of ultrathin membrane-polysulfone support composites in fiberglass tubes has the advantage of in situ maintenance. Because of the inert nature of the polysulfone support, it is possible to remove the membrane from the tube and regenerate a new membrane without removing the support liner from the module tube. The regeneration process involves circulating a solution containing 1% NaOH and 10% methanol through the tube for one hour. The tube is then washed with 1% citric acid at 800 psi for five minutes. The new membrane is cast on the polysulfone according to the method described earlier.

Two successive membrane regenerations were carried out, on-site, using sand-filtered Webster water as described in the previous section. Fig. 7 compares the reverse osmosis performance of the original and regenerated membrane in two single five-foot tubes. The membranes were cast under conditions to produce thicker membranes than used in the long term Webster test. The membranes were successfully regenerated twice and exhibited rejections above 90 percent and initial fluxes above 20 gfd. The high flux decline for the original and regenerated membranes was due to the lack of turbulence promoters and low flow rates which produced an increased rate of iron oxide fouling. The reason for the unusual flux behavior of the second regenerated membranes is unknown.

Wet-Dry Cycling

The capability of drying the composite ultrathin cellulose acetate-polysulfone support membrane without serious reduction in water flux is important for reverse osmosis applications where the modules must be moved from one site to another. The annealing of this composite membrane by drying indicated its potential for wet-dry cycling.

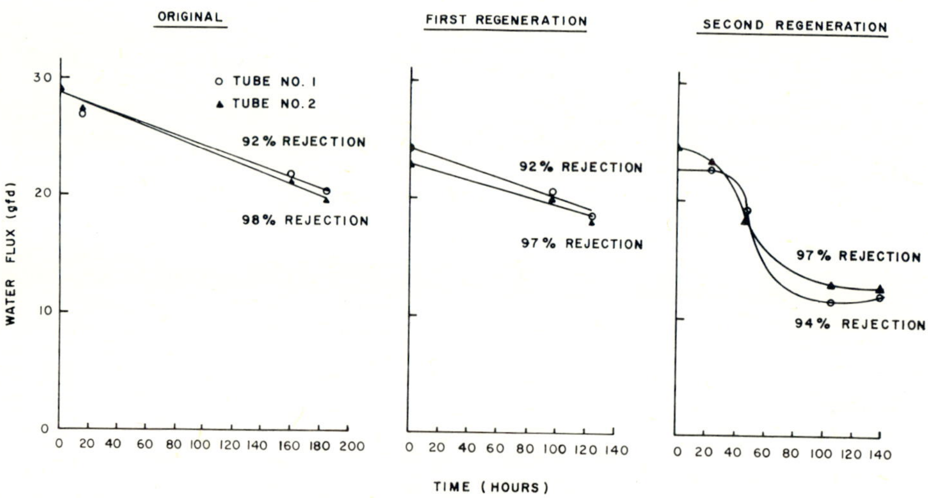

TEST CONDITIONS

TEMPERATURE	10–11° C
PRESSURE	800PSI
FLOW RATE	3/4 GAL / MIN
FEED	SAND FILTERED WELL WATER ≈2000 Mg/L DISSOLVED SOLIDS
TUBE	1/2" DIAMETER FIBERGLASS (5 FEET IN LENGTH)

Fig. 7 - Reverse Osmosis Performance of Original and Regenerated Ultrathin Cellulose Acetate Membranes in Single Tubes

Reverse osmosis tests to determine the affect of wet-dry cycling were carried out using American Standard TM 5-8 modules (eight five-foot tubes per module). To obtain complete drainage of the brine from the modules prior to the drying cycles, the top and bottom end plates of the headers were provided with drain plugs. The details of the test facility can be found in reference 12. Testing of the modules consisted of three reverse osmosis runs of 100 hours each, using a 0.5% sodium chloride feed at a pressure of 600 psi, a feed flow rate of 3 gpm, and a temperature of 25°C. After each run, the modules were drained and held for at least 50 hours in a chamber at no greater than five-percent relative humidity.

The reverse osmosis performance for wet-dry cycling of a typical module containing the ultrathin membrane-support composite is plotted in Fig. 8. For three cycles the salt rejection was 95% or above. After the first 100 hours of reverse osmosis testing the water flux was 25.3 gfd. The decline in water flux between the first and second service cycle was 8.3 gfd and between the second and third service cycle was 3.0 gfd.

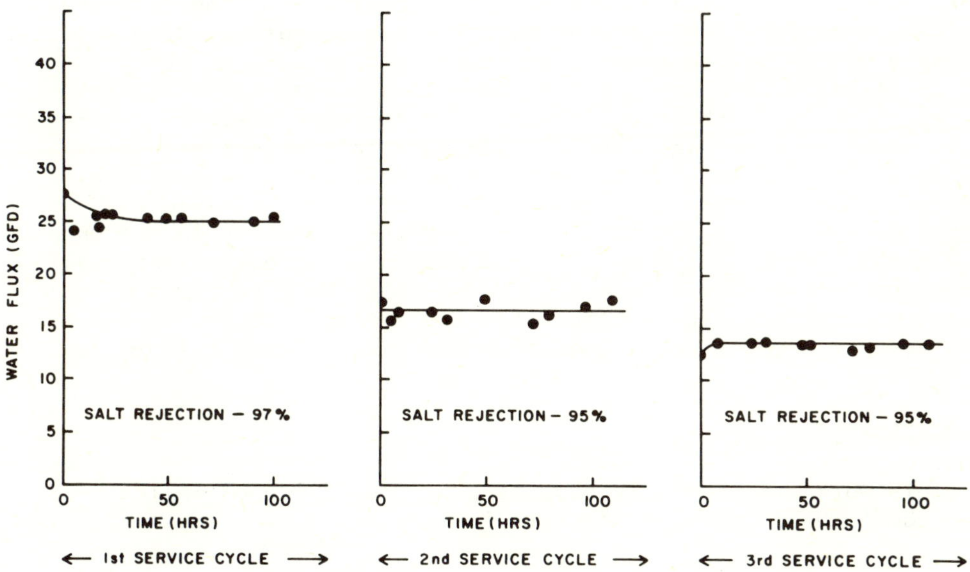

Fig. 8 - Reverse Osmosis Performance of American Standard TM 5-8 Modules Containing Ultrathin Membrane-Polysulfone Composites During Three Wet-Dry Service Cycles

This decline in the water flux on wet-dry cycling was attributed to increased hydrophobicity of the polysulfone support after being exposed to the severe conditions of the drying cycle. However, the flux decline after the second drying was significantly less than that after the first drying. Apparently once the polysulfone support film is thoroughly dried in the first cycle, repeating the process has a smaller effect. It would be expected that the flux decline would become insignificant after further drying cycles, with the ultimate flux probably being above 10 gfd. Preliminary testing has indicated that the incorporation of hydrophilic additives, such as sodium dodecyl sulfate, into the polysulfone support reduces the hydrophobicity, resulting in lower flux declines between service cycles.

CONCLUSIONS

Composites of ultrathin cellulose acetate barriers and polysulfone support liners in commercially available fiberglass tubes exhibited stable and predictable reverse osmosis performance for water desalination. The fabrication of the tubular polysulfone microporous support and ultrathin membrane is simple and does not require difficult conditions such as high temperature. The reverse osmosis properties can be varied by controlling the membrane fabrication variables of casting solution concentration, draw rate, viscosity of the polyacrylamide solution, and annealing. Long term performance of these tubular composites, both in the laboratory and on-site, with brackish waters has demonstrated its durability combined with high flux and adequate salt rejection. Flux decline due to fouling can be reversed in most cases by membrane cleaning. Removal and replacement of the ultrathin cellulose acetate membranes without manipulation of the composite in the tube and without damage to the polysulfone supports is feasible and may be an important economic factor in large scale reverse osmosis water desalination. These composites are amenable to drying between reverse osmosis operations without serious loss in flux.

ACKNOWLEDGMENT

This work was partially supported by the Office of Saline Water, U. S. Department of the Interior.

REFERENCES

1. C. E. Reid and E. J. Breton, J. Appl. Polymer Sci., 133, 1 (1957).

2. S. Loeb and S. Sourirajan, Advan. Chem. Ser. 38, 117 (1963).

3. L. T. Rozelle, J. E. Cadotte, and D. J. McClure, J. Appl. Polymer Sci., Applied Polymer Symposia No. 13, 61 (1970).

4. J. E. Cadotte, L. T. Rozelle, R. J. Petersen, and P. S. Francis, J. Appl. Polymer Sci., Applied Polymer Science Symposia No. 13, 73 (1970).

5. L. T. Rozelle, J. E. Cadotte, R. D. Corneliussen, and E. E. Erickson, Office of Saline Water Research and Development Progress Report No. 359, U. S. Government Printing Office, Washington, D. C., October, 1968.

6. Millipore Corp. Catalog, MF-67, p. 24.

7. Emil Ott, Cellulose and Cellulose Derivatives, Interscience Publisher, Inc., New York, Vol. V, Part III (1955).

8. J. L. O'Toole, Modern Plastics Encyclopedia, 1969-1970, Vol. 46, No. 10A (October 1969).

9. L. T. Rozelle, J. E. Cadotte, and D. J. McClure, Office of Saline Water Research and Development Progress Report No. 531, U. S. Government Printing Office, Washington, D. C., June, 1970.

10. H. K. Lonsdale, U. Merten, and R. L. Riley, J. Appl. Polymer Sci., 9, 1341 (1965).

11. Public Health Drinking Water Standards - 1962, Federal Register 27:2152, U. S. Government Printing Office, Washington, D. C. (1962).

12. L. T. Rozelle, J. E. Cadotte, W. L. King, A. J. Senechal, and B. R. Nelson, "Development of Ultrathin Reverse Osmosis Membranes for Desalination", Office of Saline Water Research and Development Progress Report, U. S. Government Printing Office, Washington, D. C., in press.

PREPARATION, MORPHOLOGY, AND TRANSPORT PROPERTIES OF COMPOSITE REVERSE OSMOSIS MEMBRANES FOR SEAWATER DESALINATION

R. L. Riley, G. Hightower, and C. R. Lyons
Gulf General Atomic Company
San Diego, California 92112

INTRODUCTION

The modified cellulose acetate membrane, first prepared by Loeb and Sourirajan[1] in the 1960s, is clearly the starting point of the modern reverse osmosis process. Development of the thin-film composite membrane began after it was shown by electron microscopy that the asymmetric cellulose acetate membrane is a two-layered structure consisting of a thin, dense semipermeable barrier, about 2000 Å thick, supported by a finely porous substructure[2-4]. Because the membrane is effectively very thin, high water fluxes are attained.

It was known that there were deficiencies in the asymmetric membrane. Hydrolysis of the cellulose acetate represents an inherent limitation to membrane life; selectivity is inadequate for single-pass seawater desalination; and the membrane is subject to a characteristic decline in water flux with time at high pressure [5,6]. Each of these limitations has serious implications on the design and performance of reverse osmosis systems. It was also apparent that the Loeb-Sourirajan method for the preparation of asymmetric membranes would not be generally applicable to other polymer systems, a conclusion that subsequent research has amply demonstrated.

A quantum improvement in membrane properties will most likely result from a more general method of membrane preparation. We have previously described such an approach in which composite membranes, capable of single-pass seawater desalination, have been prepared by laminating a very thin film of a semipermeable material to a finely porous support membrane[7,8]. It has been demonstrated by this method that the thin semipermeable barrier, typically 800 Å thick, can be prepared virtually free of imperfections, from materials other than cellulose 2.5-acetate and with thicknesses that can be reproducibly varied and controlled. With this method of preparing membranes, the thin film and the porous support can be individually optimized for their specific functions.

TRANSPORT PHENOMENOLOGY

The transport of water and simple salts through the thin cellulose acetate semipermeable barrier of the composite membrane can be adequately described by the solution-diffusion model of membrane transport[9]. The phenomenological equation for non-coupled water flow is

$$J_1 = \frac{D_1 c_1 \overline{V}_1}{RT} \frac{(\Delta P - \Delta\pi)}{\Delta x} \equiv A(\Delta P - \Delta\pi) , \qquad (1)$$

where J_1 is the water flux, D_1 and c_1 are the diffusion coefficient of water and the water content, $\overline{V}_1$ is the partial molar volume of water in the membrane, $\Delta P - \Delta\pi$ is the net pressure difference across the membrane (applied pressure less osmotic pressure), R and T have their usual significance, and Δx is the thin-film thickness. The product D_1c_1 is the water permeability. For a given membrane at constant temperature, the water flux is just proportional to the net pressure difference, and we have defined the proportionality factor, A, as the "membrane constant."

A similar phenomenological equation exists for solute flow:

$$J_2 = D_2 \frac{\Delta c_2}{\Delta x} = D_2 K \frac{\Delta \rho_2}{\Delta x} \equiv B\Delta\rho_2 , \qquad (2)$$

where the subscript 2 denotes the solute, c_2 is the solute concentration in the membrane, and K is the solute distribution coefficient relating c_2 to the concentration of salt in the external solution, ρ_2. The product D_2K is the solute permeability. For a given membrane, the ratio of the salt flux to the salt concentration difference is a constant B, which we have defined as the "solute permeation constant."

Equation (1) can be used to predict water flux through a membrane of known thickness, and equation (2) can be used to predict solute flux; from these two equations the semipermeability or solute rejection can be derived. Defining solute rejection as

$$\text{Solute Rejection} = \frac{\rho_2' - \rho_2''}{\rho_2'} , \tag{3}$$

where ' and '' denote the pressurized feed solution and the permeate, respectively, we obtain, assuming diffusive flows only,

$$\text{Solute Rejection} = 1 - \frac{\rho_1'' J_2}{\rho_2' J_1} . \tag{4}$$

This equation expresses the fact that the permeate concentration is determined only by the water and solute fluxes during reverse osmosis. We then have

$$\text{Solute Rejection} = \left[1 + \frac{D_2 K R T \rho_1''}{D_1 c_1 \overline{V}_1 \ (\Delta P - \Delta\pi)} \right]^{-1} . \tag{5}$$

The rejection is thus independent of thin-film thickness and is determined by the relative permeabilities and the applied pressure.

The permeability of cellulose acetate membranes to water and sodium chloride has been measured and found to depend strongly on the degeee of acetylation[9]. For the material normally used to prepare modified reverse osmosis membranes, 39.8%-acetyl cellulose acetate, having a degree of substitution of 2.5, the permeabilities at 25°C are $D_1c_1 = 2.6 \times 10^{-7}$ g/cm-sec and $D_2K = 3.4 \times 10^{-11}$ cm^2/sec. Based on the intrinsic permeabilities and equation (5), the expected salt rejection at an applied pressure of 102 atm with a 3.5 wt-% sodium chloride brine at 25°C is 99.76% for an imperfection-free membrane. In practice, seawater rejections observed experimentally with Loeb-Sourirajan type membranes range from 97% to 99%. The rejection is clearly less than the predicted value. This has been generally observed and is attributed to the presence of membrane imperfections. In spite of these imperfections, the salt rejections achieved in practice are quite adequate for recovery of potable water from most brackish water sources. For seawater desalination to be practical, it is necessary to have reasonably high water recoveries. This makes it necessary for the membrane to reject about 99.5% of the sodium chloride to yield potable water in a single pass.

MEMBRANE PREPARATION

The composite membrane is structurally similar to the Loeb-Sourirajan membrane in that it possesses a very thin semipermeable barrier supported by a finely porous matrix. An important difference, however, is the method of preparation. Briefly stated, an

800-Å thin film of Eastman Chemical Co. 432-130B cellulose triacetate (43.2% acetyl, 2.83 degree of substitution) was formed directly upon a masking layer of polyacrylic acid (PAA) deposited onto the finely porous surface of a cellulose nitrate-cellulose acetate (CN-CA) support membrane. The masking layer, a 3-μ-thick water-soluble coating of 50,000 molecular weight PAA (Rohm and Haas Co., Acrysol A-1), was applied to the finely porous glossy surface of the CN-CA support membrane by airless spray from a 2.5 wt-% solution in equal volumes of ethanol and water. The support membrane was preheated and maintained at 85°C for spraying. The thin semipermeable film was formed directly upon the surface of the PAA masking layer by dipping into a 0.5 wt-% solution of cellulose triacetate in chloroform; the vertical withdrawal rate was 1.9 cm/sec. During withdrawal, a stream of dry air was passed over the surface of the membrane.

The porous support membrane was prepared by casting a solution containing 7.8 wt-% cellulose nitrate (DuPont Co. DHA35E), 1.3 wt-% cellulose acetate (Eastman E383-40), 53.7 wt-% acetone, 19.9 wt-% absolute ethanol, 13.3 wt-% n-butanol, 1.9 wt-% glycerin, and 0.5 wt-% surfactant (Rohm and Haas Triton X-100) onto a glass plate[10,11]. The membrane was allowed to gel under controlled air flow, temperature, and humidity conditions. Increased relative humidity, increased concentration of diluent in the casting solution, or decreased temperature led to increased surface porosity in the membrane. Electron photomicrographs have shown that the membrane is asymmetric, with a very thin skin on the air-dried surface that ranges from about 5% to 15% in porosity, supported by a graduated finely porous substrate of interconnected pores. The thickness of the support ranges from 100 to 125 μ.

Several properties appear to be essential in the porous support for supporting the thin film of the composite membrane. The pores on the surface in contact with the thin film must be very small, with the maximum diameter preferably considerably smaller than the thickness of the thin film; the surface porosity should be high to minimize flow resistance; and the support should be stable against creep and chemical or physical deterioration under high compressive loading for long periods. In addition, solvent compatibility is an important consideration. The support must be tolerant, and preferably indifferent, to the solvent from which the thin film is applied.

Several measurements of the support properties are used to characterize the membrane. The water permeability is determined in a low-pressure system at 2.7 atm with distilled water. Typical membrane constants of the support suitable for supporting an 800-Å thin film range from 100 to 5000 x 10^{-5} g/cm^2-sec-atm. In addition, electron photomicrographs are taken of the finely porous surface to determine the pore characteristics, i.e., pore size, pore distribution, and total surface porosity[12].

Either a water soluble or a highly water permeable masking layer is necessary to fill or coat over the pores of the finely porous surface of the support prior to forming the thin film to avoid intrusion of the cellulose acetate thin-film solution into the pores. Polyacrylic acid has been found to perform well as this layer, since it is insoluble in chloroform, the solvent from which the thin film is formed. Analysis of the permeate through composite membranes in reverse osmosis tests for organic carbon shows that the retention of PAA is a function of molecular weight and the surface pore size of the support membrane. Masking layers prepared from 50,000 molecular weight PAA are not retained by the support membrane normally used after the composite membrane is placed in operation. Higher molecular weight (1 to 2 million) PAA masking layers are totally retained by the support.

The thin cellulose acetate film was applied to the masked surface of the support by dip-coating in a dilute chloroform solution at 25°C. The thickness of the thin film is a function of the concentration of the solution from which the PAA-coated support membrane is withdrawn, the withdrawal rate, and the viscosity and surface tension of the solution[13]. The thickness of a liquid film left on a substrate withdrawn vertically from solution is given by

$$\Delta x_w = \frac{0.93\ (\mu v)^{2/3}}{\sigma^{1/6}(\rho g)^{1/2}}\ , \tag{6}$$

where μ, σ, and ρ are the viscosity, surface tension, and density, respectively, of the thin-film solution; v is the withdrawal rate; and g is the gravitational constant. The thickness of dry polymer film left when the solvent evaporates is given by the product of the wet thickness and the solution concentration. The surface tension and density of the pure solvent, chloroform, were assumed. Thus, $\sigma = 26$ dynes/cm and $\rho = 1.50$ g/cm^3. The predicted dry thickness, Δx, becomes

$$\Delta x = 0.016\ (\mu w)^{2/3} c\ , \tag{7}$$

where c is the solution concentration expressed as g polymer/g solution.

Thin-film thicknesses are presented in Fig. 1 as functions of both withdrawal rate and solution concentration for 43.2%-acetyl cellulose acetate at 25°C. The thin-film thicknesses were determined gravimetrically by floating a known area of thin film

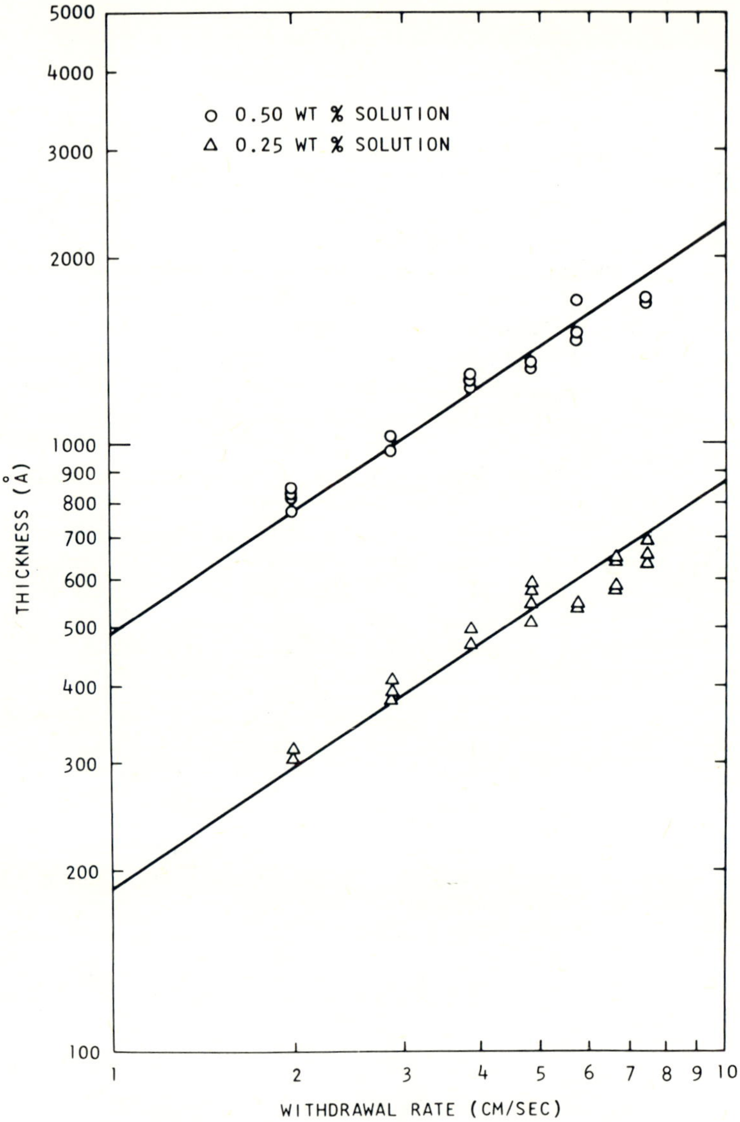

Fig. 1 - Thickness of Thin Films of 43.2%-Acetyl Cellulose Acetate Versus Withdrawal Rate and Concentration

off a PAA-coated glass substrate onto a water surface, drying, and weighing; a dry density of 1.3 g/cm^3 was assumed. The lines drawn through the data are the "theoretical" lines calculated from equation (7).

Useful films, as thin as 350 Å, have been successfully prepared by this method from cellulose acetate. However, when the thin-film thickness is less than 200 Å, the salt rejection of the composite membrane decreases. The electron photomicrographs in Fig. 2 show both 350- and 200-Å thin-film composite membranes at various stages of development. The striking difference between photomicrographs (c) and (d) is due to the change in continuity

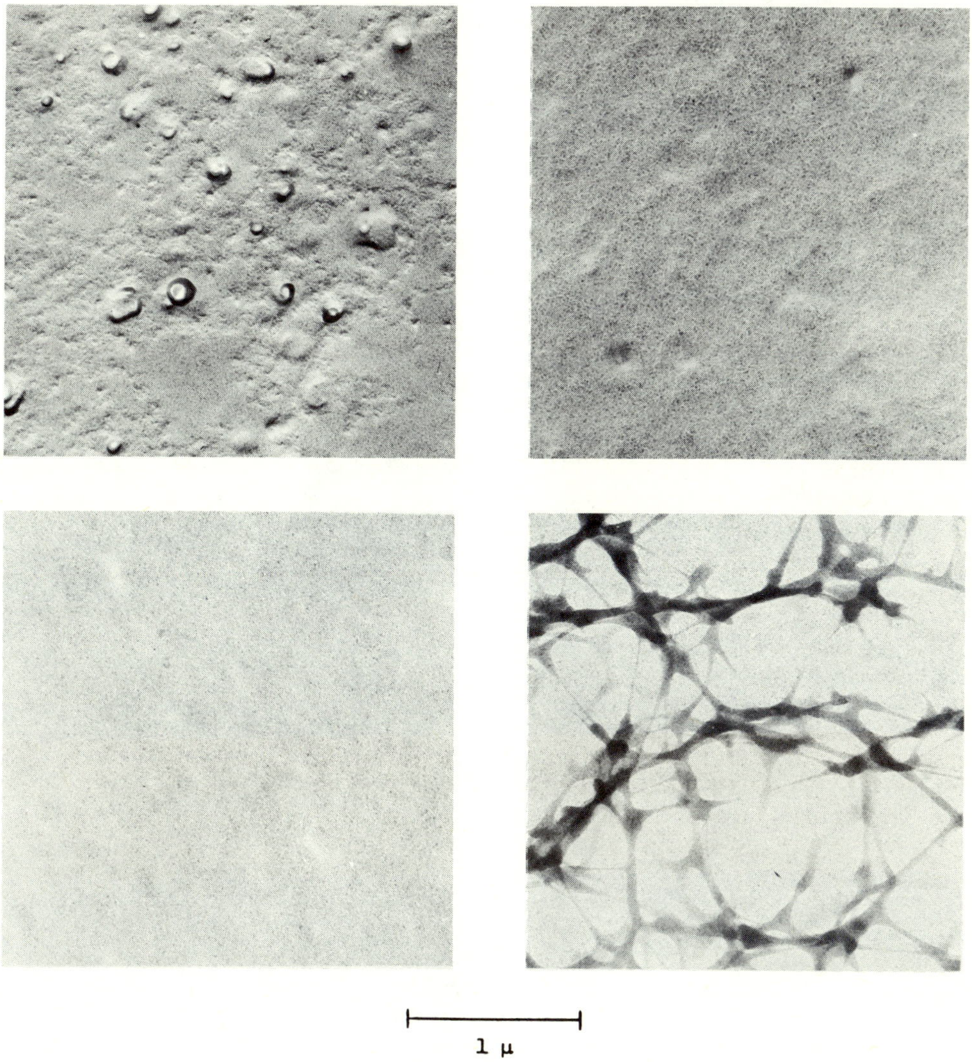

Fig. 2 - Electron Micrographs of Preshadowed-Carbon Membrane Replicas: (a) Finely Porous CN-CA Support Surface, (b) After PAA Coating, (c) PAA-Coated Support Surface Having a Cellulose Acetate Film Nominally 350 Å Thick, and (d) As Above, With a Cellulose Acetate Film Nominally 200 Å Thick (30,000 magnification)

of the thin film; the 350-Å thin film is continuous and free of imperfections, while the 200-Å thin film failed to form a continuous film.

REVERSE OSMOSIS TESTING

The composite membranes were tested at 25°C and at an applied pressure of 102 atm. A 3.5 wt-% sodium chloride brine was passed over the surface of the membrane at a velocity of 150 cm/sec. The membrane constant, A, as defined by equation (1), was calculated from the water throughout rate, and the salt rejection was calculated from the salt content of the feed and permeate in accordance with equation (3). The dependence of water flux on time was calculated according to a power law,

$$\log J_1 = -m \log t \ , \tag{8}$$

where the constant m (the flux-decline slope) depends on membrane type, brine composition, pressure, and operating temperature.

The morphology of the supporting membrane and the composite structure was determined by electron microscopy both before and after reverse osmosis testing[12]. The internal cross-sectional structure and the finely porous surface of the supports were examined both before and after composite membrane preparation; the supports examined after composite preparation had also been subjected to long term reverse osmosis testing. The internal structure of the supporting membrane was scanned in cross section to include the finely porous skinned surface. The thin-film surface of the composite membrane was also observed before and after testing. In addition, the size and form of the fouling contaminant deposited on the thin-film surface of the composite membranes during testing were examined. The electron photomicrographs of palladium-shadowed carbon replicas of the various membrane structures are shown in Figs. 3-6.

Results obtained for composite membranes prepared on supporting membranes of different permeabilities are presented in Table 1. The duration of the reverse osmosis test was 143 hr.

The excellent salt rejections, typical of composite membranes, are clearly sufficient for desalinating seawater in a single pass. As with asymmetric cellulose acetate membranes, maximum rejection is attained after several hours of operation. There is little, if any, dependence of salt rejection on support permeability. The expected salt rejection, based on the measured salt permeability of cellulose triacetate (0.33×10^{-11} cm^2/sec), is 99.94%.

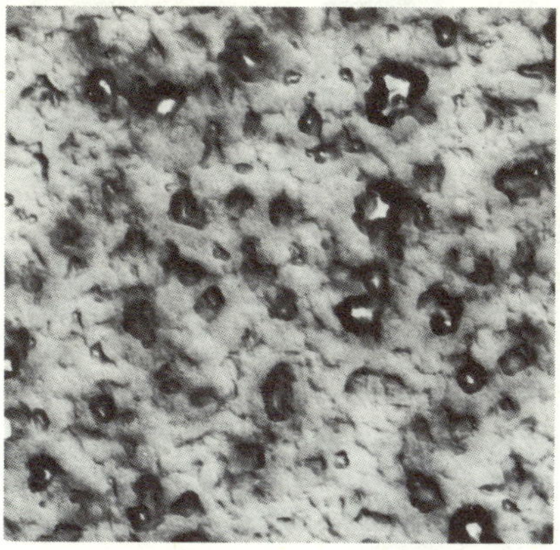

30,000x

1 μ

Finely Porous Surface of CN-CA Support Membrane

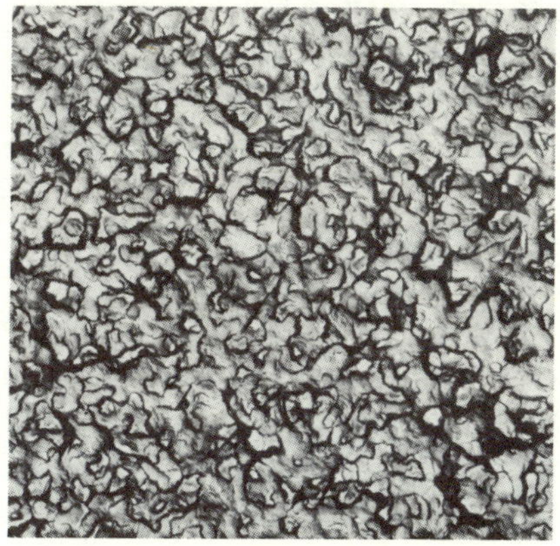

6,000x

1 μ

Bottom Surface of CN-CA Support Membrane

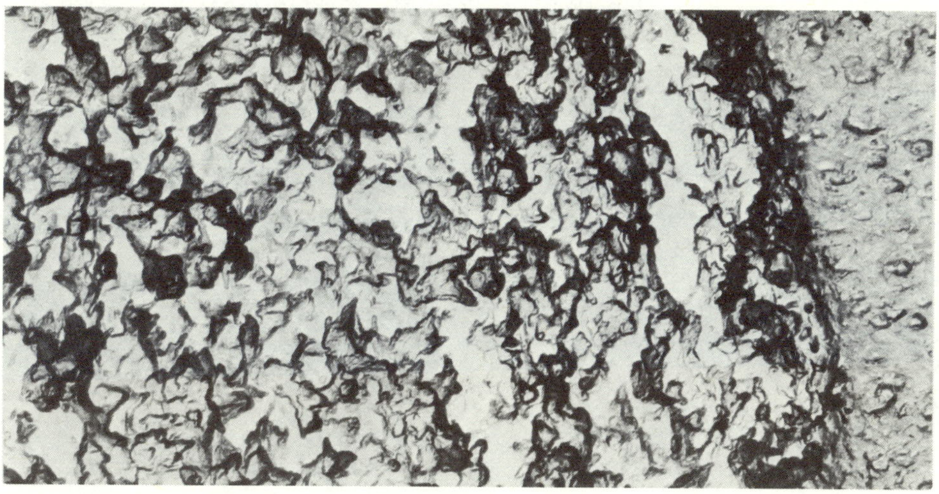

30,000x

1 μ

Internal Support Structure Including Finely Porous Surface of CN-CA Support Membrane

Fig. 3 - Electron Photomicrographs of the CN-CA Support Membrane Before Composite Preparation (Membrane Constant = 140×10^{-5} g/cm^2-sec-atm)

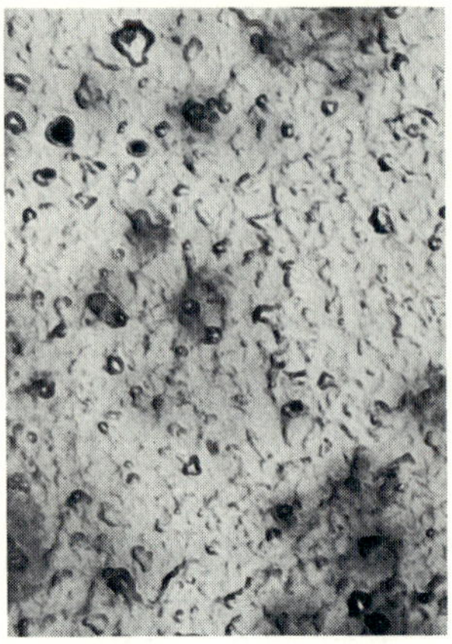

30,000x

1 μ

Finely Porous Surface of CN-CA Support Membrane

30,000x

1 μ

Finely Porous Surface of CN-CA Support Membrane with Thin Film (Fouling Contaminants Removed)

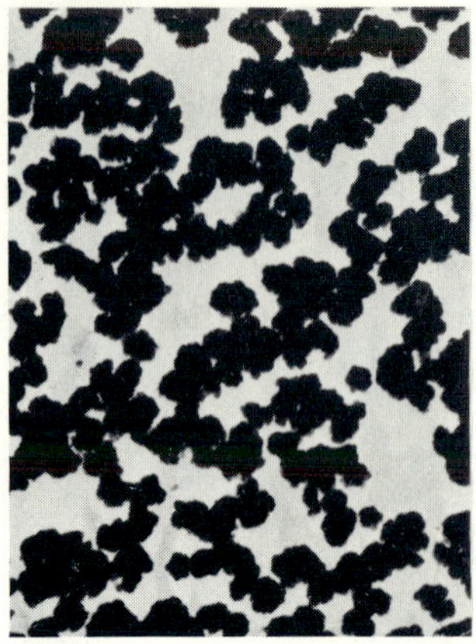

30,000x

1 μ

Finely Porous Surface of CN-CA Support Membrane with Thin Film (Fouling Contaminants Not Removed)

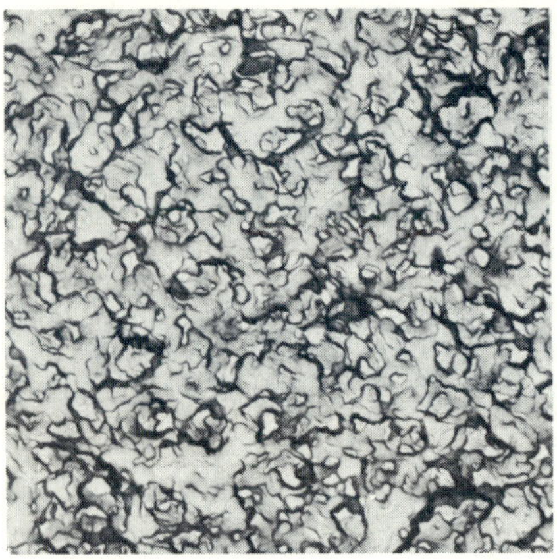

6,000x

1 μ

Bottom Surface of CN-CA Support Membrane

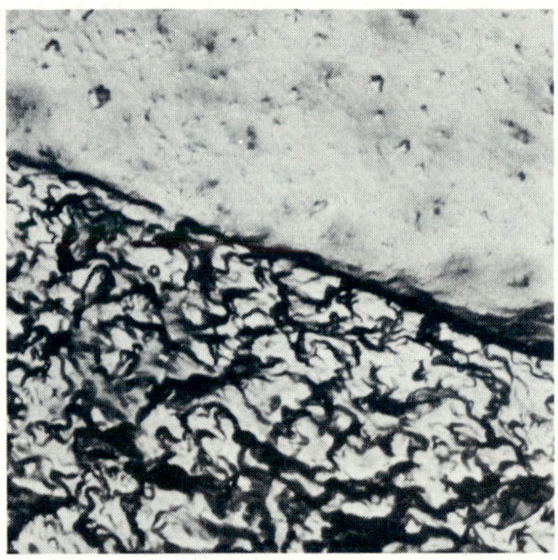

30,000x

1 μ

Internal Structure Including Finely Porous Surface of CN-CA Support Membrane

Fig. 4 - Electron Photomicrographs of the CN-CA Support Membrane after Composite Preparation and Testing (Support Membrane of Fig. 3)

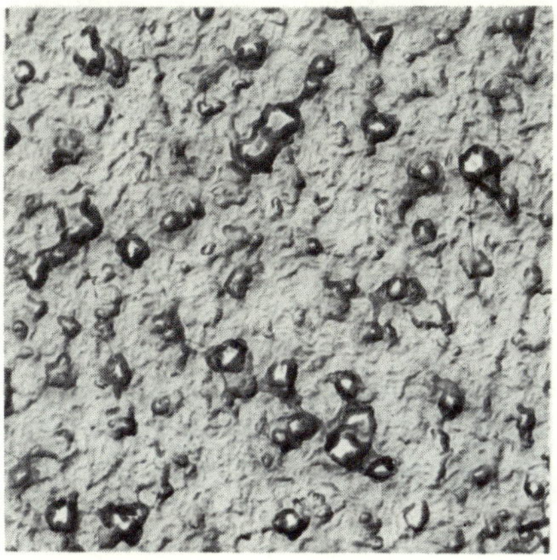

30,000x

1 μ

Finely Porous Surface of CN-CA Support Membrane

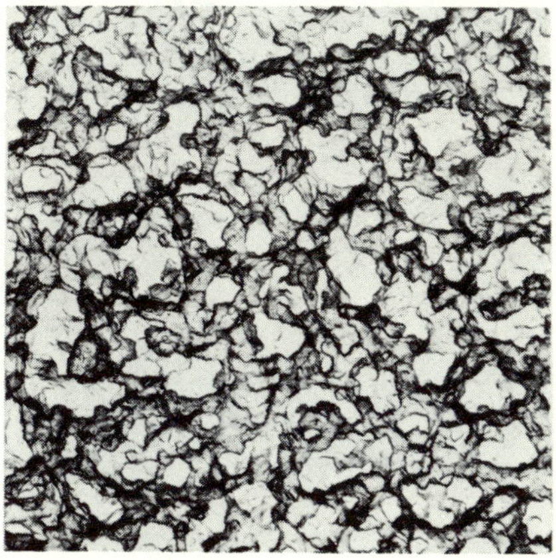

6,000x

1 μ

Bottom Surface of CN-CA Support Membrane

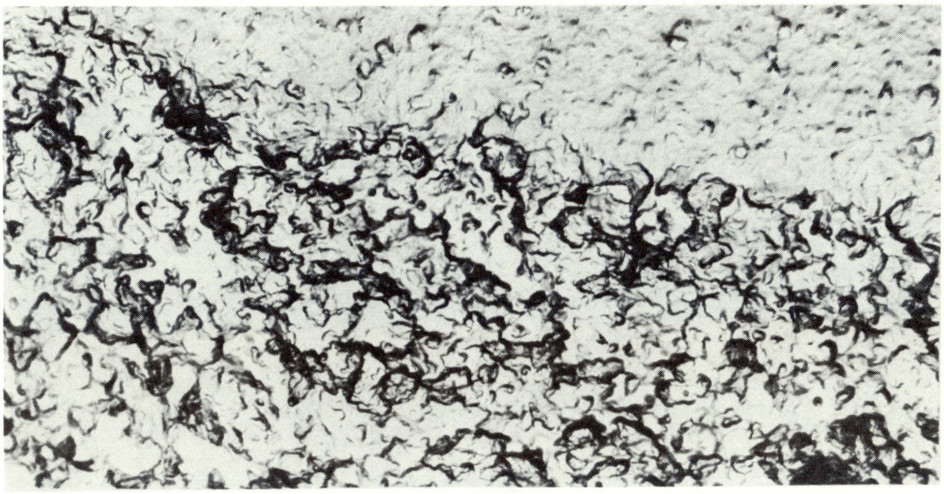

30,000x

1 μ

Internal Structure Including Finely Porous Surface of CN-CA Support Membrane

Fig. 5 - Electron Photomicrographs of the CN-CA Support Membrane Before Composite Preparation (Membrane Constant = 3500×10^{-5} g/cm^2-sec-atm)

30,000x

1 μ

Finely Porous Surface of CN-CA Support Membrane

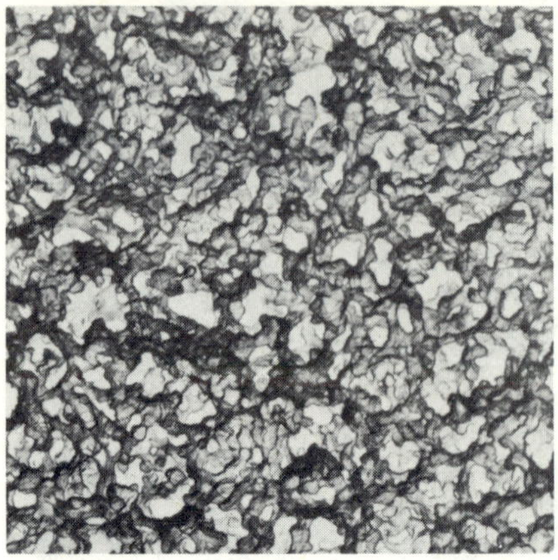

6,000x

1 μ

Bottom Surface of CN-CA Support Membrane

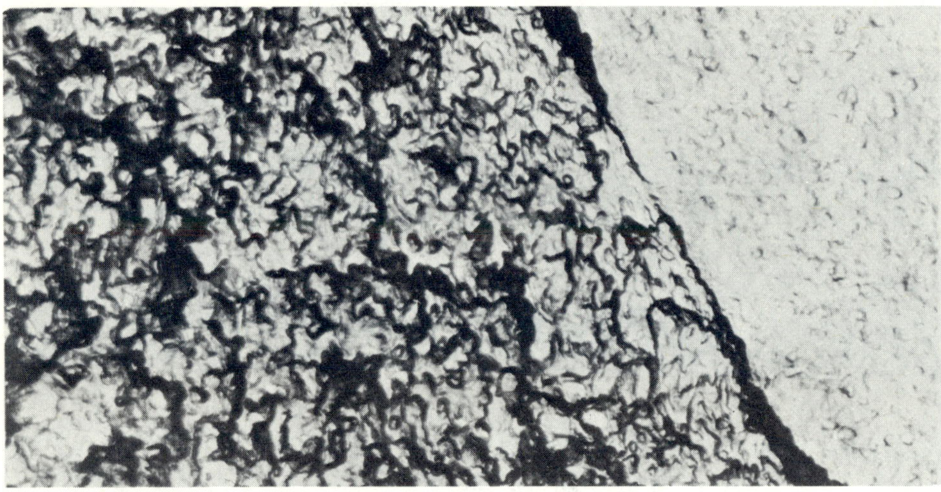

30,000x

1 μ

Internal Structure Including Finely Porous Surface of CN-CA Support Membrane

Fig. 6 - Electron Photomicrographs of the CN-CA Support Membrane After Composite Preparation and Testing (Support Membrane of Fig. 5)

TABLE 1

REVERSE OSMOSIS RESULTS FOR COMPOSITE MEMBRANES

Membrane Constant, A, of CN-CA Support at 2.7 Atm (10^{-5} g/cm^2-sec-atm)	Observed Membrane Constant (10^{-5} g/cm^2-sec-atm)		Observed A / Calculated A (143 Hr)	Water Flux (gal/ft^2-day)		Salt Rejection (%)		Flux-Decline Slope After 143 Hr
	0.5 Hr	143 Hr		0.5 Hr	143 Hr	0.5 Hr	143 Hr	
140	0.46	0.35	0.39	7.2	5.7	97.6	99.63	-0.052±0.004
140	0.57	0.42	0.46	8.9	6.8	98.0	99.53	-0.056±0.004
3500	0.44	0.35	0.39	7.1	5.7	95.4	99.68	-0.046±0.002
3500	0.52	0.41	0.45	8.3	6.7	95.1	99.73	-0.048±0.003

The observed water flux was significantly less than that calculated based on the thin-film thickness and the water permeability of cellulose triacetate (1.0×10^{-7}g/cm-sec). Furthermore, the water flux was independent of the membrane constant and the surface porosity of the support. A summary of the surface pore properties of the support, determined from electron photomicrographs both before and after composite preparation and testing, is presented in Table 2. The pore size distributed (in terms of number of pores per square micron vs. pore diameter) and pore area for the support membranes both before and after composite preparation and testing are given in Figs. 7-10.

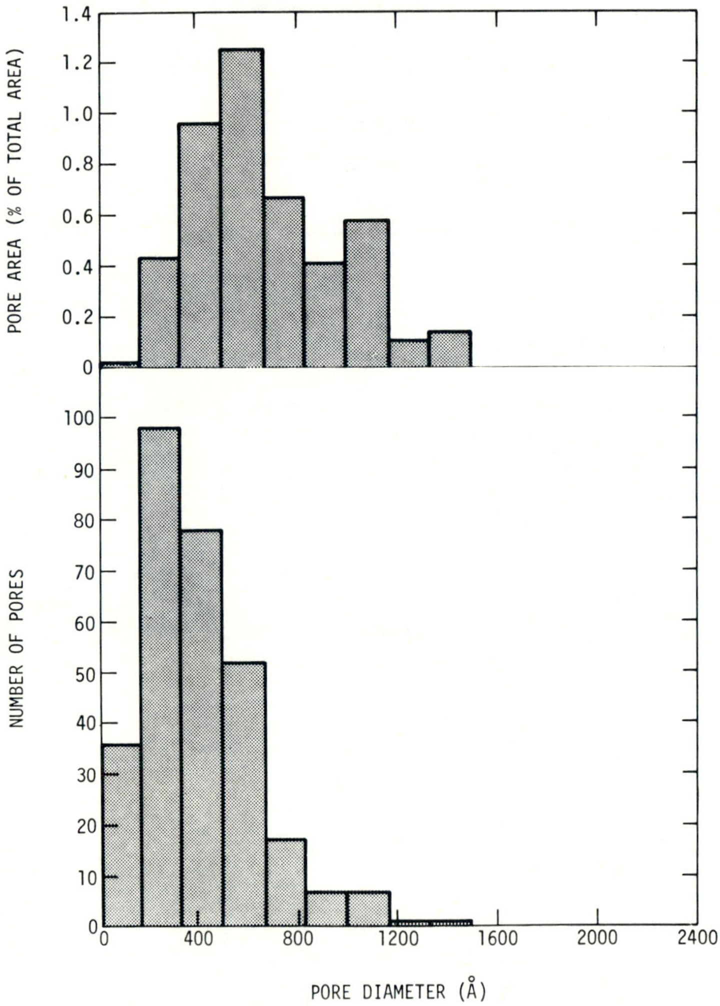

Fig. 7 - Pore-Size Distribution and Pore Area for Support Membrane Before Composite Preparation (Membrane Constant = 140×10^{-5} g/cm^2-sec-atm)

TABLE 2

SUMMARY OF SURFACE PORE PROPERTIES OF POROUS CN-CA SUPPORT MEMBRANES

Support History	Membrane Constant at 2.7 Atm (10^{-5} g/cm^2-sec-atm)	Surface Porosity (%)	Mean Pore Radius (Å)	Mean Half-Distance Between Pores (Å)	Number of Pores Per Square Micron
Before composite preparation and reverse osmosis testing	140	4.58	225	1000	28
After composite preparation and reverse osmosis testing	140	5.64	225	950	31
Before composite preparation and reverse osmosis testing	3500	14.03	500	1300	36
After composite preparation and reverse osmosis testing	3500	14.38	500	1300	21

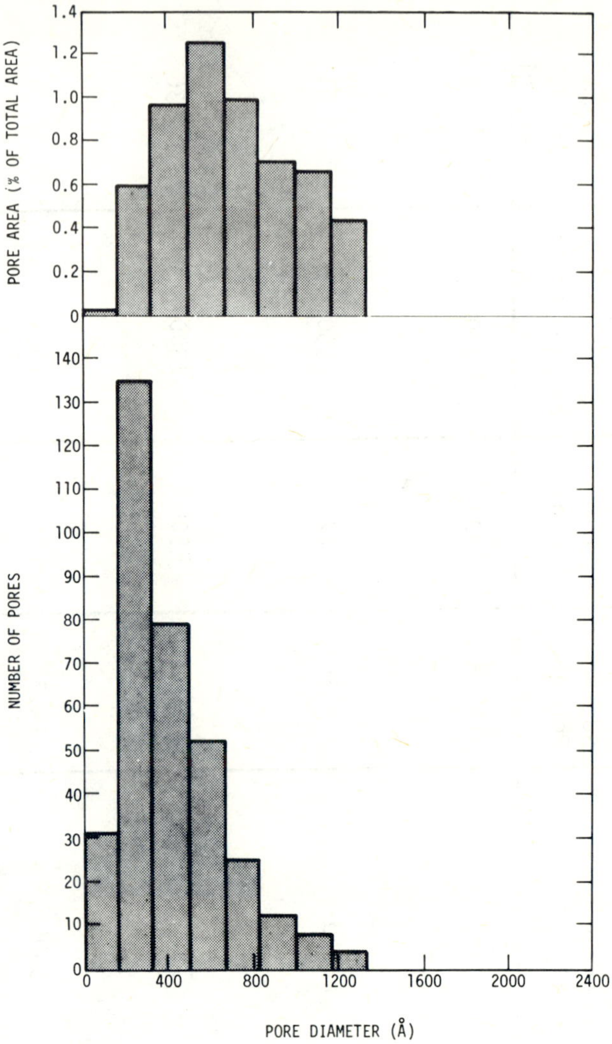

Fig. 8 - Pore Size Distribution and Pore Area for Support Membrane After Composite Preparation and Reverse Osmosis Testing (Support Membrane of Fig. 7)

The pore properties of the support that are most important to the water flux of composite membranes are the mean distance between pores and the number of pores per square micron. The initial observed composite flux was only about 50% of that expected. The magnitude of the reduction has been previously evaluated quantitatively for a wide range of thin-film thicknesses and pore characteristics[14].

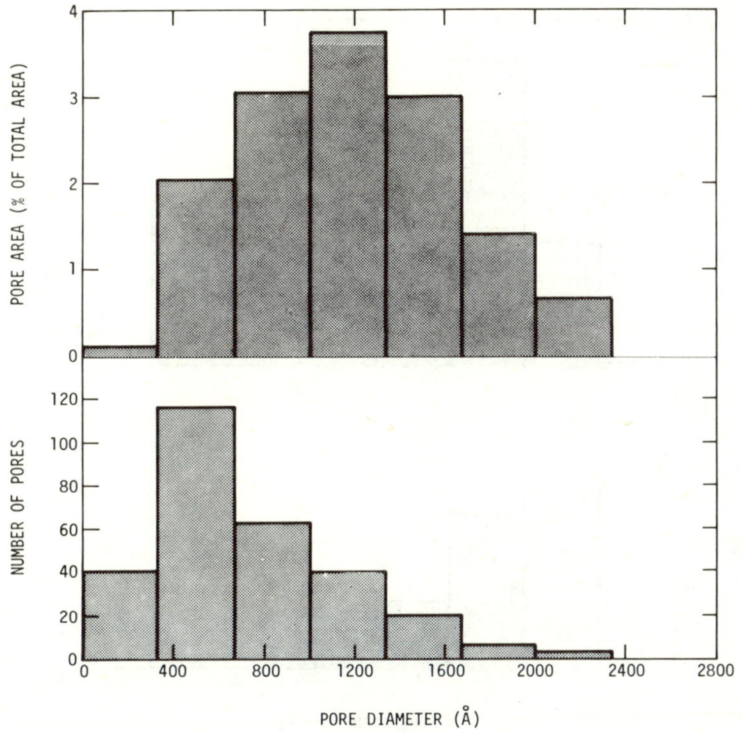

Fig. 9 - Pore-Size Distribution and Pore Area for Support Membrane Before Composite Preparation (Membrane Constant = 3500 x 10^{-5} g/cm^2-sec-atm)

The surface pore properties of the supports were unchanged after long-term testing in the composite membrane form. This suggests that the compaction or flux decline phenomenon is not due to an effective increase in the thin-film thickness resulting from a change in surface porosity.

A fouling material found attached to the thin-film surface of the composite after the reverse osmosis test is believed to be a major factor contributing to the flux decline. This material was spherically shaped, about 1000 Å in diameter, and opaque to the electron beam (see Fig. 4). It covered about 80% of the thin-film surface and was easily removed by washing. The material is believed to be a corrosion product of the all-stainless-steel system.

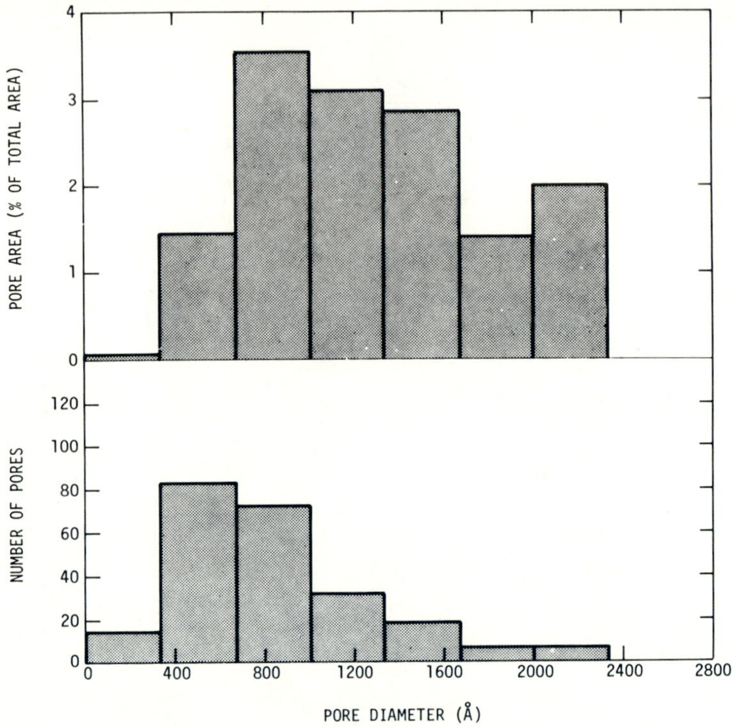

Fig. 10 - Pore-Size Distribution and Pore Area for Support Membrane After Composite Preparation and Reverse Osmosis Testing (Support Membrane of Fig. 9)

The internal structure of the CN-CA support was compacted after the test. The permeability and thickness of the support were reduced by approximately 80% and 40%, respectively. As a result, a detrimental increase in the resistance to flow could also be expected in this area of the membrane.

SUMMARY

The composite membrane approach offers a new route to the preparation of high flux asymmetric membranes for reverse osmosis. The membrane exhibits excellent semipermeability and appears to be virtually free of imperfections when the thin-film thickness exceeds 200 Å. The results presented here suggest that composite membranes have an inherently good flux stability. The observed flux decline is attributed more to a fouling phenomenon than to a change in the structural properties of the membrane itself. Desalination of seawater to potable levels can be achieved in a single pass.

REFERENCES

1. S. Loeb and S. Sourirajan, Advan. Chem. Series, 38, 117 (1963).

2. R. L. Riley, H. K. Lonsdale, C. R. Lyons, and U. Merten, J. Appl. Polymer Sci., 11, 2143 (1967).

3. R. L. Riley, H. K. Lonsdale, L. D. LaGrange, and C. R. Lyons, Office of Saline Water Research and Development Progress Report No. 386, U. S. Government Printing Office, Washington, D. C., May 1968.

4. R. L. Riley, J. O. Gardner, and U. Merten, Science, 143, 801 (1964).

5. U. Merten, (Ed.), Desalination by Reverse Osmosis, M.I.T. Press, Cambridge, Massachusetts, 1966.

6. K. D. Vos, F. O. Burris, Jr., and R. L. Riley, J. Appl. Polymer Sci., 10, 825 (1966).

7. R. L. Riley, H. K. Lonsdale, and C. R. Lyons, in Proceedings of the 3rd International Symposium on Fresh Water from the Sea, Dubrovnik, Yugoslavia, September 13-16, 1970, Vol. 2, p. 551; J. Appl. Polymer Sci., 15, 1267 (1966).

8. H. K. Lonsdale, R. L. Riley, C. E. Milstead, L. D. LaGrange, A. S. Douglas, and S. B. Sachs, Office of Saline Water Research and Development Report No. 577, U. S. Government Printing Office, Washington, D. C., Oct. 1970.

9. H. K. Lonsdale, U. Merten, and R. L. Riley, J. Appl. Polymer Sci., 9, 1341 (1965).

10. A. Goetz, U. S. Patent 2,926,104 (1960).

11. R. Zsigmondy, Z. Angew. Chem., 30, 398 (1926).

12. R. L. Riley, U. Merten, and J. O. Gardner, Desalination, 1, 30 (1966).

13. V. G. Levich, Physicochemical Hydrodynamics, Prentice Hall, Englewood Cliffs, New Jersey, 1962.

14. H. K. Lonsdale, R. L. Riley, C. R. Lyons, and D. P. Carosella, Jr., "Transport in Composite Reverse Osmosis Membranes," Chapter 6 in Membrane Processes In Industry and Biomedicine, M. Bier (Ed.), Plenum Press, New York 1971.

X-RAY STUDIES OF ULTRATHIN DESALINATION MEMBRANES

S. Krishnamurthy and D. McIntyre
Institute of Polymer Science
The University of Akron
Akron, Ohio 44304

INTRODUCTION

The importance of reverse osmosis as a novel and effective method of desalination was successfully demonstrated by membranes made from cellulose acetate, the most widely used material in desalination. Two types of membranes have been successfully fabricated: low flux[1] and high flux membranes[2,3].

Low flux membranes are cast from a solution of cellulose acetate and solvent with the subsequent solvent removal by evaporation. The flux and salt rejection are not altered when either surface of the membrane is exposed to the salt solution at high pressure in reverse osmosis, clearly indicating a symmetric nature of the membrane towards flow.

High flux membranes are usually cast from a mixture of polymer, solvent, and nonsolvents through a series of fabricating steps involving casting, evaporation, leaching, and annealing. High salt rejection with reasonable flow rate of desalinized water is realized only when the air dried surface of the membrane faces the high pressure salt solution. High flow rates with poor salt rejection occur when the casting-surface side faces the salt solution[4]. The asymmetric nature of a high flux membrane is the result of an upper dense layer of 2000 Å thickness, on the air dried surface, and an underlying porous substructure formed on the side in contact with the casting surface[5].

Early work on the elucidation of membrane structure, based on electron microscopy[5], dye staining techniques[6-8], and transport studies[4], confirmed the asymmetric nature of the high flux membranes. Further, these studies pointed out that effective desalination occurred essentially through the osmotic skin of the asymmetric membrane.

Extremely thin films (from 600 Å and above) were used in desalination studies[9,10] as a means of improving the flux of desalinized water under reverse osmosis conditions because of their shorter transport paths. These ultrathin films offered an easy means to vary the thickness of desalination membranes unlike the uncontrollable and undefined osmotic skin of the asymmetric membrane.

The performance of these membranes is comparable to that of the high flux Loeb-Sourirajan type membranes. These observations suggest that either the "skin", or the ultrathin membrane by itself, contains the molecular clue concerning the structure of the membrane necessary for effective desalination.

Because of the difficulties involved in handling these ultrathin membranes, only a few characterization studies, very limited in scope and conclusions, have been made to date[9]. As a consequence, thick films[11,12] (15 μ and greater) formed by complete solvent removal in the evaporation of the cellulose acetate-acetone casting solution have been used to characterize the structure of the thin osmotic skin (∿ 2000 Å) of the high flux membrane. Such studies cannot be expected to highlight differences between the skin and the supporting layer.

Techniques have been developed for making and handling both ultrathin and thick films. The ultrathin membranes discussed in this work are approximately 1000-3000 Å thick and should be comparable to the osmotic skin of Loeb-Sourirajan membranes. Only the characterization by X-ray diffraction is presented in this paper.

EXPERIMENTAL

Materials

The polymers used in this investigation were cellulose acetate (CA) E-398-3 and cellulose triacetate (CTA) EXP-4412A, both supplied by Eastman Chemical Products, Inc. Reagent grade acetone, methylene chloride, and methanol were used as solvents in the preparation of solutions. Hexamethylene tetramine was used as a standard to determine instrumental broadening in the X-ray diffraction studies.

Methods

Ultrathin Film Preparation. The following three methods of preparation were employed in casting ultrathin films: the Carnell method[13], casting on a water surface, and casting on a mercury surface.

Ultrathin films of CA and CTA were prepared from dilute solutions of CA in acetone (4 wt %) and CTA in a 90:10 (volume) mixture of methylene chloride and methanol (2.2 wt %) by the Carnell dipping method. In general, the procedure described by Carnell in his original article[13] was used in preparing the clean glass plates. This consisted of cleaning the glass plates with nonabrasive detergents, followed by washing in running water, and drying with lint-free cloth. The dried glass plates were immersed in a suitable solvent such as 2-butanone for at least half an hour and then air-dried for a short time before being introduced into the dilute solution for casting. In the Carnell method, 4" wide x 12" long x 1/4" thick float glass plates with chamferred edges were used as the casting substrate. A variable speed drive mechanism, shown in Fig. 1, was used to immerse the glass plates into the

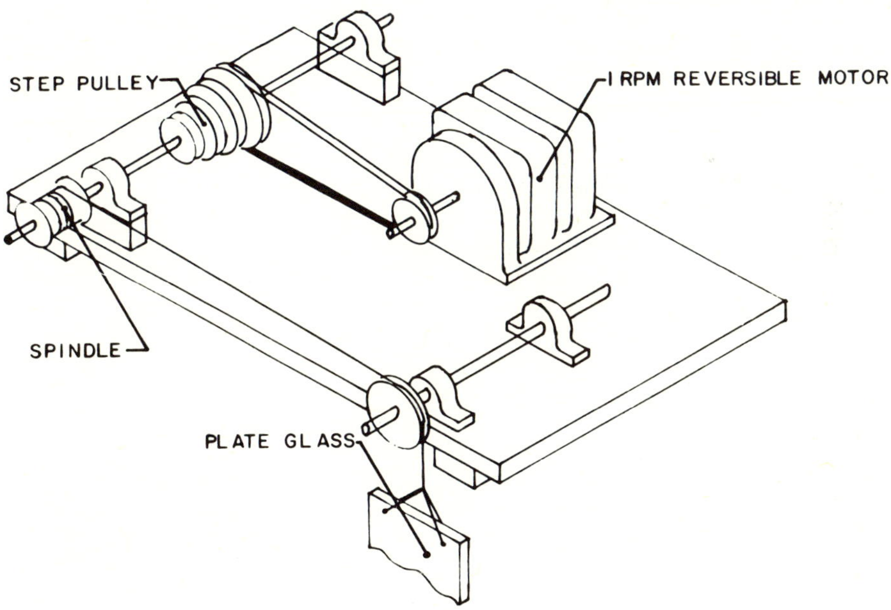

Fig. 1 - Ultrathin Film Caster

polymer solution contained in a tall, one gallon Dewar flask, and then to subsequently withdraw them. The films were cast at a dipping and withdrawing speed of 1.2 mm/second and were allowed to air dry for 30 to 300 seconds. After air drying, the glass plate with the deposited film was introduced into a glass tray containing ice-cold distilled water. By scoring the edges of the glass plate with a sharp tool, the film could be released from the plate to the surface of the water.

In the casting of ultrathin films on a water surface, the polymer solution (∿ 4 wt %) was smeared along the length of a glass stirring rod and the rod was swept across the water surface. Uniformity in the smearing of the solution and the drawing of it across the water surface were highly essential in obtaining good quality films.

Dilute solutions of polymers (∿ 0.2 to 0.4 wt %) were used in the casting of films on a mercury surface. A measured volume of the dilute polymer solution was placed on the mercury surface inside a glass ring that floated on the mercury surface. The liquid film was allowed to dry at ambient conditions for two or more hours.

Thick Film Preparation. Thick films of cellulose acetate and cellulose triacetate were prepared from concentrated solutions (20 to 25 wt %) by casting on a glass surface using an adjustable doctor knife. The film was allowed to dry at ambient conditions before peeling it from the glass plate.

Sample Preparation and Treatment. The ultrathin films prepared by the methods described above were extremely thin, less than 3000 Å in thickness. Because of the thickness, multilayered samples had to be used to obtain meaningful information in the X-ray diffraction studies. Multilayered samples of ultrathin films were prepared by rolling the wet ultrathin film on to a suitable support material and subsequently drying at ambient conditions and later under vacuum. Thick films were also dried under vacuum before use in X-ray diffraction experiments.

Ultrathin and thick film samples were annealed in hot air to study the effect of annealing on their structure. Annealing was often carried out with multilayered samples as attempts to remove single layers of ultrathin films annealed on the glass plates in hot air were not successful. To avoid the curling up of the multilayered film during annealing, the film was sandwiched between glass microscope slides or aluminum sheets sprayed with a silicone mold-release agent. At annealing temperatures under 100°C, two Mylar sheets surrounded the stacked ultrathin films to prevent adhesion and these were in turn sandwiched between glass or aluminum sheets. In all cases, the sandwich was wrapped in aluminum foil and annealed in a hot air oven provided with good temperature control. Annealing was carried out for 4 to 8 hours and the sample was allowed to cool back to room temperature over a period of 2 to 4 hours.

Sample annealing under pressure was carried out in a hydraulic press provided with heaters and temperature control. The multilayered sample was sandwiched between two aluminum sheets sprayed with a mold-release agent, and then inserted between the plates in the hydraulic press. The desired pressure and temperatures were maintained for 4 to 8 hours. The sample was cooled to room temperature over a 2 to 4 hour period. The stacking and orientation of multilayered samples used for X-ray studies is shown in Fig. 2. Orientation Ⓐ was utilized in the recording of the diffraction patterns presented here, unless noted otherwise.

X-Ray

Symmetrical Reflection Geometry. The wide angle X-ray diffraction studies were made with CuKα nickel-filtered radiation generated at 40 KV and 20 mA with an Ultrastable Phillips generator (Model PW 1310). The intensity of scattered radiation was measured using a proportional counter. Continuous scanning at goniometer speeds of 1/8° (all values in this paper refer to

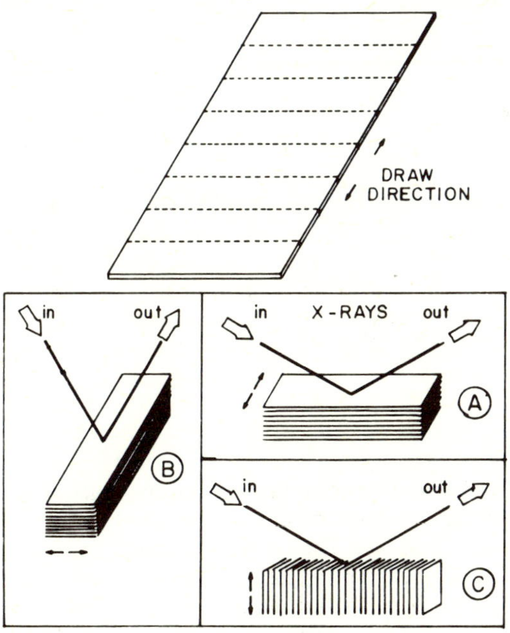

Fig. 2 - Stacking and Orientation of Films for X-Ray

degrees of 2θ) per minute was carried out. At peak areas, step scanning over 1/100° intervals was carried out manually. In all the step scannings, counting of the scattered intensity of radiation was performed for a period of 100 seconds. All of the above mentioned details apply to scattering done in the symmetric reflection mode. The scattering was performed from 4 to 36° and at room conditions in the wide angle X-ray investigation.

In this work, an alternative slit system suggested by Pollack and Alexander[14] was used. This system, besides improving the intensity of scattered radiation, eliminated a systematic error in counting. Accordingly, a larger sample holder with a bigger port was used and the antiscatter slit was removed from the system. For improving the scattered intensity in the higher angle region (12 to 36°), a 1° divergence slit was used instead of the 1/2° divergence slit used in the 6 to 12° interval. In the 6 to 12° and 12 to 36° ranges, the 0.006'' receiving slit was used.

Normal Transmission Geometry (Flat Plate Camera). In this method a small sample specimen was attached to the revolving sample holder with adhesive tape. In this pinhole collimator system, the diffraction pattern of the material was recorded on a photographic plate by the forward reflection method. After exposing the film to CuKα nickel-filtered radiation generated at 40 KV and 20 mA for periods of 6 to 24 hours, the film was developed using normal dark room procedures.

RESULTS

Unannealed Films

The diffraction pattern from 6 to 36° obtained by continuous scanning is presented in Fig. 3 for an "as cast" ultrathin cellulose acetate-acetone film. In this figure, the appearance of multiple peaks around 9.5 and 28.5° is evident. There is only a slight indication of the presence of a peak at 19°. Broad amorphous peaks around 8 to 10° and 16 to 20° are present in all these patterns.

Annealed Films

The scattering patterns recorded at room temperature of hot air annealed, stacked ultrathin films show the growth of sharp peaks at 9.5, 19.0 and 28.5°. These are shown in Figs. 3 and 4. On annealing at 96°C the multiple peaks at 9.5 and 28.5° were still present but showed increased intensity. The development of a peak at 19.0° is clearly evident in the annealed samples.

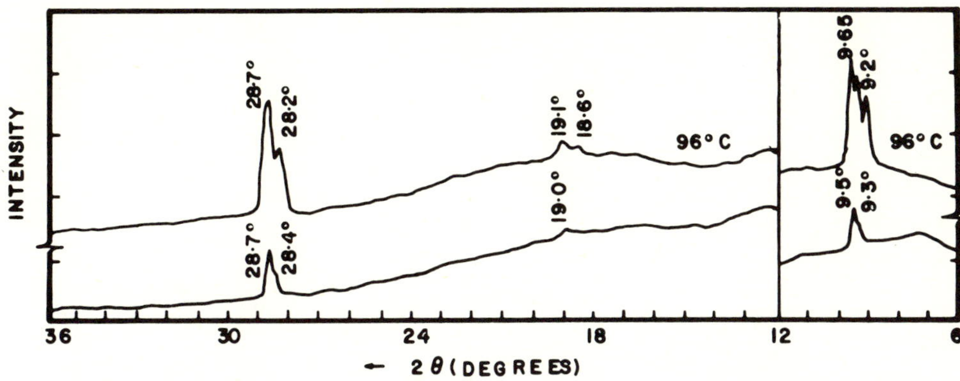

Fig. 3 - X-Ray Diffraction Patterns of Ultrathin CA Films Annealed at Room Temperature and 96°C

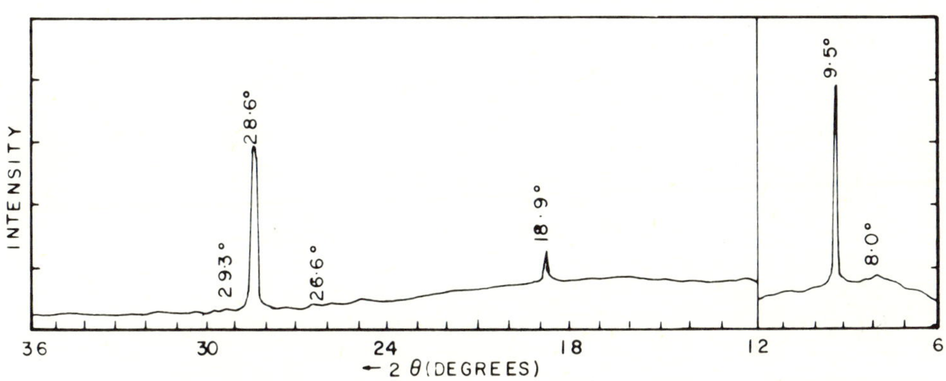

Fig. 4 - X-Ray Diffraction Pattern of Ultrathin CA Films Annealed at 180°C

The samples annealed at 180°C invariably show high intensity single peaks, at 9.5, 19.0 and 28.5° of scattering angle. Besides these "ultrathin peaks" a small peak (of very poor intensity) at 26.6 or 26.7° is found in the 180°C annealed sample.

In Fig. 5 the diffraction patterns of ultrathin film samples of CTA annealed at room temperature, 96, and 180°C are presented. The multiple peaks shown at 9.5 and 28.5° in the X-ray diffractogram of the room temperature annealed film reduce to single peaks in the annealed sample's diffraction patterns. As in the case of ultrathin CA films the peak intensities increase with increasing

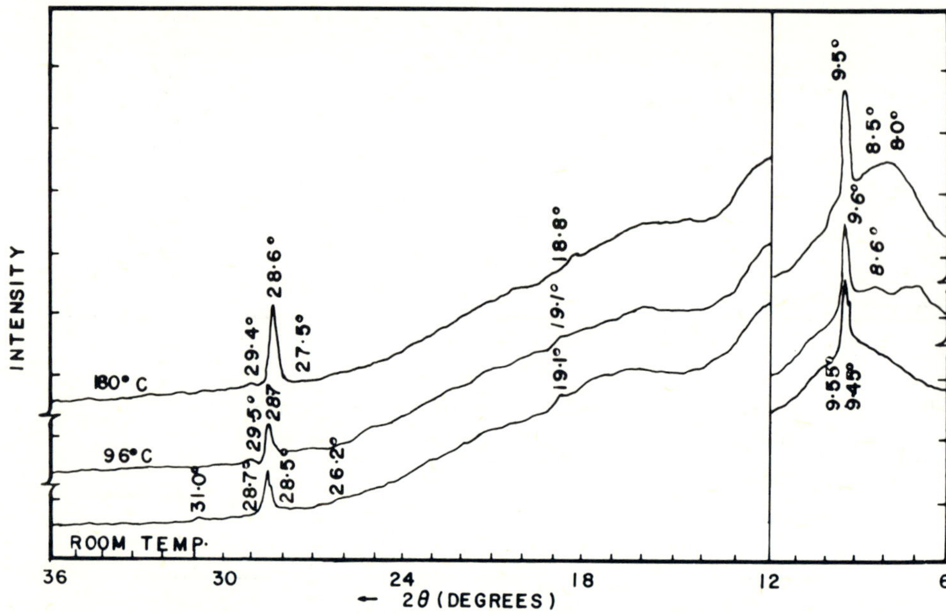

Fig. 5 - X-Ray Diffraction Patterns of Annealed Ultrathin CTA Films

annealing temperature. Though the development of the CTA II crystalline peaks is evident at 96°C, only after annealing at 220°C does full development of the peaks take place. The "ultrathin" structure is still present as shown by the sharp peaks at 9.4 and 28.5° in Fig. 6. Here again the presence of a peak at 19.0° is obscured by the CTA crystalline peaks in the vicinity.

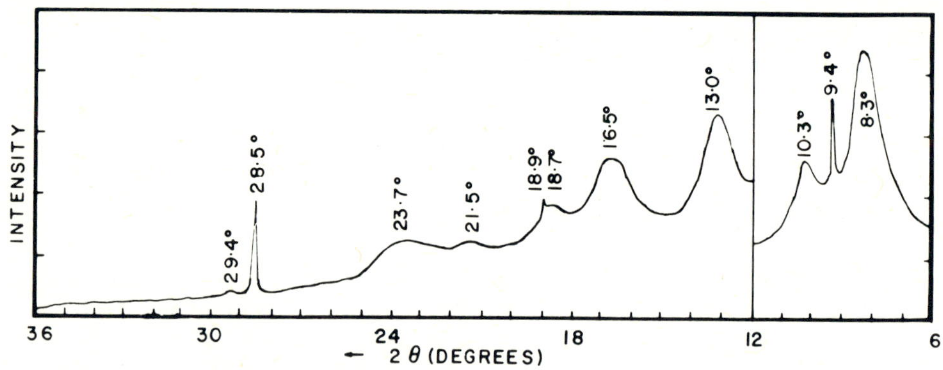

Fig. 6 - X-Ray Diffraction Pattern of Ultrathin CTA Films Annealed at 220°C

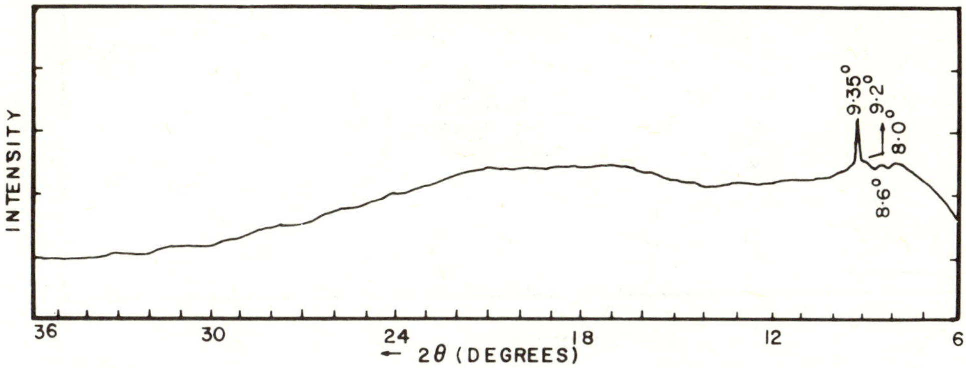

Fig. 7 - X-Ray Diffraction Pattern of Ultrathin CA Films. Ⓒ Orientation.

In Fig. 7 the diffractogram of the sample exposed to X-radiation in the Ⓒ orientation is shown (refer to Fig. 2). The diffractograms for samples annealed at 96 and 180°C were identical to the one annealed at room temperature. Only one peak around 9.3° was present in the diffraction patterns. The diffraction patterns obtained in the Ⓐ and Ⓑ orientations were essentially identical except for a small decrease in the intensity of the peaks in the Ⓑ orientation.

The samples were continuously rotated along the X-ray beam axis while the flat plate diffraction patterns were taken. The angular positions of the peaks in the flat plate pattern for an ultrathin CA film annealed at 180°C were 7°21', 18°45', 25°50', and 35°10'.

Pressed Films

Pressing at 1500 psi with simultaneous heating at 180°C resulted in the development of the conventional CTA II peaks together with the "ultrathin peaks" at 9.5 and 29.5° as shown in Fig. 8. These two peaks were of reduced intensity compared to those of the annealed but unpressed sample. The presence or absence of a peak at 19° cannot be confirmed because of the conventional crystalline peaks around this angular region.

Ultrathin Films Versus Thick Films

The results of X-ray diffraction (not shown here) of thick films of CA cast from acetone solutions show only two broad amorphous peaks in the 8 to 10° and 18 to 20° regions. The "ultrathin

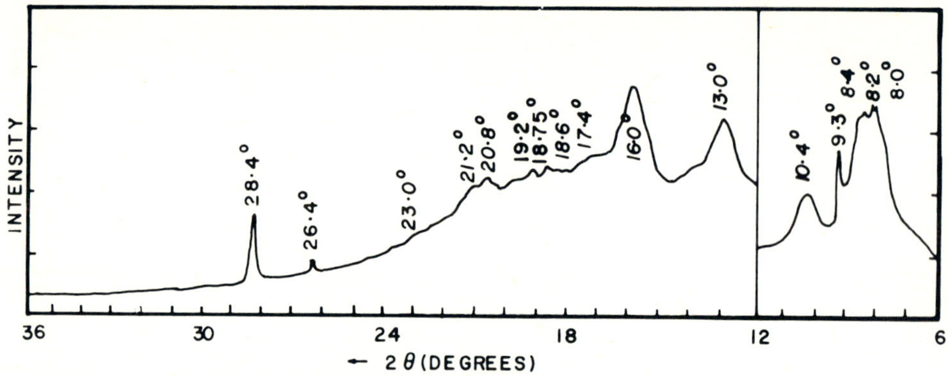

Fig. 8 - X-Ray Diffraction Pattern of Ultrathin CA Films Annealed at 180°C Under 1500 psi

peaks" at 9.5, 19.0, or 28.5° were completely absent. Annealing thick films (40- and 100 μ-thick individual layers) at 180°C does not result in the formation of either the CTA II peaks or the "ultrathin peaks." On annealing multilayers of thick films at 180°C and 1500 psi pressure, a poor development of the CTA II structure pattern is evident. Extremely small "peaks" appear at 9.5 and 28.5° whose intensities are negligibly small (see Fig. 9).

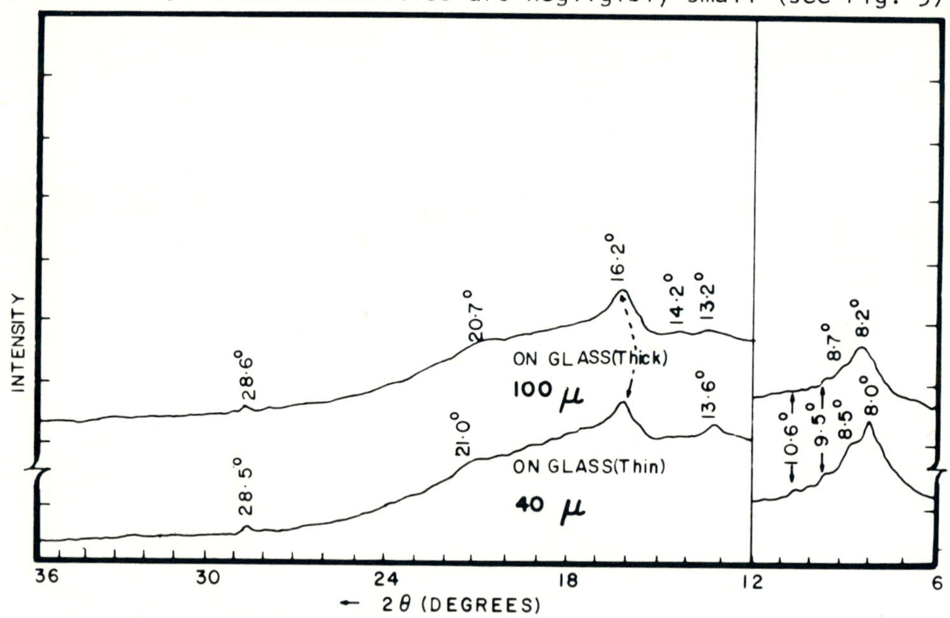

Fig. 9 - X-Ray Diffraction Patterns of Thick CA Films, Annealed at 180°C Under 1500 psi

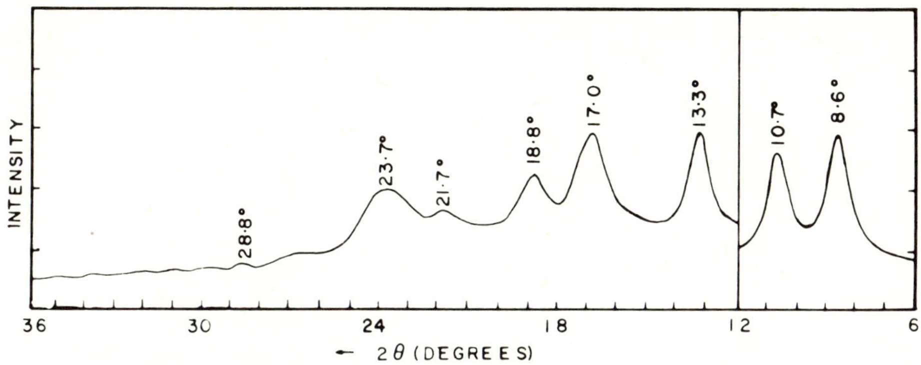

Fig. 10 - X-Ray Diffraction Patterns of CTA Films Annealed at 280°C

The diffraction pattern of a single layer of thick CTA film annealed at 280°C (close to the decomposition temperature of the material) is shown in Fig. 10. The CTA crystalline peaks appearing in this diffractogram compare well with those reported for CTA II[15,16,17]. Also the peak positions reported for ultrathin films of CA and CTA (shown in Figs. 6 and 8) correspond well with those of CTA II shown in Fig. 10.

Effect of Different Casting Surfaces

The X-ray diffraction patterns of ultrathin films of CA and CTA cast on a water surface are identical to those cast on a glass surface by the Carnell method, both in the "as cast" state and in the annealed state, and are not shown here. In Fig. 11 are shown the diffraction patterns of mercury-cast ultrathin cellulose acetate films annealed at room temperature, 96, and 180°C. The "as cast" film exhibits very small peaks (slightly above the noise level) at 9.5° and multiple peaks around 28 to 30°. Intensification of the peaks and reduction of the multiple peaks to single sharp peaks take place with annealing as shown in the diffraction patterns.

DISCUSSION

The 's' Spacing and Crystal Geometry

The X-ray diffraction patterns of unannealed multilayered ultrathin CA and CTA films shown in Figs. 3 and 5 reveal the presence of multiple peaks at 9.5 and 28.5°, the peak angles bearing a 1:3 relationship in these regions within the experimental

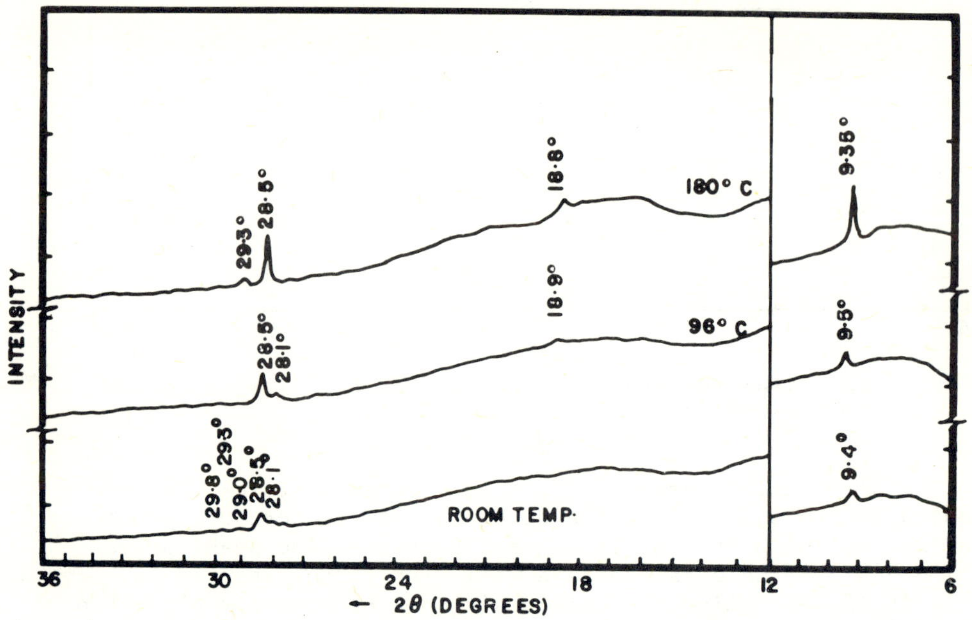

Fig. 11 - X-Ray Diffraction Patterns of Annealed Ultrathin CA Films Cast on Mercury

error. The peaks at 9.5, 19.0, and 28.5° are sharp and intense in the annealed sample (180°C), and the angles bear a 1:2:3 relationship.

The presence of these three main peaks with a definite but simple ratio of the angular positions (and, therefore, in their reciprocal spacings) suggests that the material is arranged in some well defined way. The 1:2:3 ratio of the angular position of the peaks corresponds within experimental error to a 3:1.5:1 ratio in the respective Bragg spacings of regularly spaced scattering units. For example in mica (aluminum silicate hydrate), which has a hexagonal crystal geometry, the three main crystal peaks are at 9.14, 4.57 and 3.04 Å with the peak intensities in the ratio of 40:50:100[18]. In the 180°C annealed ultrathin CA film the three main peaks are at 9.45, 4.77, and 3.15 Å and their respective peak intensities are in the ratio of 100:10:50 taking the amorphous diffraction background as the base line. Although the intensity ratios of these peaks will be slightly different if the instrumental background is used as the base line there is no reason to suspect that the peaks are the result of a mineral impurity from a chemical viewpoint. In spite of the absence of other peaks, a hexagonal crystal geometry was considered in the ultrathin CA films in analogy with the mica system.

Among other possible crystal structures, the tetragonal crystal cell system proposed for hydroxypropyl cellulose[19] was considered. The X-ray diffraction pattern of hydroxypropyl cellulose was reported to have seven meridional reflections corresponding to the ratios of the angular positions of 1:2:3:4:5:6:7. Based on this report, attempts were made to locate peaks, if any, at 4.75 and 33.25° as well as at other multiples of 4.75° of 2θ. The experimental X-ray diffraction patterns did not indicate the presence of any other pronounced sharp peaks besides the three main peaks.

The possible crystal geometry which CA might assume under suitable conditions was also considered. The CTA II crystal structure has been shown to be the most probable crystal structure for secondary CA[20,21]. But the peak positions reported for CTA II do not correspond to any of these ultrathin film peak positions. This observation definitely rules out the possibility of attributing these "ultrathin peaks" to previously studied bulk polymer crystalline structures.

Other ordered systems besides crystalline systems were considered. Among these are the lipid-water systems and liquid crystals [22,23]. It is customary to express the X-ray diffraction scanning angle in terms of the reciprocal spacing, s, where $s = 2(\sin\theta)/\lambda$. For the ultrathin films investigated in this work the peak positions correspond to s values in the ratio of 1:2:3. This regularity in the s spacing is similar to that reported by Luzzati, et al. [22,23]. They have expressed the peak positions in the lipid-water system as $s = \ell/d$ with $\ell = 1,2,3,\ldots$ etc. Most of the s spacings reported in the lipid-water system have values of 0.1 $Å^{-1}$ or less, whereas in the ultrathin films they start around 0.1 $Å^{-1}$ and become larger. Luzzati, et al., have discussed in detail the significance of these spacings and conclude that a well developed order exists in the lipid-water system. This order exists on a large scale, i.e., on the scale of a whole molecule, the maximum s value being less than 0.1 $Å^{-1}$. If an order exists in the ultrathin films, the order seems to be on a small scale because the smallest s value is not less than 0.1 $Å^{-1}$. That is, the ordering may exist over distances of the order of a repeating segment of the polymer.

Extent of Ordering

In order to determine the extent to which the ordered phases occur, integral half-width measurements of the peaks were employed. Usually the integral half-width peak measurements are used in calculating the size of the crystallites. The crystallite size and the half width are related through the Debye-Scherrer equation [24-28]:

$$D = K\lambda/(\beta \cos \theta),$$

where β is the corrected half width. In general the half width of a crystalline peak is broadened due to several factors including instrumental broadening. The instrumental broadening was determined by using a pressed sample of hexamethylene tetramine.

Step scanning in the intervals of 1/100° of 2θ was performed on ultrathin CA films cast from acetone and annealed at 180°C. Corrections due to the absorption of the specimen were incorporated in the step scanning diffraction patterns according to the procedure given by Keating and Warren[29]. Crystallite sizes measured by using the Debye-Scherrer equation with and without corrections were not significantly different and are around 1000 Å. If corrections due to instrumental broadening are to be incorporated in the observed half widths, the sizes obtained would be even larger. This suggests that these peaks may be due to some ordered phase that extends over macroscopic areas. However, the contribution to the total intensity of scattering from these ultrathin peaks is around 10%. In order to be positive that these ultrathin peaks are not due to any foreign impurities in the films, electron probe microanalysis was performed on these ultrathin films[30]. According to the analysis that was sensitive to elements above fluorine, the total amount of foreign matter in these films is less than 0.1%. Another alternative for the large size of the ordered region may be due to large single crystals[31]. In the absence of electron micrographs on these ultrathin films, this possible structure cannot be discussed further.

CONCLUSIONS

Ultrathin films of cellulose acetate have a different X-ray diffraction pattern than thick films of cellulose acetate. The differences may be divided into two categories:

1. The ultrathin films of CA, unlike the thick films, can crystallize in the cellulose triacetate II structure very readily when pressure is applied and the temperature is above 180°C. Thick and ultrathin cellulose triacetate films exhibit high crystallinity that develops at different temperatures. Thick films of CTA develop crystallinity on annealing to 280°C, whereas ultrathin films develop high crystallinity at 220°C. Both these features are consistent with the idea that the ultrathin film represents a more ordered polymer system than the thick film.

2. The major difference is the presence of very sharp peaks in ultrathin films. These peaks represent Bragg spacings in a ratio of 1:1.5:3 and are dependent on the orientation of the film. The amount of this ordered material is about

10%, but the ordering extends over a large area. The spacings do not correspond to the crystal spacings known for bulk cellulose triacetate, although these distances represent inter-atomic distances that are close to the expected inter-chain distances. The peaks do not disappear upon heating nor do different casting surfaces totally eliminate them. Electron probe microanalysis for elements above fluorine shows less than 0.1% total impurity. Therefore, these spacings can only be considered as an ordered polymer phase. The effective molecular packing required for a good desalination membrane may be this ordered phase.

ACKNOWLEDGMENTS

This work was performed under grants from the Office of Saline Water, U. S. Department of Interior, Washington, D. C. The cellulose triacetate sample EXP-4412A was kindly provided by Louis D. Moore, Jr. of Tennessee Eastman Co. Electron microprobe analysis was performed by B. Kersch through the courtesy of G. C. Berry at Carnegie-Mellon University.

REFERENCES

1. C. E. Reid and E. J. Breton, J. Appl. Polymer Sci., 1, 133 (1959).

2. S. Loeb and S. Sourirajan, Advan. Chem. Ser. 38, 117 (1963).

3. S. Manjikian, Ind. Eng. Chem. Prod. Res. Develop., 6, 23 (1967).

4. S. Loeb, UCLA Dept. of Engineering, Rept. No. 63-32 (1963).

5. R. L. Riley, J. O. Gardner, and U. Merten, Science, 143, 801 (1964).

6. B. Keilin, Office of Saline Water Research and Development Progress Report No. 84, U. S. Government Printing Office, Washington, D. C., 1963.

7. B. Keilin, Office of Saline Water Research and Development Progress Report No. 117, U. S. Government Printing Office, Washington, D. C., 1964.

8. W. Banks and A. Sharples, Office of Saline Water Research and Development Progress Report No. 143, U. S. Government Printing Office, Washington, D. C., 1965.

9. P. S. Francis, Office of Saline Water Research and Development Progress Report No. 177, U. S. Government Printing Office, Washington, D. C., 1966.

10. H. K. Lonsdale, R. L. Riley, C. E. Milstead, L. D. LaGrange, A. S. Douglas, and S. B. Sachs, Office of Saline Water Research and Development Progress Report No. 577, U. S. Government Printing Office, Washington, D. C., 1970.

11. H. K. Lonsdale, U. Merten, R. L. Riley, K. D. Vos, and J. C. Westmoreland, Office of Saline Water Research and Development Progress Report No. 111, U. S. Government Printing Office, Washington, D. C., 1964.

12. H. K. Lonsdale, U. Merten, and R. L. Riley, J. Appl. Polymer Sci., 9, 1341 (1965).

13. P. H. Carnell, J. Appl. Polymer Sci., 9, 1863 (1965).

14. S. S. Pollack and L. E. Alexander, J. Chem. Eng. Data, 5, 88 (1960).

15. B. S. Sprague, J. L. Riley, and H. D. Noether, Text. Res. J., 28, 275 (1958).

16. S. Watanabe, M. Takai, and J. Hayashi, J. Polymer Sci., C28, 825 (1968).

17. W. J. Dulmage, J. Polymer Sci., 26, 277 (1957).

18. ASTM, STP 48-M2, "Index to the Powder Diffraction File", 1963.

19. R. J. Samuels, J. Polymer Sci., A2 (7), 1197 (1969).

20. R. W. Work, Text. Res. J., 19, 381 (1949).

21. D. W. Tanner, R. D. Patel, G. C. Berry, and T. G. Fox, Carnegie-Mellon Univ. personal communication.

22. R. P. Rand and V. Luzzati, Biophys. J., 8, 125 (1968).

23. T. Gulik-Kyzywicki, E. Rivas, and V. Luzzati, J. Mol. Biol., 27, 303 (1962).

24. L. E. Alexander, X-Ray Diffraction Methods in Polymer Science, Wiley, New York, 1969, p. 423.

25. R. Bonart, R. Hosemann, and R. L. McCullough, J. Appl. Polymer Sci., 4, 199 (1963).

26. C. H. MacGillavry and G. D. Rieck, International Tables for X-Ray Crystallography, Vol. III, The Kynoch Press, Birmingham, England, 1968, p. 318.

27. D. R. Buchanan, R. L. McCullough, and R. L. Miller, Acta Cryst., 20, 922 (1966).

28. S. Kavesh and J. M. Schultz, J. Polymer Sci., A2 (8), 243 (1970).

29. D. T. Keating and B. E. Warren, Rev. Sci. Inst., 23, 519 (1952).

30. G. C. Berry, Carnegie-Mellon Univ. personal communications.

31. G. N. Patel and R. D. Patel, J. Polymer Sci., A2 (8), 47 (1970).

POROUS TUBULETS FOR DESALINATION BARRIERS

Bernard Baum, R. A. White, and W. H. Holley, Jr.
DeBell & Richardson, Inc.
Hazardville, Enfield, Connecticut 06082

INTRODUCTION

One economic requirement of the reverse osmosis process is that the operating energy be conserved by providing a high ratio of membrane surface area to volume of reverse osmosis module. One approach to achieving this high ratio is to coat desalinating films or membranes onto fine porous tubulets, which can be closely packed in great numbers into modules, to give a high surface area-to-volume ratio.

The purpose of this research was to produce thermoplastic compounds which, when extruded into fine tubulets and leached in water, would serve as rigid porous supports for reverse osmosis membranes. The technique for producing these porous materials was to compound various water-soluble liquids or polymers into thermoplastics and extract with water to produce the desired porosity.

Target requirements for the porous tubulets were a porosity in the range of 40 to 80%, pore size of 1000 to 5000 Å (0.1 to 0.5 μ), and a water flux of 100 to 1000 gal/ft^2-day (gfd) at 40 psi. This high flow is essential to permit the membrane to reject the salts without developing excessive back pressure across the support. The small pore size is necessary to provide adequate support for the desalination barrier without risk of rupture of the barrier.

EXPERIMENTAL TECHNIQUES

Flat Films

Formulations of rigid thermoplastic and water-soluble additives were compounded on a steam-heated, differential-speed, two-roll mill. To evaluate these formulations, flat films were compression molded from the milled compounds and were leached by tumbling in distilled water. Samples were dried at 40°C under vacuum for 18 hours and weight loss (percent leachables extracted) gave a rough measure of porosity. Water flow rate through the porous materials was determined by passing distilled water through the films at 40 psig pressure in a pressure filter funnel.

Promising samples were tested for pore size by forcing mercury into the samples under pressure. Curves of cumulative pore volume (volume of mercury) versus pore diameter (a function of the pressure necessary to force the mercury into the sample) yielded average pore diameters and the volume distributions of pore diameters. A typical intrusion curve is presented in Fig. 1 for a material with an average pore size of 0.35 μ.

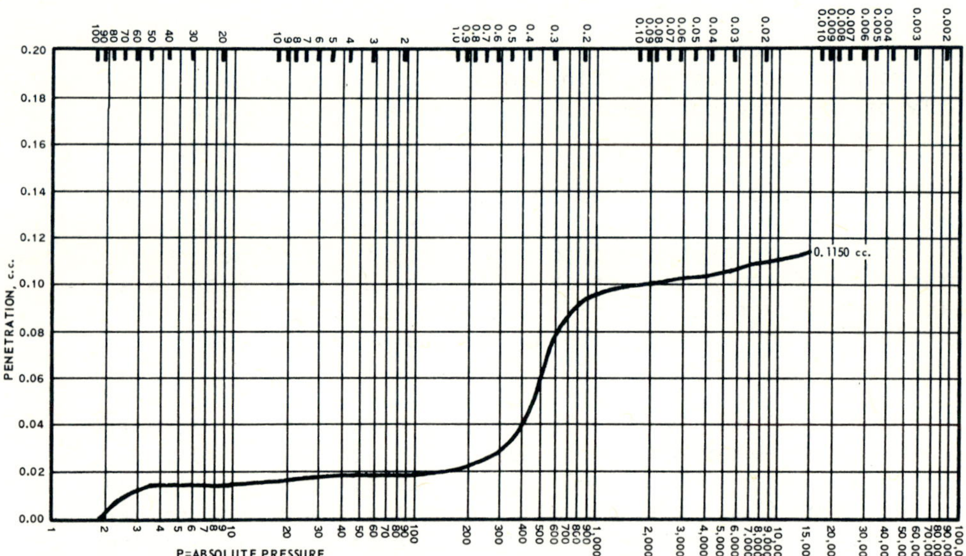

Fig. 1 - Volume Penetration Vs. Pressure or Pore Size. Formulation 82653. The Upper Scale Gives The Pore Diameter in Microns. Measured Pore Volume: 0.15 cc/g

Scanning electron photomicrographs gave a clearer understanding of pore size and shape at the surface.

Tubulets

Tubulets were prepared from either milled stock or from dry blended materials. Extrusion trials were made with a 1" Killion extruder using an annular die measuring 0.050" inside diameter and fitted with a 0.028" mandrel. Some of the tubulets were drawn down to 0.030"-0.035" OD with 0.008"-0.010" wall. As with flat films, tubulets were leached in water, dried, and tested for pore size by mercury intrusion porosimetry.

A conventional test loop with two test cells was used to determine flow rate of the porous tubulets (Fig. 2). Porous tubulets were supported within perforated 1/2" OD by 16" acrylic tubes by sealing both ends in an epoxy potting compound (Figs. 3 and 4). The top end of the tubulet was sealed by encapsulating it within the epoxy potting compound while the bottom end protruded through the epoxy seal to facilitate collecting the product water. The tubulet test cells were then positioned in the test cell housing (Fig. 5). Water flow rates were determined by circulating water at 40 psig and collecting the effluent in a graduated cylinder (Fig. 6). Approximate pore size was established by determining if latex emulsion particles of 1 μ or 0.2 μ diameter would pass through the tubulet or porous film.

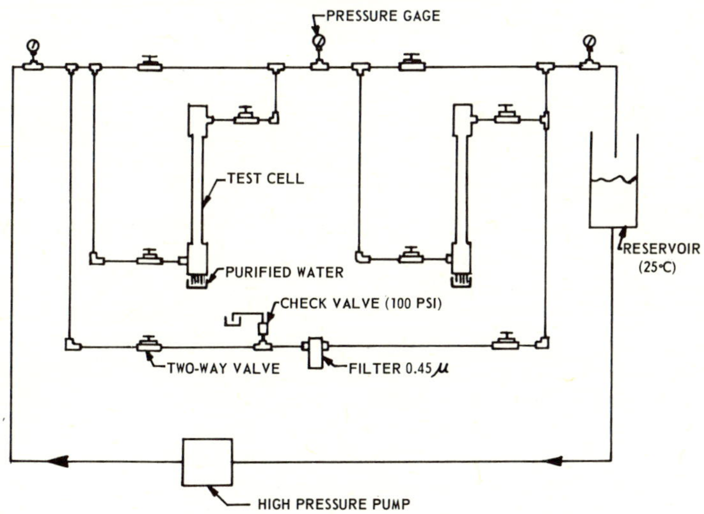

Fig. 2 - Tubulet Test Cell System

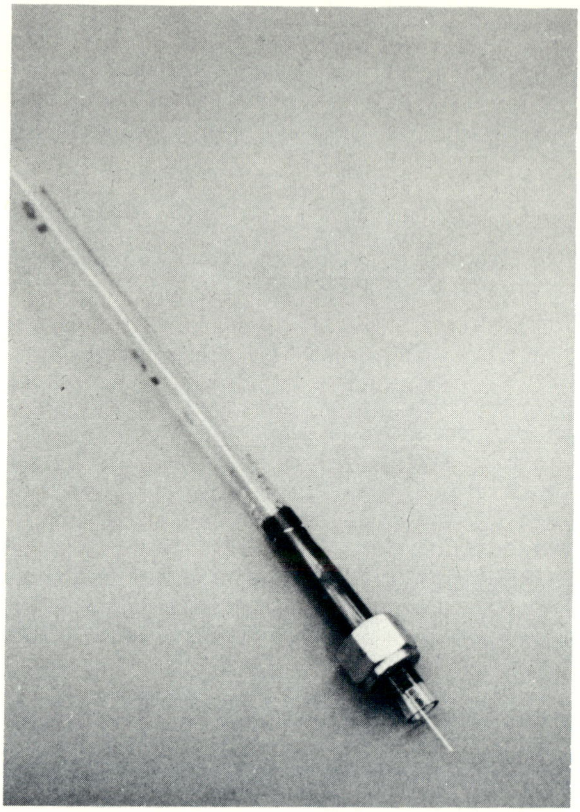

Fig. 3 - Tubulet Test Cell

POROUS MATERIAL DEVELOPMENT

Preparation and Properties of Flat Films

To develop a porous material, numerous thermoplastics and water-soluble additives were screened. Rigid polymer systems included low and medium molecular weight polyvinyl chloride (PVC), vinyl chloride/vinyl acetate copolymer, vinyl chloride/propylene copolymer, blends of vinyl chloride/propylene copolymer and vinyl chloride homopolymer, styrene/acrylonitrile copolymer (SAN), polystyrene (PS), and polypropylene (PP).

PVC Homopolymer. To produce an easily extractable, porous PVC, low and medium molecular weight polyvinyl chlorides were blended with processing aid, stabilizer, surfactant, and finally water-soluble polyvinyl methyl ether (PVME) on a two-roll mill. The formulations were:

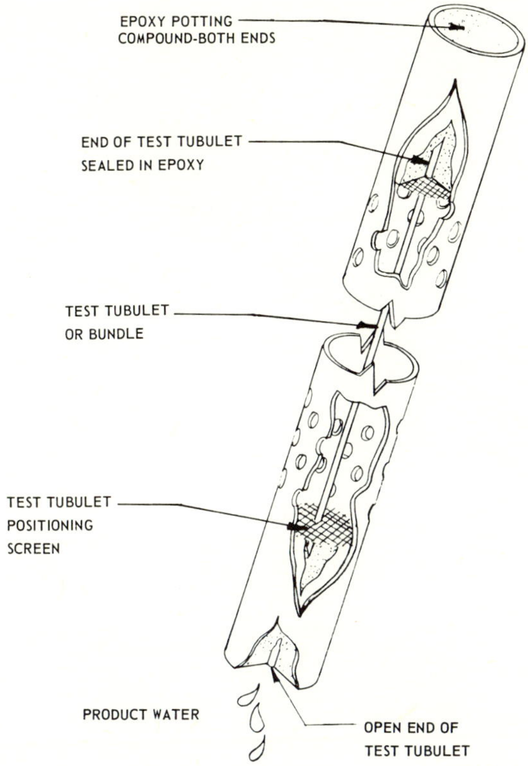

Fig. 4 - Enlarged View of the Tubulet Test Cell

Formulation 82653 (Polyvinyl Chloride Homopolymer)

Material	Manufacturer	Function	Quantity (Parts Solids)
Geon 103 EP or Rucon B-20	B. F. Goodrich Hooker Chemical	PVC homopolymer resins	100
Gantrez M-154	GAF Corporation	PVME[(1)]	15[(2)]
Triton X-100	Rohm & Haas	Water-soluble surfactant[(1)]	10
Acryloid K120N	Rohm & Haas	Processing aid	3
Thermolite 831	M&T Chemicals	Stabilizer	2

[(1)] Water-leachable additives (channel formers).
[(2)] Thirty parts of a 50% water solution.

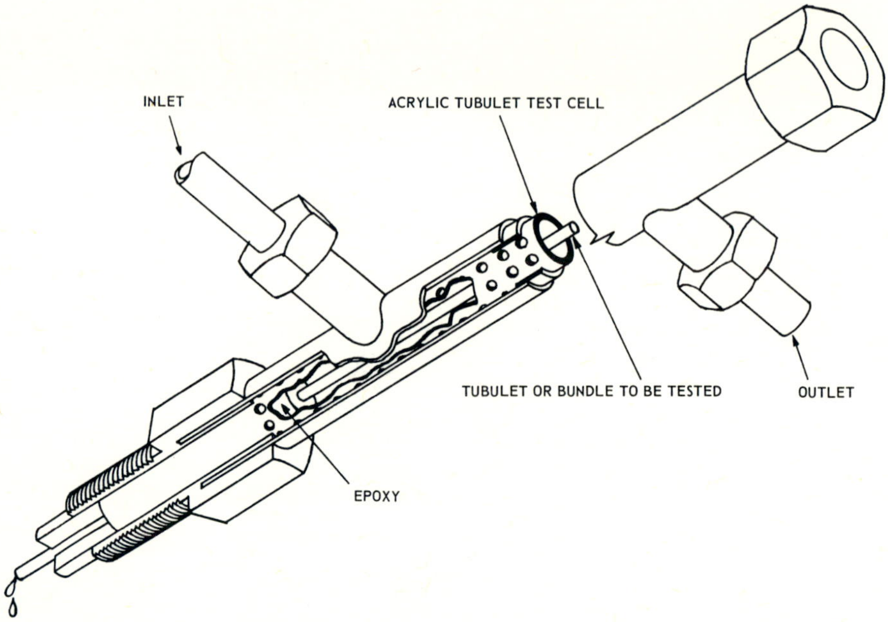

Fig. 5 - Tubulet Test Cell Assembly

Inherent water permeabilities, normalized for a 5-mil film, were two orders of magnitude greater than that of a cellulose acetate asymmetric membrane (Table 1). The latex pore-sizing test showed an average pore size of less than 0.2 μ for the Geon 103 EP, but the pores in the Rucon material were too large with this particular formulation. Mercury intrusion porosimetry on the Geon compound indicated an average pore size of about 0.35 μ; 80% of the pores fell between 0.05 μ and 0.08 μ. Scanning electron photomicrographs show the surface porosity (Fig. 7).

In order to improve the water flux through the leached Geon 103 EP/PVME formulation (82653), a higher molecular weight PVME (Gantrez M-155) was substituted for Gantrez M-154. This formulation, which was very similar to 82653, produced variable results. One batch gave essentially 100% extraction of leachable material and 1200 gfd water flux at 40 psig. Another batch allowed only 61% extraction of leachable ingredients and 53 gfd. One source of this variability was apparently the processing temperature.

To study this phenomenon, samples were compression molded at various temperatures. Increasing compression molding temperature resulted in a decreasing percent leaching of water extractables

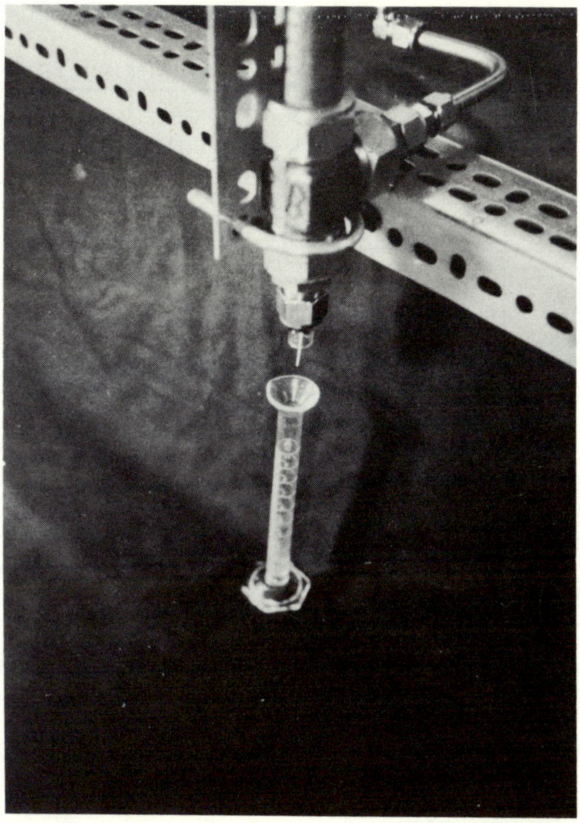

Fig. 6 - Collecting Effluent During Water Flow Test

(Fig. 8) with a sharp drop-off at about 187°C. Tensile strength, elongation, and modulus increased with compression molding temperature. Modulus seemed to level off at about 180°C. It is obvious that processing temperature is critical in achieving a proper balance of physical properties and degree of porosity.

Other Materials. Porous films prepared with similar water-leachable additives from a blend of vinyl chloride/propylene copolymer and PVC homopolymer had a water flow rate of 925 gfd and an average pore size of 0.65 μ (Table 2, Fig. 9). Styrene/acrylonitrile copolymer in a similar formulation gave a flux of 2700 gfd at 40 psig (Table 3). Various compounds were formulated with polystyrene and polypropylene, but all had very low water flow rates (Table 4).

Extrusion of Tubulets

Extrusion Conditions. PVC Formulation 82653, when compounded on a two-roll mill and extruded, was straw-colored, indicating

TABLE 1

PROPERTIES OF LOW VS MEDIUM MOLECULAR WEIGHT PVC FILMS

Composition		Material Extracted by Water Leaching	Water Flow Rate through Film (gfd)	Passes 1 μ Latex	Normalized Permeability for 5-Mil Sheet[2]
Low molecular weight PVC PVME Triton X-45	100[1] 15 10	3.7	23	Yes	8.0×10^{-4}
Medium molecular weight PVC PVME Triton X-100	100[1] 15 10	10	180	No	3.0×10^{-3}
Cellulose acetate asymmetric membrane		-	-	-	1×10^{-5}

(1) Formulation also contained 3 phr Acryloid K120N and 2 phr Thermolite 831.

(2) g/cm^2-sec-atm at 1500 psig and 25°C with 3.5% sodium chloride.

---

slight degradation. To minimize this effect, the long temperature/pressure chamber in the die throat was removed. The resulting decrease in volume within the die reduced the residence time of the plastic material in the extruder, reducing the possibility of material decomposition. To further reduce degradation, a "dry blend" approach followed by extrusion compounding was substituted for the relatively lengthy milling procedure. The extrudate was smooth by both methods.

The blend of PVC homopolymer and vinyl chloride/propylene copolymer gave an extrudate that was very rough and had low strength. The SAN (styrene/acrylonitrile copolymer) extruded without melt fracture and could be drawn down to as thin as 5 mil OD tubulets. The SAN tubulets split longitudinally on leaching, however, and were unsatisfactory.

Physical Properties of Tubulets. In order to compare the tensile properties of PVC Formulation 82653, compounded by milling, against the same formula compounded by dry blending and extrusion, 10-mil sheets of both were prepared by compression molding. Both

Fig. 7 - Surface of a Porous PVC Film - 1100X

the ultimate elongation and tensile strength of unextracted sheet molded from the milled compound were higher than that of the unleached sheet made from dry blend (Table 5). The water-leached sheet has similar tensile strength and somewhat lower percent elongation in comparison to the unleached sheet.

Tubulets extruded from the dry blend showed a marked decrease in elongation on drawing down to 0.035" OD from a 0.050" ID die (Table 6), indicating that a tougher tubulet could be obtained by extruding with no draw-down through a smaller extrusion die.

Flow Rate and Pore Size of PVC Porous Tubulets. Tubulets approximately 30 and 60 mils in diameter were extruded from both the PVC homopolymer (82653) milled compound and from dry blend. The water flow rates through the 30-mil leached tubulets extruded from dry blend were 3 to 5 times higher than those extruded from milled stock (Table 7).

TABLE 2

PROPERTIES OF PROPYLENE-VINYL CHLORIDE COPOLYMER AND PROPYLENE-VINYL CHLORIDE COPOLYMER/VINYL CHLORIDE HOMOPOLYMER BLENDS

Material	Parts	% Extract-ables in Sample	% of Extract-ables Leached	Water Flow Rate[(1)] (gfd)
VCl/propylene copolymer PVC homopolymer Dowfax 2A1 PVME	50 50 50 15	53.8	67	247
VCl/propylene copolymer PVC homopolymer Dowfax 2A1 PVME	50 50 11.25 25	25.1	76/87	925
VCl/propylene copolymer Dowfax 2A1 PVME	100 15 15	22	79	7.0
VCl/propylene copolymer Dowfax 2A1 PVME	100 35 15	32.6	100	180
VCl/propylene copolymer PVME Triton X-100	100 15 10	19.7	6.0	0
VCl/propylene copolymer Polyox WSR-301[(2)]	100 25	20	-	0

(1) Normalized for a 5-mil thickness at 40 psig.

(2) Polyethylene glycol (Union Carbide).

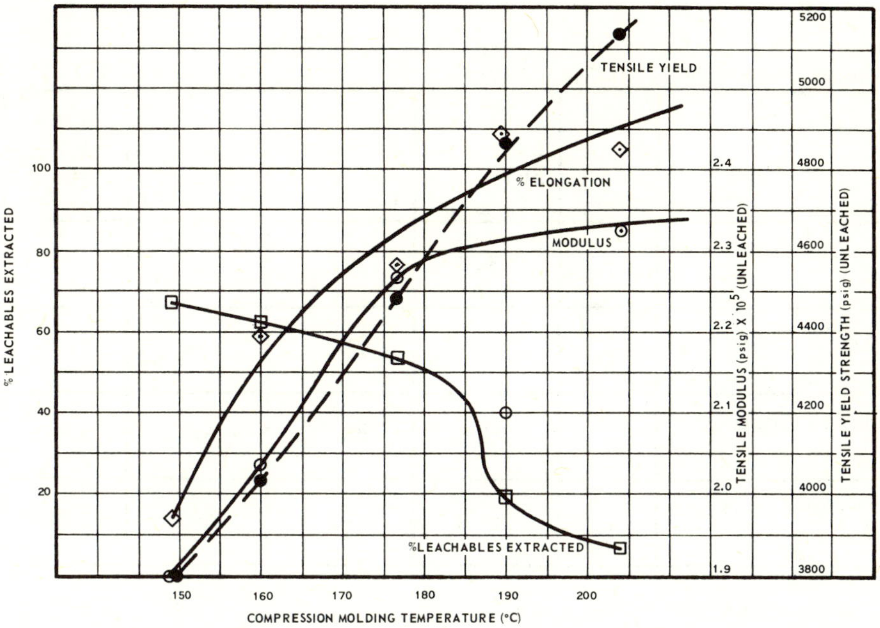

Fig. 8 - Properties Vs. Molding Temperature of PVC/PVME Films

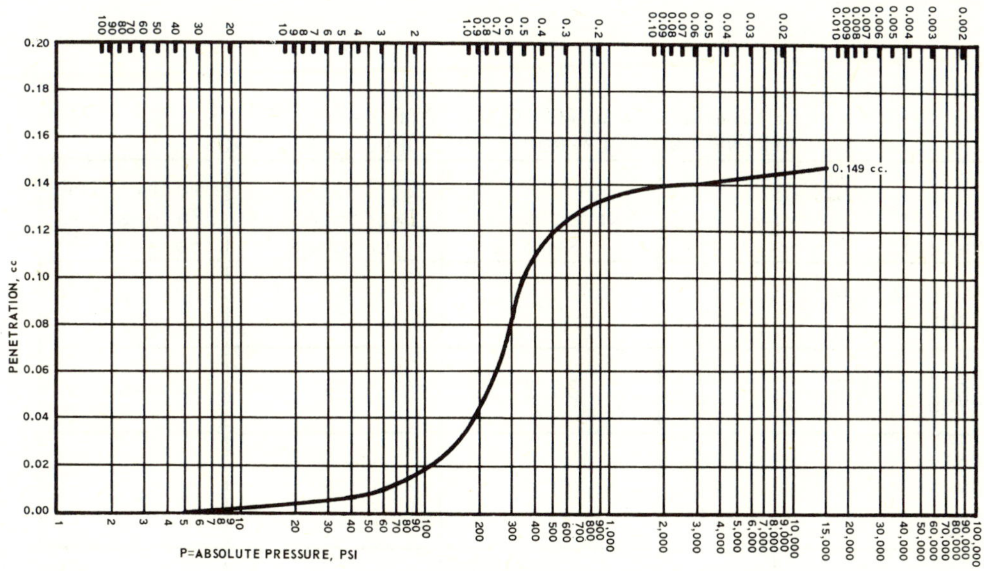

Fig. 9 - Volume of Mercury Penetration Vs. Pressure or Pore Size for PVC Homopolymer/Copolymer Blend. The Upper Scale Gives the Pore Diameter in Microns. Measured Pore Volume: 0.27 cc/g.

TABLE 3

LEACHING AND WATER FLOW DATA FOR POROUS FILMS OF STYRENE/ACRYLONITRILE COPOLYMER RESIN

Material	Parts	% Extractables in Sample	% of Extractables Leached	Passed Latex[1]	Water Flow Rate[2] (gfd)
Styrene/AN Copolymer PVME	100 15	13.0	7.7	-	-
Styrene/AN Copolymer PVME	100 50	33.3	77	Yes	5200
Styrene/AN Copolymer PVME	100 40	28.5	80	No	1000
Styrene/AN Copolymer PVME[3]	100 40	28.5	-	No	2700

(1) 0.2 μ average particle size.

(2) Normalized for a 5-mil thickness at 40 psig.

(3) Lower molecular weight.

TABLE 4

LEACHING AND WATER FLOW RATE DATA FOR POLYSTYRENE AND POLYPROPYLENE POROUS MATERIALS

Material	Parts	% Extractables in Sample	% of Extractables Leached	Water Flow Rate at 40 psig (gfd)
Polystyrene PVME	100 25	20	38	0
Polystyrene PVME	100 50	33.3	72	0
Polystyrene PVME Surfactant	100 25 15	28.5	72	0
Polypropylene PVME	100 25	20	20.7	0

TABLE 5

EFFECT OF DRY BLENDING AND LEACHING ON TENSILE PROPERTIES

Description of Sample	Not Leached: Tensile Break (psi)	Not Leached: Ult. Elong. (%)	Leached: Tensile Break (psi)	Leached: Ult. Elong. (%)
Compression molded from milled stock	3150	33	3510	20
Compression molded from dry-blended and extruded material	2610	15	-	-

TABLE 6

EFFECT OF DRAW-DOWN ON TENSILE PROPERTIES OF TUBULETS

Sample Dimensions: OD (in.)	Sample Dimensions: ID (in.)	Not Leached: Tensile Break (psi)	Not Leached: Ult. Elong. (%)	Leached: Tensile Break (psi)	Leached: Ult. Elong. (%)
0.035 (1)	0.015	3260	8	4176	5
0.055	0.022	2990	46	4110	15

(1) Drawn down from 0.055 inch.

TABLE 7

PROPERTIES OF POROUS PVC TUBULETS

OD (in.)	Wall Thick. (in.)	Water Flow Rate (gfd)	Passes 0.2 μ Latex	Mercury Porosimetry: Avg. Pore Size (μ)	Mercury Porosimetry: Range (3) (μ)
0.031 (1)	0.0085	3-8	No	0.12	0.05-0.16
0.033 (2)	0.010	20-30	No	-	-
0.060 (1)	0.012	400-1100	Yes	0.45	0.03-1.3
0.055 (2)	0.016	4-6	No	-	-

(1) Extruded from milled compound.
(2) Extruded from dry blend.
(3) Range in microns between which 80 volume percent of pores fall.

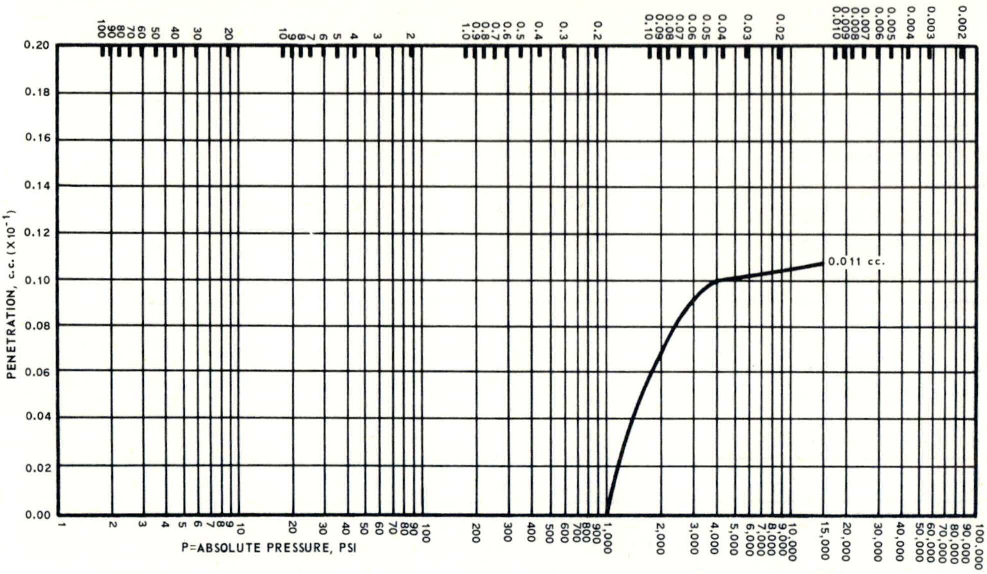

Fig. 10 - Volume Penetration Vs. Pressure or Pore Size for 0.030'' OD Porous PVC Tubulets. The Upper Scale Gives the Pore Diameter in Microns. Measured Pore Volume: 0.13 cc/g.

Mercury intrusion porosimetry for the 30-mil tubulets extruded from milled stock indicates that the pores are small, explaining the low water flow (Fig. 10). Scanning electron photomicrographs of a 0.030'' diameter tubulet (Fig. 11), when compared with those for a flat porous film of the same formulation (Fig. 7), show a definite orientation of the surface character of the tubulet, presumably due to the drawing of the tubulet during extrusion. Drawing allows easy control of pore diameter but causes elongation of pores and thereby reduces water flow rate; thus it may be necessary to extrude through a smaller die so that drawing or stretching can be minimized.

The flow rates for the 60-mil tubulets from milled stock were in the range of 400 to 1100 gfd, within the target range of 100-1000 gfd. The large pore diameters as indicated by mercury intrusion porosimetry (Fig. 12) explain the high flow. The low flow through the 55-mil tubulets may be the result of excessive wall thickness (0.016'' OD).

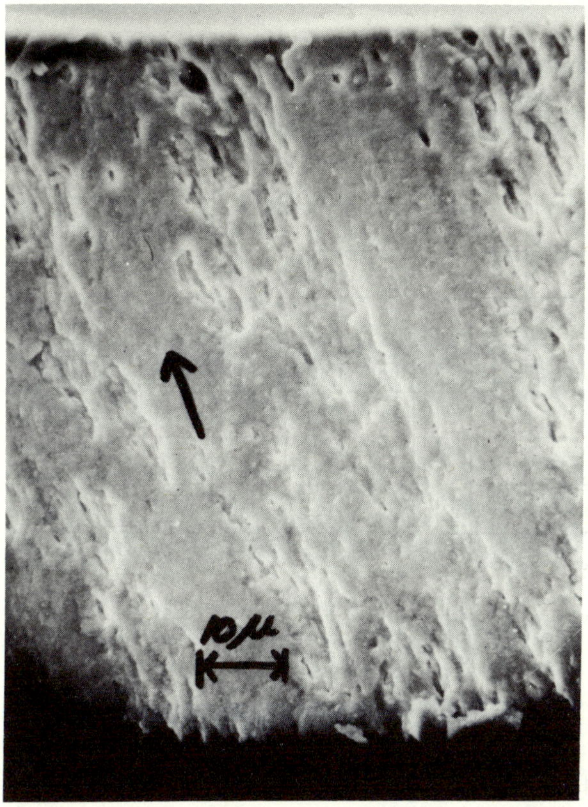

Fig. 11 - Surface of a 0.030 Inch Diameter Porous PVC Tubulet - 1100X

CONCLUSIONS

A number of thermoplastic polymers were evaluated as the matrices for porous materials to be used as supports for reverse osmosis desalination membranes. Formulations consisted of a rigid hydrophobic polymer blended with a water-soluble polymer and surfactant. Porosity was produced by extraction of the water-soluble ingredients. In flat sheet form, porous membranes were prepared from several homopolymers, copolymers, and homopolymer/copolymer blends which exhibited water permeability in the range of interest.

A polyvinyl chloride homopolymer compound was successfully extruded from both dry blend and milled stock into approximately 30- and 60-mil OD tubulets. After aqueous extraction, the 30-mil tubulets allowed a water flux of 20 to 30 gal/ft^2-day at 40 psig.

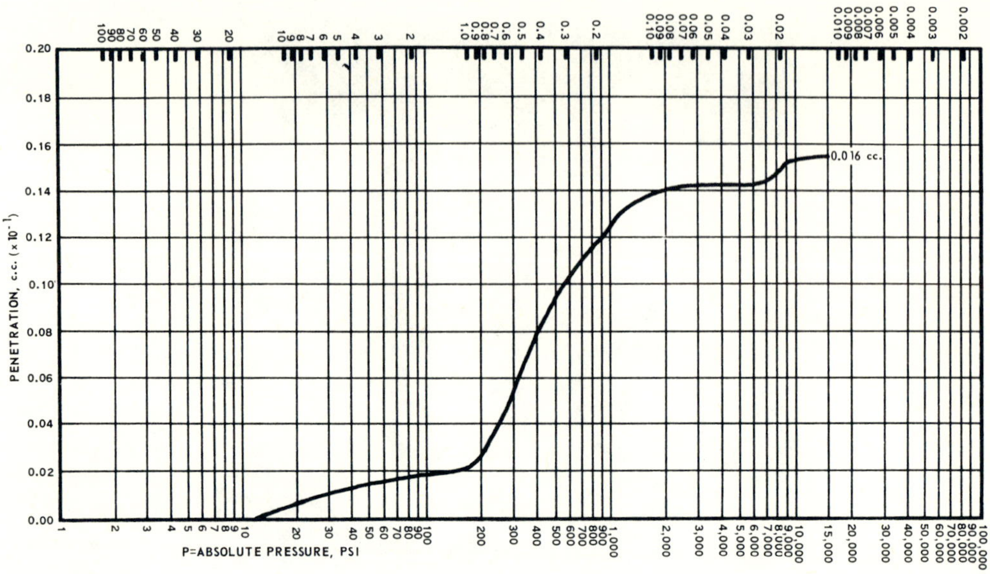

Fig. 12 - Volume Penetration Vs. Pressure or Pore Size for 0.060" OD Porous PVC Tubulets, Formulation 82653. The Upper Scale Gives the Pore Diameter in Microns. Measured Pore Volume: 0.17 cc/g.

Permselective coatings are currently being developed for these tubulet supports. The resulting composite membranes will be used for brackish and/or seawater desalination.

ACKNOWLEDGMENT

This work was carried out under contract with the Office of Saline Water, U. S. Department of the Interior We gratefully acknowledge the advice and assistance of Dr. H. E. Podall and Dr. L. M. Kindley, Polymer and Biophysics Division, Office of Saline Water.

VI. SUMMARY AND FUTURE TRENDS

FUTURE TRENDS OF OSW MEMBRANE RESEARCH

W. S. Gillam
Assistant Director, Research
Office of Saline Water
U.S. Department of the Interior
Washington, D.C.

INTRODUCTION

The development of reverse osmosis as a process for desalting brackish and seawater originated with the research of Reid and Breton at the University of Florida in 1954-1956. These investigators showed that cellulose acetate film is capable of separating salt from water by reverse osmosis, although at a very slow rate. Soon after this, Loeb and Sourirajan at the University of California developed a unique process for making an asymmetric or "skinned" cellulose acetate membrane. The water flux and salt rejection of this membrane were sufficiently high to make the reverse osmosis process an attractive candidate for desalination. Just ten years later, small capacity commercial reverse osmosis units were available to the public.

During this period, a number of workers made significant contributions to the fundamental understanding of the reverse osmosis process. Merten of Gulf General Atomic, Incorporated, published on his theory of reverse osmosis desalination and Kedem of the Weizmann Institute of Science analyzed the process phenomenologically in terms of irreversible thermodynamics. During this same time period other researchers were investigating various practical configurations for making reverse osmosis a viable technique for removing salt from water. Mahon of Dow Chemical Company did some of the original research on cellulose acetate hollow fibers. Merten of Gulf General Atomic, Incorporated, tested preliminary designs of the spiral wound modules a few years later. Rapid progress has been made in reverse osmosis membrane technology, so that today reverse osmosis equipment, membranes, and systems of different designs are commercially available in plant capacities up to several million gallons per day. The configurations used include not only improved versions of the original tubular design, but also spiral wound modules, and more recently, the hollow fibers of the duPont Company.

At present, it may be said that much of the research and development supported by the OSW on the reverse osmosis process for brackish water desalination has been successfully taken over by industry, so that in the future we will probably see this area emphasized to a lesser degree by OSW, Research.

RECENT ACCOMPLISHMENTS OF OSW, RESEARCH IN REVERSE OSMOSIS

Brackish Water Membranes

Research on cellulose acetate (CA) blend membranes and crosslinked cellulose acetate methacrylate (CAM) membranes led to significant improvements in these materials for brackish water desalination. The optimized blend membrane has been successfully incorporated into a spiral wound module which is now commercially available.

Asymmetric ethyl cellulose membranes were prepared in flat sheet and tubular form as candidates for reverse osmosis desalination of alkaline brackish water. Because of their stability at high pH, these membranes are reasonably effective for rejecting boron (as the sodium salt of boric acid) as well as sodium chloride.

Hollow fibers have been prepared from asymmetric cellulose acetate using several methods of fabrication. With cellulose acetate of different degrees of substitution, fibers have been prepared by melt spinning, solution spinning, and thermal gelation techniques. All of these show promise for brackish water desalination, and research in these areas is continuing.

A tubular composite membrane system was developed for brackish water and was advanced to the engineering development stage for conversion into a practical regenerable module. The system consists of an ultrathin cellulose acetate barrier deposited in situ on a tubular microporous polysulfone support film within a porous fiberglass support tube. A significant feature of this system is that it enables in situ regeneration of the ultrathin desalination barrier, without affecting the relatively inert polysulfone substrate.

Dynamically-formed hydrous zirconium oxide polyacrylic acid dual layer membranes have been demonstrated for brackish water feeds. Such membranes retain good rejections and reasonable fluxes under conditions of relatively high water recover (> 80%). Flux decline appears not to be excessive in runs of 200-400 hours. Membrane regeneration can be accomplished by in situ chemical procedures. Some of this work has now advanced to the engineering-development stage for field testing.

Seawater Membranes

Because the requirements are much more severe, progress in the area of desalting seawater by reverse osmosis has been slower. Nevertheless there have been notable achievements in developing membranes for possible use in this application.

One of the significant accomplishments has been the development of a blend cellulose acetate for seawater desalination by annealing at a high temperature to increase salt rejection. First generation membranes of this type have been advanced to the engineering-development stage for further evaluation and development into practical tubular and spiral wound modules.

Crosslinked CAM membranes can also be formulated and fabricated to give high salt rejections at reasonable fluxes. These membranes have shown outstanding flux stability in flat sheet form. Development effort is being pursued to fabricate practical seawater modules using these membranes.

Melt spun cellulose triacetate hollow fibers with rejections satisfactory for seawater desalination have been prepared on a laboratory scale. The flux is reasonably high for a hollow fiber configuration, but problems of flux decline rate have yet to be overcome. New composite ultrathin cellulose triacetate barriers on a specially designed porous support are being scaled up for incorporation into spiral wound modules for operation at or above 1000 psi. Laboratory studies have clearly shown their potential as high performance seawater membranes, with sodium chloride rejections of 99.7% or more.

Finally, aromatic polyamide and poly(amide-hydrazide) membranes have been prepared and tested on a seawater feed in tubular and flat sheet form with encouraging results.

Basic Studies

Basic studies relating to membranes and the reverse osmosis process continue to provide the foundation for building and developing the technology of reverse osmosis.

The simple solubility-diffusion model developed some years ago provides a reasonable approach to the problem of membrane transport. Refinements of this model and extension to explain the mechanism of salt rejection and water transport on a molecular basis have been the subjects of a number of studies. Progress has been made, but exact quantitative correlations with the properties and structure of the polymer or membrane require further research.

Qualitative guidelines relating polymer structure to efficacy of desalination suggest that a candidate reverse osmosis membrane should have: (1) some hydrophilic character; (2) polar sites for binding (reversibly) and facilitating the movement of water; and (3) a fairly regular molecular structure so that there is tight packing.

In recent years significant progress has been made in elucidating the mechanism of forming asymmetric membranes. The use of the solubility parameter approach provides a more scientific basis for selecting solvent systems for casting asymmetric membranes. A simple microscopy technique has been developed for rapidly examining the precipitation and formation of asymmetric membranes.

For some years the gradual irreversible decline in flux with time of an asymmetric membrane has been attributed to "compaction" of the membrane. It seemed logical to assume this involved compression of the porous substructure due to viscoelastic creep with resulting loss of effective porosity. However, recent studies suggest that flux decline in many cases results primarily from changes in the "skin" or the region between the skin and the porous substructure. There is also increasing evidence that simple fouling of the membrane surface may account for at least part of the decline in flux which is observed during reverse osmosis operation.

Improvements in the Technology of Forming Membranes and Novel Concepts for Supports

Research and development of a tubular module concept, supported by OSW over the past few years, has served to catalyze the commercialization of a new and improved desalination system. The system is based on a continuous process for producing one inch diameter cellulose acetate membranes which are braided with a flexible fiber support in fifty-foot coiled modules.

In another development, the practical feasibility of a module design amenable to in situ membrane regeneration was demonstrated in a laboratory unit. This system is comprised of bunches of one-quarter inch diameter porous tubes within a pressure shell, coated externally with cellulose acetate. In addition to the in situ membrane regeneration feature, other potential advantages are: (1) reduced tendency to foul; (2) ease of assembly and disassembly; and (3) ability of the membrane to withstand greater pressure and pressure fluctuations than in systems in which the membrane is inside the support. A somewhat similar system has been developed which consists of bundles of very small diameter microporous tubulets, with the desalination barrier coated on the outside of each tubulet within a high pressure shell. The most promising support developed thus far is a porous polyvinyl chloride tubulet of 60 mils outside diameter. A number of formulations and techniques are being evaluated for coating the tubulets with salt rejecting barrier materials.

A new reverse osmosis membrane fabrication process, which has shown promise in small scale laboratory tests, has recently been reported. This is a dry process that eliminates the necessity for immersion and annealing steps, and appears to offer significant advantages over conventionally prepared membranes with regard to quality control, wet-dry stability, and ease of shipping and storage.

Further improvements in the "wet" methods of membrane fabrication have also been made. A procedure known as thermal gelation is used to simultaneously gel and anneal the membrane in one step. A variation in the casting technique called "hydrocasting" has been developed for forming an asymmetric membrane inside small diameter tubules. Hydrocasting involves blowing a plug of polymer casting solution with a bubble of a gas through the tubule, followed by cold water or an aqueous salt solution for immediate gelation.

FUTURE REVERSE OSMOSIS RESEARCH

Change of Emphasis

Under the Saline Water Conversion Act of 1971, the Office of Saline Water is charged with the responsibility of conducting research on conversion of saline and other chemically contaminated water into water suitable for beneficial consumptive uses. Thus, renovation or upgrading of waste water along with removal of salt may be of increasing interest. Because of the success of the reverse osmosis process in brackish water desalination, and the interest which industry has shown in development of this technology, the OSW will probably concentrate more of its desalting research support on seawater. The key to many of the problems of reverse osmosis desalination of sea water resides in the ability to design polymers which meet the requisite membrane properties. Increased research emphasis on non-cellulosic polymers is predicted as an area which might lead to reverse osmosis membranes with properties superior to those of cellulosics for desalination of sea water.

Sea Water Desalination

The research effort of OSW to uncover new and improved polymeric barriers for rejection of different solutes will be accelerated. Although the primary goal initially will be the development of membranes suitable for seawater desalination, the same membranes will also be investigated for use in rejecting organic solutes which could be present in a variety of waste-water feeds. For rejection of both organic and inorganic solutes, non-cellulosic materials having a potential for improvement in chemical stability, mechanical strength, and rejection capabilities will be emphasized.

For both one-stage and two-stage seawater reverse osmosis desalination processes, there is a need for supports that can perform in the range of 1000-1500 psi. Continued research in this direction is contemplated for the future.

Waste Water Renovation

In the area of renovating chemically-contaminated water, there exists a vast potential for reverse osmosis applications. To realize this potential a new look into membranes and their capabilities for handling a variety of waste water feeds is planned. For this purpose, a significant effort to develop membranes with selective rejection characteristics will be initiated. Such membranes would permit the retention of certain solutes, such as boric acid or heavy metals, while allowing passage of more desirable or harmless solutes. Applications might be foreseen for treatment of irrigation return flow waters and heavy metal-containing brines.

Another approach to membranes which may have a role in wastewater renovation is the composite membrane. Although considerable research on this concept has already been done, further exploratory work is desirable. These membranes differ from the conventional "skinned" membranes in that the ultrathin barrier and microporous support film are discrete materials and may consist of entirely different polymers--each tailored for its respective function. This type of composite membrane provides a unique opportunity for improved hydraulic permeability, without sacrifice of rejection of different solutes and without compromising the requisite mechanical properties.

Membrane Fouling and Degradation

The problem of membrane fouling and degradation is inherent to all membrane processes operating on natural waters containing colloidal matter and/or sparingly soluble inorganic or organic substances. To facilitate the development of techniques to minimize such effects, a better understanding is required of the underlying physical chemical parameters associated with the adhesion of the foulant or scale to the membrane. Although OSW Research has recognized the need for work in this area, and has undertaken several programs in past years, more emphasis on this problem is planned for the future. For some types of feed water and module configurations the cost of pretreatment may be the major factor in determining the cost effectiveness of the reverse osmosis process. An expansion of effort is planned in the field of feed water chemistry including membrane interactions and pretreatment. Not only is it desirable to understand membrane-feed interactions, but practical methods of pretreatment are required which are tailored to specific feed waters and reverse osmosis module configuration.

Further research is also planned to assess the contribution of fouling to the flux decline observed with most reverse osmosis membranes. It is hoped that these studies will resolve the question of just how much of the decline in flux is due to fouling and how much is due to membrane "compaction."

Other Applications of Reverse Osmosis

Divalent ions are rejected from aqueous solution by most reverse osmosis membranes more easily than univalent ions. Since it is the divalent ion species which contributes to hardness in water as well as scaling and fouling problems, reverse osmosis may be considered as a process for water softening.

During previous years, OSW, Research has collaborated with the Department of Defense (DOD) and the National Aeronautics and Space Administration (NASA) on special applications of RO to problems of mutual interest. Joint programs with both of these agencies are in progress and are planned for continuation in the future. The possible application of reverse osmosis to renovation and recycling water in a space vehicle is a problem of current interest to NASA. OSW and NASA are jointly supporting programs to develop membranes and reverse osmosis systems for this application. A key feature of this program is the requirement for membranes which can operate at a sterilization temperature of about 75°C. One can foresee future requirements for such membranes for processing industrial waste effluents at elevated temperatures.

Specific membrane property requirements have been of interest to DOD and have been jointly investigated by OSW, Research and DOD. For example, programs aimed at development of modules which can operate after exposure to freezing temperatures have been recently completed. Currently, research is being conducted to develop reverse osmosis membranes which retain their properties after wet-dry cycling. Certainly OSW will continue to collaborate with other agencies on reverse osmosis programs of mutual interest.

Mechanism of Water Diffusion and Salt Rejection

Finally, support of the more basic aspects of reverse osmosis will be continued. Although the main objectives of future membrane research will be the development of practical reverse osmosis membranes and systems, basic research programs aimed at explaining the mechanism of water transport and the underlying molecular factors responsible for solute rejection will receive some support for an indefinite time. It is from these basic programs that new ideas, new concepts, and basic information will come, which will enable predictions of membrane and reverse osmosis system performance, and furnish guidelines for advancing reverse osmosis technology.

INDEX